IET MATERIALS, CIRCUITS AND DEVICES SERIES 60

IP Core Protection and Hardware-Assisted Security for Consumer Electronics

Other volumes in this series:

Volume 2 **Analogue IC Design: The current-mode approach** C. Toumazou, F.J. Lidgey and D.G. Haigh (Editors)
Volume 3 **Analogue–Digital ASICs: Circuit techniques, design tools and applications** R.S. Soin, F. Maloberti and J. France (Editors)
Volume 4 **Algorithmic and Knowledge-based CAD for VLSI** G.E. Taylor and G. Russell (Editors)
Volume 5 **Switched Currents: An analogue technique for digital technology** C. Toumazou, J.B.C. Hughes and N.C. Battersby (Editors)
Volume 6 **High-frequency Circuit Engineering** F. Nibler *et al.*
Volume 8 **Low-power High-frequency Microelectronics: A unified approach** G. Machado (Editor)
Volume 9 **VLSI Testing: Digital and mixed analogue/digital techniques** S.L. Hurst
Volume 10 **Distributed Feedback Semiconductor Lasers** J.E. Carroll, J.E.A. Whiteaway and R.G.S. Plumb
Volume 11 **Selected Topics in Advanced Solid State and Fibre Optic Sensors** S.M. Vaezi-Nejad (Editor)
Volume 12 **Strained Silicon Heterostructures: Materials and devices** C.K. Maiti, N.B. Chakrabarti and S.K. Ray
Volume 13 **RFIC and MMIC Design and Technology** I.D. Robertson and S. Lucyzyn (Editors)
Volume 14 **Design of High Frequency Integrated Analogue Filters** Y. Sun (Editor)
Volume 15 **Foundations of Digital Signal Processing: Theory, algorithms and hardware design** P. Gaydecki
Volume 16 **Wireless Communications Circuits and Systems** Y. Sun (Editor)
Volume 17 **The Switching Function: Analysis of power electronic circuits** C. Marouchos
Volume 18 **System on Chip: Next generation electronics** B. Al-Hashimi (Editor)
Volume 19 **Test and Diagnosis of Analogue, Mixed-signal and RF Integrated Circuits: The system on chip approach** Y. Sun (Editor)
Volume 20 **Low Power and Low Voltage Circuit Design with the FGMOS Transistor** E. Rodriguez-Villegas
Volume 21 **Technology Computer Aided Design for Si, SiGe and GaAs Integrated Circuits** C.K. Maiti and G.A. Armstrong
Volume 22 **Nanotechnologies** M. Wautelet *et al.*
Volume 23 **Understandable Electric Circuits** M. Wang
Volume 24 **Fundamentals of Electromagnetic Levitation: Engineering sustainability through efficiency** A.J. Sangster
Volume 25 **Optical MEMS for Chemical Analysis and Biomedicine** H. Jiang (Editor)
Volume 26 **High Speed Data Converters** Ahmed M.A. Ali
Volume 27 **Nano-Scaled Semiconductor Devices** E.A. Gutiérrez-D (Editor)
Volume 29 **Nano-CMOS and Post-CMOS Electronics: Devices and modelling** Saraju P. Mohanty and Ashok Srivastava
Volume 30 **Nano-CMOS and Post-CMOS Electronics: Circuits and design** Saraju P. Mohanty and Ashok Srivastava
Volume 32 **Oscillator Circuits: Frontiers in design, analysis and applications** Y. Nishio (Editor)
Volume 33 **High Frequency MOSFET Gate Drivers** Z. Zhang and Y. Liu
Volume 38 **System Design with Memristor Technologies** L. Guckert and E.E. Swartzlander Jr.
Volume 39 **Functionality-Enhanced Devices: An alternative to Moore's law** P.-E. Gaillardon (Editor)
Volume 43 **Negative Group Delay Devices: From concepts to applications** B. Ravelo (Editor)

IP Core Protection and Hardware-Assisted Security for Consumer Electronics

Anirban Sengupta and Saraju P. Mohanty

The Institution of Engineering and Technology

Published by The Institution of Engineering and Technology, London, United Kingdom

The Institution of Engineering and Technology is registered as a Charity in England & Wales (no. 211014) and Scotland (no. SC038698).

First published 2019

The Institution of Engineering and Technology
Michael Faraday House
Six Hills Way, Stevenage
Herts, SG1 2AY, United Kingdom

www.theiet.org

British Library Cataloguing in Publication Data
A catalogue record for this product is available from the British Library

ISBN 978-1-78561-799-7 (hardback)
ISBN 978-1-78561-800-0 (PDF)

Typeset in India by MPS Limited
Printed in the UK by CPI Group (UK) Ltd, Croydon

Contents

About the authors xv
Acknowledgments xvii
Foreword by Thomas M. Coughlin xix
Foreword by Pradipta Patra xxi
Foreword by Swarup Bhunia xxiii
Foreword by Peter Corcoran xxv
Preface xxvii
List of acronyms xxxi
List of notations xxxvii

1 Introduction to IP core protection and hardware-assisted security of consumer electronics **1**
1.1 Consumer electronics and security perspectives 1
1.2 Hardware-assisted security and IP core protection 3
1.3 Intellectual property (IP) cores/hardware 4
1.3.1 Utility of IP cores in CE devices 4
1.3.2 Why security and protection of hardware/IP cores? 5
1.3.3 Traditional forms of IP protection not enough? 6
1.4 IP core protection and hardware-assisted security of CE device—DSP core 6
1.4.1 Security and protection methodologies available for IP core/hardware 7
1.4.2 Different IP core protection and hardware-assisted security mechanisms: advantages and disadvantages 12
1.4.3 HLS (architectural synthesis) as design backbone for implementing security algorithms for DSP IP cores 13
1.5 Hardware-assisted media protection 15
1.6 Physical unclonable functions 16
1.7 Organization of the book 17
1.8 Conclusions 19
1.9 Exercises 19
References 20

2 Security in consumer electronics and internet of things (IoT) **23**
2.1 Internet of things (IoT) – a broad overview 23
2.1.1 IoT – architecture 25
2.1.2 IoT – driving technology 27

2.1.3 IoT – applications 28
2.1.4 IoT – challenges 30
2.2 Security, privacy, IPR in IoT, and consumer electronic systems – a big picture 31
2.2.1 IoT security – attacks and countermeasures 32
2.2.2 Trustworthy consumer electronic systems 34
2.2.3 Hardware-assisted security and protection 35
2.2.4 Different aspects of security and privacy 37
2.2.5 Different aspects of intellectual property (IP), ownership right, or copyright protection 38
2.3 Memory security 39
2.3.1 Memory security attacks 39
2.3.2 Memory security solutions 40
2.4 Radio-frequency identification (RFID) security 43
2.4.1 RFID security attacks 43
2.4.2 RFID security solutions 44
2.5 Near-field communications (NFC) security 46
2.5.1 NFC security attacks 47
2.5.2 NFC security solutions 48
2.6 Smart transportation security 50
2.6.1 Smart car security 51
2.6.2 UAV or drone security 55
2.7 Smart healthcare security 58
2.7.1 Smart healthcare security attacks 59
2.7.2 Smart healthcare security solutions 60
2.8 Firmware 62
2.8.1 Firmware attacks 62
2.8.2 Firmware solutions 63
2.9 Blockchain technology 64
2.9.1 Blockchain – overview 65
2.9.2 Blockchain – application 66
2.9.3 Blockchain as a security framework 67
2.9.4 Blockchain – issues 68
2.10 Conclusions 69
2.11 Exercises 69
References 70

3 Trojan security aware DSP IP core and integrated circuits 79
3.1 Introduction 79
3.2 Types of hardware Trojans 81
3.2.1 Trojan features 81
3.2.2 Benefit of Trojan security at higher abstraction level 84
3.2.3 Threat model 85
3.3 Hardware Trojan in a 3PIP module 86
3.3.1 Example of a hardware Trojan 86
3.3.2 Trojan detectability in a 3PIP module at RTL/lower levels 87

3.4 Selected Trojan security approaches 88
3.4.1 Trojan security approaches for DSP cores 88
3.4.2 Trojan security approach for combinational/sequential circuits 94
3.5 Trojan security aware DSP IP core 95
3.5.1 Definition 96
3.5.2 Goal 96
3.5.3 Formulation 97
3.5.4 Models 97
3.6 Design process of Trojan secured DSP IP core 99
3.6.1 Deriving the CDFG of a DSP core 99
3.6.2 Generating the DMR of the CDFG 108
3.6.3 Trojan secured scheduling of DMR CDFG 108
3.7 Analysis of case studies/test cases 113
3.7.1 DSP applications and system setup for the case studies 113
3.7.2 Security analysis 114
3.7.3 Design cost analysis 115
3.7.4 Comparative perspectives 116
3.8 Conclusion 117
3.9 Exercises 118
References 119

4 IP core and integrated circuit protection using robust watermarking 123
4.1 Introduction 123
4.2 Selected watermarking approaches 124
4.3 Design process of watermarked IP core/hardware 128
4.3.1 Problem formulation 128
4.3.2 Design process of single-phase watermarked IP core/hardware 129
4.3.3 Design process of triple-phase watermarked IP core/hardware 148
4.3.4 Desired properties of IP core watermark 161
4.3.5 Possible cases of dishonest claim of IP core/hardware ownership and its resolution 162
4.4 Analysis on case studies 163
4.4.1 Security analysis of triple-phase watermark for DSP IP cores 163
4.4.2 Design cost analysis of triple-phase watermark for DSP IP cores 164
4.5 Conclusion 166
4.6 Exercises 168
References 169

5 Symmetrical protection of DSP IP core and integrated circuits using fingerprinting and watermarking 171
5.1 Introduction 171
5.1.1 Background on watermark and fingerprint 173

5.1.2 Threat model 173
5.1.3 Benefits of protection at higher abstraction 173
5.2 Fundamentals of IP core protection 174
5.2.1 Overview on non-symmetric IP core protection techniques 174
5.2.2 Overview on symmetric IP core protection techniques 175
5.3 Symmetrical IP core protection for DSP core 175
5.3.1 Problem formulation 177
5.3.2 Symmetrically protected design—area evaluation model 177
5.3.3 Symmetrically protected design—delay evaluation model 177
5.3.4 Symmetrically protected design—cost evaluation function 177
5.3.5 Encoding rules of buyer fingerprint and seller watermark for DSP IP cores 178
5.3.6 Multi-variable signature embedding process 180
5.3.7 Signature detection process 181
5.3.8 Desirable properties of signature 182
5.4 Case study of symmetrical IP core protection 182
5.4.1 Demonstration of fingerprinting constraints embedding process 183
5.4.2 Demonstration of watermarking constraints embedding process 186
5.5 Analysis of case studies for DSP cores 187
5.5.1 Analysis of embedding cost, security metric on DSP Cores symmetrical protection 188
5.5.2 Comparative study between symmetrical and non-symmetrical technique 190
5.6 Conclusion 194
5.7 Exercises 194
References 195

6 Computational forensic engineering for resolving ownership conflict of DSP IP core 199
6.1 Introduction 199
6.1.1 Overview of forensic engineering 200
6.2 Computational FE technology 202
6.3 IP core feature extraction algorithms 204
6.3.1 Feature extraction rules 204
6.3.2 IP core validation 213
6.3.3 Important characteristics of customized CFE 214
6.4 Analysis on case studies 214
6.4.1 Results of the customized CFE approach 215
6.5 Conclusion 222
6.6 Exercises 222
References 223

7 Structural obfuscation of DSP cores used in CE devices **227**
7.1 Introduction 227
7.1.1 Threat model 230
7.1.2 Benefits of providing security at higher design abstraction level 230
7.2 Obfuscation for IP core protection—a broad view 231
7.2.1 Code obfuscation techniques 231
7.2.2 Logic obfuscation techniques 231
7.2.3 Structural obfuscation techniques 232
7.3 Compiler transformation-driven structural obfuscation 233
7.3.1 Formulation and evaluation models 235
7.3.2 Multistage high-level transformation techniques 236
7.4 Low-cost structural obfuscation for DSP IP core 242
7.4.1 Overview on PSO 242
7.4.2 Movement of particle 242
7.4.3 Terminating condition of PSO 243
7.5 A case study for multistage structural obfuscation 243
7.6 Analysis of case studies 246
7.6.1 Result of multistage structural obfuscation 246
7.6.2 Comparative study and discussion 248
7.7 Conclusion 250
7.8 Exercises 251
References 251

8 Functional obfuscation of DSP cores used in CE devices **255**
8.1 Introduction 255
8.2 Attack scenarios and threat model 256
8.2.1 Possible attack scenarios 256
8.2.2 Threat model 259
8.3 Selected functional obfuscation approaches 260
8.4 Design of functionally obfuscated DSP core 262
8.4.1 Formulation 262
8.4.2 Low-cost obfuscation method for DSP core 262
8.5 Security of functionally obfuscated DSP core design 266
8.5.1 Keyspace 266
8.5.2 Security analysis 267
8.5.3 Countermeasures against attacks 267
8.6 Optimization engine for functional obfuscation of DSP cores 273
8.6.1 Particle encoding 273
8.6.2 Particle fitness 273
8.6.3 Updating particle 274
8.7 Analysis of case studies 275
8.7.1 Security analysis 275
8.7.2 Overhead analysis 277

8.7.3 Comparative analysis 279
8.8 Conclusion 282
8.9 Exercises 283
References 283

9 Obfuscation of JPEG CODEC IP core for CE devices 287
9.1 Introduction 287
9.2 Overview of JPEG compression and decompression 289
9.2.1 DCT-based JPEG image compression process 290
9.2.2 DCT-based JPEG image decompression process 293
9.3 Design process of structurally obfuscated JPEG IP core 293
9.3.1 Threat model, problem formulation, and optimization framework 293
9.3.2 Constructing non-obfuscated DFG for JPEG compression 294
9.3.3 Generating structurally obfuscated JPEG compression IP core 296
9.3.4 Generating structurally obfuscated JPEG decompression IP core 298
9.4 Implementation of JPEG CODEC IP core 299
9.4.1 Designing obfuscated JPEG compression IP core 299
9.4.2 Designing obfuscated JPEG decompression IP core 302
9.4.3 End-to-end JPEG CODEC through designed hardware/IP core 302
9.5 Analysis on case studies 308
9.6 Conclusion 312
9.7 Exercises 312
References 313

10 Advanced encryption standard (AES) and its hardware watermarking for ownership protection 317
10.1 Introduction 317
10.2 AES algorithm 318
10.2.1 Overview of AES 318
10.2.2 AES algorithm—description and custom hardware design 318
10.3 AES digital watermarking 326
10.3.1 AES watermark encoding 327
10.3.2 Process of embedding watermark in AES 329
10.3.3 Signature detection 329
10.4 Case study of a watermarked AES hardware 330
10.5 Conclusion 330
10.6 Exercises 334
References 334

11 Hardware approaches for media and information protection and authentication **337**
11.1 IP Protection—a broad overview 337
11.1.1 Digital rights management 338
11.1.2 Copyright protection of multimedia—a brief history 339
11.1.3 Hardware versus media protection 342
11.2 General framework for copyright protection 343
11.2.1 The encoder 343
11.2.2 The decoder 344
11.2.3 The comparator 344
11.3 Types of digital watermarks 344
11.3.1 Spatial versus frequency domain watermarking 345
11.3.2 Based on multimedia objects 345
11.3.3 Based on human perception 346
11.3.4 From applications point of view 346
11.3.5 Based on embedding techniques 347
11.3.6 Hardware-based watermarking systems 347
11.4 Applications of digital watermarks 347
11.4.1 Copyright protection 347
11.4.2 Ownership assertion 348
11.4.3 Authentication and integrity verification 348
11.4.4 Fingerprinting 348
11.4.5 Usage control 348
11.4.6 Broadcast monitoring 348
11.4.7 Content labeling 349
11.4.8 Misappropriation detection 349
11.4.9 Anti-counterfeiting 349
11.4.10 UAV safety 349
11.4.11 Medical signals authentication 350
11.5 Desired characteristics of watermarks 350
11.5.1 Perceptibility 350
11.5.2 Robustness 350
11.5.3 Tamper resistance 351
11.5.4 Bit rate 351
11.5.5 Modifiability, multiplicity, cascadability, and orthogonality 351
11.5.6 Scalability 351
11.5.7 Unambiguity and universality 352
11.5.8 Pixel alteration and human intervention 352
11.5.9 Reliability 352
11.5.10 Blindness 352
11.5.11 Security 353
11.5.12 Real-time operation 353
11.5.13 Cost and complexity 353
11.5.14 Energy consumption 353

11.5.15 Integrability 354
11.5.16 Characteristics specific to a watermark 354
11.6 Technical challenges for watermarking 355
11.6.1 Properties of visual signals 356
11.6.2 Properties of the human visual system 357
11.6.3 How much watermark signal to add and where? 357
11.6.4 Spread spectrum communications 358
11.7 Hardware-based approaches for watermarking 358
11.7.1 Image watermarking hardware systems 359
11.7.2 Video watermarking hardware systems 372
11.7.3 Secure better portable graphics (SBPG) 378
11.7.4 Trust cam 379
11.8 Dynamic watermarking in smart car or UAV 381
11.9 Medical signals authentication 382
11.10 Side-channel information leakage attacks and countermeasures 383
11.10.1 An encryption hardware 383
11.10.2 Side-channel analysis attacks 384
11.10.3 Side-channel attack countermeasures 387
11.11 Attacks on watermarks and watermarking systems 388
11.11.1 Removal and interference attacks 389
11.11.2 Geometric attacks 389
11.11.3 Cryptographic attacks 389
11.11.4 Protocol attacks 389
11.12 Limitations of watermarks and watermarking 390
11.13 Conclusion 391
11.14 Exercises 391
References 392

12 Physical unclonable functions (PUFs) 403
12.1 Introduction 403
12.2 PUF: Principle 406
12.3 Properties or characteristics of PUFs 407
12.3.1 Uniqueness 407
12.3.2 Reliability (correctness) 409
12.3.3 Randomness (uniformity) 409
12.3.4 Correlation (bit aliasing) 409
12.3.5 Power consumption 410
12.3.6 Speed 410
12.4 Classification of PUFs 410
12.4.1 Device-based PUFs 411
12.4.2 Security-based PUFs 411
12.5 Ring oscillator–based PUFs 412
12.6 Reconfigurable or dynamic PUFs 415
12.7 SRAM-based PUF 419
12.8 Memristor-based PUFs 420

12.9 Diode-based PUF 424
12.10 Carbon-based PUFs 426
12.10.1 CNT-based PUF 427
12.10.2 Graphene-based PUF 428
12.11 Microprocessor-based PUF 429
12.12 Magnetic PUF 430
12.13 Practical implementation of PUF 432
12.14 PUF: case study applications 433
12.15 PUF: issues 438
12.16 Conclusion 440
12.17 Exercises 440
References 441

Appendix A **447**

Appendix B **459**

Index **493**

About the authors

Dr. Anirban Sengupta is an Associate Professor in Computer Science and Engineering at Indian Institute of Technology (IIT) Indore, where he directs the research lab on "CAD for Consumer Electronics Hardware Device Security and Reliability." He holds a Ph.D. and M.A.Sc. in Electrical and Computer Engineering from Ryerson University, Toronto (Canada), and is a registered Professional Engineer of Ontario (P.Eng.). He has been an active researcher with strong publications in the emerging areas of "Hardware Security," "IP Core Protection," "Privacy and Digital Rights Management for Electronics Devices," "Forensic Engineering for Security." He is recipient of honors like IEEE Distinguished Lecturer by CE Society in 2017, IEEE Computer Society TCVLSI Editor Award in 2017, IEEE Consumer Electronics Society Best Paper Award in 2019 International Conference on Consumer Electronics (ICCE), Las Vegas, IEEE Computer Society TCVLSI Best Paper Award in iNIS 2017 and IETE Best Research Award 2018 by IETE Sub-Center Indore. He is a Chairman and Editor-in-Chief of IEEE Computer Society – TCVLSI and General Chair of 37th IEEE International Conference on Consumer Electronics 2019, Las Vegas. He is the Deputy Editor-in-Chief of *IET Computers and Digital Techniques* which is a Journal that has nearly 40 years of history. He is also the Founder Chairman of IEEE Consumer Electronics Society Chapter – IEEE Bombay Section. He is also a recipient of "Visvesvaraya Faculty Research Fellow" by Digital India Corporation, Ministry of Electronics and IT.

He is an author of more than peer-reviewed 182 publications and patents. He is an author of a book from IET on "IP Core Protection and Hardware-Assisted Security for Consumer Electronics." This book is the first of its kind in the scientific community. He also currently serves in 14 editorial positions as Senior Editor, Associate Editor, Editor and Guest Editor of several IEEE Transactions/Journals, IET and Elsevier Journals including *IEEE Transactions on Aerospace and Electronic Systems (TAES)*, *IEEE Transactions on VLSI Systems*, *IEEE Transactions on Consumer Electronics* (to join from Jan 2019), *IEEE Access Journal*, *IET Journal on Computer & Digital Techniques*, *IEEE Consumer Electronics*, *IEEE Canadian Journal of Electrical and Computer Engineering*, *IEEE VLSI Circuits and Systems Letter* and *Elsevier Microelectronics Journal*. He further serves as Guest Editor of *IEEE Transactions on VLSI Systems*, *IET Computers and Digital Techniques* and *IEEE Access Journal*. He is the General/Conference Chair of 37th IEEE International Symposium on Consumer Electronics (ICCE) 2019, Las Vegas, Technical Program Chair of 36th IEEE International Conference on Consumer Electronics (ICCE) 2018 in Las Vegas, 15th IEEE International Conference on Information Technology (ICIT) 2016, 3rd IEEE International Symposium on Nanoelectronic and Information Systems (iNIS) 2017 and 2019 IEEE International Symposium on VLSI (ISVLSI) in Florida.

More than a dozen of his IEEE publications have appeared in "Top 50 Most Popular Articles" with few in "Top 5 Most Popular Articles" from IEEE Periodicals. His patents have been cited in various industry patents of IBM Corporation, Siemens Corporation, Qualcomm, Amazon Technologies, Siemens Aktiengesellschaft (Germany), Mathworks Inc, Ryerson University and STC University of Mexico multiple times. His professional works have received wide media coverage nationally and internationally such as in IET International News (UK), Times of India, Central Chronicle, DBPOST News, Free Press Journal, Dainik Bhaskar, etc. He has supervised more than 15 candidates including several graduated Ph.D. candidates all of whom are/were placed in academia and industry. He has successfully commissioned special issues on "Hardware Security, Privacy, IP protection" in IEEE TVLSI, IET CDT, IEEE Access as well as IEEE CEM, which have received wonderful response from scientific community. Further due to high quality special issue in IEEE Access Journal in 2016, he has been conferred award from IEEE Access for his outstanding contribution. He is also a recipient of "Best Research Paper Award 2017" by Indian Institute of Technology Indore. He has been awarded highest rating "Excellent" by expert committee of Department of Science and Technology (DST) based on the performance (output) in externally funded project in 2017. His research ideas have been awarded competitive funding by various prestigious agencies such as Department of Science and Technology (DST), Council of Scientific and Industrial Research (CSIR) and Department of Electronics and IT (DEITY). He held external affiliation as "Honorary Chief Scientist" at VividSparks IT Solutions Pvt Ltd, besides his regular affiliation at IIT-I, between 2016 and 2018.

Dr. Saraju P. Mohanty is a Professor at the University of North Texas. His research is in "Smart Electronic Systems" which has been funded by National Science Foundations, Semiconductor Research Corporation, US Air Force, and Indo-US Technology Forum. He received IEEE-CS-TCVLSI Distinguished Leadership Award in 2018 for outstanding services to the IEEE, and to the VLSI research community. He has been recognized as an IEEE Distinguished Lecturer by the Consumer Electronics Society (CESoc) in 2017. He was conferred the Glorious India Award in 2017 for his exemplary contributions to the discipline. He received Society for Technical Communication (STC) 2017 Award of Merit for his outstanding contributions to *IEEE Consumer Electronics* Magazine. He was the recipient of 2016 PROSE Award for best Textbook in Physical Sciences and Mathematics from the Association of American Publishers for his *Mixed-Signal System Design* book published by McGraw-Hill in 2015. He was conferred 2016–17 UNT Toulouse Scholars Award for sustained excellent scholarship and teaching achievements. He is the EiC of the *IEEE Consumer Electronics* Magazine (CEM). He serves as the Chair of Technical Committee on VLSI, IEEE Computer Society. He has received four best paper awards and has delivered multiple keynote talks at various International Conferences. He authored 280 research articles, three books, and invented four US patents. His Google Scholar h-index is 29 and i10-index is 89. More about his biography, research, education and outreach activities can be obtained from his website: http://www.smohanty.org

Acknowledgments

We would like to thank our family and friends for the support and encouragement throughout the execution of the book project. We would also like to thank the Indian Institute of Technology (IIT) Indore (India) and University of North Texas (UNT), Denton (USA), for their support in executing this work. We would also like to thank Ph.D. candidates Dipanjan Roy and Deepak Kachave at IIT Indore for their inputs. We would like to thank the Ph.D. candidate Venkat P. Yanambaka at UNT for all his help.

Special thanks to Dr. Swarup Bhunia at the University of Florida (USA), Dr. Peter Corcoran at the University of Galway (Ireland), Dr. Tom Coughlin at Coughlin Associates, California (USA), and Dr. Pradipta Patra at Samsung Electronics Research (India) for writing the foreword for this book.

Foreword by Thomas M. Coughlin

Consumer electronic (CE) devices are a ubiquitous part of our everyday lives. These devices are increasingly connected to the Internet and depend upon web-based services. Because these devices capture and contain a lot of personal data and can be used to monitor and track consumers, the security of these devices (from both hardware and data perspectives) has become of paramount importance. As the growth in what is often called Internet of things (IoT), including several smart hardware devices as the backbone, creates loopholes for hardware threats as well as an increasing number of targets for consumer data and privacy invasion, we need strong hardware and data protection and intellectual property (IP) core ownership security built into these devices at the most basic level.

At the same time that consumer content must be protected, IP cores in these devices as well as the authenticity of the supply chain used to manufacture and distribute consumer products have become a major concern and are intimately tied to hardware and content security. Older techniques for protecting hardware security using today's commoditized system-on-chip (SoC) network connected devices with IP cores from multiple sources are insufficient to meet the needs of modern consumer products and copyrighted content on these devices.

IP core protection and hardware-assisted security for consumer electronics address the general issue of protecting hardware (especially DSP cores) of CE as well as data on CE, including new demands created by IoT enabled products. The book also addresses the potential threats from intervention in the CE design supply chain and how these threats can be circumvented. These supply chain security solutions may include hardware watermarking, hardware fingerprinting, symmetrical IP core protection, hardware metering, computational forensic engineering for IP core protection, and various forms of hardware obfuscation.

The book explores and addresses many different types of security attacks that can be launched against CE products, including power and timing "side channel" attacks, as well as backdoor and brute-force attacks. Methods are discussed for defeating/blocking Trojan horse attacks as well as reverse engineering attacks.

The book discusses the latest methods for active and passive IP core security of DSP cores used in CE device and protection mechanisms for the buyers of ICs and incorporated IP that includes "watermarking" and "fingerprinting" as well as structural obfuscation, functional obfuscation, hardware AES, watermarked AES, and physically unclonable functions (PUFs). The book also explores the use of forensic engineering for resolving ownership issues for reusable digital signal processing (DSP) IP cores. In addition to these direct security concerns, the book discusses the trade-offs of energy consumption, security and performance for CE hardware.

The first of its kind on CE security, this book should be read by designers, programmers, or engineers involved in the design of CE devices as well as for services using these devices.

Thomas M. Coughlin
IEEE Fellow,
Chair, IEEE Consumer Electronics Future Directions Committee

Foreword by Pradipta Patra

We are delimited by consumer electronics (CE) in our everyday life. It is hard to envision a smart life without smart devices currently. However, a smart CE device is valuable as long as its security is intact. Security is integral to the successful deployment of a smart CE product. The term "security" encompasses hardware security, hardware assisted security, data security, media security, etc. At the heart of every security protocol is the underlying hardware which is the lifeline of every application. In the current world of cybersecurity, securing and protecting hardware (and intellectual property (IP) cores) as well as securing data/information are paramount.

For the first time authors in this book present an all-inclusive comprehensive book that encompasses **security of CE devices from hardware security perspective as well as data/media perspective**. The book covers a wide spectrum of modern security topic ranging from fundamentals of IP core protection and hardware-assisted security, Trojan security of DSP cores used in CE hardware, hardware watermarking of DSP /IP cores/IC, symmetrical DSP/IP core protection using watermarking and fingerprinting, computational forensic engineering, structural obfuscation of DSP cores, including a case study on JPEG CODEC, functional obfuscation of DSP cores, hardware AES, watermarking, IoT security, PUF, and media protection.

To the best of my knowledge, there is no book that presents the details of **multiple hardware-assisted security topics** under one umbrella. The book covers various state-of-the-art security attack details possible on DSP IP core as well as recommends interesting solution to them, which have industry applicability. The attack and solutions described in the book can be deployed for hardware security and IP core protection of DSP cores used in CE devices such as digital camera, camcorder, smart phone, smart watch, etc. The book can be extremely interesting for industry practitioners, researchers (both beginners and advanced level), graduate students/scholars, and any academicians.

The book also provides conceptual questions and exercise for each chapter. This book provides a great opportunity to challenge the understanding for each security topic and refine a reader's understanding level. The book can also be easily adopted in for any graduate-level course (master's and Ph.D.) at an institute/university. Additionally, it can also serve as designer's handbook who is eager to learn designing secured DSP hardware core. The book provides several examples of industry-level practical design procedure on secured DSP cores such as digital filters, JPEG, DCT, and FFT. End-to-end design flow for devising secured DSP IP cores is described in each chapter, in the context of different threat models.

To my knowledge, this is the first book ever on CE device IP core protection and hardware-assisted security. It is a must-read book for a hardware designers, FPGA programmers, or CAD tool designers, as well as researchers and academicians involved in the design of CE devices.

Pradipta Patra, Ph.D.
Power Innovation Lab, Frontier Research Team
Samsung Research,
Samsung Electronics, Seoul, South Korea

Foreword by Swarup Bhunia

Digital signal processing (DSP) and multimedia applications affect every spheres of modern society. These applications work on discrete physical signals (e.g., image, video, speech, and physiological signals), analyze their behavior, and recognize the patterns of significance. Over the years, they have made pervasive and profound impact in the consumer electronics (CE) and other industrial sectors. With unprecedented growth of smart and sensor-enriched Internet of things (IoT) edge devices, which interface with the physical world through sensors and actuators, the role of DSP to analyze signals and extract meaningful information from them has become ever more significant. We observe a growing trend of incorporating autonomy and smartness in IoT devices as well as in complex system of systems (e.g., autonomous vehicles). In this context, we expect to see the increasing role of DSP and its integration with artificial intelligence to make complex reliable decisions without or with little human intervention.

Implementation of DSP and multimedia algorithms has taken various paths, ranging from purely software implementation to implementation in a domain-specific processor, i.e., DSP, a reconfigurable platform (e.g., FPGA), to purely custom hardware implementation (e.g., ASIC), and any combination of these. In all possible implementation platforms, traditionally, designers have aimed at optimizing the power-performance-area to achieve target quality of service (QoS). Security issues associated with these implementations and ways to address them have not been at the forefront of efficient realization of wide variety of DSP and multimedia algorithms. However, security for DSP/multimedia applications has emerged as a key design parameter as variety of security issues involving the hardware itself or the content being processed have surfaced. These issues include tampering and piracy of DSP/multimedia cores, while the protection approaches include fingerprinting, watermarking, and secure hardware design.

The book, for the first time to my knowledge, provides a compendium of chapters covering, rather comprehensively, the security issues and protection approaches in hardware/software implementation of DSP/multimedia functions, and touches on both current practices and recent innovations. The comprehensive coverage of DSP and hardware security concepts with relevant background material and description of current industrial practices are the key distinctive features of this textbook.

Some of the salient features of the book include the following:

- The book encompasses hardware-assisted security approaches for DSP/multimedia cores with comprehensive coverage for today's CE devices.
- It includes emerging attack vectors such as Trojan attacks in DSP/multimedia core.

- It covers promising countermeasures for various attacks, e.g., robust watermarking for DSP/multimedia cores, symmetrical protection for DSP/multimedia cores that combine fingerprinting with watermarking and structural and functional obfuscation for DSP/multimedia cores.
- It presents forensic analysis techniques on DSP/multimedia cores to examine and provide proof of ownership with case studies on JPEG CODEC.
- It includes chapters which focus on the protection of digital media and the role of secure hardware design in IoT security.
- It provides several examples of practical design procedure in industry for secured DSP cores, such as digital filters, JPEG, DCT, and FFT.
- It depicts an end-to-end design flow for devising secured DSP IP cores.

The book can serve a valuable resource for a wide readership consisting of students, researchers, and practitioners. It can be an essential read for a hardware designers, FPGA programmers, CAD tool designers, as well as researchers and academicians involved in the design of secure CE devices. It can be easily adopted in as a graduate-level elective course at an institute. Additionally, it can also serve as a handbook for a designer, who needs to understand the security issues and realize secure hardware while implementing DSP/multimedia functions. I believe the book would greatly help the readers to gain knowledge on this topic, motivate them for advanced research, and effectively do the security analysis and security design.

It is my great pleasure to see the book getting published. I wish it all the success in fulfilling the authors' mission.

Dr. Swarup Bhunia,
Preeminence Professor and Steven Yatauro Faculty Fellow,
University of Florida, Gainesville, FL, USA,
ACM Distinguished Speaker

Foreword by Peter Corcoran

We live in a time of continual change and nowhere is this more evident than in the consumer electronics (CE) industry. Nearly 6 years ago, I wrote an editorial for *IEEE Consumer Electronics Magazine* entitled "Change and the Consumer Electronics Industry" (http://bit.ly/CEchange). A key point of this article is that if you want to work in this industry you have to embrace change and go with it. Technology is constantly adapting to the market and the needs of the consumer and this was never truer than it is today.

Today, we live in a world of shortening product cycles, increasing NRE costs, and, unless your organization owns a global technology platform or manufactures' high-value consumer device platforms, there are enormous challenges in building sustainable revenue from our engineering works. Add to these day-to-day concerns the constant struggle to keep up with an ever-changing technology landscape and new technical challenges.

Among these challenges we now find that hardware design is the "new" software are now being built in the same way that complex software products like operating systems were built in the 1990s—constructed from the work and contributions of different engineering teams with specialized skills and integrated onto a single platform—these increasingly complex "system-on-chip" designs that are the mainstay of today's advanced CE products, from smartphones, to TVs, from autonomous vehicles to immersive AR/VR headsets.

With this growth in system complexity, it becomes challenging for smaller engineering companies to build their own integrated circuits and for many of our business model becomes that of the fabless IP supplier. We painstakingly build specialized hardware designs to license into larger SoC platforms owned by global corporations. And we face the same challenges faced by smaller software vendors in the 1990s—how can we prevent the theft of our engineering designs by copying from one IC design to another? After all they are just patterns of registers and transistors, albeit highly complex and sophisticated ones embodying many thousands of hours of engineering effort.

In this context, I am delighted to introduce this new book addressing the increasingly important topic of *IP Core Protection and Hardware-Assisted Security for Consumer Electronics*. This textbook is targeted to this increasingly important design challenge for today's CE engineers.

This text starts with the fundamentals of IP core protection and proceeds to develop a number of alternative and complimentary approaches including a number of detailed case studies. The text examines Trojan security approaches, robust watermarking and symmetrical approaches that combine fingerprinting with watermarking. Forensic analysis techniques are presented to examine and provide proof of ownership, and several advanced techniques based on design and functional obfuscation methodologies are presented together with case studies on JPEG CODEC and DSP cores. Additional chapters consider the protection of digital media and the role of secure hardware design in IoT security.

This text appears at a very timely juncture when CE engineers are facing increasing challenges. It provides a detailed and comprehensive coverage of this increasingly important field and will be of great benefit to both practicing engineers looking to update their knowledge in this field and equally to graduate students working on advanced hardware R&D projects.

Dr. Peter Corcoran
IEEE Fellow and Emeritus Editor-in-Chief,
IEEE Consumer Electronics Magazine
National Inventors Hall of Fame, Ireland

Preface

The recent rapid proliferation of smart city and Internet of things (IoT) framework has only become a reality owing to the expansive developments in the field of consumer electronics (CE), cloud computing, and high-speed networking infrastructure. CE include smart mobile devices that are responsible for combining multiple functions into one device, providing a phone, camera, video recorder, alarm clock, personal music player, paging system, TV remote control, etc. It also acts as gateway for some other devices. It provides user interface to access, manage, and control our things. It is a disruptive technology. Cloud computing has evolved a set of sophisticated infrastructure for storage, messaging, security, content, and connectivity. In terms of connectivity technology, we now have more sophisticated and reliable Wi-Fi connections due to the tremendous growth in computing infrastructure and computing power augmented with the exponential growth of network data for next few years.

However, each of these modules as part of IoT or smart city is solely dependent on the security and protection of the underlying hardware system. For example, smart mobile devices in CE, hardware infrastructure in cloud computing, and system framework for reliable and trusted connectivity technology would need built-in hardware security technologies to protect against unauthorized access to personal information, detect/correct malicious hardware, data integrity, etc. There could be several types of cyber attackers depending on the probability of occurrence and potential damage caused, such as (1) amateur attackers whose probability of occurrence is high but potential damage caused is low, (2) petty criminals whose probability of occurrence is not as high but its potential damage could be higher than amateur attackers, and (3) terrorists activists/state-sponsored attackers whose probability of occurrence is low but potential damage caused could be extremely severe. Thus, in the world of CE and IoT, security, privacy, and protection of hardware and its information are critically imperative. It is universally acknowledged that for IoT, security of underneath hardware is pivotal for reliable functioning. The underlying hardware may have been affected by several threat factors. An adversary may generate secret exist that leaks crucial data such as encryption key employed in secure transmission line, the maker could intervene the design by implanting hardware Trojans, or implant artifacts with recognized dependability susceptibilities. Current generation designs assimilate intellectual property (IP) blocks from manifold vendors, mass-produced and verified by diverse companies worldwide. Accordingly, numerous access points exist for hardware to be attacked. For a trusted or secured hardware design, defense of intellectual property cores is of ultimate significance.

Electronics design industry in future is heading for a paradigm shift toward secured, reliable, and low-cost CE hardware from the conventional ones. Reading this book will enable the readers to push their boundaries of knowledge to dive into the some emerging hardware-assisted security and IP core protection aspect of CE devices. In this book, we present for the first time "intellectual property core protection and hardware-assisted security" for CE. We define "hardware-assisted security" as the security/protection of hardware cores. Intellectual property cores of CE systems and hardware are designed to secure/protect the information processed by CE systems. The term "security" encapsulates a broad theme that covers many aspects including hardware security, protection, privacy, trustworthiness, and intellectual property protection and information security. The system security may refer to the security of the system (e.g., a specific CE device) that handles the data or information. Though there have been textbooks focusing on system security, hardware defense against malicious Trojan, however no textbook encompasses details of wide-spectrum hardware threats/attacks and its hardware-assisted defense mechanism. In this book, explicit focus has been made on hardware (IP core) Trojan, IoT security, hardware watermarking, hardware structural obfuscation, hardware functional obfuscation, hardware encryption, watermarking encryption hardware, forensic engineering, JPEG core, memory attacks, security, privacy, IP right in IoT and CE systems obfuscation, IP core fingerprinting, PUF, and several others.

Broadly, the theme of this book includes the following:

1. Security aspects in IoT
2. Hardware security/protection against Trojans
3. Hardware security/IP core protection against reverse engineering attacks
4. Symmetrical IP core protection of CE hardware
5. Active and passive IP core protection by functional and structural obfuscation
6. Forensic engineering-based protection of CE hardware
7. Protection mechanisms of IC/IP buyer
8. Protection mechanisms of IC/IP seller by watermarking and fingerprinting
9. Hardware-assisted approaches for media and information security
10. PUF-based security and protection methods of CE hardware

Chapter 1 presents an "Introduction to IP core protection and hardware-assisted security of consumer electronics." The salient features of this chapter include providing a security perspective, defining hardware-assisted security, providing background on IP cores and its significance, necessity of security and protection of DSP kernels—with details on high-level synthesis framework and finally ending with hardware-assisted media protection.

Chapter 2 presents "Security in consumer electronics and Internet of things." The salient features of this chapter include IoT architecture, challenges, and its applications; the big picture on security, privacy, IP right in IoT and CE systems; memory security, radio frequency identification (RFID) security; near field communications (NFC) security; smart healthcare and transportation security, firmware security, and finally ending with blockchain technology.

Chapter 3 presents "Trojan security-aware DSP IP core and integrated circuits." The salient features of this chapter include types of hardware Trojan, hardware

Trojan in a 3PIP module, Trojan security approaches, Trojan security-aware DSP kernel/IP core design process, and finally ending with analysis on case studies.

Chapter 4 presents "IP core and integrated circuit protection using robust watermarking." The salient features of this chapter include introduction to hardware (IP core) watermarking, description of selected IP core watermarking approaches, design process of a watermarked IP core, properties of a hardware watermarking, and finally ending with analysis on case studies.

Chapter 5 presents "Symmetrical protection of DSP IP core/IC using fingerprinting and watermarking." The salient features of this chapter include background on watermark and fingerprint, its associated threat model, fundamentals of IP core protection, symmetrical IP cores protection for DSP kernels, case study on symmetrical IP core protection, and finally ending with analysis on different DSP kernel test cases.

Chapter 6 presents "Computational forensic engineering (CFE) for resolving ownership conflict of DSP IP core." The salient features of this chapter include overview on CFE for IP core protection, IP core feature extraction algorithms, important characteristics of customized CFE, and finally ending with analysis on case studies.

Chapter 7 presents "Structural obfuscation of DSP cores used in CE devices." The salient features of this chapter include introduction to structural obfuscation, its associated threat model, relative perspective of logic, code, and structural obfuscation techniques, compiler transformation-driven structural obfuscation, design process of low-cost structurally obfuscated IP core, case study on multistage structural obfuscation, and finally ending with case studies on DSP kernels.

Chapter 8 presents "Functional obfuscation of DSP cores used in CE devices." The salient features of this chapter include attack scenarios and threat model, selected functional obfuscation approaches for DSP IP core, design of functional obfuscated DSP IP core, security of functionally obfuscated DSP core design, optimization engine for functional obfuscation of DSP kernels, and finally concluding with analysis on DSP kernel case studies.

Chapter 9 presents "Obfuscation of JPEG CODEC IP core for CE devices." The salient features of this chapter include the overview of JPEG compression and decompression process, design process of structurally obfuscated JPEG IP core, its associated threat model, formulation, optimization framework, implementation of JPEG CODEC IP core, and finally concluding with design analysis.

Chapter 10 presents "Advanced encryption standard (AES) and its hardware watermarking for ownership protection." The salient features of this chapter include introduction to the AES algorithm, its description and custom hardware design, AES digital watermarking, process of embedding watermark in AES hardware, and finally ending with the design analysis.

Chapter 11 presents "Hardware approaches for media and information protection and authentication." The salient features of this chapter include introduction to digital rights management (DRM), copyright protection of multimedia, hardware versus media protection, general framework for copyright protection, types of digital watermarks, applications of digital watermarks, desired characteristics of watermarks, hardware-based approaches for watermarking, dynamic watermarking in smart car, side-channel information leakage attacks and countermeasures, attacks on watermarks and watermarking systems, and finally ending with the limitations of watermarking.

Chapter 12 presents "Physical unclonable functions (PUFs)." The salient features of this chapter include principles of PUFs, properties or characteristics of PUFs, classification of PUFs, description of different types of PUFs such as ring oscillator based, static random access memory based, memristor based, diode based, magnetic based, practical implementations of PUF, case study of PUF, and finally ends with discussion on its issues and challenges.

This book aims to present novel solutions for any security/protection problems related to hardware used in CE. Authors believe that there is no book that presents details of numerous hardware-assisted security topics under one canopy. The theme is important for researchers and graduate students in different areas of specialization, as it encompasses overlapping contents of EDA-CAD, hardware design security, VLSI design (high-level synthesis, RTL, gate-level synthesis, etc.), IP core protection, optimization using evolutionary computing, system-on-chip design, and finally application-specific processor/hardware accelerator design for CE applications. All the aforesaid topics are encapsulated in the proposed theme where researchers, practitioners, industry experts, CAD/design engineers, system architects, and students are expected to show interest in reading.

The book offers theoretical exercise for each chapter. This presents abundant prospect to test the understanding for each security topic and hone readers understanding. The book is prepared keeping in mind that can be easily integrated to any master's and Ph.D. level course as well as senior bachelor's course. The book provides materials to train senior-level engineering students on hardware-assisted security topics. Furthermore, it also serves as designer's handbook for a designer who is eager to learn designing secured hardware, especially IP core for DSP kernels. Each chapter also includes end-to-end design flow for secured IP core architecture from the perspective of different threat/attack models.

Dr. Anirban Sengupta, Senior Member IEEE
IEEE Distinguished Lecturer
Associate Professor
Computer Science and Engineering
Indian Institute of Technology Indore
Email: asengupt@iiti.ac.in

Dr. Saraju Mohanty, Senior Member IEEE
IEEE Distinguished Lecturer
Professor
Computer Science and Engineering
University of North Texas, USA
Email: saraju.mohanty@unt.edu

List of acronyms

Acronyms	Expansion
3PIP	Third-party IP
A	
AMS-SoC	Analog/mixed-signal system on a chip
AES	Advanced encryption standard
ASIC	Application-specific integrated circuit
ARF	Autoregression lattice filter
ASAP	As Soon As Possible
ALAP	As Late As Possible
ASCII	American Standard Code for Information Interchange
AP	Anti-parallel
B	
BCD	Binary-coded decimal
BFOA	Bacterial foraging optimization algorithm
BER	Bit error rate
BPG	Better portable graphics
BCH	Bose, Chaudhuri, and Hocquenghem
C	
CE	Consumer electronics
CODEC	Compression and decompression
CFE	Computational forensic engineering
CDFG	Control data flow graph
CU	Crypto unit
CIG	Colored interval graph
CS	Control step
CAD	Computer-aided design

CNF	Conjunctive normal form
CPTWG	Copy Protection Technical Working Group
CPA	Correlation power analysis
CRP	Challenge response pair
CNT	Carbon nanotube
CNT-FET	Carbon nanotube field-effect transistor
CED	Concurrent error detection
CD	Compact disc
CMOS	Complementary metal-oxide semiconductor

D

DSP	Digital signal processing
DCT	Discrete cosine transformation
DSE	Design space exploration
DIT	Decimation in time
DFG	Data flow graph
DoS	Denial of service
DMR	Dual modular redundancy
DWT	Discrete wavelet transformation
DRM	Digital rights management
DVD	Digital video disc
DPCM	Differential pulse code modulation
DST	Discrete sine transform
DPA	Differential power analysis
DES	Data encryption standard
DPR	Dynamic partial reconfiguration
DSP-IP	Digital signal processing-Intellectual property

E

EZW	Embedded Zero Wavelet
EPR	Electronic patient record
EEG	Electroencephalogram
EM	Electro-migration
ECU	Electronic control unit

F

FFT	Fast Fourier transform
FIR	Finite impulse response
FPGA	Field-programmable gate array
FL	Framing logic
FSM	Finite state machine
FE	Forensic engineering
FU	Functional unit
FT	Fourier transform
FoM	Figures of Merit
FFO	Feed-forward MUX PUF overlap structure
FFC	Feed-forward MUX PUF cascade structure
FFS	Feed-forward MUX PUF separate structure

G

GPS	Global positioning system
GSM	Global System for Mobile communication
GDS	Graphic Data System
GF	Galois Field
GOP	Group of picture
GNRFET	Graphene nano-ribbon field-effect transistor

H

HLS	High-level synthesis
HWT	Haar wavelet transformation
HDL	Hardware description language
HIPAA	Health Information Privacy and Security Act
HITECH	Health Information Technology for Economic and Clinical Health
HD	Hamming distance
HVS	Human visual system
HEVC	High-efficiency video coding
HCI	Hot carrier injection
HRS	High resistive state
HRNG	Hardware random number generation
HDA	Helper data algorithm

I

IoT	Internet-of-things
IP	Intellectual property
IPR	Intellectual property rights
IC	Integrated circuit
ID	Distinctive identifier
IDCT	Inverse discrete cosine transformation
IoE	Internet of Everything
IDS	Intrusion Detection Systems
IoMT	Internet of Medical Things
IoHT	Internet of Health Things
ISCAS	International Symposium on Circuits and Systems
IZZ	Inverse zigzag
ILB	IP functional locking blocks

J

JPEG	Joint Photographic Experts Group
JND	Just noticeable distortion

L

LoRaWAN	Long range wide area network
LUT	Lookup table
LT	Logic transformation
LU	Loop unrolling
LICM	Loop-invariant code motion
LFSR	Linear feedback shift register
LSB	Least significant bit
LRS	Low resistive state

M

MPEG	Moving Picture Experts Group
MSE	Mean square error
MND	Minimally noticeable distortion
ME	Motion estimate
MC	Motion compensation
MTJ	Magnetic tunnel junction

N

NFC	Near field communication
NIST	U.S. National Institute of Standards and Technology
NBTI	Negative bias temperature instability

O

OCR	Office for Civil Rights
ORF	One-way random functions

P

PUF	Physically unclonable functions
PIN	Personal identification number
PSO	Particle swarm optimization
PSO-DSE	Particle Swarm Optimization-Design space exploration
PSNR	Peak signal-to-noise ratio
POK	Physically obfuscated key
PBTI	Positive bias temperature instability
PMOS	P-type metal-oxide-semiconductor
PCM	Phase change material
PRNG	Pseudo-random number generation

R

RT	Register-transfer
RFID	Radio-frequency identification
RF	Radio-frequency
RTL	Register transfer level
RE	Reverse engineering
ROE	Redundant operation elimination
RFL	Robust functional locking
RF-Ob	Robust functional obfuscation
RLE	Run-length encoding
RC	Round coefficient
ROI	Region of interest
RF	Radio frequency
RO	Ring oscillator
RPR	Runtime partial reconfiguration

S

SM	Sensor module
SSH	Secure shell
SSL	Secure socket layer
SSI	Signal strength intensity
SoC	System-on-chip
SCPA	Semiconductor Chip Protection Act
SAT	Satisfiability
SoO	Strength of obfuscation
SNR	Signal-to-noise ratio
SDC	Secure digital camera
SBPG	Secure better portable graphics
SVM	Support vector machine
SRAM	Static random access memory
SHIC	Super high information content
SAC	Strict avalanche criterion
SHIC	Super-high information content
SEM	Scanning electron microscopy

T

THT	Tree height transformation
TDDB	Time-dependent dielectric breakdown
TC	Thermal cycling

U

UAV	Unmanned aerial vehicles
UPC	Universal product code

V

VoIP	Voice over Internet protocol
VHDL	Very-high-speed integrated circuit hardware description language
VLIW	Very long instruction word

W

WT	Wavelet transform

Z

ZZ	Zigzag

List of notations

Chapter 2

M	= memory
V	= contents of memory
A	= addressed of memory
$M[R, A]$	= memory at the address A is read and it will return the data that is associated at A
$M[W,A,V]$	= the value V is written into the address A
T	= time instance
$M[W,A,V,T]$	= the value V is written into the address A at time T
$M[R,A,T]$	= memory at the address A is read and it will return the data that is associated at A at time T
V_A	= the value at the address location A at a time T_A
V_A^*	= the value at the address location A at a time T_B

Chapter 3

En	= enable of BCD Adder
P	= number of counters
Q^*	= incorrect output
Q	= correct output
O	= primary output
A and B	= two primary inputs
S	= sum of adder
U or UF	= unrolling factor
I	= total number of iterations
$I_1, I_{2,}, \ldots, I_n$	= total number of nested loops
P	= particle encoding
R_n	= resource array
UF_1	= the unrolling factor value for outermost nested loop
UF_n	= the unrolling factor value for innermost nested loop
M_v	= vendor allocation mode

V_1, V_2	= two distinct vendors
$N(R_D)$	= the number of resources of resource-type R_D
U_n	= unrolling factor of nth loop of n-times nested loop-based CDFG
X_i	= optimal design solution
A_T^{DMR} and T_E^{DMR}	= the total area and total execution delay of the DMR design
A_{cons} and T_{cons}	= user-specified area and execution delay constraints
$A(R_i^{V_j})$	= the area of resource type R_i corresponding to vendor-type v_j
$N(R_i^{V_j})$	= the number of instance of resource type R_i corresponding to vendor-type v_j used in the design
R_i	= resource type
C_{body}^{DMR}	= number of control steps required to execute the loop body of Trojan-secured DMR CDFG
C_{first}^{DMR}	= number of control steps required to execute the first iteration of Trojan-secured DMR CDFG
Δ	= delay of a single control step
$D(R_i^{V_j})$ and $D(R_i'^{V_j})$	= the delay of resource type R_i corresponding to vendor-type v_j of original and resource type R_i' corresponding to vendor type v_j of duplicate unit
cs	= the control step number where the R_i is allocated
$C_f(X_i)$	= the design cost of Trojan-secured DSP IP core with design solution X_i
A_{max}^{DMR} and T_{max}^{DMR}	= the maximum area and execution in the design space
w_1 and w_2	= user-specified area and delay weightage factor
$x(i)$	= the input signal in a DCT function
$X(m)$	= the output signal of a DCT function
N	= number of data points in a DCT function
C_1 to C_7	= the DCT coefficients in ascending order (C_1 indicates the maximum positive DCT coefficient value and C_7 indicates the minimum positive DCT coefficient value)
k_1 to k_8	= DCT coefficients
$x[0]$ to $x[7]$	= input signals primary inputs to a DCT function
$X[0]$ to $X[7]$	= output signal of a DCT function
$x(n)$	= a sequence of N values in a DFT function
"N"	= represented as an integer power of 2
$X(k)$	= complex-valued sequence
W_N	= the complex-valued phase factor

$G(k)$ and $H(k)$	= the two $N/2$-point DFTs of even and odd sequences
$A(k)$ and $B(k)$	= the two $N/4$-point DFTs of even and odd sequences corresponding to $G(k)$
$C(k)$ and $D(k)$	= the two $N/4$-point DFTs of even and odd sequences corresponding to $H(k)$
M	= filter order of FIR
h_i	= FIR coefficient
$x(n)$	= input impulse
$y(n)$	= output impulse
$N(R_{\mathrm{D}})$	= the number of resource of Dth dimension of the particle
"α"	= a random value within the range of maximum and minimum possible value of resources
$V^{+}{}_{d_i}$ and V_{d_i}	= new and current velocity of ith particle in dth dimension
R_{d_i}	= resource value of ith particle in dth dimension
$R_{d_{\mathrm{lb}_i}}$	= local best position of ith particle in dth dimension
$R_{d_{\mathrm{gb}}}$	= global best of dth dimension
ω	= the inertia weight
b_1 and b_2	= social and cognitive factors
r_1 and r_2	= random number between 0 and 1
$R^{+}_{\mathrm{d}_i}$	= the updated resource value of ith particle in dth dimension
V^{+}_{d}	= new velocity
maximum and minimum limit of velocity	= $+V^{\max}\mathrm{d_i}$ and $-V^{\max}\mathrm{d_i}$
au	= area unit (transistor)
ns	= nano seconds
$I_{\max}$	= terminating criterion of PSO
Δ	= maximum iterations for which global best solution is checked before PSO is terminated

Chapter 4

$M_i(n)$	= ith sub-mark or mathematical function
C_i	= the clock cycle number during which the sub-mark is sent to a "temporally free" IP output port
M	= filter order of a digital FIR
h_i	= FIR coefficient
$x(n)$	= input impulse

$y(n)$ = output impulse

$H(e^{j\omega})$ = transfer function

δ_1 and δ_2 = parameters specifying allowable pass-band and stop-band of a digital filter

N = order of a digital IIR Butterworth filter

Ω_c = cutoff frequency

$H(s)$ = transfer function

B_0B_1 = parameters of an IIR Butterworth filter

$H(z)$ = represents the transfer function of the filter in frequency domain

"α", "β", "γ", "'i", "I", "T", and "!" = signature digits of a watermark

P_c = probability of coincidence metric

"w" = watermark strength/size

$N(R_i)$ = number of functional unit of type i

D = the type of functional unit

C = number of colors used in colored interval graph before embedding watermark

T_t = tamper tolerance ability

"v" = number of variables used in encoding

$C_f(x_i)$ = the watermarked design cost for user provided resource configuration X_i

L_{total} and A_{total} = total watermarked design latency and area

L_{max} and A_{max} = maximum design latency and area based on the available resources in the module library

w_1 and w_2 = weightage parameters for latency and area

Chapter 5

X_i = optimal design solution

$N(R_D)$ = the number of resource type R_D

A_{total} and L_{total} = total area and latency of the symmetrically protected design

A_{cons} and L_{cons} = user-provided area and latency constraints

$A(R_i)$ = area of resource R_i

A(buffer) = area of a buffer

N(buffer) = number of buffers

A(mux) = area of multiplexers

N(mux) = number of multiplexers

L_{total} = total design latency of symmetrically protected reusable IP core

$total_{cs}$ and $delay_{cs}$	= total number of control step of the design and execution delay of each control step
$C_f(X_i)$	= design cost of the symmetrically protected reusable IP core with resource configuration X_i
A_{max} and L_{max}	= maximal area and latency in the design space
w_1 and w_2	= user-provided area and latency weightage factor
c	= the number of colors required
W	= the size of watermarking constraints
P_c	= probability of coincidence
F	= size of fingerprint constraints

Chapter 6

A_n	= finite set of algorithms/tools in CFE
P	= problem
S	= solution
n	$= 1, \ldots, i$
k	= number of clusters such that each cluster contains solutions that are closely matched in CFE
IP_{Vn}	= vendor competing for ownership of an IP core
n	= is the number of vendors claiming ownership of IP core
IP_{ID}	= IP core whose ownership is to be awarded/IP core under test
$CS_S(FU_i)$ and $CS_S(FU_j)$	= starting control step of ith and jth FU
$CS_E(FU_i)$ and $CS_E(FU_j)$	= ending control step of ith and jth FU
CS	= control step
MUL_En_1	= multiplier enable
$CS_S(FU_i)$ and $CS_E(FU_i)$	= starting and ending control steps of ith FU
$CS_E(N)_1$ and $CS_E(N)_2$	= the ending control step of data sets 1 and 2
μ	= match percent

Chapter 7

A_{cons}	= user-provided area constraint
L_{cons}	= user-provided latency constraints
X_i	= optimal design solution
$N(R_D)$	= number of resource-type R_D
A_{total}^{obf} and L_{total}^{obf}	= total area and latency of the structurally obfuscated design
$A(R_i)$	= area of resource R_i

A(buffer)	= area of a buffer
N(buffer)	= number of buffers
A(mux)	= area of multiplexers
N(mux)	= number of multiplexers
D	= total number of dimensions
${L_{\text{body}}}^{\text{obf}}$	= single iteration execution delay of the loop body of obfuscated CDFG
I	= total loop iteration (loop count)
${L_{\text{first}}}^{\text{obf}}$	= delay to execute the first iteration of obfuscated CDFG
UF	= unrolling factor
$C_f(X_i)$	= design cost of the structurally obfuscated DSP IP core with resource configuration X_i
${A_{\text{max}}}^{\text{obf}}$ and ${L_{\text{max}}}^{\text{obf}}$	= maximum possible area and latency in the design space
w_1 or $Ø_1$ and w_2 or $Ø_2$	= user-specified area and latency weightage factor
V_{d_i} and $V_{d_i}^{+}$	= current and new velocity of ith particle in dth dimension
X_{d_i}	= resource value or unrolling factor of ith particle in dth dimension
$X_{d_{\text{lb}_i}}$	= local best position of ith particle in dth dimension
$R_{d_{\text{gb}}}$	= global best of dth dimension
ω	= the inertia weight
b_1 and b_2	= social and cognitive factors
r_1 and r_2	= random numbers
X_{d_i}	= new resource value or unrolling factor of ith particle in dth dimension
g_{best}	= current global best in the population
l_{best}	= own best location of a particle
I_{max}	= maximum number of predefined iteration counts
Δ	= number of iterations to check if no improvement in global best is found
P	= population size in PSO
P^{obf}	= power of obfuscation
n_i	= unique number of modified nodes due to ith stage transformation method
n_T	= total number of nodes of the original (before applying any transformation method) DSP circuit

Chapter 8

μ	= a random integer number used in PSO during functional obfuscation
T_{ILB}	= total number of ILB structure/template available to insert in the design
K_{s}	= number of possible key-space for a locked netlist through ILBs
“b”	= the number of key-bits per ILB
m	= the output bit-width of each functional unit
f	= the number of functional units used to design the datapath
P_i	= ith encoded particle in the particle population
$n(R_{\mathrm{d}})$	= the number of resources of type R_{d}
d	= the total number of resource types available
“α”	= a random integer number within the range of maximum and minimum resource value of resource type R_{d}
$C_f(P_i)$	= normalized weighted cost/fitness of a particle P_i
$\emptyset_1$ and $\emptyset_2$	= user-provided weightage factor for area/power and delay components, respectively
P^{FO} and D^{FO}	= the total power consumption and execution delay of a functionally obfuscated (FO) design
${P_{\max}}^{\mathrm{FO}}$ and ${D_{\max}}^{\mathrm{FO}}$	= the maximum possible power (estimated as power consumption of maximum number of hardware resources available in the module library)
${D_{\max}}^{\mathrm{FO}}$	= maximum possible delay (estimated as delay utilizing minimum number of hardware resources) of FO design
$V_{d_i}^{+}$ and V_{d_i}	= are new and current velocity of ith particle in dth dimension
R_{d_i}	= is the resource value of ith particle in dth dimension
$R_{d_{\mathrm{lb}_i}}$	= is the local best position of ith particle in dth dimension
$R_{d_{\mathrm{gb}}}$	= is the global best of dth dimension
ω	= is the inertia weight
b_1 and b_2	= are social and cognitive factors
r_1 and r_2	= are random numbers between 0 and 1
$R_{\mathrm{d}_i}^{+}$	= is the updated resource value of ith particle in dth dimension
d	= number of dimensions
SoO	= strength of obfuscation

Chapter 9

T or $T(i, j)$	= 2D-DCT coefficient matrix
“N”	= the size of the DCT matrix

"i" and "j"	= the *i*th row, *j*th column component of the 2D-DCT matrix
$C_1, \ldots, C_7$	= DCT coefficients
X	= transformed matrix after DCT processing
M	= input image matrix
D	= matrix $(d_{11}, d_{12}, \ldots, d_{88})$ transforms the rows of the "M" block
Q	= quantization matrix
X'	= 2D quantized 8×8 block
X''	= decompressed 8×8 block
O	= IDCT block of X'' block
E	= matrix $(e_{11}, e_{12}, \ldots, e_{88})$ transforms the rows of the X'' block
$Cf(X_i)$	= hardware design cost for design solution Xi
${L_{total}}^{JPEG}$ and ${A_{total}}^{JPEG}$	= the total design latency and total hardware area of the obfuscated JPEG IP core
${L_{max}}^{JPEG}$ and ${A_{max}}^{JPEG}$	= the maximum possible design latency and hardware area
w_1 and w_2	= weightage factor of two design parameters (latency-area)
$A_1, \ldots, A_8, \ldots Q_1, \ldots, Q_8$	= the primary inputs are represented through
$i_1, i_2, \ldots, i_{135}$	= dependent inputs
i_{136}	= represents the output
SoO	= strength of obfuscation
PSNR(A,B)	= Peak signal-to-noise ratio for two images A and B of size $M \times N$
MAX_A	= the maximum pixel value in image A
MSE(A, B)	= mean square error between images A and B

Chapter 10

K	= bit-key length
n_r	= number of internal rounds
i	= index indicating the current round
A	= input data size
B	= output encrypted data
$X_0, \ldots, X_{15}$	= input byte to S-box
$Y_0, \ldots, Y_{15}$	= output byte from S-box
$Z_0, \ldots, Z_{15}$	= output from Mix Column diffusion layer
$K_0, \ldots, K_{15}$	= keys

k_i	= subkey
$O_0, \ldots, O_{15}$	= output from key addition layer
$P(x)$	= modular reduction function
RC	= round coefficient
$M_0, \ldots, M_3$	= inputs to XOR blocks in Key Scheduler at round 0
$M_4, \ldots, M_7$	= outputs from XOR blocks in Key Scheduler at round 1
"α" and "β"	= embedding signature digits during AES hardware watermarking
S_0, S_1	= select lines of multiplexer in Mix Column structure
$S_0^{'}$, $S_1^{'}$	= select lines of de-multiplexer in Mix Column structure

Chapter 11

I	= original input image
W	= watermark input
I'	= output watermarked image
E	= encoder function
D	= decoder function
J	= image which needs to be challenged
W'	= the watermark recovered by the decoder
δ	= threshold
$C_\delta\ (W, W')$	= correlation between W and W'
w_i	= watermark sequence
$F_k(u, v)$	= the DCT coefficient of the *k*th image block
α_k	= scaling factor
$F_k^{\,i}(u, v)$	= watermarked DCT coefficients
K	= key
I_w	= watermarked image
I_N	= the neighborhood image pixel gray value
α_i	= global scaling factor
C_1, C_2, C_3, C_4	= linear regression coefficients used to linearize cubic root
i_{Wk}	= *k*th watermarked image block
β_k	= embedding factor
μ_{kI}	= normalized mean gray value of *k*th image block
μ_I	= normalized standard deviation of the gray values of the *k*th image block
σ_{kI}	= normalized mean gray value of the image
C_I	= frequency components

C_{Iw} = frequency components of the watermark image
C_{Iwk} = frequency components of *k*th block of the watermark image
ε = probability of error

Chapter 12

i, j = two different chips
R_i, R_j = response generated by chips i and j
HD = hamming distance
n = length of response
K = key-bit length
R'_i = response generated by the chip i in a different environment
x = number of different samples
S = reliability
t_{inv} = delay of an inverter
V_{DD} = supply voltage
C_L = the load capacitance
W = effective width of the channel
L = effective length of the channel
m = mobility of carriers
$V_{D_{SAT}}$ = saturation source-to-drain voltage
t_{ox} = oxide thickness
e_{ox} = oxide dielectric permittivity
V_T = threshold voltage
f_o = oscillation frequency of the ring oscillators
N_{inv} = number of inverters
$C(n, k, t)$ = binary linear code
n, k = length of the dimension of code
t = maximum number of errors that can be corrected
p_0 and p_1 = probabilities of binary source of outputs
p_{max} = maximum probability
$p_{A_{max}}$ = maximum probability of A
$p_{B_{max}}$ = maximum probability of B
$M(t)$ = memristance of the memristor
R_0 = maximum resistance offered
ΔR = the difference between the maximum and minimum memristance offered

η = applied voltage polarity
μ = mobility
D = device thickness
$T_{\mathrm{wr,min}}$ = minimum threshold of time
V_{word} = voltage of the word lines
V_{bit} = voltage of the bit lines
N_{m} = noise margin

Chapter 1

Introduction to IP core protection and hardware-assisted security of consumer electronics

This chapter presents an overview of the book and discussions on topics that could serve as an introductory part toward the overall content of the book. It is assumed that the readers will have some subject knowledge on topics such as very large scale integration (VLSI), embedded systems, hardware design flow, etc., before adopting the section of this book.

1.1 Consumer electronics and security perspectives

Consumer electronics (CE), embracing high-end devices ranging from digital camera, multi-spectral camera, IP TV, smart tablets, night vision camera to smart meter, accompanied by data and communication know-how makes emerging smart cities and internet-of-things (IoT) a reality. In today's world of CE and IoT, security, privacy, and protection of hardware and its information are increasingly becoming critically imperative. It is universally accepted that for IoT, security of underneath hardware is pivotal for its correct operation. The underlying hardware is affected by several threat factors. An adversary may generate secret exist that leaks crucial data such as encryption key employed in secure transmission line, the maker could meddle the design by implanting 'hardware Trojans,' or inserting artifacts with recognized dependable susceptibilities. Current generation designs and assimilates intellectual property (IP) blocks from manifold vendors, which are mass-produced and verified by diverse companies worldwide. Accordingly, numerous access points exist for any typical hardware and are prone to be attacked. For a trusted hardware design, defense of IP cores is of utmost importance and significance (Sengupta and Kundu, 2017; Sengupta *et al.*, 2018).

A typical CE system (Figure 1.1) has digital signal processing (DSP), embedded processor, memory, system software, and application software. Significantly, various forms of attacks can occur to this CE system, i.e.: system security, information security, system privacy, information privacy, network security, firmware security, system trustworthiness, hardware IP rights, and information copyrights. The origins of all these attacks vary and are different, for example, they can be remote or local. They can be either attacked by software or hardware. They can attack either software

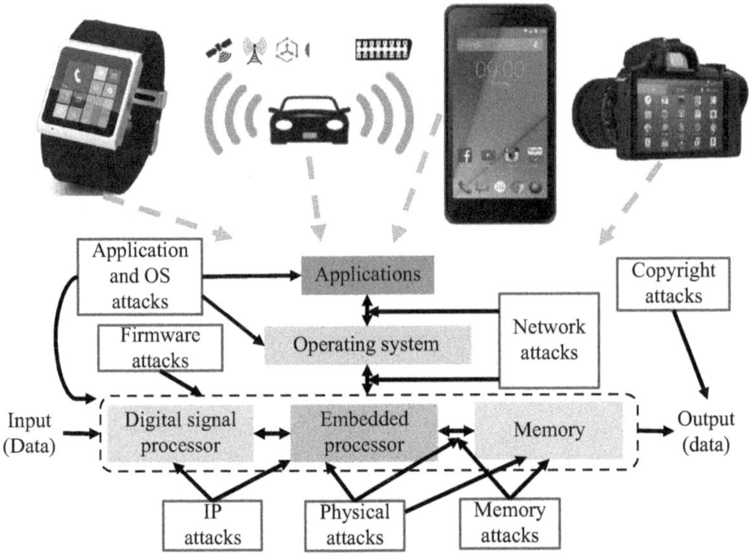

Figure 1.1 Consumer electronics components and security vulnerability

or hardware components of the CE system. Thus, it is a quite challenging and difficult task for any system design engineer to ensure that the CE system that is being designed is completely full-proof. There is no single solution that can work for (1) overall system for all types of attacks and (2) for all types of hardware (e.g., digital, analog, and radio-frequency components). Thus, the protections from intellectual property rights (IPR) of both the hardware components and the information (including media) that a CE system processes needs different solutions. The chapter corresponding to this section will discuss the details of these attacks and explore the corresponding possible solutions for their mitigation. Moreover, many of these attacks and solutions will be discussed throughout the different chapters of this book. The hardware-assisted security is defined as the 'security provided by hardware' for: (1) information being processed, (2) hardware itself, and (3) overall system (Sengupta *et al.*, 2018; Mohanty 2018a, 2018b). The hardware-assisted security solutions may be in different forms such as: (1) additional hardware components used for security, (2) hardware design modification performed, and (3) system design modifications performed (Mohanty, 2018a, 2018b). Overall, a system can be considered as trustworthy, if it incorporates security, privacy, and IP protection solutions during the design phase while obtaining the components from trusted vendors and trusted manufacturing units.

It is a well-known fact that the CE system is a heterogeneous combination of various types of hardware, software, and firmware and can be conceptualized as analog/mixed-signal system on a chip (AMS-SoC) (Mohanty, 2015). CEs or CE system are essentially AMS-SoCs, and it includes almost everything, e.g., smart watch, smart camera, smart phone, smart car, smart healthcare device, etc. (see Figure 1.1). In the hardware-digital-hardware, analog hardware, RF hardware, and mixed-signal hardware are needed for various signal processing and communications that happens in the CE system. Memory is needed for buffer processing and storage. The software are system software, application software, and middleware. The security vulnerability can arise from any of these components of a typical CE system (Mohanty, 2018a, 2018b). This section will present a brief overview, and detailed discussions are presented in the corresponding chapters. The following sections discuss about hardware design flow and more specifically at the architecture level.

1.2 Hardware-assisted security and IP core protection

Currently, design age of CE devices comprises complex data/control intensive hardware such as DSP cores in the form of IP cores which depend on the integrated circuit (IC) supply chains globally. To enhance the design efficiency and improve the design speed, the use of IP cores for DSP applications in CE devices, normally provided by a third-party vendor, has turned into a typical exercise in the CE industry. Nonetheless, surging threats to security is prevalent and rising piracy concerns make global supply chains vulnerable (Sengupta, 2015, 2016). Figure 1.2 shows the hardware threats in a typical system-on-chip design cycle comprising third-party IP (3PIP) cores. The first location is during the process of importing 3PIP cores from outside vendors and second location is at the foundry where copying of design netlist is performed. Thus, security is mandatory in these two locations. Furthermore, Figure 1.3 shows the various elements in a chip design cycle which are trustworthy and untrustworthy by nature. Other forms of hardware threats such as infringement of IP core ownership, IP core buyer ownership, etc.,

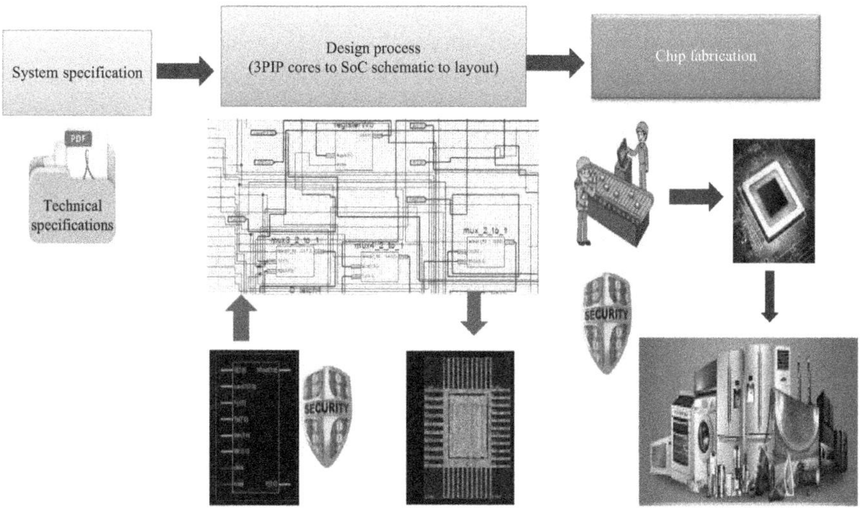

Figure 1.2 SoC design and fabrication including 3PIP cores indicating possible hardware vulnerabilities (Sengupta, 2017)

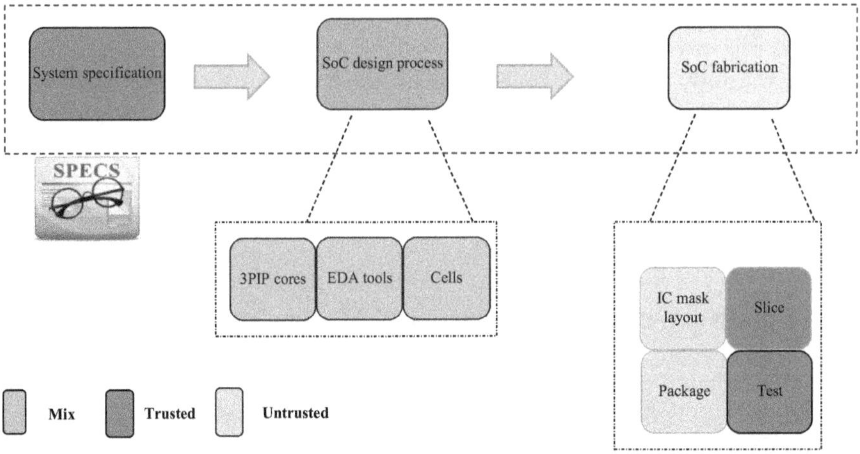

Figure 1.3 Chip design flow used in CE devices indicating trustworthy and untrustworthy elements (Sengupta, 2017a, 2017b)

are also possible which will be discussed in the latter half of this chapter and other chapters.

In this book, we define 'hardware-assisted security' as the security/protection of hardware/IP core for DSP and multimedia applications in CE devices or information by a hardware/system of CE devices. The term 'security' encapsulates a broad theme that covers many aspects including hardware security, protection, privacy, trustworthiness, and IP core protection and information security. The terms 'security' depicts the security of a system (such as CE device) that handles the data or information. In a CE-based structure, defense of its hardware and IP cores is very essential. Therefore, the practice of using secured IPs becomes significant. In this age of smart cities, growing CE hardware design complexity is outpacing the designer yield, thus, resulting into bigger undertakings. Additionally, the present group of CE designs is enormously reliant on global IP supply chains. To upsurge the design productivity and to reduce the design time, the usage of IP cores often supplied by a third-party vendor has turned into a vital and *de-facto* practice in the industry. It is estimated that top counterfeits could have an impact of $300B on the semiconductor market. However, increasing threats to security and swelling piracy concerns do affect the global supply chains. Thus, information security that covers the security of data, information, and multimedia which are handled all the time by CE devices becomes important (Sengupta *et al.*, 2018; Mohanty 2018a, 2018b).

1.3 Intellectual property (IP) cores/hardware

1.3.1 Utility of IP cores in CE devices

The express intensification of silicon technology resulted into terrific upsurge in design complexity of CEs devices. In the realistic operation of multi-billion dollar foundries, more and more applications are unified on a single silicon die. Synchronously, demands of market have directed bigger design start-ups, reduced design time, and growing time-to-market burden. To defy these difficulties, IP cores, especially for complex DSP and multimedia applications, have become a pragmatic solution. An IP core once corroborated in a design with a specific process technology can be employed to other designs that use the identical process technology. In turn, this eradicates the necessity to embrace such a core in most of the system-level simulation testing, which in turn, decreases the intricacy of inclusive system design and verification. To provide a simple illustration, once a JPEG CODEC core or wavelet transformation core used in image compression is validated for a specific silicon process, it can be assimilated into SoC designs of other CE devices easily (Sengupta, 2016).

1.3.2 Why security and protection of hardware/IP cores?

Demands of sophisticated CE devices, including complex DSP/multimedia cores, have led to extensive usage of reusable IP cores. The swelling reuse of such complex IP cores inexorably uncovers them to perils of counterfeiting. Since an IP core can represent several years of design, R&D, and validation, a critical concern is how to safeguard this asset? and how can a counterfeit of a core be identified/prevented?

To the best of our knowledge, it is a fact that the privileges of the actual IP owner can be battered both deliberately and inadvertently. This can be conceivable by an abuse of the license treaty by deploying the IP core in numerous devices, when a solo license has been obtained. There can be a high enticement for a foundry or even to other IP-core makers to integrate a high-quality design into other system components where the actual licensing rule can prevents this. As the systems grow composite, there is a rising trend for IP core makers to sub-license other IP cores for including that in their designs, and once a design process has been finished, it is alluring to reuse it in other supplementary products that may not encompass the actual licensing. Another impending threat to IP cores is that of piracy, where by dishonest means or reverse engineering, direct stealing/copying of the IP for reuse without authorization are done. As a result of the IP theft, the attacker possibly may even claim the entitlement of the IP. Thus, a method of annulling dishonest claims of IP ownership is highly desirable. For the makers of the IP core or a whole-chip design, it is both problematic and costlier to demonstrate that their IP in the product has been employed unlawfully. In case of silicon chips, one of the realistic methods of identifying an IP core is to acquire an example of the product under test for sending to a professional laboratory for analysis. Furthermore, injection of any form from the third-party invasion in the hardware-design process may cultivate and result in serious security apprehension for CE hardware. Thus, safeguarding CE/IoT hardware can be realized as a protection of IP cores against dishonest entitlement of ownership. Another version of threat called ‘hardware Trojan’ is an act of intentionally injecting malicious logic into a design (such as IP core hardware) by an adversary in the vendor house, which will target detection/correction policies affecting security protocol. Both the hardware vulnerabilities, mentioned above, may occur in any of the design abstract levels. Pre-processing or post-processing style approach is defined based on the level at which security is employed, which is often driven by complexity of the DSP core. For example, in case of complex DSP/multimedia cores comprising gates in the order of 15K–50K, security approach during architectural (high-level) synthesis is more beneficial and fruitful (Sengupta and Kundu, 2017).

1.3.3 Traditional forms of IP protection not enough?

For the last several years, various methodologies for the protection of such cores through patents, copyrights, trademarking, etc., have been employed. Yet, these approaches are not established adequately to shield the worth of IP cores used in CEs devices. The primary reason is the design costs of implementation; in addition, liability of proof to exhibit infringement is also too expensive. Besides, when it comes to the protection of reusable IP cores in CE industry, most of the IP core vendors trust on a combination of communal conviction of their business collaborators and privacy during operation.

1.4 IP core protection and hardware-assisted security of CE device—DSP core

As discussed earlier, at the core of a CE device, several DSP cores used for data intensive or control intensive functions are present. These DSP cores are IP cores in the form of SoC platform which are responsible for data processing of large chunks of data. For example, digital video cameras almost always have a second built-in microprocessor, dedicated to video calculations. This process is called as digital signal processor (DSP)—its role is to perform repetitive mathematical tasks in real time. In simpler terms, while our smart phone's main microprocessor is checking to see if we have any incoming calls, second processor will simultaneously run the email in the background, manage Wi-Fi signal, check when video is coming through the lens, etc., and all such calculations are taken care by second dedicated processor, DSP core. A thematic example is shown in Figure 1.4. As evident from this figure, the multimedia/DSP core (JPEG compression/decompression IP core) plays a significant role in the successful operation of a CE device. Similarly, other DSP cores such as discrete cosine transformation (DCT) core, fast Fourier transform (FFT), digital filters, etc., also play a critical role in CE device design. For example, FIR filter helps to improve the audio quality and the cellular infrastructure in the CE device. Protection of such complex DSP/multimedia cores relies on the high-level synthesis (i.e., architectural synthesis) as backbone design framework. This is because such complex DSP cores that comprise 15K–60K gates never initiates the design process from lower abstraction level. Furthermore, selecting/controlling the architecture of these DSP cores is seamlessly possible through high-level synthesis framework as it includes design space exploration (DSE) module (driven through intelligent evolutionary algorithms). This is not possible if these designs have to start from register-transfer (RT)/gate-level/lower levels.

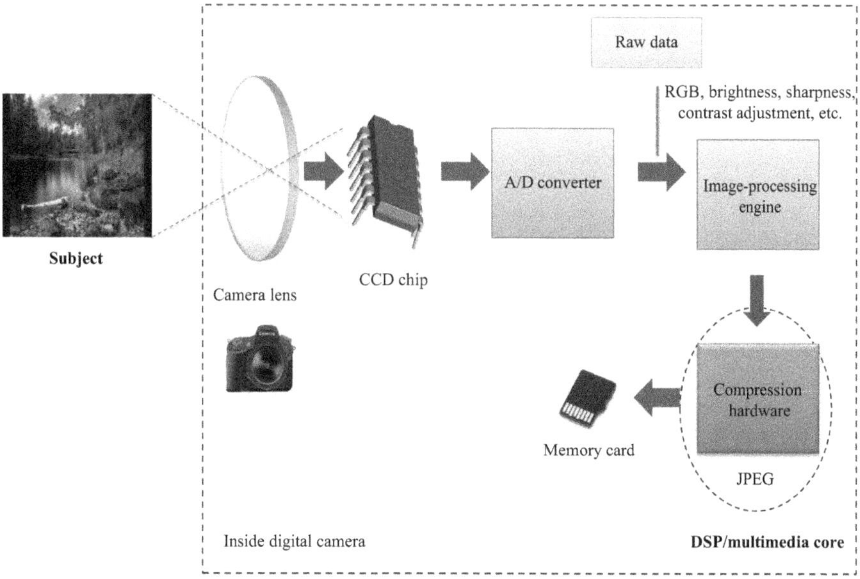

Figure 1.4 DSP/multimedia core in CE device: digital camera

A protection/security mechanism of such complex cores is thus intertwined with the high-level synthesis design process. For such hardware, protection is mostly applied during higher levels of abstraction (i.e., during high-level synthesis). On the contrary, security/protection mechanism of less complex hardware such as adder, encoder, decoder, etc., can be performed at lower levels such as RT level/gate level, etc. Figure 1.5 shows the classification of different hardware-assisted securities and IP core protection mechanisms at higher and lower levels of abstraction. As discussed, for DSP/multimedia cores such as infinite impulse response (IIR) filter, FIR, JPEG, etc., protection/security algorithms are applied at higher levels of abstraction which enables the constraints of the security algorithms to propagate to the lower levels also through RT level/gate level synthesis.

1.4.1 Security and protection methodologies available for IP core/hardware

Some of the recognized protection mechanisms used extensively for DSP IP cores of CE designs that depend on the threat model/vulnerability are watermarking, fingerprinting, hardware metering, computational forensic engineering, Trojan security, structural obfuscation, functional obfuscation, physically unclonable functions (PUF), and patents/copyrights (Alkabani *et al.*, 2007; Castillo *et al.*, 2007; Christof and Jan, 2009; Hong and Potkonjak, 1999; Koushanfar *et al.*, 2005; Marvel, 2005; Ni *et al.*, 2006; Sengupta and Roy, 2018a, 2018b). Figure 1.6 shows an overview of threat models along with its hardware security countermeasures for CE device.

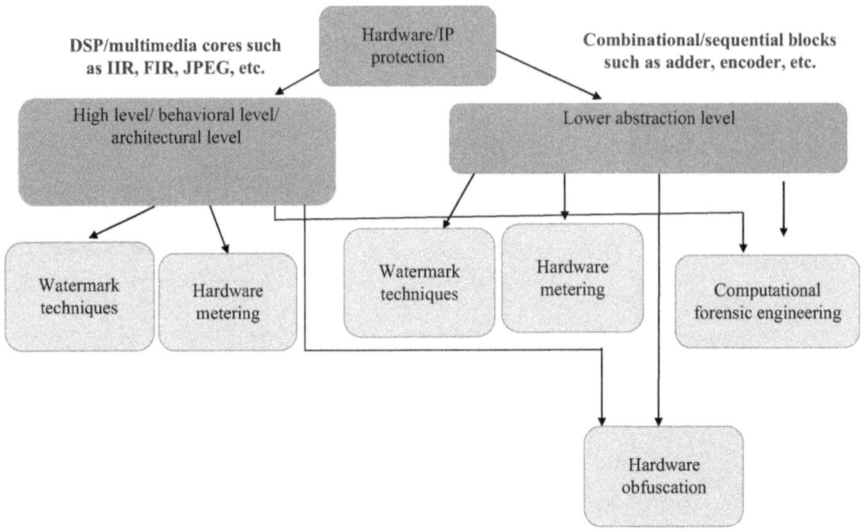

Figure 1.5. Hardware protection mechanisms

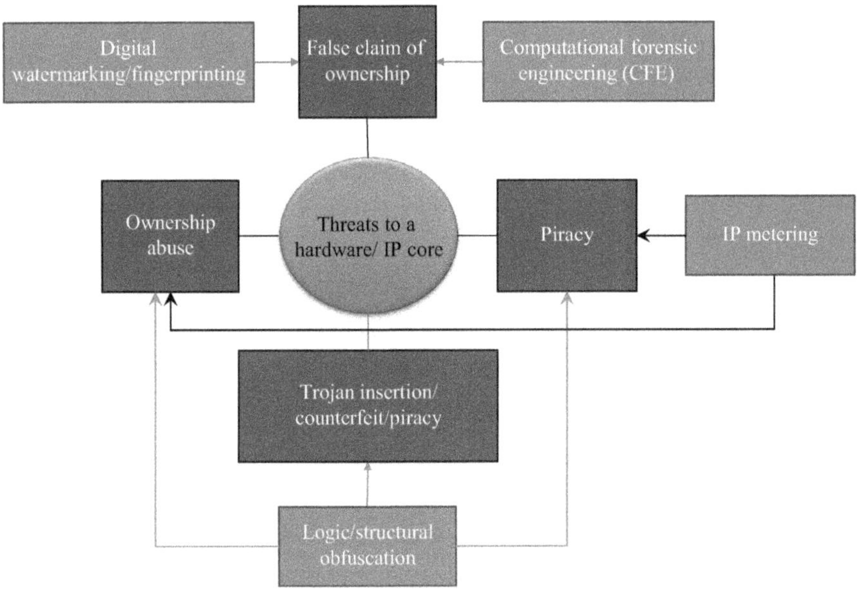

Figure 1.6 Overview of some common hardware threats and its countermeasures

(i) IP core/hardware watermarking: This hardware protection mechanism is employed when the hardware threat anticipated to be false claim of ownership. This threat model is basically because of ownership abuse or false claim of IP core ownership or piracy. Although watermarking protects against the aforesaid threats, it can only detect and not prevent the threats. In other words, watermarking approach is capable of providing passive form of protection to an IP core. More details on this protection mechanism are presented in Chapter 4.

(ii) Computational forensic engineering (digital forensics) for IP cores: This hardware protection mechanism is employed when the hardware threat under question is ownership abuse or false claim of IP core ownership or piracy. It has the same threat model as watermarking and provides the detection of the legal owner in case of ownership conflicts. The objective of computational forensic engineering (CFE) is to recognize the individual that produced the IP. It examines certain properties of a given IP and measures the probability that a particular individual has produced it. The CFE method identifies from a group of synthesis tool, the one which is employed to make a specific IP design. It is to be noted that this approach is also capable of providing passive forms of protection to an IP core. More details on this protection mechanism are presented in Chapter 6.

(iii) Hardware metering: This is the method of monitoring an IC. Some of the effective methods are to physically identify serial numbers on chips or storing unique identifiers in its memory (Alkabani and Koushanfar, 2007; Roy *et al.*, 2008). The taxonomy of hardware metering is shown in Figure 1.7. Hardware metering can be broadly classified into two categories: non-functional and functional. In non-functional identification method, a unique ID is created that is separate from the chip functionality. However, it can have the following drawbacks:

1. Vulnerability to cloning and/or removal;
2. Once chip is tagged, foundry can copy same tag on other chips or simply remove tag so chip cannot be traced;
3. Possibility to overproduction;
4. Foundry can produce multiple chips with same tag;
5. Out of millions of chips, probability of finding two matching tags is small.

Non-functional identifiers in hardware metering are sub-classified into two types: (1) reproducible and (2) unclonable. Reproducible identifiers are a unique ID stored inside a chip package, die, or a chip memory. They can be in the form of

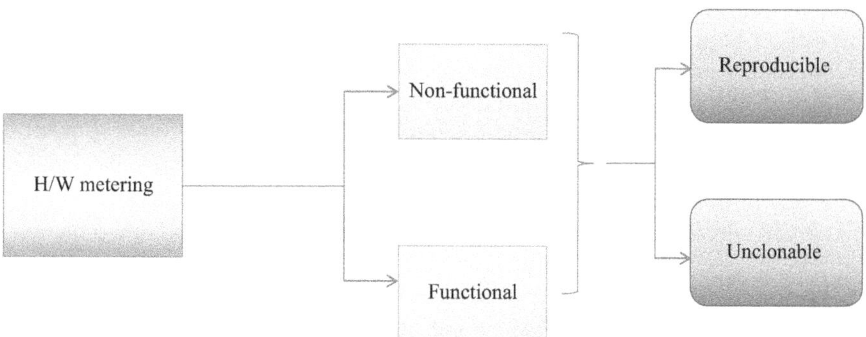

Figure 1.7 Taxonomy of hardware metering methods

indented serial number or digitally stored serial number. Its advantages are that it does not depend on randomness and are easy to track/identify. Disadvantages are that it is easy to clone/modify and easy to overproduce. On the other hand, unclonable non-functional identifiers can use extrinsic or intrinsic methods for metering. For example, intrinsic method may use IC leakage, power, timing, path signatures, etc., which are all unique due to process variations. This method does not need any additional logic and can be used as such on existing designs. On the contrary, extrinsically unclonable non-functional identifiers uses additional logic such as PUF which is a structure comprising ring of oscillators (RO) capable of generating random values due to process variations. PUFs are impacted by power supply noise, temperature, delay, etc.; therefore, the values will keep on changing constantly. The other class of metering method is functional identifier-based metering. This method links identifiers to chip's internal functional details during the synthesis process. Some of the properties are:

- Each chip's function gets a unique signature;
- Additional states added generate same output;
- Function unchanged from input to output;
- Internal transactions unique to each chip.

The above methods discussed so far in the context of hardware metering are all passive metering methods. There exists an active metering method which uses external-locking techniques to lock IC. Some of the properties of active metering methods are:

- This technique tries to allow IP owner to have control over;
- Uses key encryption to lock correct functionality of a chip;
- At the gate level, XOR gates are placed on selected non-critical paths;
- Requires that every chip be activated with an external key. Only IP owner can generate key (Alkabani and Koushanfar, 2007; Roy *et al.*, 2008).

(iv) IP core/hardware fingerprinting: This hardware protection mechanism is employed when the hardware threat anticipated is false claim of ownership by the IP seller/vendor or illegal redistribution of extra copies without the knowledge of the IP buyer. The threat model is unlawful redistribution/resell of IP core copies in the market by a dishonest IP seller (Hong and Potkonjak, 1999). Thus, IP core fingerprinting protects against the ownership abuse of IP buyer. Although fingerprinting protects against the aforesaid threats, it can only detect and not prevent the threats. In other words, fingerprinting approach is capable of providing passive forms of protection to an IP core. More details on this protection mechanism are presented in Chapter 5.

(v) IP core/hardware obfuscation and encryption: This hardware protection mechanism is employed when the hardware threat anticipated is IP netlist theft/counterfeit/piracy through reverse engineering. There are two types of hardware obfuscation methodologies: (1) structural obfuscation and (2) functional obfuscation. In structural obfuscation, the goal is to enhance the reverse engineering complexity to make it more challenging for an attacker to discover the actual functionality of an IP core from its transformed structure. On the contrary, in functional obfuscation, the goal is to disable the IP core through 'keys' such that normal functionality of the obfuscated design can only be enabled by the successful application of a single pre-determined key or a series of covert keys at the input; otherwise, an IP core design will operate in a mode, which will exhibit incorrect functionality. Both these types of obfuscation methods aim to provide protection against IP infringement and Trojan insertion. Structural obfuscation is regarded as a 'passive' form of obfuscation as it does not affect the functionality of the IP core, while only making syntactic changes, i.e., modifying the circuit description in a soft form such that it becomes challenging for an attacker to understand the internal working of an IP core. A major shortcoming of the passive approach is that it does not change the black-box functionality of a design, and hence unable to avert potential usage of an IP as black-box in a design. However, since it does not add extra-locking circuitry to the IP core, the chances of design overhead is lower than active approaches which makes it an equally attractive choice in CE industry. More detail on this protection mechanism is presented in Chapters 7 and 8.

(vi) Trojan security/detection of an IP core: Hardware Trojan in the context of IP cores is a malicious alteration of logic inside IP core that causes to change its functionality such that under rare event condition, the IP core produces wrong output. This hardware protection mechanism is employed to safeguard against hardware Trojan secretly hidden inside a 3PIP cores used in SoC integration. Due to globalization of IC design chain, involvement of third-party IP cores play a crucial role in the mass production of CE devices in stricter time to market pressure. Trojan security of DSP IP cores can be achieved through dual modular redundancy technique during system-on-chip integration running concurrently with an optimization framework. This enables detection of hardware Trojan that causes functional change. More detail on this protection mechanism is presented in Chapter 3.

(vii) Physically unclonable functions: PUF is a 'digital fingerprint' that serves as a unique identity for the hardware. It acts as hardware security primitive due to its complexity and the unpredictability between its outputs and the applied inputs. PUFs are based on physical variations which occur naturally during semiconductor manufacturing and which make it possible to differentiate between otherwise identical hardware. Because of the random process variations, no two ICs even with same layout are identical. There are various types of PUFs, such as: (1) *Latch PUFs* where the output changes due to intrinsic property of the latch; (2) *SRAM PUF* where the initial power up values of the SRAM cells are random; (3) *Ring of oscillator (RO) delay-based PUF* where the path delay is unpredictable due to manufacturing variation; and (4) *Arbiter PUF* where the path delay is unpredictable due to physical randomness. In arbiter PUF, each challenge selects a unique pair of delay gates. Random process variation determines which path is faster (Potkonjak and Goudar, 2014). More detail on this protection mechanism is presented in Chapter 12.

(viii) Hardware-based encryption: Another hardware protection mechanism for securing classified data used in CE device is hardware-based encryptions. Hardware-based encryption process can be used for securing data storage and transfer, as well as for DSP IP core protection in the context of CE devices. Advanced encryption standard (AES) (Paar and Pelzl, 2009) is the most robust encryption process that can be used as the hardware-based encryption of classified data as well as in DSP IP core protection from external attacks. More detail on this protection mechanism is presented in Chapter 10.

(ix) Patent and copyright: A patent is an authorized privilege sanctioned by the government of a specific dominion which on submission safeguards the IP of the maker and thus can only be sold and used by the approval of the owner. More explicitly, the owner is given the constitutional privilege to thwart others from commercially abusing the invention. It is the owner's right to report the case of any contravention under civil law (Sengupta, 2015).

IP protection that relates to printed publications as well as in any form of public communication such as TV broadcasting, and hardware systems for storing, is called 'Copyright.' It shields the maker in terms of privileges in creative works against fraud who aim to plagiarize (Sengupta, 2015).

1.4.2 Different IP core protection and hardware-assisted security mechanisms: advantages and disadvantages

1.4.2.1 Hardware metering vs. hardware watermarking

In hardware watermarking, a secret signature is chosen which is implanted on the original design. In IP metering, a distinctive identifier (ID) is injected into the product for validating any unlicensed copy made. In case of watermarking, a design-inspection step is required during signature detection process. IP metering also includes a design-inspection step as the ID is accessible by reading some registers or trivial I/O access. Furthermore, implanting a watermark is certainly secure for cases of false claim of ownership or IP piracy as only the owner can evidently validate the enactment of a secret watermark. Similarly, IP metering is also effective for identifying the IP piracy and for reckoning the likely measure of piracy, which makes it extremely significant in court proceedings to endorse assessments of compensations.

Design-overhead may follow due to implanting a concealed signature in the original IP core, in case of hardware watermarking. This also depends on the type of signature selected and its size. But, in case of IP metering, the design-overhead is non-existent or nominal, as a trivial area of the IP-core is re-configured for each designed/manufactured hardware. IP metering only comprises modification in configuration in a tiny segment of the design.

In the context of DSP IP cores, watermarking techniques are mostly implanted before the RT level synthesis, i.e., as a pre-processing or during in-synthesis, i.e., signature is implanted during the execution of the high-level synthesis process. On the contrary, IP metering is generally executed post-layout or during manufacturing, through memory programming or indenting serial number on device body.

During watermark, care should be exercised so that the signature insertion does not damage the functionality of the design and hampers its quality. There is a clear tradeoff in case of watermarking with respect to robustness vs. design overhead. In other words, larger size watermark is certainly more robust but may incur more design overhead. None of these parameters are present for metering (Sengupta, 2015).

1.4.2.2 Hardware obfuscation vs. hardware watermarking

Hardware obfuscation is completely different than hardware watermarking of IP core. In hardware obfuscation, the circuit description of the IP core is deliberately obscured either through encryption or by introducing unreadability in the circuit structure, so that the reverse engineering process becomes significantly harder. On the contrary, hardware watermarking introduces ownership information in a concealed manner within the IP core design itself, so that when required, legal ownership of the rightful owner can be proved. Although the threat models of both these protection mechanisms are slightly different, hardware watermarking can be used in unification with hardware obfuscation. In an obfuscated design, watermarking can be effective in providing a next line of defense against unlicensed copying efforts. Both these protection methods may incur design overhead, if not properly implanted/performed.

1.4.2.3 Computational forensic engineering vs. hardware watermarking

Computational forensic engineering for IP core protection employs a set of digital forensic evidences to infer the rightful owner. It is a signature-free approach, unlike hardware watermarking for IP core protection. Furthermore, as CFE does not employ signature for resolving IP ownership conflicts, it has zero chance of quality degradation on the original design. Additionally, no extra care is needed for CFE as it does not disturb the functionality of the design. Yet, the general process to execute CFE is more composite than watermarking as it warrants cognizance on the various phases of statistics collection, feature extraction, clustering, and validation. Thus, CFE incurs more time complexity when compared with watermarking in case of DSP IP cores.

1.4.3 HLS (architectural synthesis) as design backbone for implementing security algorithms for DSP IP cores

As DSP/multimedia IP cores depend on high level synthesis (HLS) design framework as its backbone, the security/protection algorithms are mostly in-synthesis by nature, i.e., the security constraints/techniques are performed during the high-level synthesis itself. Before reading the subsequent chapters that deal with the details of the security methodologies for DSP cores, it is important to review and study the high-level synthesis design process in this chapter itself. This sub-section will present an overview of the generic HLS design/tool flow for DSP/multimedia cores. The design flow shown in Figure 1.8 is broadly applicable to any modern HLS methodology/HLS tool such as Vivado HLS (Vivado User Guide, 2018; Xilinx, 2018). The HLS flow can be used to design and secure any complex IP core such as FIR filter, IIR Butterworth filter, high/low-pass digital filter, DCT, inverse DCT, decimation in time, FFT, Haar wavelet transformation, JPEG compression, JPEG decompression, band pass filter, MPEG motion vectors, H.264, etc.

A typical HLS design flow accepts a DSP application in the form of C-code and converts it into a graphical representation called control data flow graph through parsing. Along with this application, HLS also accepts the technology library of modules based on which the design evaluation is performed. The technology library information is based on a particular technology scale and comprises details of silicon area, delay, power, etc., of a unit module. Once a control data flow graph (CDFG) is obtained after parsing the C-code, it then undergoes a series of transformations to convert it into an optimized form. These series of transformations can be compiler driven or Boolean function driven by nature. This optimized CDFG accepts HLS pragmas/directives to pass the information to the DSE system in HLS. The first step of DSE system is operation scheduled, which also accepts its own set of HLS pragmas/directives as well as scheduled configuration to assign time (cycle) stamp to each operation. Examples of directives at this step include chaining, multi-cycling, data pipelining, functional pipelining, loop unrolling, etc. This is followed by resource allocation and binding phase which also accepts its own set of directive for executing spatial constraint to functional resources. Binding enables sharing/reuse of functional hardware resources for multiple operations.

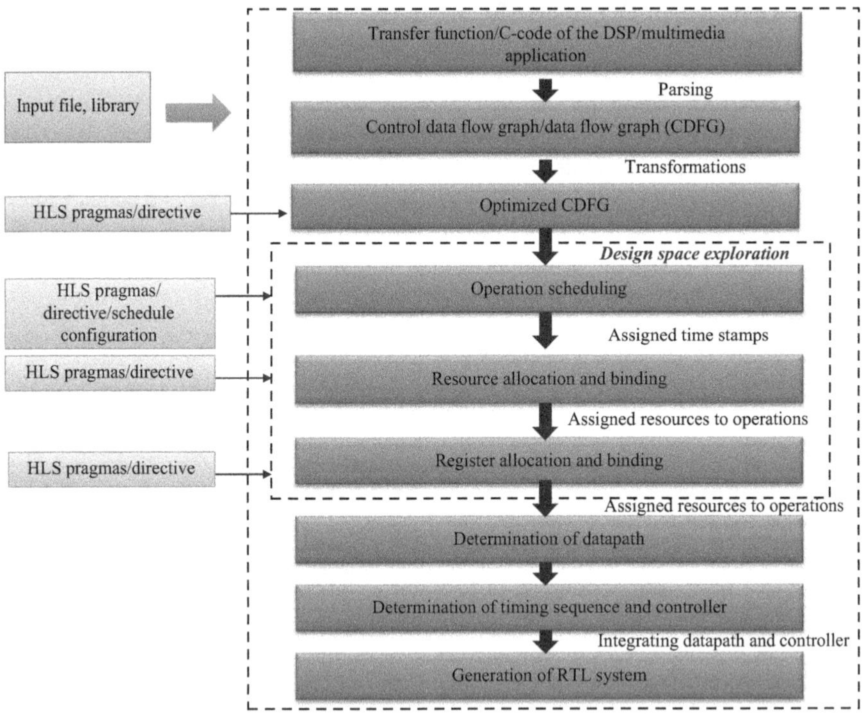

Figure 1.8 Overview of generic HLS design/tool flow

Subsequently, register allocation is performed through its own set of HLS directive to obtain an optimal number of register sets required for implementing the application in hardware. Once these phases are over, then the determination of datapath is performed by analyzing the I/O connectivity, multiplexers, demultiplexers, etc., based on the datapath obtained, and the timing sequence for the generation of control path is determined. This yields cycle-by-cycle execution details of each hardware resource. Both the datapath and controller are described in hardware description language (HDL) and finally integrated to form the RTL system schematic/HDL description of the DSP/multimedia application. As the security algorithms/features of DSP IP cores are integrated during one or more HLS synthesis steps, hence understanding the HLS design framework is crucial. More details on HLS design process are available in Sengupta and Mohanty (2016).

1.5 Hardware-assisted media protection

The idea of ownership right, copyright, or IPR that a CE system can be subjected to is depicted in Figure 1.9. As far as ownership right is concerned, the common factor of hardware, software, overall CE system, or media piracy is the loss of revenue of the actual or legal owner. But, these ownership issues/attacks are diverse, and correspondingly solutions to be deployed may differ. The hardware IP core protection aspects which are results of hardware piracy have been discussed in previous sections. Overall, system piracy or cloning is also a problem in the CE area, due to which multiple fake CE systems such as cameras, USB drives, hardware components for smart cars, fake batteries, are available in black market. These may be cheap, and user may think that he/she got a good deal, without having a

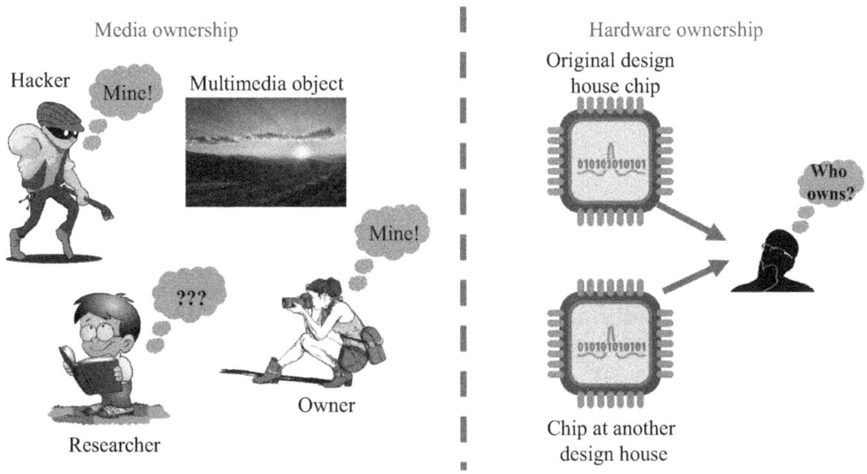

Figure 1.9 Concept of ownership right, copyright, or intellectual property (IP) rights

knowledge that these would have lots of wrong implications. For example, a cloned/fake hardware may be providing a backdoor entry to the CE system that uses such hardware. Imagine, the case of a smart car, a hacker through this back door can get access to the smart car and take control of it. This can lead to fatal consequences, e.g., an attacker/hacker can create accidents. Similarly, a fake/cloned battery can be a safety hazard for the CE system as a bad battery can explode and catch fire, examples of all such several case scenarios are available in the Internet. Use of pirated software also can have potential backdoor for the hackers to take control over the system. Every CE system processes data, information, video, audio, and game in every second of its active state (Mohanty, 2017a, 2017b). It is estimated that the sensors, social networks, web pages, image and video applications, and mobile devices generate more than 2.5 quintillion bytes (i.e., 10^{18} bytes) data per day (Mohanty *et al.*, 2016). This will be further complicated as smart cities initiatives around the world take momentum in which IoT with billions of connective devices process data throughout the day and night. It is rightly said that ***it is not oil rather it is data that is the new gold***. In terms of media piracy video, audio have impact on the entertainment industry, online streaming companies, and various-related industry in terms of revenue loss. In the chapter corresponding to this ownership matters, discussions will be presented on the hardware-assisted solutions for media protection. Both software-based and hardware-based solutions are possible, but hardware provides advantages in terms of real-time performance and protection right from the source-end to consumer-end.

1.6 Physical unclonable functions

It is a fact that most of the security, privacy, and IP rights solutions are based on some forms of encryption, scrambling, hashing, and watermarking. These mechanisms rely on some form of random number or pseudorandom number. It is a fact that a general-purpose processor, which is a deterministic machine, computes the next instruction based on a program counter that cannot generate true random numbers. Also, the keys random/pseudorandom numbers typically are stored in the memory or drives for possible use in future. Hacking a system is a matter of time only, and it is possible for a hacker to get access to it and consequently access to the system and information that it contains. Primitive PUF hardware generates a random number on the fly as shown in Figure 1.10 and completely eliminate the need of storing keys on a hard-drive, thus making it secure (Yanambaka *et al.*, 2017, 2018; Joshi *et al.*, 2017). A PUF relies on the variations at the physical level of the IC which is originated due to the nanoscale semiconductor manufacturing. Thus, PUF makes good use of the process variations instead of a designer trying to reduce the impact of the process variations on the IC. The output generated by a particular PUF is unique to it, and hence serves like a signature of it. The PUF chapter will discuss on topics such as PUF principles, various types of PUFs, and applications of PUFs.

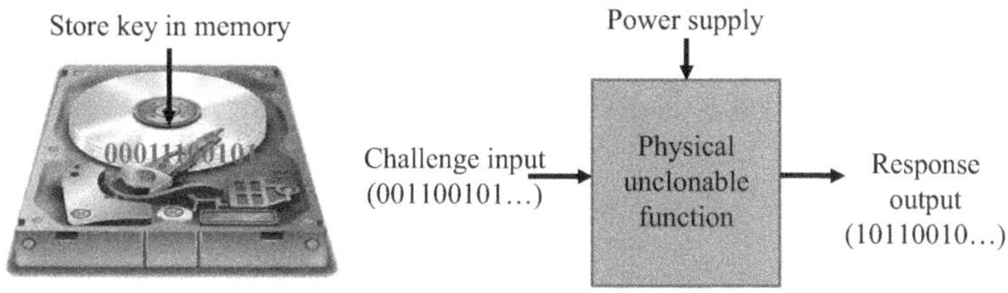

(a) Physical unclonable function (PUF) does not need to store key

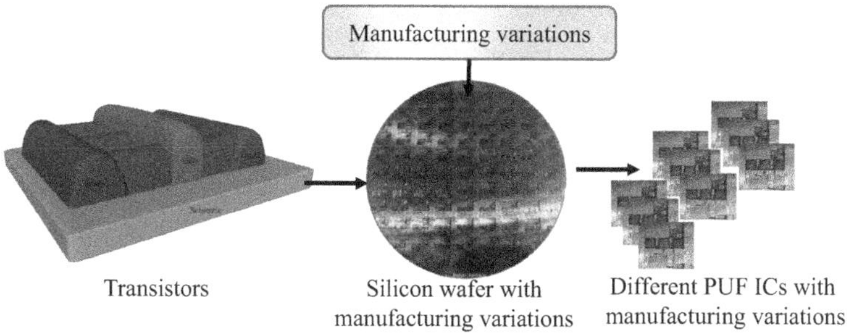

(b) Physical unclonable function (PUF) relies on nanoscale semiconductor variations

Figure 1.10 Concept of physical unclonable function (PUF)

1.7 Organization of the book

This section summarizes the roadmap of the book in terms of overview of the organization of the expected content.

Chapter 2 presents the broad overview as well as some specifics on security, privacy, and ownership protection in the context of CEs and IoT. It presents a broad perspective of IoT and CE including its driving technology and challenges. It further discusses about security and privacy in IoT/CE systems with an emphasis on the attacks and its countermeasure, memory security, radio frequency identification (RFID) security, smart healthcare, and smart transportation security.

Chapter 3 presents the details on hardware Trojan security aware DSP IP core/IC. The chapter includes introduction to the concepts, followed by features of the Trojan-secured methodology—that includes threat model, description of features, and advantages of performing Trojan security at higher abstraction level. The next sections of this chapter comprises an example of hardware Trojan inside 3PIP module, discussion on selected Trojan security approaches, goal/models/problem formulation of a Trojan security aware DSP IP core, and the detailed description of its design process. Subsequently, the implementation and results are presented, followed by conclusion and questions and exercise for the learners and practitioners.

Chapter 4 presents the details on how to protect a DSP IP core/IC using robust watermarking. The chapter includes introduction to watermarking, description of selected watermarking approaches, design process of watermarked

IP core/hardware—including threat model, explanation on single-phase and multi-phase watermarks, followed by discussion on desirable properties of a watermark and threat scenarios. Subsequently, details on implementation and results are provided—that includes security analysis and design cost/overhead analysis. The chapter end with a conclusion of the concepts learnt and some exercise for the learners/practitioners.

Chapter 5 presents the details on symmetrical protection of DSP IP core/IC using fingerprinting and watermarking. The chapter includes introduction to the problem and concepts on symmetrical protection, description of background on fingerprinting and watermarking, followed by threat models and benefits of providing symmetrical protection at higher design abstraction level. The next sub-sections focus on the fundamentals of IP core protection, process of implanting symmetrical IP core protection, and finally finishing with a motivation example of this approach. Subsequently, implementation and results are presented, followed by conclusion and exercise questions for learners and practitioners.

Chapter 6 presents the forensic engineering for resolving IP core ownership problem. The chapter includes introduction to the concept, followed by the description of the customized CFE approach for IP core protection. The next sub-sections focus on IP core feature extraction algorithms, followed by implementation and results. Finally, the chapter concludes with an inference, followed by exercise questions for learners and practitioners.

Chapter 7 presents the structural obfuscation of DSP IP cores used in CE devices. The chapter includes the introduction to the concept, followed by the description of its threat model and benefits of obfuscation at higher abstraction layer. The next sub-sections focus on the fundamentals of obfuscation for IP core protection, process of structural obfuscation, and generation of low-cost obfuscated IP core design. Subsequently, a motivational example is presented, followed by the details of implementation and results. The chapter ends with a conclusion and exercise questions for learners and practitioners.

Chapter 8 presents the description of functional obfuscation of DSP cores used in CE devices. The chapter encompasses the following—introduction to the concepts, discussion on selected functional obfuscation approaches, followed by the explanation of the design process of the functionally obfuscated DSP core and security view point—analysis and attack scenarios—of the functionally obfuscated DSP core. The next sections of this chapter presents optimization engine for deriving the low-cost/optimal functionally obfuscated DSP core architecture, implementation and results, and finally ending with the conclusion and questions and exercise for learners and practitioners.

Chapter 9 presents the details on a real-world complex application—obfuscation of JPEG CODEC IP core for CE devices. The chapter includes introduction to the concepts, overview of JPEG compression/decompression process, followed by design process of an obfuscated JPEG IP core design—including threat models and detailed design steps. The next sub-sections focus on the implementation of the design, presentation of the results and conclusion. Finally, the chapter ends with exercise questions for learners and practitioners.

Chapter 10 presents the details on AES and its hardware watermarking for ownership protection. The chapter starts with the introduction to the concepts, followed by the description of AES algorithm—encompassing overview of AES, custom hardware design, and its digital watermarking. Next, results are presented for watermarked AES hardware, followed by conclusion and exercise questions for learners and practitioners.

Chapter 11 presents the details on the protection and authentication of media content with emphasis on the hardware vs. media protection, digital rights management (DRM), copyright protection of multimedia as well as general framework of multimedia watermarking, its applications, challenges, attacks, and countermeasures.

Chapter 12 presents the details on the PUF with an emphasis on its properties, classification, and discussion on different types of PUF based on security, structure, and fabrication. It also presents the details on the implementations of PUF, challenges, and other issues on PUF.

1.8 Conclusions

This chapter introduced 'hardware-assisted security and IP core protection' in the context of CE systems. Various types of security threats are briefly summarized from an overall system point of view. It also presented an overview of different IP core protection and hardware-assisted security techniques, with its pros and cons. Through the help of a real-life CE design, the significance of DSP core was explained along with the different hardware threats possible on DSP IP cores. The chapter also summarized the roadmap of this book—including the expected contents of each chapter from a reader's perspective.

1.9 Exercises

1. What is consumer electronics? Name some its modern applications?
2. What is the role of IP core blocks in CE device design?
3. Explain the SoC design flow and indicate location of hardware vulnerabilities?
4. What is 3PIP?
5. Define 'hardware-assisted security'?
6. What are the components of a SoC design process?
7. What are steps inside SoC fabrication process?
8. What are the advantages of IP cores in CE device design?
9. Why traditional forms of security and IP core protection are not enough for reusable IP cores?
10. Explain the DSP component inside a camera system?
11. Explain the threat models for watermarking and fingerprinting?
12. How do you counter (prevent) Trojan insertion inside an IP core design?
13. Explain drawbacks of IP metering?
14. Explain the threat model of CFE? How is it different from fingerprinting?

15. What are the different types of hardware metering techniques?
16. Explain the different types of PUFs?
17. Compare obfuscation and watermarking?
18. What are the HLS directives applied during operation scheduling in HLS?
19. Explain the design flow of a generic HLS system?
20. What are the phases inside design space exploration system?
21. How do you get an optimized CDFG of a DSP IP core?
22. What is the difference between multi-cycling and chaining?
23. Explain the difference between datapath and control path in HLS?
24. Explain the different types of scheduling algorithm employed during HLS?
25. What is the difference between allocation and binding?

References

Alkabani Y., Koushanfar F., and Potkonjak M., 'Remote activation of ICs for piracy prevention and digital right management,' in Computer-Aided Design, 2007. ICCAD 2007. IEEE/ACM International Conference on, San Jose, CA, USA, November 2007, pp. 674–677.

Alkabani Y. M. and Koushanfar F.,'Active hardware metering for intellectual property protection and security," In Proceedings of 16th USENIX Security Symposium on USENIX Security Symposium (SS'07), Niels Provos (Ed.). USENIX Association, Berkeley, CA, USA, 2007, Article 20, 16 pages.

Castillo E., Meyer-Baese U., García A., Parrilla L., and Lloris A., 'IPP@ HDL: Efficient intellectual property protection scheme for IP cores,' *Very Large Scale Integration (VLSI) Systems, IEEE Transactions on*, 2007, vol. 15 (5), pp. 578–591.

Hong I. and Potkonjak M., 'Behavioral synthesis techniques for intellectual property protection,' in *Proceedings of the 36th annual ACM/IEEE Design Automation Conference*. ACM, 1999, pp. 849–854.

Joshi S., Mohanty S. P., and Kougianos E., 'Everything you wanted to know about PUFs', *IEEE Potentials Magazine*, 2017, vol. 36 (6), pp. 38–46.

Koushanfar F., Hong I., and Potkonjak M., 'Behavioral synthesis techniques for intellectual property protection,' *ACM Transactions on Design Automation Electronic Systems*, 2005, vol. 10 (3), pp. 523–545. [Online]. Available: http://doi.acm.org/10.1145/1080334.1080338.

Marvel L., 'Information hiding: Steganography and watermarking,' in Optical and Digital Techniques for Information Security, ser. Advanced Sciences and Technologies for Security Applications, B. Javidi, Ed. Springer, New York, 2005, vol. 1, pp. 113–133. [Online]. Available: http://dx.doi.org/10.1007/0-387-25096-4.

Mohanty S. P., 'Security and energy tradeoffs in consumer electronics', Keynote Abstract, 3rd Zooming Innovation in Consumer Electronics International Conference (ZINC), 2018, Novi Sad, Serbia, 31st May 2018a.

Mohanty S. P., 'Smart cities – demystified', Keynote Abstract, 2nd International Conference on Man and Machine Interfacing (MAMI), 2017a, Bhubaneswar, India.

Mohanty S. P., 'Internet of things (IoT) – demystified', Keynote Abstract, 16th International Conference on Information Technology (ICIT), 2017b, Bhubaneswar, India.

Mohanty S. P., Nanoelectronic Mixed-Signal System Design. McGraw-Hill Education, New York, 2015, no. 0071825711.

Mohanty S. P., 'Energy and security tradeoffs in CE systems', Panel, IEEE International Conference on Consumer Electronics, Las Vegas, 2018b.

Mohanty S. P., Choppali U., and Kougianos E., 'Everything you wanted to know about smart cities', *IEEE Consumer Electronics Magazine (CEM)*, 2016, vol. 5 (3), pp. 60–70.

Ni Z., Shi Y.-Q., Ansari N., and Su W., 'Reversible data hiding,' *Circuits and Systems for Video Technology, IEEE Transactions on*, 2006, vol. 16 (3), pp. 354–362.

Paar C., and Pelzl J., 'Understanding cryptography – A textbook for students and practitioners', Springer-Verlag, Berlin Heidelberg, eBook ISBN 978-3-642-04101-3, Number of Pages: XVIII, 372, November 2009.

Qu G., and Potkonjak M., 'Intellectual property protection in VLSI designs', Springer US, eBook ISBN 978-0-306-48717-0, Number of Pages: XIX, 183, February 2003.

Potkonjak M. and Goudar V., 'Public physical unclonable functions', *Proceedings of the IEEE*, 2014, vol. 102 (8), pp. 1142–1156.

Roy J. A., Koushanfar F., and Markov I. L., 'EPIC: Ending piracy of integrated circuits,' *2008 Design, Automation and Test in Europe*, Munich, 2008, pp. 1069–1074.

Sengupta A., 'Protection of reusable IP core at architectural level', *IEEE VLSI Circuits & Systems Letter*, 2015, vol. 1 (2), pp. 14–17.

Sengupta A., 'Protection of IP-core designs for CE products', 2016, *IEEE Consumer Electronics Magazine*, vol. 5 (1), pp. 83–88.

Sengupta A., 'Hardware vulnerabilities and its effect on CE devices: Design-for-security against Trojan', 2017a, *IEEE Consumer Electronics*, vol. 6 (3), pp. 126–133.

Sengupta A., 'Hardware security of CE devices: Threat models and defence against IP Trojans and IP piracy', 2017b, *IEEE Consumer Electronics*, vol. 6 (1), pp. 130–133.

Sengupta A. and Mohanty S. P., 'High-level synthesis of digital circuits in the nanoscale, mobile electronics era', IET Book: Nano-CMOS and Post-CMOS Electronics: Circuits and Design, (Eds: Saraju P Mohanty& Ashok Srivastava), Invited Book Chapter, e-ISBN: 9781785610004, pp. 219–261, June 2016.

Sengupta A. and Roy D., 'Multi-phase watermark for IP core protection,' *2018 IEEE International Conference on Consumer Electronics (ICCE)*, Las Vegas, NV, 2018a, pp. 1–3.

Sengupta A. and Roy D., 'Reusable intellectual property core protection for both buyer and seller,' *2018 IEEE International Conference on Consumer Electronics (ICCE)*, Las Vegas, NV, 2018b, pp. 1–3.

Sengupta A. and Kundu S., 'Securing IoT hardware: Threat models and reliable, low-power design solutions', *IEEE Transactions on Very Large Scale Integration (VLSI) Systems*, 2017, vol. 25(12), pp. 3265–3267.

Sengupta A., Mohanty S. P., and Rose G., 'Hardware-assisted design for security and protection of consumer electronics', *IET Computers and Digital Techniques*, Accepted, June 2018.

Xilinx Vivado Design Suite User Guide – https://www.xilinx.com/support/documentation/sw_manuals/xilinx2018_1/ug910-vivado-getting-started.pdf, Last accessed on January 2018.

Xilinx – Vivado HLS – https://www.xilinx.com/products/design-tools/vivado/integration/esl-design.html, Last accessed on February 2018.

Yanambaka V. P., Mohanty S. P., and Kougianos E., 'Making use of semiconductor manufacturing process variations: FinFET-based physical unclonable functions for efficient security integration in the IoT', *Springer Analog Integrated Circuits and Signal Processing Journal*, 2017, vol. 93 (3), pp. 429–441.

Yanambaka V. P., Mohanty S. P., and Kougianos E., 'Making use of manufacturing process variations: A dopingless transistor based-PUF for hardware-assisted security', *IEEE Transactions on Semiconductor Manufacturing (TSM)*, 2018, vol. 31 (2), pp. 285–294.

Chapter 2

Security in consumer electronics and internet of things (IoT)

In a typical consumer electronics generation, diverse forms of attacks exist, e.g., system security, information security, information privacy, system trustworthiness, hardware IP protection, information copyright protection. Each of the attacks have their own threats and serve different purposes. Objective of this chapter is to differentiate each of these attacks in the context of internet of things (IoT). IoT is a very hot, trending topic of research, where different devices, sensors, and computers are connected to a network, collect data, and process for different purposes. A network of devices and sensors connected to the cloud can complete many unimaginable tasks, and without the existence of cloud, almost all task would have been very difficult to achieve and complete. Since, most connectivity occurs in cloud, there is always a probability of security and privacy threats. Because all types of the different devices are connected to the internet, attackers pose a huge threat to the system. For example, if any electronic system that is responsible for home care, healthcare, running an entire city, etc., are attacked, then the consequence of this threat will be a major catastrophe.

This chapter is organized as follows: Section 2.1 presents an introduction and broad overview of IoT; Section 2.2 presents a broad view of security, energy, and cost trade-offs in consumer electronic systems; Section 2.3 presents different security threats and solutions to the memory modules of the system; Section 2.4 provides different threats and solutions to the radio-frequency identification (RFID) modules; Section 2.5 provides security flaws and how to mitigate them in an near-field communication (NFC) module of the system; Section 2.6 discusses different security flaws in the smart vehicles, the autonomous cars, and the unmanned aerial vehicles (UAVs); Section 2.7 discusses different ways on how an attacker can attack a healthcare device and provides solutions for some of the attacks; Section 2.8 discusses the threats on firmware of the devices; and Section 2.9 discusses the emerging technology of blockchain, its advantages, and issues. The conclusion and future directions are presented in Section 2.10.

2.1 Internet of things (IoT) – a broad overview

Communication has evolved in the world at a very fast and rapid pace. Several solutions have been proposed for communications, short or long distance.

Technological growth also helped this in developing high-performance computing devices, making it a necessity to look for newer inventions. One of the such major creation is 'internet of things' and this has become an integral part of everyday life. There are many definitions for IoT that differs depending upon the individual's perspective. In an IoT environment, all the devices or 'THINGS' are connected to a network and exchange data among themselves or send to the cloud (Mohanty *et al.*, 2016). The concept of '3-I' in the context of IoT is shown in Figure 2.1, which includes. 'Interconnection,' 'Intelligence,' and 'Instrumentation.' These three are the important aspects of an IoT environment, as illustrated in the figure pertaining to the smart city environment. Interconnection is the communication platform that provides the network for the devices and Internet connection that helps in transmitting the data among themselves or to the cloud services, where post-processing take place. Intelligence is the brain of the entire environment, and Instrumentation refers to various devices and sensors that are present in the environment. There are various devices present in an IoT environment such as microcontrollers, single board computers, and sensors. Each of them should have a unique IP address to be considered as a THING for the IoT (Mohanty, 2018).

An IoT can be defined as Any TIME connection, Any THING connection, and Any PLACE Connection as shown in Figure 2.2 (IEEE, 2015). IoT is not localized

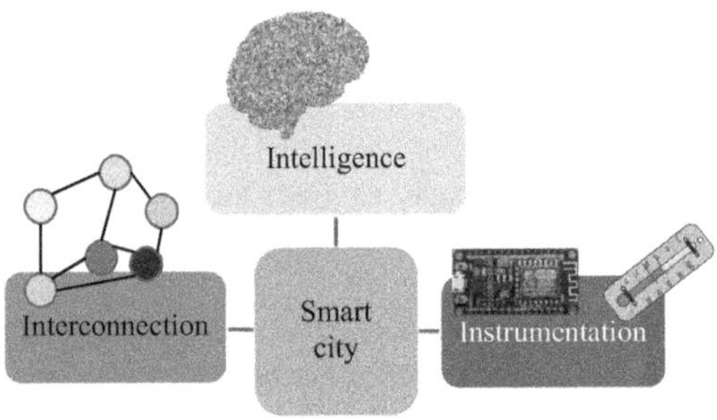

Figure 2.1 Smart cities 3 'I's

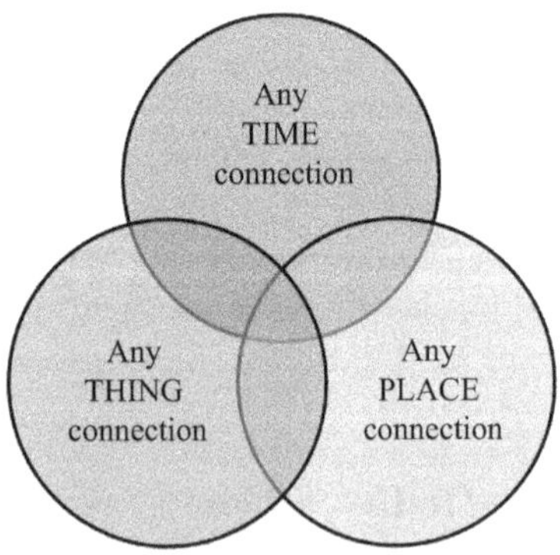

Figure 2.2 Definition of internet of things

to one location, but it is omnipresent and can be felt anywhere. Some of the IoT devices can be deployed in places where they are not monitored continuously. Its role is to monitor a parameter and transmit the data over the air (Yanambaka *et al.*, 2016a). In such cases, it should be continuously connected to the internet which needs 'Any PLACE Connection,' 'Any TIME Connection,' and 'Any THING Connection.' With the technological advancements, various devices and board are attaining communication capabilities which can help in implementing an IoT environment at a very low cost.

2.1.1 IoT – architecture

The structure of an IoT solution is described in its architecture. It includes various aspects including physical, the THINGS and virtual, the communication protocols. The IoT architecture is shown in Figure 2.3. As shown in the figure, the IoT architecture contains three different aspects which are connected to each-other by Internet. The THINGS comprise of the physical components which include the sensors, actuators, and various IoT devices. The sensors and actuators are connected to the IoT devices which help gather the data. The IoT devices are smart devices which can process the data that is collected from the sensors. Various sensors and actuators are connected using wired or wireless connectivity. Different levels of abstractions of an IoT model are shown in Figure 2.4.

There are various devices that communicate with the cloud and other IoT devices in different ways as shown in Figure 2.5. Some of the sensors and devices communicate directly to the cloud and transmit the data. Such devices will have high performance and can also do some of the data processing before uploading the data to the cloud. There are gateway devices which act as an intermediary devices between them and the sensors. These play a major role in data acquisition. After preprocessing the data, it is then sent to the cloud. Various cloud applications are available which can provide an interface for the end-user of the data to interact with it.

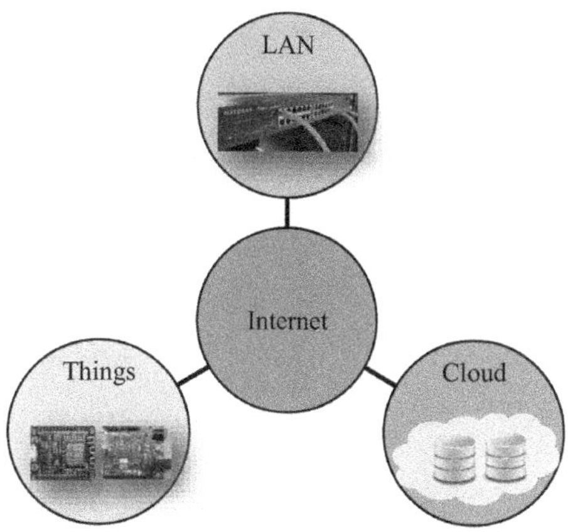

Figure 2.3 Architecture of internet of things

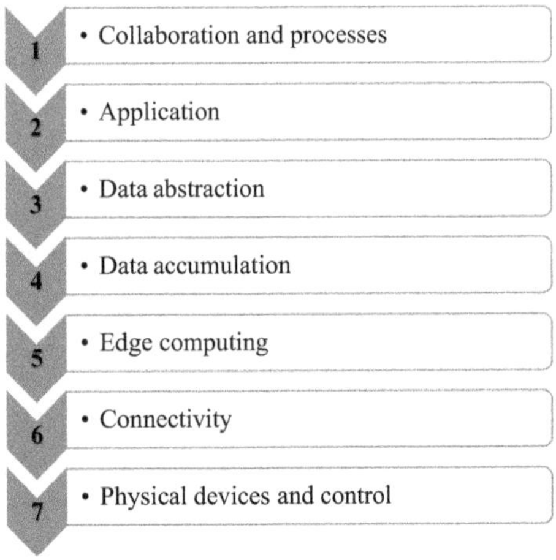

Figure 2.4 Architecture levels of internet of things

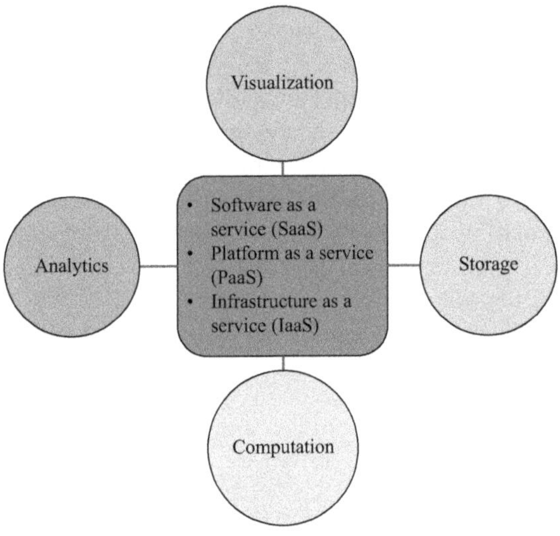

(a) Internet of things—cloud

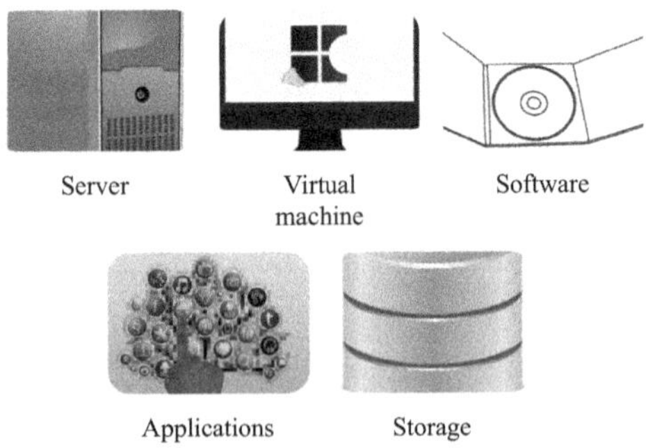

(b) Internet of things—various aspects

Figure 2.5 Various aspects of internet of things

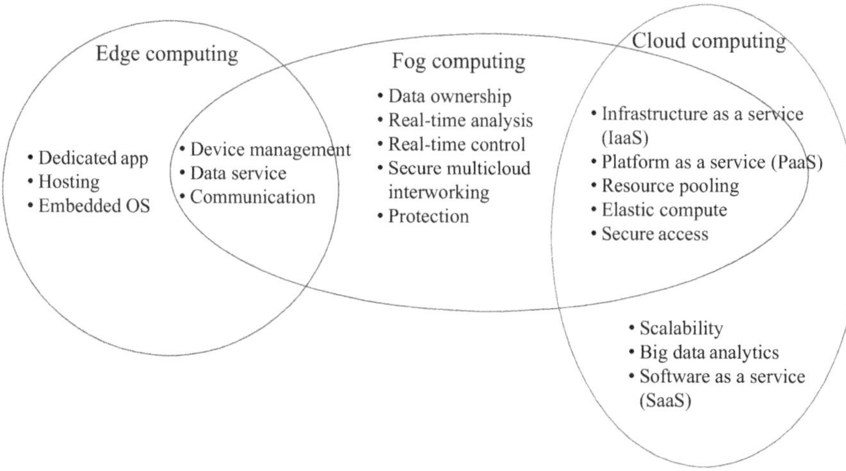

Figure 2.6 Fog computing versus edge computing versus cloud computing

Using the application programming interface (API), the client will have much exploratory choices and process the data further if needed.

IoT, fog computing, and edge computing are three different concepts that provide the functions to control and transfer the data to the cloud storage as presented in Figure 2.6. The cloud computing is extended to the edge; thereby, giving it the secure control and management of software, domain specific hardware, and standard compute. While edge computing is the management and control of the standalone device either individually or by software in the fog domain.

2.1.2 IoT – driving technology

In an IoT environment, every THING will have a unique IP address as discussed above. Every sensors would not have the capability of communication. In such cases, the sensors combined with a device that has the capability of communication will become the THING that is necessary in an IoT environment. There are various aspects to IoT. It has spread its wings to various domains like smart city, smart healthcare, smart transportation, smart home, and so on. As shown in Figure 2.7, various sensors are connected to different single computer boards or microcontrollers which have the capability of having a unique IP address and transmit the data over internet to form a THING. In the figure, the sensor is connected to a smart watch which will collect the data from the sensor and transmit the same to the cloud or other devices.

Various sensors and devices are the driving technologies for IoT. There are various sensors and devices for IoT, and those are called as smart devices. One of the main concerns of these devices is security (Sivaraman *et al.*, 2018). Research is being conducted for a very long time in this aspect of IoT driving technology, various IoT smart devices, and their security (Almajali *et al.*, 2018; Sivaraman *et al.*, 2018; Surendran *et al.*, 2018; Valea *et al.*, 2018). Another aspect of driving

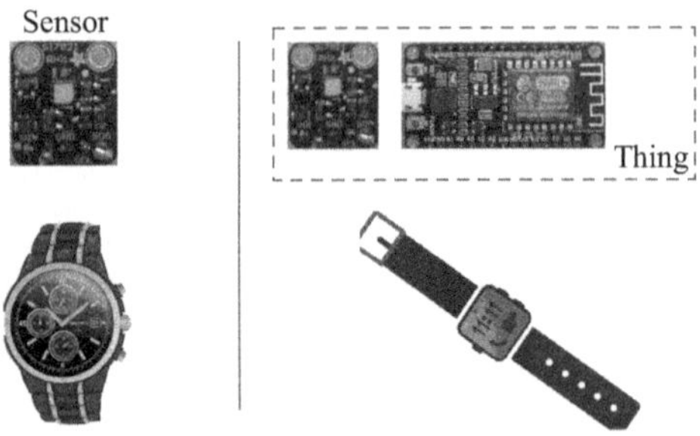

Figure 2.7 Driving technologies of internet of things

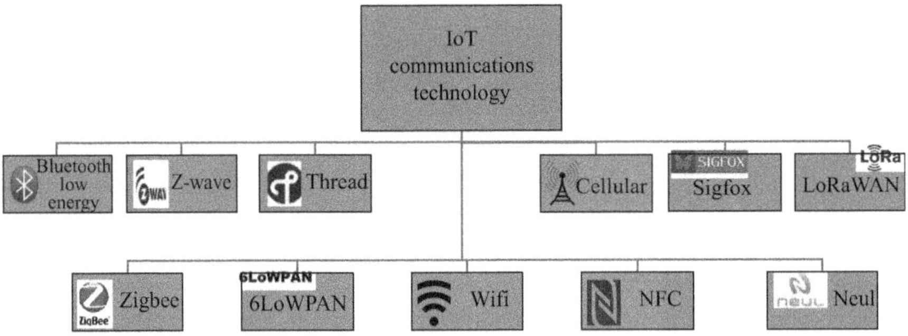

Figure 2.8 Communications of internet of things

technologies in an IoT architecture is the communications. Figure 2.8 shows various communication technologies that are used in the IoT environment (RS Components, 2015). Most of the components are using low-power Bluetooth and WiFi components in the single-board computers and boards for communication, but other communication protocols are also used in IoT. Low-power Bluetooth is not suitable for data transfer of large files, but can be used for smaller data packets. Zigbee is another protocol that is widely popular in the industrial scenario. It offers a low-power operation, robust security, and scalability. For home environments, Z-wave is employed. 6LowPAN is an Internet Protocol-based communication technology, unlike Bluetooth, operates over the network defining the encapsulation and header compression. WiFi and cellular are the options that are explored where ever they can be employed. Unlike other protocols, NFC does not have a wide range and also not capable of sending high chunks of data (RS Components, 2015). Long-range wide area network (LoRaWAN) is getting popular due to its low-energy consumption and long range.

2.1.3 IoT – applications

INTERNET and THINGS are the two main aspects of IoT architectures. Based on them, two different architectures can be proposed. When the Internet is given the

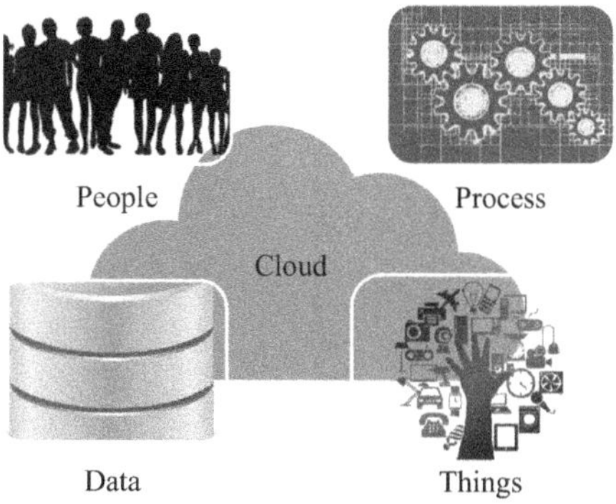

Figure 2.9 Internet of everything

main focus and network aspects are given a priority in the IoT architecture, it is called as an internet-centric architecture (Gubbi *et al.*, 2013). With the cloud at the center of the whole architecture, the full potential can be brought out. With the introduction of cloud, the entire system becomes highly scalable and also makes it simple and flexible to divide the costs associated. When the smart devices or 'THINGS' are the main focus of the architecture, it is called as an object-centric architecture (Lopez *et al.*, 2012). Internet of everything (IoE) as depicted in Figure 2.9 will advance IoT to even more broader scope with human and process integrations.

The smart devices or THINGS have been around for a very long time (Carter *et al.*, 2008; Lee and Ke, 2018; Luo *et al.*, 2009; Obuchi *et al.*, 2018). With RFID becoming the leading technology, many smart devices are taking advantage. Developments in technology and high-performance devices have been a great advantage in developing IoT applications (Mohanty, 2017; Mohanty *et al.*, 2016). Many areas use the concept of IoT to perform actions with more accuracy and acquire more data that was not at all possible before. Figure 2.10 shows different applications of IoT. Smart healthcare is a growing area which takes advantage of smart devices to develop medical devices for improving the quality of life, diagnosing, and providing treatment to various diseases and health conditions (Catarinucci *et al.*, 2015; He *et al.*, 2018; Sundaravadivel *et al.*, 2016). There are various components of a smart city, which includes infrastructure, buildings, energy, transportation, technology, education, governance, and citizens (Mohanty *et al.*, 2016). Economy, governance, society, and environment are different themes of a smart city. Each of the components and themes uses the concept of IoT in a very efficient way to improve the quality of living in a smart city. In a smart home environment, controlling various devices securely can be achieved through IoT (Al-Kuwari *et al.*, 2018; Lee *et al.*, 2017; Son *et al.*, 2016). IoT can also be implemented for various military applications where LoRaWAN, as presented in the previous section, can be used (Jalaian *et al.*, 2018). IoT in military applications is also sometimes referred as Internet of Battlefield Things or Internet of Battle Things (Bolisetti *et al.*, 2017; Farooq and Zhu, 2018; Kott *et al.*, 2016).

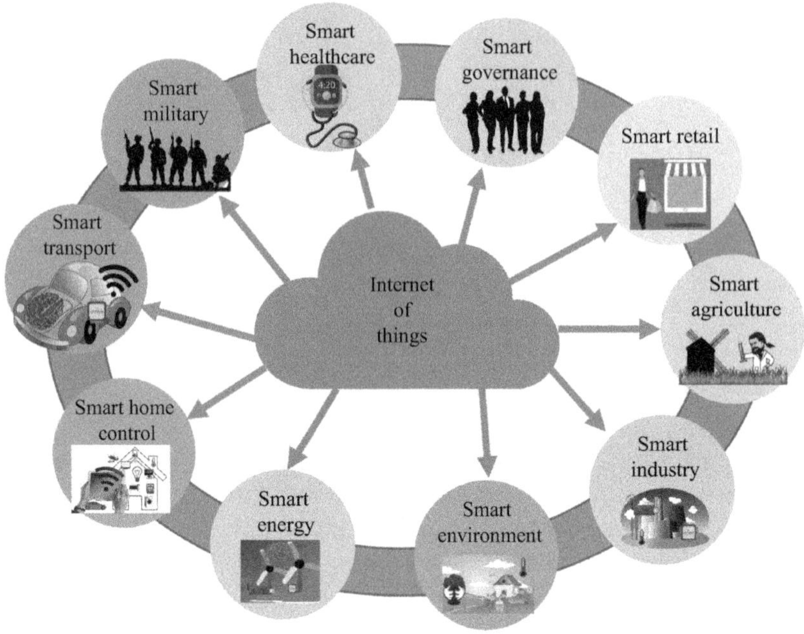

Figure 2.10 Applications of IoT

2.1.4 IoT – challenges

Technological advancements have made developing IoT applications and devices much simpler, but there are a lot of challenges to overcome in this area. Connectivity is a major issue in this regard. IoT devices or Things needs to have a unique IP address of their own which means they should be connected to internet or cloud storage. But, a high-range wireless communication requires more energy consumption and low-power communication protocols will have a transmission rate or range as a trade-off (RS Components, 2015). Sensors have improved since the beginning of IoT, but a need for better sensing devices has not yet been developed. Moreover, accuracy of the sensors has to be improved in various ways as well. Figure 2.11 shows various challenges that are present in an IoT environment (Mohanty, 2017). Various IoT architectures are proposed based on different aspects and focus of IoT (Gubbi *et al.*, 2013). Selecting an architecture based on the environmental conditions and devices being used in the environment is an uphill task and a challenge. There are various dependencies in the IoT architectures.

When various devices are connected to each other through a network, they are vulnerable to various threats. Security issues are a great challenge and addressing them should be considered as a primary priority. Various solutions are available for IoT security but with the implementation of better security solutions, power consumption is going to increase for the entire module (Yanambaka *et al.*, 2017). Collecting the data and sending it to the cloud is part of the challenge for which the

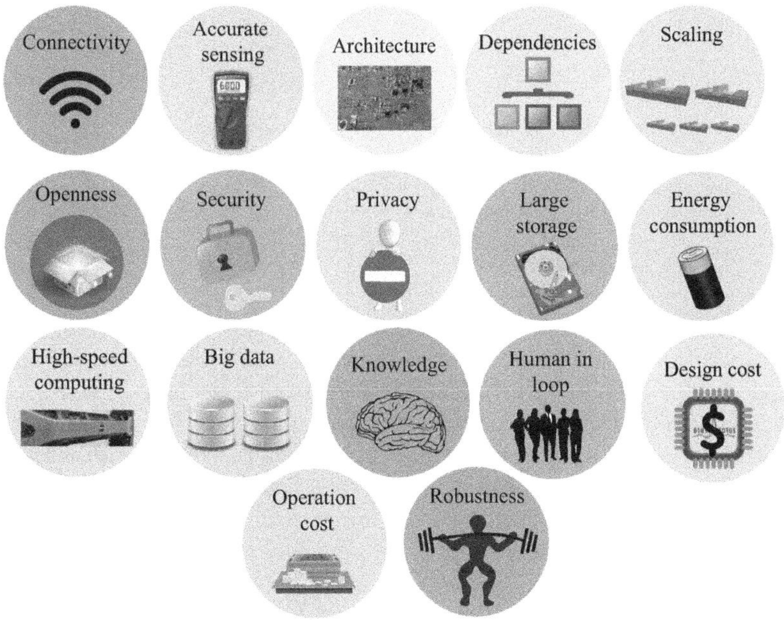

Figure 2.11 Challenges in designing IoT

later part is going to be process the data that is being collected. First, the data is sent to the cloud where it can be processed, employing high-speed servers, and storing huge amounts of data will increase the cost of design and overall IoT design. Processing the data needs high-performance computing, wherein some cases, the data should be processed in real-time that will increase the operational and overall cost of the design.

2.2 Security, privacy, IPR in IoT, and consumer electronic systems – a big picture

Security has been one of the major issues that has been around for a very long time, and it is still the main focus of the research. New devices are developed every day with the advancements in technology, and better solutions are resulted in various domains as discussed in Section 2.1. Even with the introduction of many security primitives, attackers are finding ways to hack and perform successful attacks on various secured devices. This has been a major issue since many times. Cryptography has been the solution where the main text message intended for the end-user is encrypted with a key and the end user will decrypt the text with the same key or a different private key. This has been the in practice for a very long time (Yanambaka *et al.*, 2017). The first lock and key found was 4,000-year-old, and again now in this digital era; here, the locks and keys are also digital, however the underlying concept remains the same. It may be noted that security, privacy, confidentiality, and

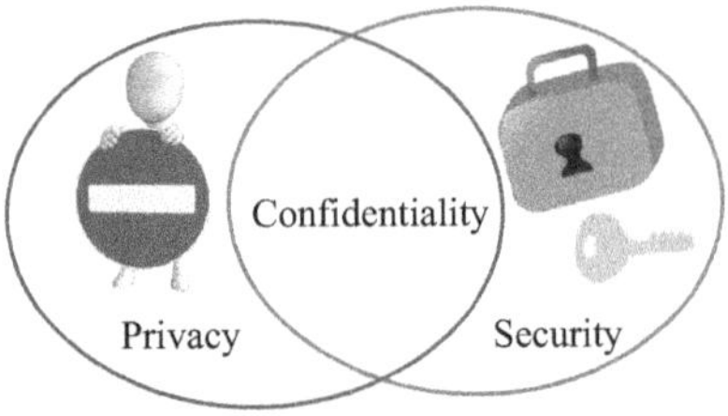

Figure 2.12 Security versus privacy

intellectual property rights (IPR) are not necessarily the same (Bhargava *et al.*, 2018; Mohanty, 2017, 2018) as depicted in Figure 2.12. Data security is typically referred to as the confidentiality, availability, and integrity of data. System security often deals with trustful, safe, normal operations of a system (e.g., smart car), etc. Data privacy is defined as the appropriate use of data as agreed for different purposes. System privacy can be making location and functions of unknown system, e. g., location of a smart car should be unknown to adversaries. Confidentiality is assured only if the information is not made available or disclosed to unauthorized individuals, entities, or processes. IPR or ownership rights deals with the preservation of ownership of the owner of a particular hardware or media which is analogous to the copyright or authorship of a book that of a publisher or author.

2.2.1 IoT security – attacks and countermeasures

The debate of security and privacy has been going on for a very long time (Notario, 2016). The definitions of privacy and security are still in debate, and the most center-point is the confidentiality part. Both security and privacy are important parts of any system. In the case of an IoT environment, these are paramount as the devices in IoT are continuously connected to a network and online. This leaves much room for the attacker to attack the system in various ways. There has been a lot of research going on in this area (Asplund and Nadjm-Tehrani, 2016; Schinianakis, 2017; Schneier, 2017; Sivaraman *et al.*, 2018; Unwala *et al.*, 2018). There are various security attacks that are happening around the world. Number of cybercrime attempts have increased, and so have been the security solutions to fight against them (Andrews, 2017; Heartfield *et al.*, 2016; Kundu *et al.*, 2018; Mazurczyk *et al.*, 2016).

Figure 2.13 shows different types of attacks that are possible on an IoT environment (Nawir *et al.*, 2016). There are many threats where unauthorized and unrecognized malicious devices can enter the network and tamper the system both externally and internally. So, therefore, identifying the users that are in the IoT environment is an important prerequisite. Research is being conducted extensively in detecting and preventing the IoT botnets (Moon *et al.*, 2015; Prokofiev *et al.*, 2018). The device should be tamper resistant. There are various types of attacks like side channel attacks and memory exploitations that can be successfully executed by tampering with the device itself (Li *et al.*, 2011; Sengupta and Kundu, 2017). Hence naturally, the IoT device should be tamper proof, so that the attacker

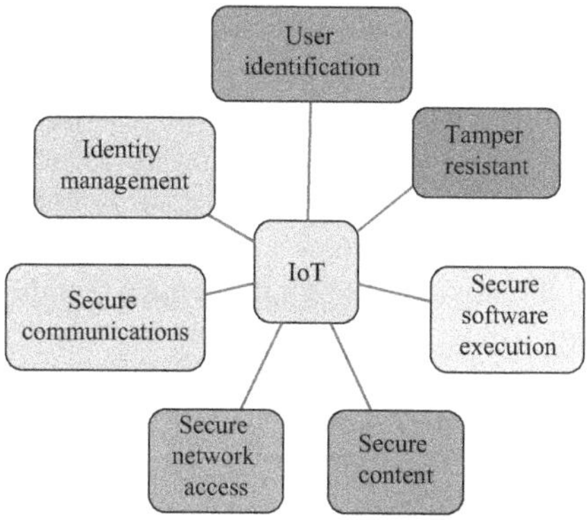

Figure 2.13 Taxonomy of security attacks in IoT

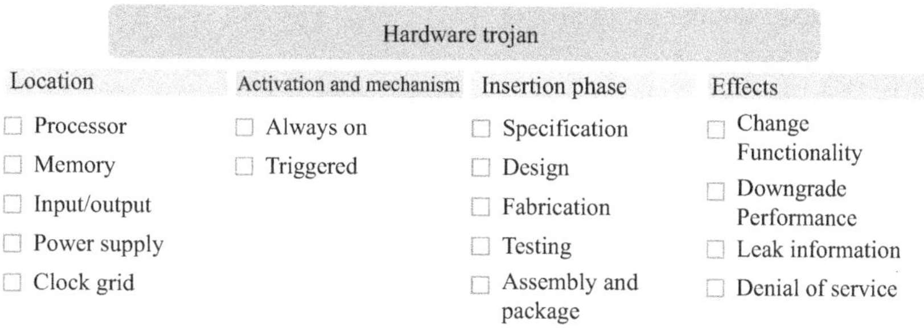

Figure 2.14 Hardware trojans

cannot perform such attacks. Software that is used in the IoT devices should be extensively scrutinized before executed on the device. This can lead to unauthorized usage of the device, data loss, or device failure. The network and the device security are of highest important. But increasing the security of the device may also increase the power consumption of the overall module. This should be avoided as the IoT devices are usually battery operated and cannot be used with a continuous power source. This makes it difficult for the development of security primitives for IoT devices. Low-power hardware security modules are being developed for these devices (Yanambaka *et al.*, 2017) so that the attacks are mitigated. Communication and device network authentication should also be given highest priority such that unauthorized devices cannot be in the network and the communication channel is not compromised (Yanambaka *et al.*, 2016a). Identity management is also a key factor in this scenario. The identity of the device should be secured such that cloning attacks are prevented.

Hardware trojans are another issue that should be dealt within the IoT environment. Figure 2.14 shows various types of hardware trojans that are currently active, the location where they will be introduced, and the phase where there is a chance of insertion (Mohanty, 2015). A hardware trojan is an unwanted and malicious design that is introduced into the main design in different phases of manufacturing by the attacker to render the entire module useless or steal the data from the module. A hardware trojan can be introduced in processor, memory, I/O devices, power supply, and clock grid. Each of these components is the important modules of the device, which play a vital role in the overall functioning of the device. Introduction of a malicious design can prove fatal to the efficient working of device. There are two different types of trojans, one which is always on when the device itself is on and the other where it can be triggered with an event or with a signal. There are many ways in which the hardware trojans can be detected and avoided (Dupuis *et al.*, 2018; He *et al.*, 2017; Sengupta, 2016, 2017; Shen *et al.*, 2018; Xue and Ren, 2018). There are many phases where the hardware trojans can be inserted as shown in Figure 2.14. Changing functionality, downgrade performance, leak information, and denial of service are some of the effects of hardware trojans. These can prove fatal for many cases where the device cannot function at the full potential in times of need.

2.2.2 Trustworthy consumer electronic systems

Besides high- and low-power consumption, any consumer electronic system should also have the properties of trustworthiness and security. Figure 2.15 shows different aspects of security, privacy, and protection (Krishnamurthy *et al.*, 2017; Mohanty, 2015). There are different layers in the system, application, operating system (OS), firmware, digital signal processor (DSP), embedded processor, memory, and radio frequency (RF) components. At each of the stages, there are different kinds of attacks that can be successfully performed on the system. Application and OS attacks are possible on the OSs and its applications. Between the OS and applications or firmware and OS, there is a network which connects them to perform operations. This network can be attacked with different kinds of

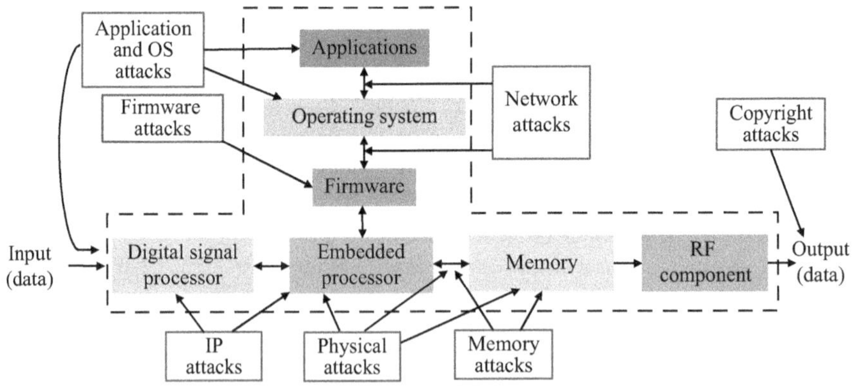

Figure 2.15 A selected attacks on a typical CE system – security, privacy, IPR

attacks: application attacks and network attacks. Some of them were discussed in the previous sections. Firmware is an integral part of the system. This acts as a bridge to the board or system to talk to other components or peripherals and describes its functionality. There are different kinds of firmware attacks possible on the devices.

Different hardware components of the device such as DSP, embedded processor, memory, and RF components, can become targets to various attacks as shown in Figure 2.15. IP attacks, side channel attacks, physical attacks, and memory attacks are some of them that are performed at the hardware level. After the output is produced, the data in the output can be subject to many copyright attacks. The RF components can also be subject to various network threats which will be discussed in the proceeding sections.

A selective attributes of CE system to be trustworthy (Krishnamurthy *et al.*, 2017; Mohanty, 2015):

1. The integrity of the data that is being processed by the system should be maintained.
2. The device will be prone to side channel and power analysis attacks once it is deployed in the field. So the data that is being processed should be secured and concealed during processing.
3. The functionality of the device should be restricted to the intended functionality. Additional operations should not be performed by the device.
4. The device should be reliable at all times and especially at times of critical operation.
5. The transparency of the device must be restricted to the owner and no one else.
6. Trusted vendors should be approached for different parts while manufacturing the device to avoid introduction of unwanted trojans.
7. Trusted fabrication facilities should be approached.

2.2.3 Hardware-assisted security and protection

With the increased security threats, software-based security alone is not going to fight against them. Hardware-assisted security plays a crucial role in mitigating the threats caused by the attackers. There are two different kinds of attacks that can be performed on the system: software-based attacks and hardware-based attacks (Chang *et al.*, 2017; Krishnamurthy *et al.*, 2017). The software attacks are performed usually from the communication channel. As IoT is connected through a network, this can easily be under threat by the attacker. These attacks are typically originated remotely and the device is affected. As the attacker need not be present at the device to perform them, they can be performed regularly and remotely. Denial of service attack, routing attacks, malicious injection, spoofing, snooping, and brute force are some of the software-based attacks.

Hardware-based attacks are typically on the device itself and most of them has to be performed at the location of the device. Some kinds of hardware attacks can also be performed remotely. In an IoT environment, the devices are deployed in areas where human interaction is very minimal and so it is highly prone to the

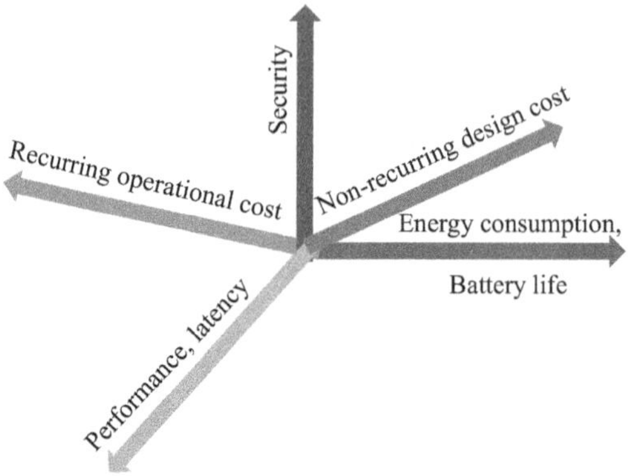

Figure 2.16 CE system security – aware design tradeoffs

attacks. Side channel attacks, backdoors, eavesdropping, inducing faults, and hardware trojans are some of the hardware attacks.

For the two kinds of attacks, software and hardware, the solutions available are also two different types: software solutions and hardware solutions (Krishnamurthy *et al.*, 2017). The software-based security solutions are easy to use, update, and also upgrade. This gives the flexibility to increase the robustness offered. As the software-based applications are dependent on the OS, they are widely supported and can be designed for cross-platform applications. The recurring operational costs are high in case of software securities. They can be used for general purpose processor but one of the main drawbacks is it will not be able to save the system against reverse engineering and other hardware attacks. In case of a hardware-security solutions, they are very fast when compared with software solutions, consumes less power, high flexibility in optimization, and low cost when an field programmable gate array or application-specific integrated circuit is used. They cannot be used to stop the software reverse engineering and are highly prone to side channel attacks.

Figure 2.16 shows the security-aware design trade-offs and describes different aspects of the design. An example of hardware-assisted security is shown in Figure 2.17. Major concerns when designing a system is low-power consumption and high performance, reliability, low cost of manufacturing and easy integration. In the design of a secure camera system, the watermarking chip is the main focus as most of the security is offered by it. The trade-offs that are present are power consumption, performance, chip area, integration into the system, and the memory. Integration of more memory into the system will increase the battery life, chip area, and design cost of the camera. Increased security will also increase the battery consumption and the other parameters as shown in Figure 2.16. Thus, a design should be designed keeping all the parameters in view and trade-offs for security, and include additional hardware components, that will perform voltage and frequency scaling like technology for security, energy, and performance optimization (Figure 2.17; Krishnamurthy *et al.*, 2017).

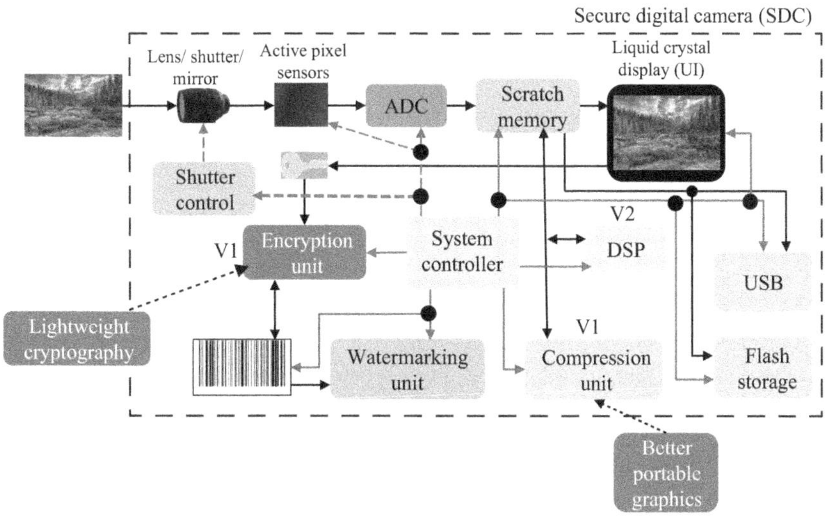

Figure 2.17 CE system design and operation tradeoffs: camera system example

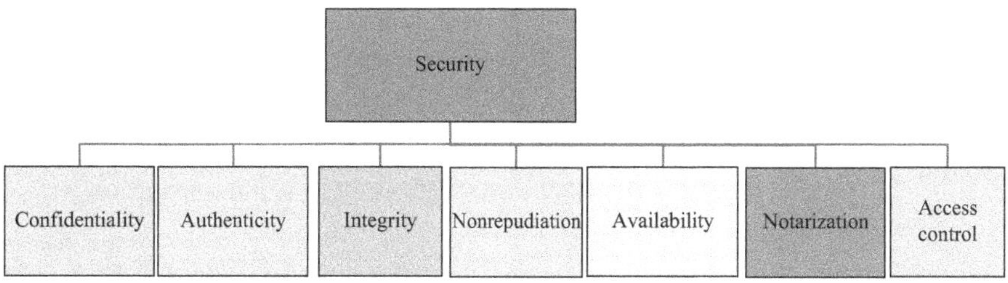

Figure 2.18 Different aspects of security

2.2.4 Different aspects of security and privacy

Security and privacy are two different aspects and are always up for a never-ending debate. The debate regarding them is going on for a very long time, and might not end anytime soon. The definitions and differences between them have not yet reached a conclusion. Different aspects of security are shown in Figure 2.18 (Mohanty, 2017, 2018). The data that are stored in the memory should be secure and confidential. It should not fall into the wrong hands and attacker should not be able to retrieve the data stored in the memory. The data that is stored should also be authentic, i.e., no unauthorized person should have tampered with the data and changed it. The integrity of the data should also be intact. The data should always be available. The storage should not be affected by the attacker and the storage must not wipe-off any of the data. Non-repudiation is another issue where the owner of the data cannot deny it. Access to the personnel should also be controlled and only authorized people should be able to access the data.

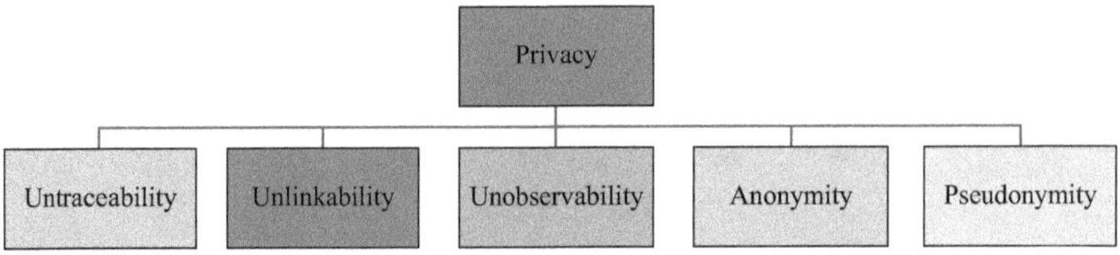

Figure 2.19 Different aspects of privacy

Different aspects of privacy are shown in Figure 2.19 (Mohanty, 2017, 2018). Privacy of anyone should not be compromised with the data that is being collected and stored. In case of a smart healthcare device, the patient's data will be collected over a period of time and sent to the doctors. The only people that should have access to the data are the patient and doctor. The cloud storage server where the data is stored should not be accessed by the attacker, and the privacy of the patient must not be compromised. In such case, the data should not reveal or lead to the patient and not be traceable. The data that is stolen from the storage should not lead to any additional details or data. The data can also be encrypted or scrambled so that the patient information or any other data cannot be retrieved from any patterns and data. Anonymity of the patient and doctor should always be maintained at the storage node.

2.2.5 Different aspects of intellectual property (IP), ownership right, or copyright protection

Ownership, IPR, and copyright protection are the main issues that are present in many different areas. Not only media but electronics are also highly susceptible to the counterfeiting (Mohanty, 2015, 2018; Mohanty *et al.*, 2017; Sengupta and Roy, 2017; Sengupta *et al.*, 2018). This has been around for a very long time. Copyrights management has been around and various laws are established such that the ownership of the entity remains intact. But attackers are finding different ways to counterfeit devices and media and sell the items on their own ownership. Figure 2.20 shows the concept of IPR, ownership and copyright protection. Consumers and researchers are having the trouble in maintaining the rights. There are cases where the user is not aware of the copyright laws and is not using the entity for their own commercial purpose. But in most of the cases of copyright infringement, the attacker takes advantage of the laws that vary from region to region and claim ownership of the device or media. Identical designs are used for the electronic components, which makes the consumer to wonder whether the hardware is legitimate or not. These are also an issue of the media where a photograph is debated for ownership by the owner and the attacker (Mohanty, 2015). Watermarking the images is a way to protect the media but this is also susceptible to many attacks where the attacker can remove the watermark and claim the ownership of the media.

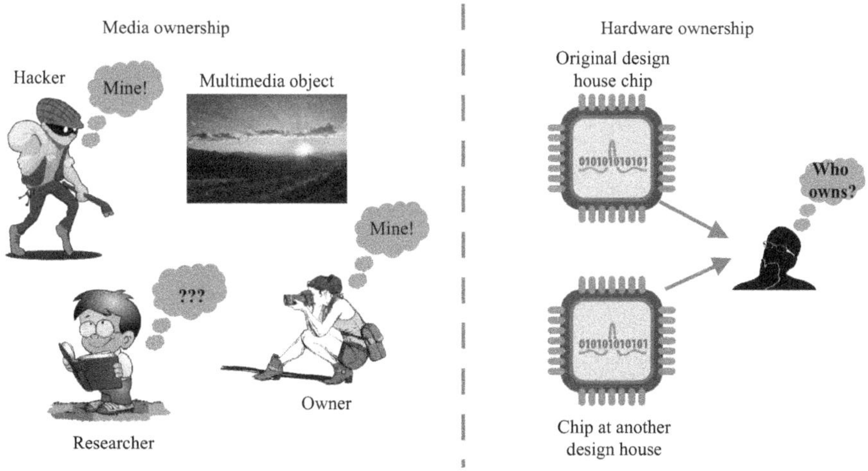

Figure 2.20 Intellectual property (IP), ownership right, or copyright – media versus IP hardware cores

2.3 Memory security

Memory is the integral part of every system. Different types of memory, primary, and secondary memories are available and are present in almost every IoT device. Non-volatile, static random access memory (SRAM), etc., are employed to improve the performance and increase the storage capabilities of the IoT devices where they will have the capabilities to store the data collected and send to the cloud at a later time. There is no IoT device without memory. But different types of memories are subject to different types of attacks. It is a highly vulnerable part of an IoT device. This section describes various kinds of attacks and solutions that are available in the market.

2.3.1 Memory security attacks

Various attacks are possible on the memory modules of the embedded systems. Figure 2.21 shows various attacks that are possible on the memory of an embedded system (Nimgaonkar *et al.*, 2013). The attacks can be broadly classified as passive and active attacks. The passive attacks may include snooping and cold boot attacks. Snooping attacks happens when an adversary is able to read and interpret confidential information in memory. The cold boot attacks, which is also a type of side channel attack, a hacker who has physical access to a computer retrieves encryption keys from a running OS after using a cold reboot to restart the computer. The attack relies on the data remanence property of random access memory (cache, SRAM, or dynamic random access memory (DRAM)) to retrieve memory contents that remain readable in seconds to minutes after the power supply has been removed. This happens as the memory chips used in the computers retain their contents for seconds to minutes after power is lost even if these are volatile, leaving the contents available for malicious or forensic acquisition. Active attacks include spoofing,

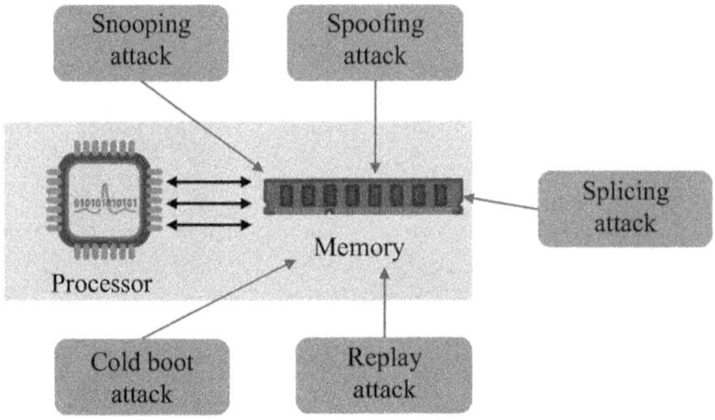

Figure 2.21 Attacks on the memory of an IoT device

Figure 2.22 Nonvolatile solutions for attacks on the memory of an IoT device

splicing, replay, and denial of service. In the active attacks, an adversary is able to change the contents of memory so it would not perform the intended functions. In spoofing attack, an adversary exchanges a block of memory with an arbitrary fake block. The block of memory at one location is replaced with another block from a different location in splicing attack. In replay attacks, an adversary stores the value of a block of memory at a given address at one time, and writes that value at exactly the same address at a different time. Replay attack is the hardest attack as it is difficult to trace and protect. The denial of service attack memory refers to a scenario, when a malicious process takes most of the memory bandwidth, leaving other processes without the required resources and thus effectively slowing them down.

2.3.2 Memory security solutions

The nonvolatile memory (a.k.a., hard-drive or solid state drive) is protected by different hardware and software methods which rely primarily on encryption as depicted in Figure 2.22. In this kind of solutions, an encryption hardware is integrated in the storage which then allows access through a password or personal identification number (PIN). Similarity, an encryption software can run in the system that uses the storage drives for data security purposes. The software then allows access to the data through password or PIN.

There are various solutions available to sense an attack and defend against it. In a typical embedded system, there are various sensors that are deployed around the entirety of the system. These different sensors are used to detect various parameters of the system. When the system is being attacked by the attacker, there will be changes in various parameters of the system like current fluctuations, increased power consumption, and thermal variations. These parameters can be constantly monitored with the help of sensors and detect any abnormal variations in them. The memory integrity verifications are performed only when these values sensed by the sensors are abnormal. In a traditional embedded system, the integrity verification is performed for all the blocks of memory which will increase the power consumption of the entire module for the preservation of data integrity in the memory.

The definition of any memory integrity verification approach can be structured as follows. Initially, the memory '*M*' is communicated by a processor. There are two attributes to the memory '*M*,' the contents of memory '*V*,' and addressed '*A*.' *M[R,A]* denotes that the memory at the address A is read and it will return the data that is associated at A. *M[W,A,V]* denotes that the value V is written into the address A. To return the value V, the write of value V should be immediately followed by address A. The writes and reads of the memory have a notion of time. In order to associate the same into the reads and writes, it should be addressed as: *M[W,A,V,T]* and *M[R,A,T]*. The association between the values that is present in the memory locations and the addresses of the locations is called as splicing attack. Let us consider an example, address A should return the value V_A and A' should return the value V'_A, when the memory is subjected to a successful splicing attack, the A will return the value V'_A. When the value that is stored in the memory location associated with an address A is changed to a random value V'_A, this is called spoofing attack. When the association between the value and the time is disoriented, it is called as replay attack. For example, let us consider V_A as the value at the address location A at a time T_A and the value of the same location at the time T_B as V_A^* and $T_A < T_B$. When a successful replay attack is performed on the memory, at a time $T > T_B$, the value that will be returned is V_A.

Implementing various sensors in a trusted environment boundary of the embedded system for monitoring various parameters of the processor have been around for a long time (McGowen *et al.*, 2006; Oh *et al.*, 2010). Research has been conducted to use the variations in power consumption and current fluctuations to monitor the attacks that are being performed on the devices. Figure 2.23 shows the implementation of such a system that can perform the memory integrity verification when the sensors detect an abnormal values of the parameters (Nimgaonkar *et al.*, 2013).

As shown in Figure 2.22, various sensors that are present in the hardware is represented as sensor module (SM). The memory bus is constantly monitored by the sensors present in the SM for the fluctuations. The memory integrity verification mechanism is present in the special cache, the 'hash cache.' The hash addresses are present in the hash cache. Hash of the memory block is written to the off-chip memory. The hash and the hash addresses are used to verify the integrity of memory. The processor in the embedded system is a trusted chip. So, the cache

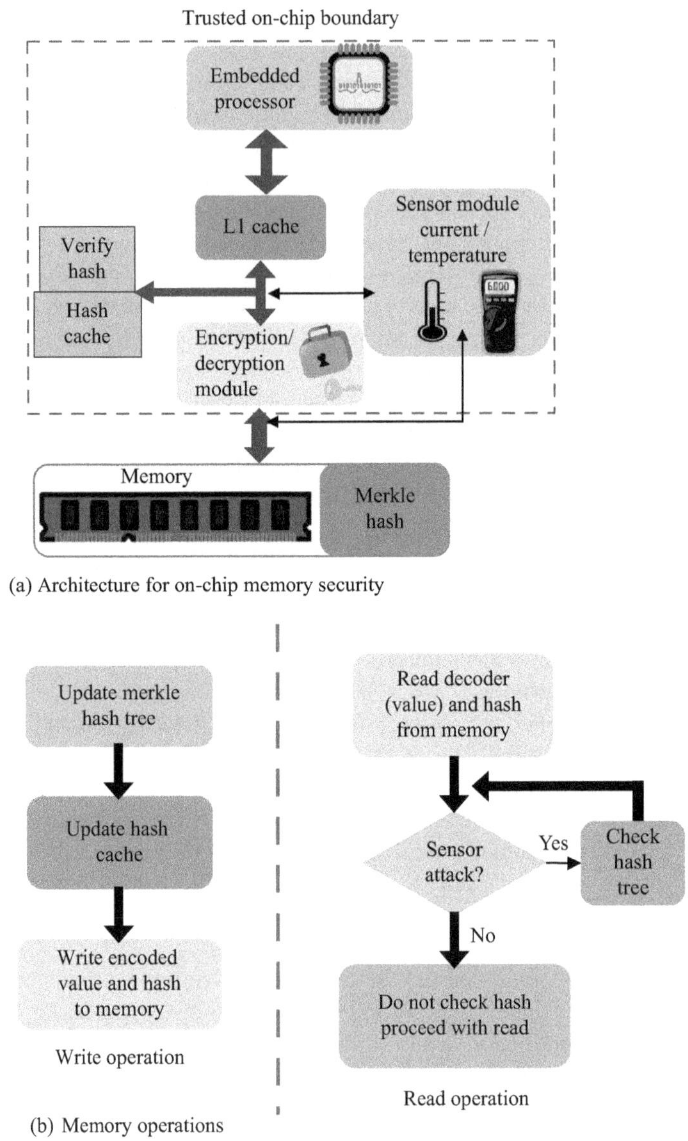

Figure 2.23 On chip memory solutions for attacks on the memory of an IoT device

and the SM are present in the processor boundary. An encryption/decryption block is present in the architecture.

The read and write operations performed on the memory are described in Figure 2.22. When data are being written to the memory, the hash address and the hash are computed by the processor and are stored in the hash cache. After storing them, the memory block is encrypted and stored. During the read operation, the memory block which is read is decrypted. During this decryption operation, the SM continuously monitors the memory bus for any abnormalities. If the voltage fluctuations in the memory bus cross a previously set threshold value, the system is considered to be under attack and in this scenario, the hash address is computed again. It is verified with the previously stored hash addresses in the cache. If there is a hit, the hash of the data itself is checked and verified using the hash stored. In case if these two did not match, the system is under attack and the data get corrupt. All the operations that are being performed by the processor are aborted. Once the voltage fluctuations in the memory bus are reduced below a threshold value, the processes stop and the threat is considered to be stopped or mitigated.

2.4 Radio-frequency identification (RFID) security

Radio-frequency identification (RFID) is an old technology and has been used in various applications. It can be said to be a starting for IoT in some ways. Tracking, security, and locating are some of the applications of RFID. In IoT, RFID is implemented in many ways. Even though the range of RFID is very less, it can be used for various purposes in IoT environments. But like every communication protocol, RFID is also vulnerable to many attacks and threats. This section discusses various security vulnerabilities and threats of RFID and solutions to some of them.

2.4.1 RFID security attacks

There are different security threats for RFID (Jayakrishnan *et al.*, 2018; Li *et al.*, 2012; Thompson *et al.*, 2014). Figure 2.24 shows various attacks on RFID (Mohanty, 2015). The attacks are categorized into three types: physical threats, channel threats, and system threats.

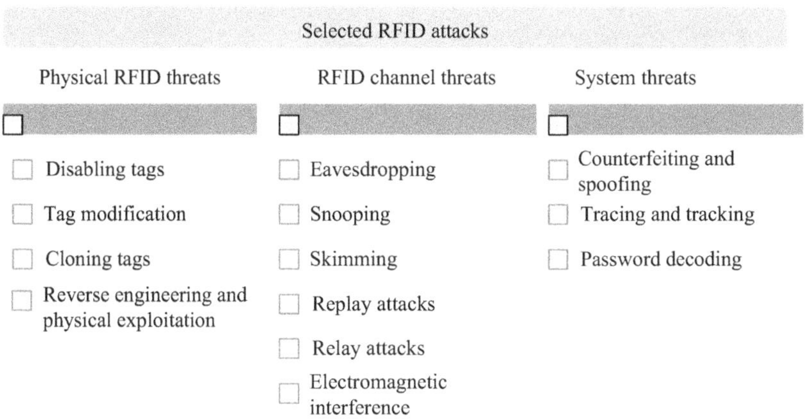

Figure 2.24 Attacks on the RFID modules of an IoT device

Physical threats are the security threats that are performed on the physical RFID device, transmitter, or receiver (Li *et al.*, 2012). Due to the low cost of the devices, the RFID tags are not tamper resistant and can be easily exploited. There is a great chance of reverse engineering the RFID tags and clone them. This allows the attacker to get the data that is stored in the tag. Every RFID tag will have a unique identification number and copying the label is not a difficult task with the device itself in hands of the attacker. The RFID tags can be easily disrupted with the electromagnetic waves which can disable them temporarily making them vulnerable. To disrupt the RFID tag, the electromagnetic radiation should be of the same frequency as that of the RFID system.

There are two components in an RFID system: the tag and the reader. The channel to communicate between them can be taken as an advantage and the vulnerabilities are exploited by the attacker. These are called as channel threats. Eavesdropping is a vulnerability of the RFID tag as the communication between the RFID tag and the reader is wireless. The information retrieved by eavesdropping can be used for different attacks like counterfeiting, copying the tag, track, etc. The transmitting power of the reader is high when compared with the RFID tag. so reading from the tag is going to be easier comparatively. To avoid the data being stolen, the data transmitted should be encrypted between the reader and the tag.

One other major vulnerability of the RFID tag is the replay attack. First, the data is collected by performing an eavesdropping attack on the RFID channel. This collected data is again replayed to the reader which is illegal and a false deception is achieved. The attacker can deceive the reader by impersonating the original tag and gain access. One time password, stamp program, and so on, are some of the solutions for these types of attacks. Man in the middle attacks, popularly known as relay attack is another security vulnerability of RFID, where the attacker can intercept the data between the reader and the tag but the attacker is far from the devices themselves. By using the devices, the data intercepted can be forwarded to another device and used for various reasons. This is called relay attack.

2.4.2 RFID security solutions

It is said that any system is as secure as its weakest element. So to secure the entire system and increase the robustness, its weakest elements should also be made secure. The two different communication channels of RFID are front-end and back-end, respectively (Vattem and Anjali, 2017). Readers collect the data from the RFID tags. This is the front-end part of the communication. Once the data is collected from the tags; the data is transmitted to the servers to process it. This is called as the back-end of the system. The back-end communication is already made secure by using secure shell and secure socket layer. This is already standardized and secure. But the front-end is prone to many issues as discussed above. This can be considered the weakest part of the RFID systems.

There are many security solutions available for RFID systems (Jayakrishnan *et al.*, 2018; Sun and Zhong, 2012). Figure 2.25 shows various solutions for the RFID attacks that can be implemented to prevent the attacker to succeed in taking

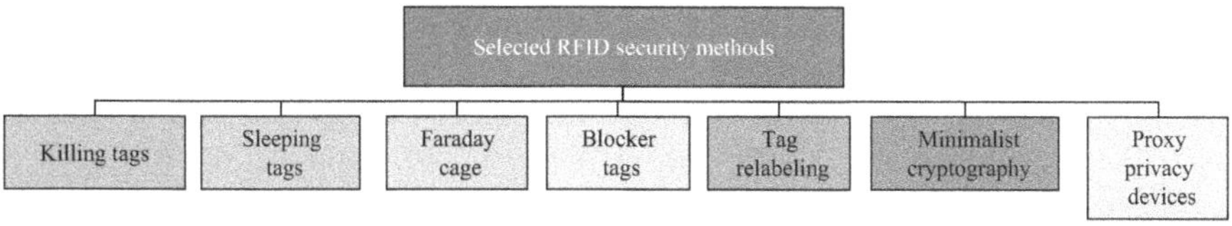

Figure 2.25 Solutions for attacks on the RFID modules of an IoT device

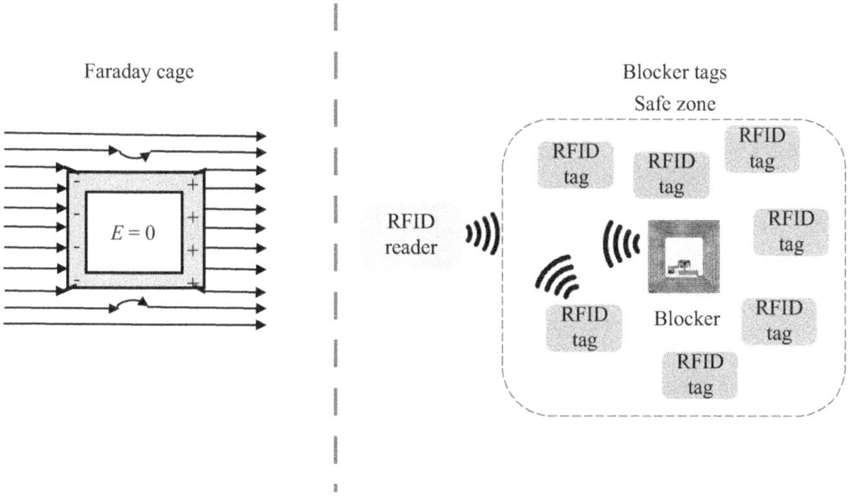

Figure 2.26 Solutions for attacks on the RFID Modules of an IoT device using Faraday cage and blocker tags

advantage of the vulnerabilities. Killing the RFID tags when a security threat is detected can be one of the ways of saving the security measures that can be implemented easily. The communication between the reader and the RFID tag can be temporarily stopped and necessary measures can be applied before continuing the communication again. The RFID tags themselves that does not contain any power source and hence the reader should be very near to the tag which needs the attacker to be near to the devices. Faraday cages can be implemented as a security measure. Faraday cages for security has been around for a very long time (Ahmadi *et al.*, 2017; Wu *et al.*, 2001). Figure 2.26 shows the structure and working of a Faraday cage. When it is implemented, the electromagnetic waves cannot pass through it as it is canceled out by the electric field generated around the Faraday cage. Cryptographic primitives like encrypting the communication between the tag and the reader can be implemented which can render the data collected by the attacker unless without the proper decryption. Blocker tags that block-out the unauthorized reader requests can be placed in proximity and can protect the authentic tags from being vulnerable.

There are also other solutions for security of RFID (Halevi *et al.*, 2013). In some cases of applications, access to the system can be blocked to certain conditions. In a scenario of a patient with an implantable medical device, access can be granted to the reader when the patient is lying down or in a specific pre-determined posture which can prevent the attacker from gathering data or tampering with the system in other conditions (Halevi *et al.*, 2013). In the case of a keyless entry car, this context can be taken as an advantage (Yang *et al.*, 2012). There are various steps for the driver before starting the car engine, for example, opening the door of the car, sitting in the driver's seat, etc. In this scenario, the posture of driver sitting in the driver's seat can be considered for starting the car's engine. Obviously, this can prevent any unauthorized usage of the vehicle by the attacker. This scenario can also be implemented for starting the car from a distance or home of the user (Yang *et al.*, 2012). The human actions and interaction with the world is an involuntary action which need not require the user to perform any additional actions for the purposes of security.

2.5 Near-field communications (NFC) security

Near-field communication, often also referred as NFC, is a short-range communication system. NFC is a contactless technology that allows data communication between the devices without internet connection. Features like Android Beam can be used in the NFC which allows for the users to transmit the data between devices. There are various applications of NFC which are shown in Figure 2.27 (Harbison, 2016). The foundation for developing NFC is RFID technology. There are some drawbacks to the NFC technology which include smaller range and low data rates.

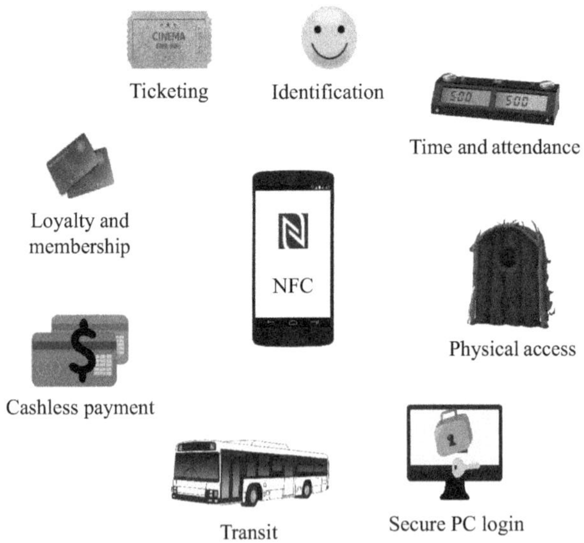

Figure 2.27 Various applications of NFC modules in IoT

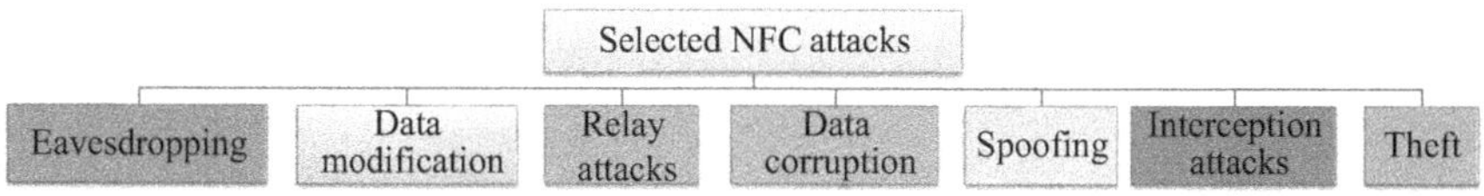

Figure 2.28 Attacks on the NFC modules of an IoT device

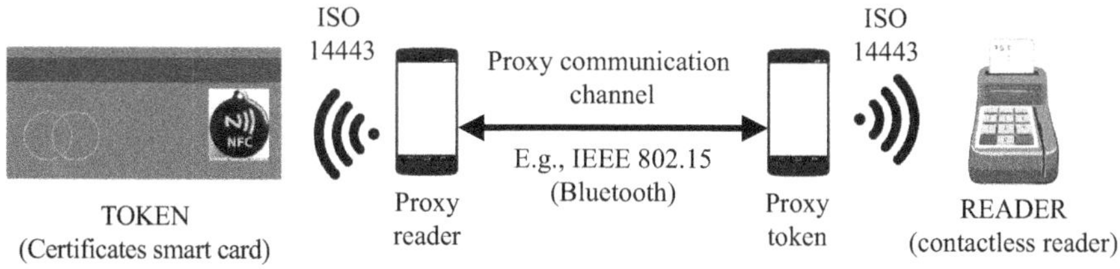

Figure 2.29 Relay attacks on the NFC modules of an IoT device

There has been a lot of research conducted in the area of NFC. There are various attacks that can be performed on NFC modules even though they are considered difficult to perform, NFC relay attack being one of them. Figure 2.28 shows various attacks that are performed on an NFC module. A trojan relay can be installed on a mobile device equipped with NFC and can initiate payments using the properties in the device of the victim (Harbison, 2016). This issue was addressed immediately but recent studies show that NFC can be used to steal the currency from various physical credit cards in the pockets of the victims. A point of sale (POS) machine which has the capability to accept payments through NFC and a mobile phone having NFC built into it are the only things needed to execute this attack (Harbison, 2016). This is possible with the NFC relay attacks which will forward the wireless communication over a large distance in contrary to the small distance of the NFC which are discussed further in the section.

2.5.1 NFC security attacks

Figure 2.29 shows the relay attack on the NFC module. As explained above, the attacks allow the data to be transmitted over a large distance. There are various parts of the attack, viz., honest prover, honest verifier, dishonest prover, and dishonest verifier. The cards of the victim and the point-of-sale terminal of the attacker are the honest prover and verifier in the scenario. The dishonest prover and verifiers are the NFC capable mobile phones of the victim and the attacker.

ISO/IEC14443 protocol of the NFC is taken as an advantage by the attacker in the relay attack. First, the attacker will forward the reader's request to the victim and the answer from the victim is sent back to the reader in real-time. In the attack scenario, the attacker should be near the victim's card and relay data over another communication channel which is sent to the second NFC reader. This secondary NFC reader has to be in the original readers' proximity. For example, in case of cashless payment, an attacker could have an accomplice covertly holding one of

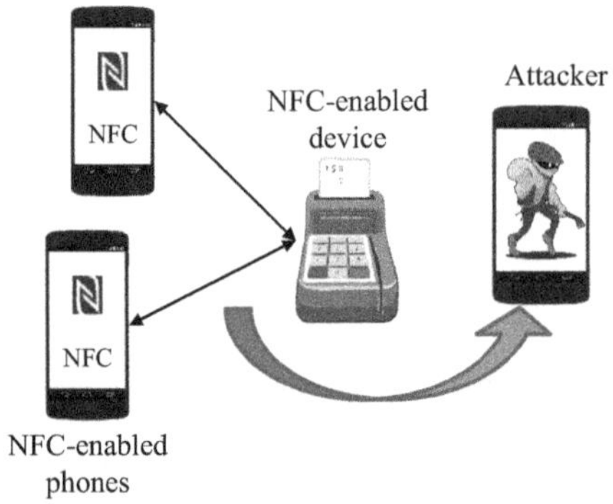

Figure 2.30 Eavesdropping attacks on the NFC modules of an IoT device

the relay devices in range of somebody else's payment card (i.e., by holding it near their wallet). Once this is done, the attacker could then use their relay device to purchase an item against the victim's card. The disadvantage to this type of attack is the time taken to transmit the data. Due to the increased distance that the data has to travel, the time taken to transmit the data will also be increased.

In case of data-modification attack, the data that is being exchanged by the two parties is captured by the attacker and modified. The device used by the attacker will inhibit the data that is exchanged through the NFC channel for a brief amount of time. Another way of interfering with the data exchange over the NFC channel is to use an RFID jammer. The data modification can be easily detected by monitoring the strength of the signal that is being transmitted.

Figure 2.30 shows the eavesdropping attack of the NFC applications. In this case, an antenna is used to monitor the traffic between two parties by the attacker. The small distance between the two devices does not reduce the possibility of this attack. In some cases of attacks, the data that is being exchanged will be corrupted rendering it as useless. Using a secure channel for transmission, the data integrity can increase.

In a spoofing attack, the attacker pretends to be a different entity and taps the device against the tag. This can be made possible if the tag is compromised allowing it to run the malicious code. Even though the distance between the sender and receiver is very small, in some cases, man in the middle attacks is also possible in NFC. The attacker can use the antennas to intercept the information and manipulate it before sending it to the original receiver.

2.5.2 NFC security solutions

Figure 2.31(a) shows the architecture of Swing-Pay, a digital card module which uses biometric authentication combined with NFC for peer-to-peer payments (Ghosh *et al.*, 2017). Various components of the Swing-Pay system are shown in the figure. For prototyping, Arduino Due board was used for the modules (Ghosh *et al.*, 2017). The prototype of Swing-Pay was presented using the off-the-shelf components (Ghosh *et al.*, 2017).

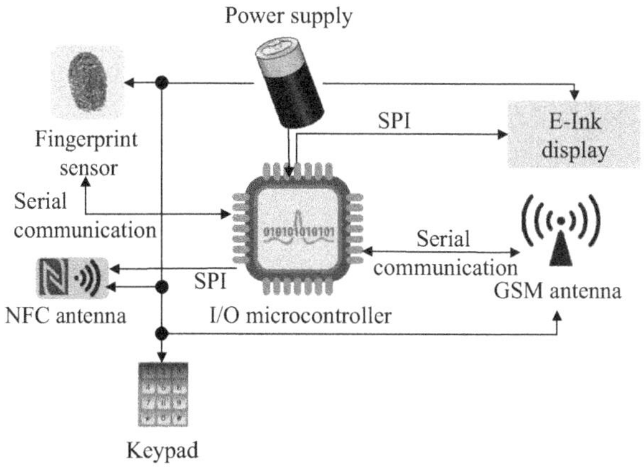

(a) Swing pay—solution for attacks on NFC

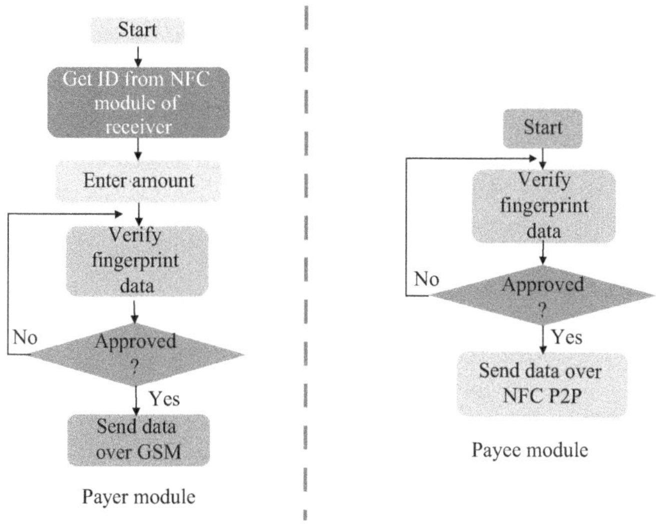

(b) Payer and payee modules of swing pay

Figure 2.31 Various aspects of internet of things

In the design, E-Ink display was used for reducing the power consumption by the modules. The displays are bistable, and the power is consumed only when a change of state is required. A global system for mobile communication (GSM) modem, Adafruit Fona was implemented in the design for communications with a small size and a wide variety of capabilities. For interaction, a capacitive touch sensor was implemented in the design. A battery, lithium ion battery, was used as a power source.

In the framework of Swing-Pay, the fingerprint of the payee is activated initially. The card will be activated once the fingerprint is authenticated. Once

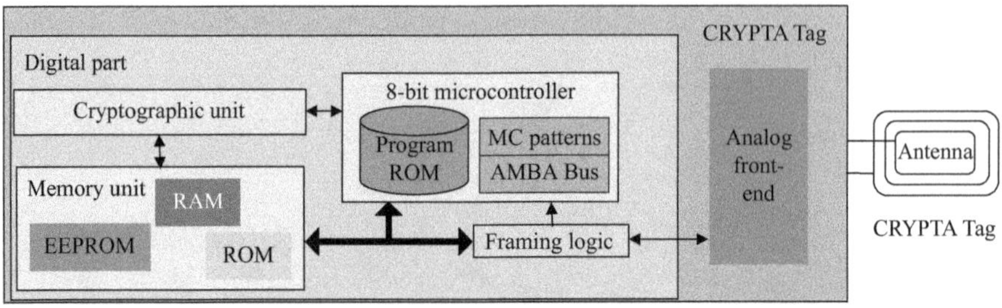

Figure 2.32 Crypta tag

payer selects pay-money, the payee will select the receive money in the respective module. The payee is authenticated using the fingerprint sensor and tap his card with the payer's card. Upon successful authentication, the Payee's ID will be transferred using NFC P2P mode to the payer module. If the authentication fails, the transaction will be blocked. Upon successful authentication, a message is sent through GSM module. In this prototype, the Android application was used and data is put in a Web Service using HTTP.

Once the data is received by the cloud server, it proceeds with the transaction. The details of bank for both payer and the payee are stored in the cloud server. Details are sent to the respective bank servers using the API. The payer's bank will transfer the money to the payee account.

A Crypta tag architecture is shown in Figure 2.32 (Plos *et al.*, 2013). One of the main components of the tag is analog front-end. This is connected to the antenna and is responsible for various tasks, extracting power supply, modulation, and demodulation of data and providing stable reset and clock signals. The digital part is attached to the analog front-end. The received demodulated data is processed in the digital part of the tag. The requested operations are performed on the data received and the responses are prepared. The digital part of the tag consists of an 8-bit Microcontroller, crypto unit (CU), framing logic (FL), and a memory unit. Internal memory stores the microcontroller program and AMBA bus is used for the communication between the FL and memory unit. Various cryptographic operations are performed on the data by the CU. An interface to the microcontroller is provided by FL and also handles some of the commands that are at lower levels.

2.6 Smart transportation security

This section discusses the security threats and solutions for smart transportation. The category of vehicles that are discussed here, include smart cars and UAV. Research is being conducted where smart vehicles are being designed to reduce human interaction in driving the vehicle on road or in the air. This means the system will be doing all the driving and the dependence on circuitry increases exponentially. They are also connected to the internet for constant updates and also communicate with the servers. This makes them vulnerable to many kinds of

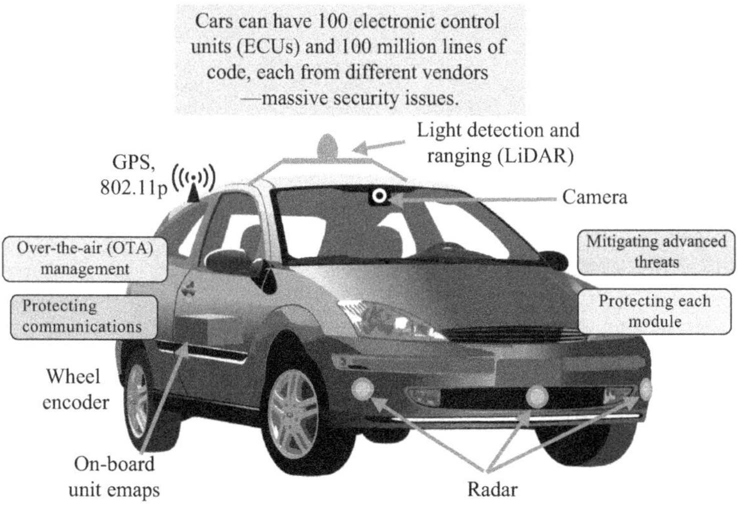

Figure 2.33 Various aspects of an autonomous car

attacks (Storm, 2015; Wu and Horng, 2017; Xiao *et al.*, 2009). If the smart car or a UAV are attacked and taken control by the attacker, the results can be catastrophic.

2.6.1 Smart car security

There are various attacks that can be performed on a smart car or an autonomous car (Petit and Shladover, 2015). Figure 2.33 shows a typical autonomous car system (Storm, 2015). Various parts of an autonomous car include global positioning system (GPS), LiDAR, on-board unit, Emaps, radar, camera, wheel encoder, etc. LiDAR is a ranging system which uses laser. It shoots laser pings which when echoed back, gives the readings from reflection of the objects that are present around the vehicle. Radars and camera are also used to detect the obstacles in the surroundings while the vehicle is in motion. Now-a-days, smart vehicles have around 100 electronic control units (ECUs) and 10 million lines of code. When such a massive system is in place there are various chances of security breaches possible. The software updates are usually Over-The-Air updates, which also means the communications of the car has to be protected. Advanced threats should also be mitigated and each module that is present in the vehicle should be secured and protected such that the attacker does not gain access to them.

2.6.1.1 Smart car security attacks

There are various threats when it comes to the smart transportation (Petit and Shladover, 2015). Figure 2.34 shows some of the attacks that a smart car is vulnerable to. GPS spoofing and jamming are considered to be high threats in a smart transportation environment. Cost of the simple jamming equipment is very less, but highly effective (Petit and Shladover, 2015). Costlier jammers do not only jam the signal but also give fake coordinates which can increase the risk. This is called GPS

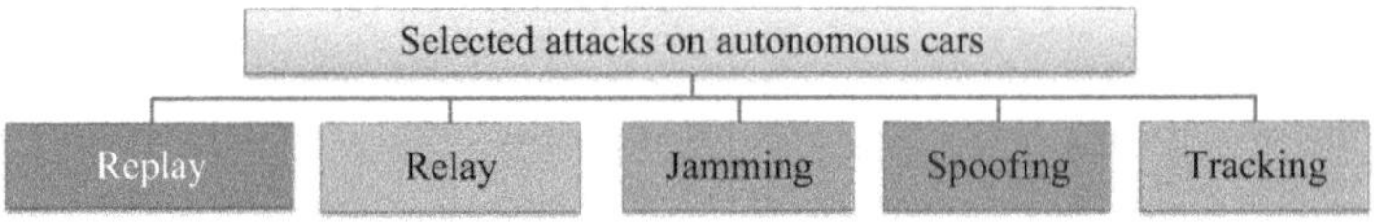

Figure 2.34 Security attacks on autonomous car

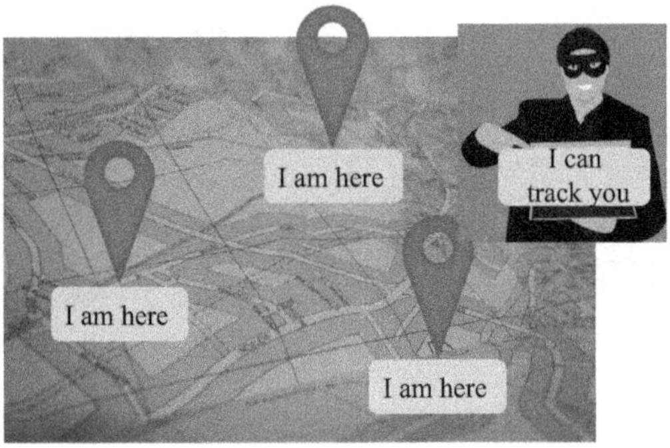

Figure 2.35 Privacy issue in an autonomous car

spoofing where false locations are given to the car. The antitheft system of a car can be easily blocked by using a GPS jammer combined with a GSM jammer. Detecting this kind of threat is highly difficult due to the environmental constraints (Petit and Shladover, 2015). Using the GPS modules, the attacker will also be able to track the location of the vehicle affecting the privacy and security of the smart vehicle. Tracking a vehicle is also possible by merely eavesdropping the communications and enabling the remote identification and tracking (Figure 2.35).

Cameras are used in the smart car to constantly monitor the environment and perform the necessary operations. It is responsible for many processes like detecting pedestrians, nearby cars, and so on. A camera lens is highly susceptible to many issues. Like any camera, the lens can be distorted using a high-intensity light which may render it useless. There are various off-the-shelf components available, infrared light sources that are able to incapacitate the camera. This may lead to under performance of the vehicle or even cause an accident due to the malfunctioning of car. There are solutions available for this issue.

When an autonomous vehicle is designed, specific decision chains are also required to be incorporated. Figure 2.36 shows the design chain of a smart car. When a human is driving the vehicle, the decisions can be made by the human, but when an autonomous vehicle is driven by the actuators and they are controlled by algorithms. This decision chain involves different sensors, perception, planning, and actuation. The sensors will collect the data from the surroundings and store it or send it for processing. Once the data is processed, the algorithms are used for planning the next operation that should be performed by the vehicle. Depending on the decision made, the appropriate actuators will be activated to perform the action.

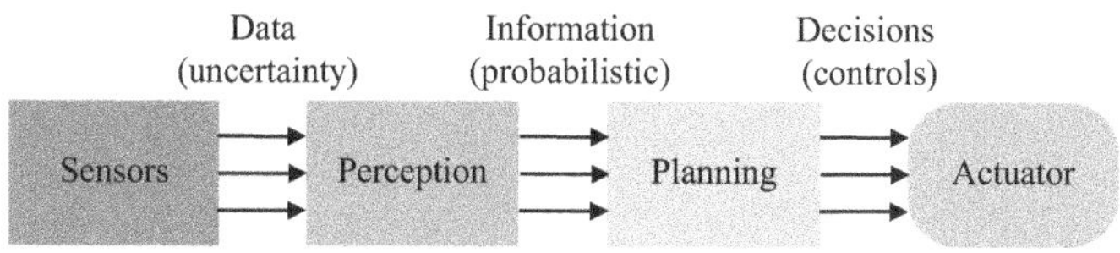

Figure 2.36 Smart car decision chain

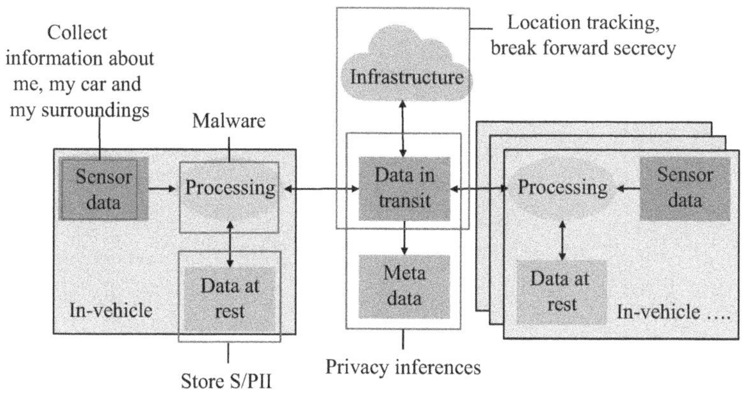

Figure 2.37 Data exchange between smart vehicles

Figure 2.37 shows the data-exchange process between different autonomous vehicles and the areas that can be attacked by the attacker (Petit *et al.*, 2014). The sensors will collect the data about the surroundings and the data is processed at the ECU in the vehicle. Data at rest is the data that is stored when the vehicle is not in motion or stopped. This data is then sent to the other vehicles and the server through the communication module. When this data is sent to another vehicle, same process repeats and data is stored as encrypted data. Each of the different layers are subject to different attacks by the attacker.

2.6.1.2 Smart car security solutions

Smart transportation is a fast-growing area with many researchers working on it. Secure components development is also one of the main focus of this area (Kong *et al.*, 2016; Pacheco *et al.*, 2016; Petit *et al.*, 2014; Satam *et al.*, 2017). Security of an autonomous vehicle cannot be compromised, or else, public can get into danger. There are many solutions available for most of the attacks performed on the smart vehicles. The cameras can be incapacitated using high-intensity IR emitters. This can be blocked by implementing the filters before the camera which can block the IR light. Securing the communication between the vehicles and to the server can help in fighting against the eavesdropping attacks. This can be achieved by implementing the hardware-assisted security which can increase the encryption strength (Yanambaka *et al.*, 2016a, 2016b, 2017).

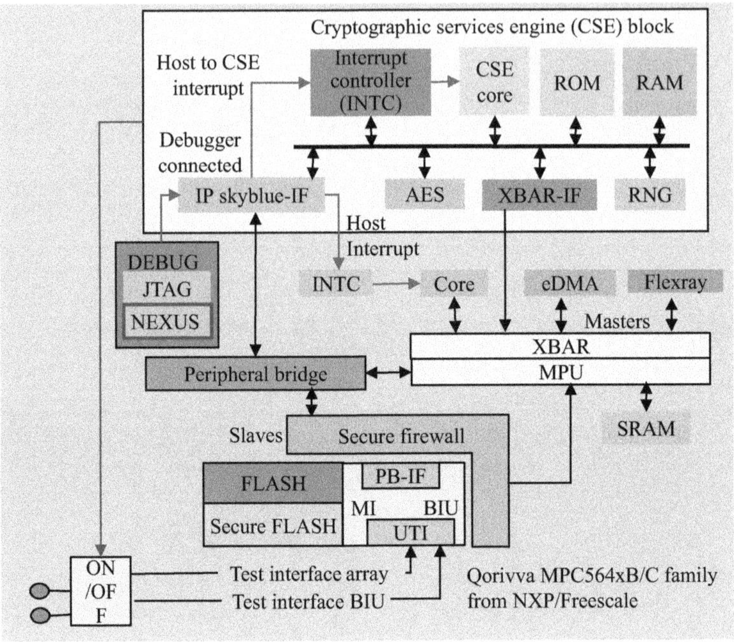

Figure 2.38 Block diagram of cryptographic services engine

There are also various frameworks that are available to develop more secure services for the smart autonomous vehicles (Ferreira *et al.*, 2013; Jin *et al.*, 2014; Soliman *et al.*, 2013). There are four different layers when designing a smart car framework (Satam *et al.*, 2017). They are the application, services, communications, and devices layers. At each of these layer, attacks can be performed. But with the abstraction of the layers, the security can be increased and entire system can be made more robust to the attacks. The sensors and other physical devices are responsible for collecting the data that is required to control the vehicle. The next layer is responsible for the communications to the other vehicles and the servers for updates. Various communications can be used like WiFi, Bluetooth, etc., as discussed in the previous sections. The services layer is responsible for performing all the middleware operations and functions that are used in smart car between the communications and applications layer. The top layer is the applications layer which has different applications that the user will need.

Various encryption modules are available for the smart vehicles as shown in Figure 2.38 and 2.39 (Freescale, 2013). A cryptographic services engine is shown in Figure 2.38, which implements the encryption of the flash modules and prevents the access to other blocks or attackers. A secure communication can be made possible between the components using the ECU and master ECU. A random number is generated and is sent to the central ECU immediately when the car is turned on. Then, the sensor will collect the data and send to the sensor ECU which are encrypted by the CSE module with the random number generated. The central

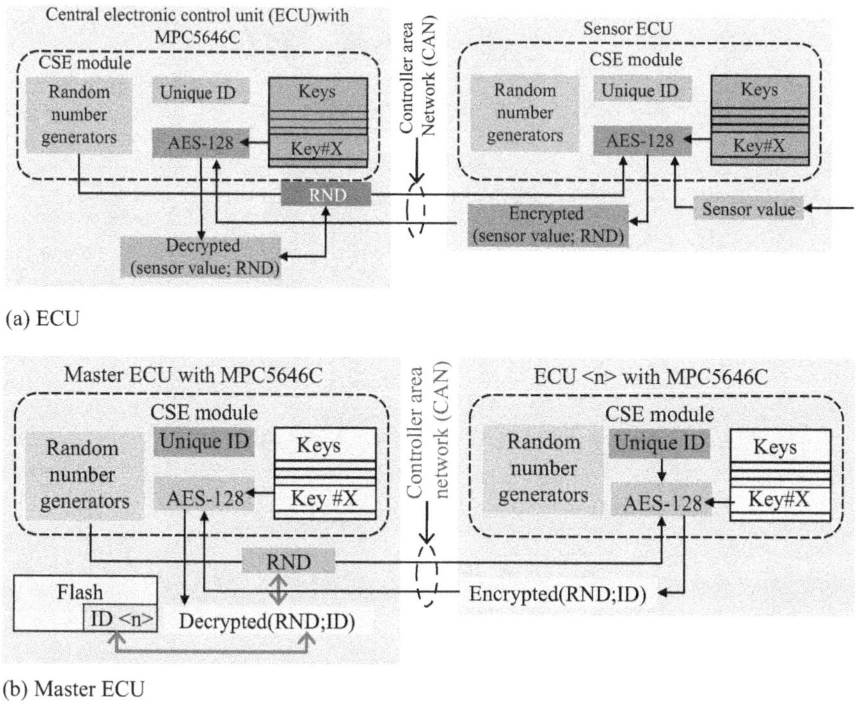

Figure 2.39 Secure communication

ECU then receives the message that is secured by the CSE. Then, the central ECU will decrypt the message using the CSE that is present in central ECU and receives the message.

2.6.2 UAV or drone security

There are many different types of UAV and drones. Drones and UAVs are used in various applications of day-to-day life from military grade to home grade (UAV, https://www.theuav.com/). These applications are designed to reduce human interaction in dangerous situations such as hostile areas and also to increase efficiency in day-to-day applications. But the drones and UAVs are susceptible to many different attacks like other smart vehicles as discussed in the previous sections. This section describes different security threats for Drones and UAVs and security measures that can protect against them.

2.6.2.1 UAV security attacks

The illustration of an UAV with selected components and attached is presented in Figure 2.40. There are various parts of an UAV that are connected to the network or a satellite that can be used to breach the security and take control of the UAV. Communications protocol connects to the satellite to receive and send data to the base station, GPS module communicates with the GPS satellites. The different

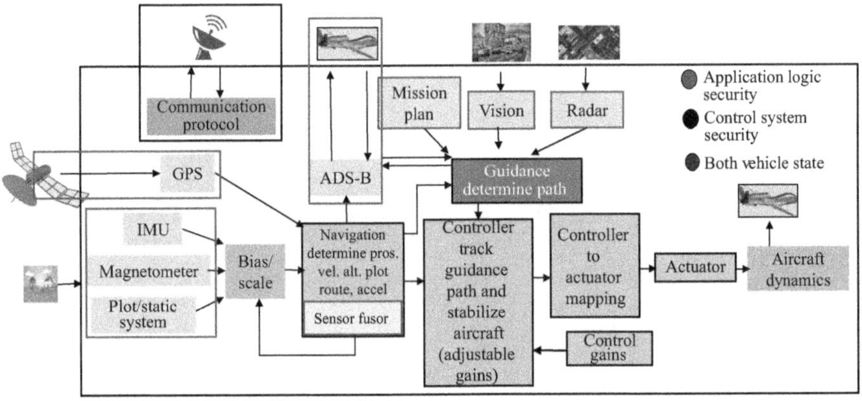

Figure 2.40 UAV security – a big picture

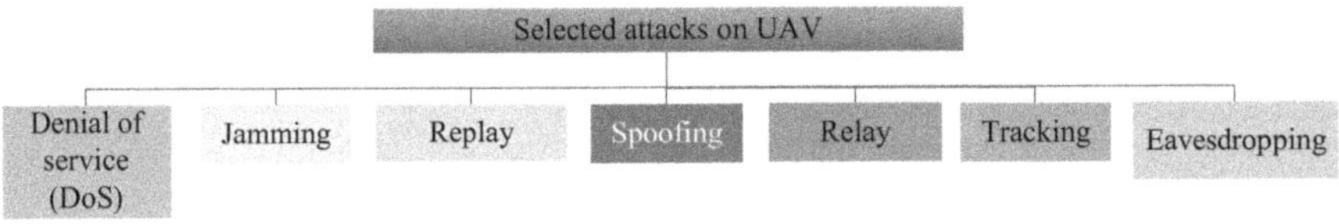

Figure 2.41 Security attacks on a UAV

logic components that are necessary for computations are shown in Figure 2.40. The control systems and security modules are shown in Figure 2.40.

Many of these components are vulnerable to various attacks originated from a computer. So the attacks can be originated from a remote source and the attacker need not be near to the devices while attacking. Different actions can be performed by the attacker like:

- Change the control rates by performing a distributed DoS (DDoS) attack.
- The data that is being collected from the sensors can be modified which can make the flight dangerous.
- The data from the drone can be analyzed for any design flaws that can be taken advantage of in a later attack.

The attacks that can be performed on a UAV can be broadly classified into two categories: integrity attacks and denial of service attacks (Sedjelmaci *et al.*, 2017). The integrity attacks compromise the functionality of the vehicle by changing and sending a false information. Figure 2.41 shows different types of attacks on a UAV. GPS spoofing is a common attack that can be performed on the UAVs receiver which can send the vehicle false coordinates which can make the flight difficult. By performing this attack, the UAV can be led to believe it is in a different location than the original location. Such attacks are performed and vehicles were compromised because of this attack (Shepard *et al.*, 2012). But to perform the attack the spoof signal should be aligned precisely with the original signal and increase the

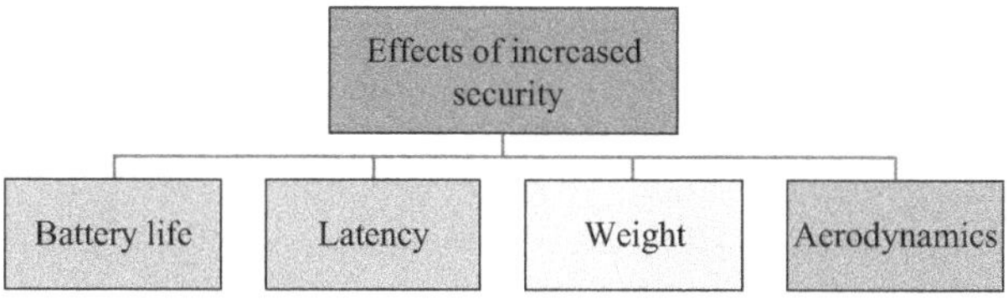

Figure 2.42 Effects of increased security

strength of the spoofing signal gradually to control the target. After gaining control, the counterfeit signal can be moved away from the original signal which controls the drone (Nighswander *et al.*, 2012; Shepard and Humphreys, 2011). Detecting this attack by the drone is highly difficult as the drone's navigation system will not raise any alarm when this attack is being performed.

UAVs can also be attacked by similar UAVs which can lead to the authentic vehicle transmitting falsified data to the receiver. A counterfeit UAV can be used to communicate with an authentic vehicle and send false data. The authentic vehicle then communicates the falsified data to the base station which can be catastrophic in some conditions. This is called the injection attack. The position of the UAV and other collision avoidance systems are handled by the ADS-B controller which is present on the UAV (Sedjelmaci *et al.*, 2016). These can be taken advantage by the attacker to perform an ADS-B attack that will send false position data to the system. This is called spoofing attack.

Another common attack that can be performed on the UAV is the denial of service attacks. This can be performed by sending huge packets of data which will disrupt the communications of the UAV rendering it useless. Jamming, gray hole, and black hole attacks are some of the DoS attacks. Jamming will jam the communications to the base station which can make it loose control and block any new commands. A gray hole and black hole attacks will initialize a node which can lure the UAV into sending the data and forward the data to the server, but then drop all the data or steal it when received at the node.

Security of a UAV can be increased but with the increase of security there can be many issues as shown in Figure 2.42. With added security modules, the weight of the entire UAV increases. When it comes to the aerial vehicles, weight plays a major role in the flight to destination. With an added weight, the aerodynamics are changed. With more components added to the system, more battery consumption is a major issue. With the more battery added, again it loops back to the aerodynamics.

2.6.2.2 UAV security solutions

Intrusion detection system is a way of stopping the attackers from getting into the system's network (Sedjelmaci *et al.*, 2017). In the delay tolerance network, the communications are sparse and hence to deploy a distributed intrusion detection system is a better approach. When a distributed intrusion detection systems (IDS) is employed, different nodes will be able to fire the IDS and then monitor various behaviors of the incoming data. Each of the UAV will be able to monitor the neighbors and secure the entire network. When a malicious UAV is detected in the network, it then sends the data regarding the UAV and the malicious UAV loses the ability to be in the network. Hence, the entire network can be secured easily. Different rules are written for detecting the GPS spoofing and jamming (Hartmann and Steup, 2013; Kim *et al.*, 2012). There are different characteristics of a spoofing attack: it generates a signal with a high strength and signal strength intensity (SSI) is very high from that of a satellite. The attacker will also send multiple signals from the same directions and from the same antenna. All of those signals will have the same signal. So while detecting such attackers, the UAV detection agent first selects all those signals and their signal strength, SSI, are determined and calculated. Then, the mean and variance of the signals are calculated and if they lie within the range of $\pm 3\sigma$, they are correct, and if not they are not correct and are from an attacker.

2.7 Smart healthcare security

IoT, the interconnection of various sensors that collect data, transmit it to the cloud, and store or process the data, is becoming an integral part of various industries including the healthcare. The idea of smart healthcare using the internet of medical things (IoMT) or internet of health things (IoHT) is presented in Figure 2.43. It is estimated that there are more connected devices on this planet than there are people (Li *et al.*, 2011) and the number is keep on increasing day-by-day. Among all the data that is being collected by the sensors, health data is one of the most sensitive information. The health conditions of the patients should be kept under high security making sure the data does not leak. If medical records of patient are leaked, it can lead to major problems like social exclusion, insurance, adverse employment, and so on.

Previous sections discussed various security flaws that can be exploited and data stolen. This infers no system is 100% secure. An IoT network means network

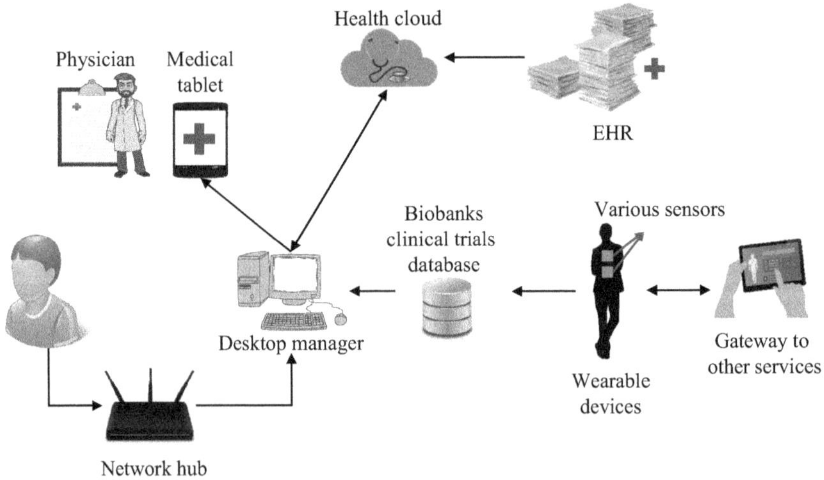

Figure 2.43 Internet of medical things (IoMT) or internet of health things (IoHT)

of connected devices and sensors. If one device is attacked and breached, the entire network will be in jeopardy. When it comes to a healthcare device, this can lead to catastrophic results where the patient's life may fall into danger. Many healthcare devices will monitor a certain parameter of the patient and deliver the drug at a preprogrammed dosage. If such devices are attacked, the configuration can be changed leading to health deterioration. Hence, the health information of the patient should be confidential.

Thoughts around health information privacy and security automatically turn to the Health Information Privacy and Security Act (HIPAA), together with the Health Information Technology for Economic and Clinical Health (HITECH) Act amendments. Administered by the Department of Health and Human Services, Office for Civil Rights (OCR), the HIPAA Rules regulate privacy, security, individual rights, and breach notification for some (but not all) health information. Moreover, they require certain (but not all) persons who process that health information to put physical, administrative, and technical safeguards in place to protect it in all forms, and to institute a security management process (including risk analysis and risk management) to protect it in electronic form.

2.7.1 Smart healthcare security attacks

There are many different risks when it comes to a smart healthcare system (Alamri, 2018; Chaudhary *et al.*, 2018; Ghoneim *et al.*, 2018; Xiao *et al.*, 2018). Figure 2.44 shows different kinds of attacks that can be performed on a smart healthcare system. Data eavesdropping is the process where the attacker will listen to the communications between the medical device and the receiver and steal the data from the communication. This can be avoided using a secure communication. Data privacy, data confidentiality, and location privacy are major issues in healthcare. There is a lot of data that is collected in the case of a medical device (Sundaravadivel *et al.*, 2016). All this data needs to be securely transmitted to the doctor. The data should not fall into the hands of an attacker. It should be confidential and privacy should be given a highest priority. Every device will be given a unique identification which is also subject to counterfeiting attacks and spoofing attacks that were discussed in the case of RFID tags. RFID tags may also be used in the medical healthcare devices which are subject to different attacks that can lead to identity threats. Access and control of the device can be taken by the attacker which will be discussed further in the section.

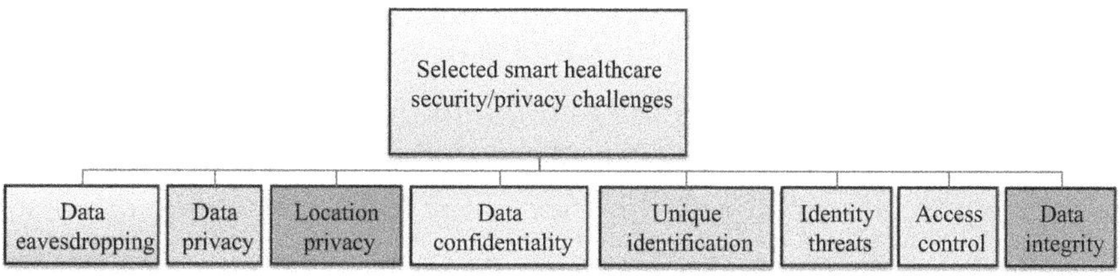

Figure 2.44 Attacks on smart healthcare systems

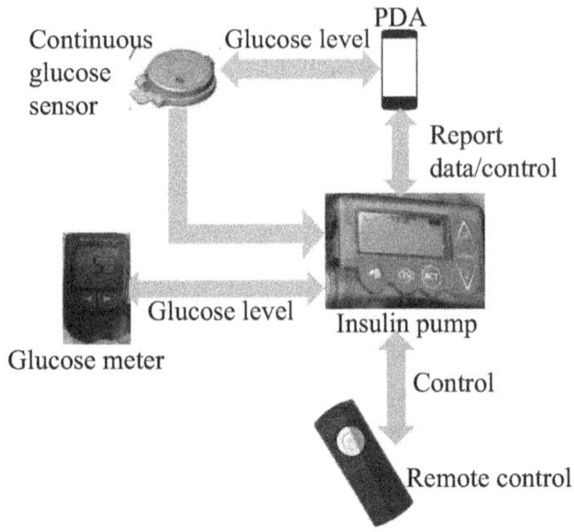

Figure 2.45 Insulin delivery system

Figure 2.45 shows the design of an insulin delivery system that is used by various diabetes patients to monitor and control the insulin levels in the body (Li *et al.*, 2011). There are various components in the system, a glucose monitor, which can monitor the levels of glucose in the blood and a pump which is used to pump glucose into the blood when necessary and a remote control. The glucose monitor will have glucose sensors, which senses the levels of glucose in blood and transmit the data to the insulin pump and to the computer when necessary. The pump will autonomously administer the insulin in prescribed doses by the doctor and is programmed. There are two different ways of administration, bolus dose, which administers quickly when the glucose levels rise quickly and basal dose which administers the dose slowly during night and between meals. These are preprogrammed into the pump and are administered accordingly. There are different ways of programming the device, buttons on pump, through a computer with a wireless connection, remote control, and wireless connection to a blood glucose monitor. Figure 2.46 shows two different kinds of attacks that are possible on the insulin pump which can affect the health of the patient. Passive attack is where the remote control is taken into control and send signal to the insulin pump (Li *et al.*, 2011). A universal radio peripheral is used in this scenario where the remote sends a malicious signal to the pump which can change the program of system and administer an unprescribed dosage of insulin. Active attack is where the same radio peripheral is used to send the signal directly to the pump and change the program of the system. Both these attacks are dangerously harmful to the patient.

2.7.2 Smart healthcare security solutions

There are many solutions available for the security threats of smart healthcare. As the smart healthcare itself is a growing field, many researchers are aiming toward securing the system (Ghoneim *et al.*, 2018; He *et al.*, 2018; Li *et al.*, 2011; Rauf *et al.*, 2018; Wang *et al.*, 2018; Xiao *et al.*, 2018). Many hardware-assisted security solutions are discussed in different sections above which can be deployed in case of smart healthcare devices as most of the components in this case are discussed

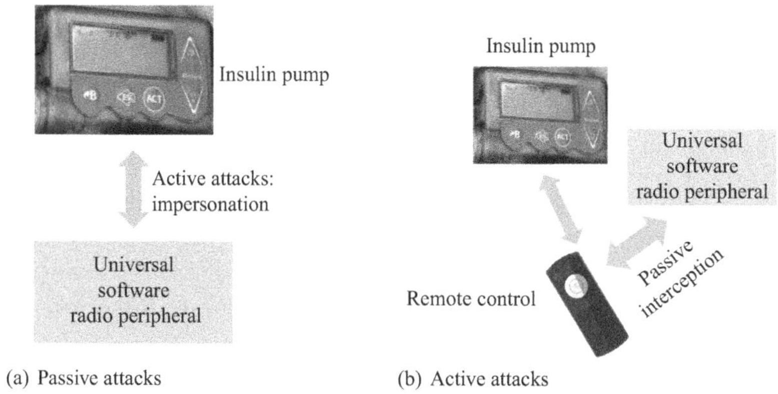

Figure 2.46 Passive and active attacks on smart healthcare

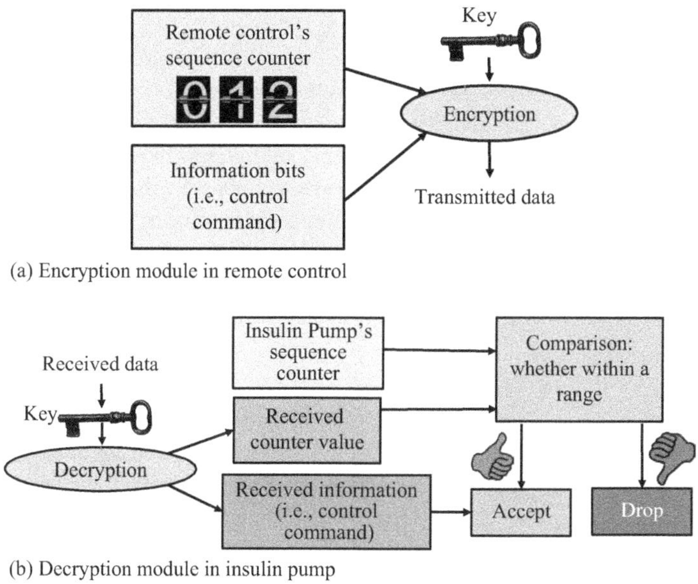

Figure 2.47 Encryption and decryption modules in an insulin delivery system

previously. Data eavesdropping, privacy, security, and location privacy can be handled by employing more secure measures to encrypt the communication. But added security modules can increase the power consumption and require more battery which is a major issue in the smart device especially healthcare (Sundaravadivel *et al.*, 2016; Yanambaka *et al.*, 2015, 2016a, 2017). So, low-power modules should be used for this purpose.

Figure 2.47 shows the ways of securing the insulin administering system (Li *et al.*, 2011). Using a different pin at a timely manner will help securing the system without adding more to the security modules already present in the system

(Alrabady and Mahmud, 2005). The remote and the insulin pump shares a communication channel which can be encrypted for a better security measure. There is a sequence counter in the remote and a key is used to encrypt the number generated by the counter. When a data is sent to the pump, the counter increased by 1 and the insulin pump will have the same key to decrypt the number shared with the pump. There may be some dropped signals between the remote and the pump which can have some difference in the numbers but if that difference is large it can be assumed the system is under attack. By employing a rolling code technique, the attacker will not be able to retrieve the pin to launch any attack. But the encryption and decryption algorithms should be more sophisticated to avoid the attacker reading the pins.

2.8 Firmware

Technology scaling has allowed us to design smaller and more powerful application devices that has network capabilities (Mohanty, 2015; Yanambaka *et al.*, 2015). These devices have allowed us to develop huge number of applications that has become an integral part of our lives. When a network capability to the device was added, this entity has given access to a whole new level of attackers. The attacker can now access the devices remotely and attack them. All the embedded systems are equipped with firmware nowadays. Firmware is the software that is embedded in the hardware and acts as a middleware between the hardware and the outer world applications or an OS, if necessary. This firmware will not be updated as often as the other applications. When updating the firmware, there can be many attacks possible which will be discussed in this section.

2.8.1 Firmware attacks

There are a growing number of attacks being performed on the firmware of the devices. The network capabilities of the devices have allowed them to be attacked from a remote location. When the firmware has been attacked, the entire hardware can be compromised. The hardware can be taken into control, or tampered with, data loss, and so on are some of the results of a successful attack. There is also a chance that the attacker can take control of the embedded system to create and send spam messages to many other devices or emails across the network (Kaspersky, 2014). Denial of service attacks are one of the most common ones that are happening (Shila *et al.*, 2016). A ThingBot can also be used to create a DDoS attack that can incapacitate a server (Kaspersky, 2014). It is a fact that the majority of these attacks are possible, because the firmware and firmware updates are insecure, low security in Internet web interfaces, insecure web applications, and so on (OWASP, 2018). There are many research facilities that are developed for the security flaws in the firmware and their updates (Choi *et al.*, 2016; Dacus and Yannakogeorgos, 2016; Danese *et al.*, 2017; Guin *et al.*, 2017).

The firmware attacks are broadly classified into two different types of attacks: static and dynamic (Figure 2.48). The static attacks are the kind where the

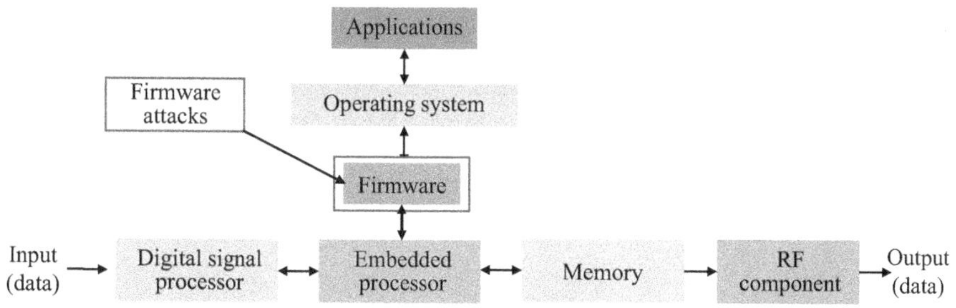

Figure 2.48 Firmware and attacks

firmware that is present in the memory of the embedded system is attacked during the update process or when it is being patched. This can be avoided by implementing a trusted environment when updating and patching the software (Greene, 2016). But using them will fail when a dynamic attack is launched on the firmware of the system.

In a dynamic attack, the attacker will divert the flow of control from the firmware to the code that he has injected and will execute it. There are two different types of dynamic attacks on the firmware of a system: code reuse and code injection. In case of a code injection attack, as the name suggests, the attacker will inject his own code into the memory of the system and will change the flow of control to execute the malicious code. Data execution prevention, Stack Canaries are some of the ways to protect the hardware against such attacks. In code reuse attacks, the attacker will try to execute the already existing code on the memory by changing the flow of control. This can compromise the functionality of the entire system. Various code snippets or gadgets can be combined and a new arbitrary logic can be developed (Shila *et al.*, 2016). This allows the attacker to avoid injecting any new code into the system.

2.8.2 Firmware solutions

There are many defenses that are being developed for firmware security as stated above (Choi *et al.*, 2016; Dacus and Yannakogeorgos, 2016; Danese *et al.*, 2017; Guin *et al.*, 2017). Commonly used security solution is the Address Space Layout Randomization (ASLR) which will randomize the address lotions of data and code segments. This will render the attacker impossible to create some of the attacks that have been discussed. Figure 2.49 shows the implementation of programming protocol of flash memory. In the process, an open channel is used between the microcontroller and the programmer (Soja, 2014). An RSA encryption scheme is used in this scenario for security. For the software image that is to be loaded onto the microcontroller, a hash is generated using SHA-256 algorithm. The hash is generated in a secure offline environment. A signature is added to the hash value which is used to identify the owner of the software. The signed hash value and the software image are transferred to the destination.

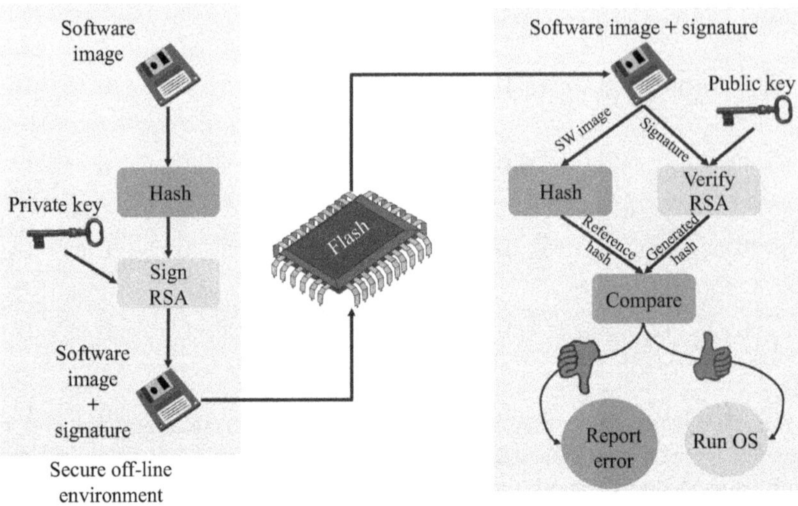

Figure 2.49 Firmware security solutions

The destination then uses the public key and verifies if the code sent to it is authentic or is tampered with. The hash value must match with the hash generated by the embedded module. If they are not matching, then the software would have been tampered, and thus, cannot be used in the system. This method does not use any power intensive mechanisms to encrypt and decrypt the entire software which uses more power and performance from the embedded system. The initially generated hash of the software image is generated in a secure offline environment which can prove very limited chances of attacks from a remote location. Access to the original software is restricted before generating the hash. All systems can be provided with a public key which can be used to authenticate the software and RSA key. This also does not demand much processing power from the embedded system.

2.9 Blockchain technology

Transactions have been around for a very long time. Especially, financial transactions and banks have been around making it easier to transfer funds between accounts. With the advent of internet, online transactions have increased lately. Huge number of financial transactions is happening in the internet for online shopping, transfer funds between accounts and overseas. An intermediary or bank is given the responsibility of transferring the funds securely from the sender to the receiver. These are accurate, responsible, legitimate, and secure. But there are questions rising lately:

- What happens when the trusted party goes rogue?
- When an attacker steals all the data, and who is held responsible?

- There is a delay between sender to intermediary and to the receiver.
- Can the intermediary be trusted for authenticity?

As an answer to all these questions, the blockchain technology was introduced by Satoshi Nakamoto with Bitcoin (Barber *et al.*, 2012; Bitcoin, 2008). It uses an open ledger or shared distributed ledger that contains the records of every transaction that occurred. There is no presence of intermediary in this scenario. This section discusses the Blockchain, its applications, and issues in implementations of blockchain.

2.9.1 Blockchain – overview

Blockchain is the concept of shared ledger where all the participants or users will have a copy of the ledger which contains all the transactions that occurred till date. There is no trusted intermediary in this case which eliminates all the questions that raised. A public key infrastructure is implemented in the case of Bitcoin which is used for the authentication of the users and control the access. Every owner will sign and authenticate the transaction with his private key. A 'block' contains multiple transactions that happened along with its hash and is time stamped. For a transaction to be approved, majority of the users or nodes has to agree to the consensus. Miners are responsible to calculate the hash value of the block according to the consensus algorithm. The hash computed will be less than a threshold that was previously agreed to avoid any issues among the users. The value is set such that the miners will be able to calculate it in a less period of time, in this case, 10 min (Li *et al.*, 2017). Figure 2.50 shows the process of blockchain. A selected characteristic of the blockchain technology is shown in Figure 2.51.

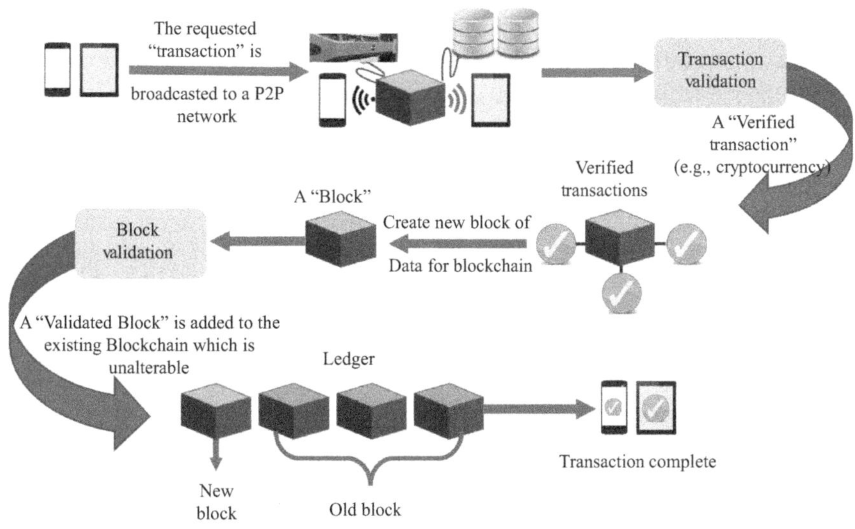

Figure 2.50 Blockchain operation and data flow

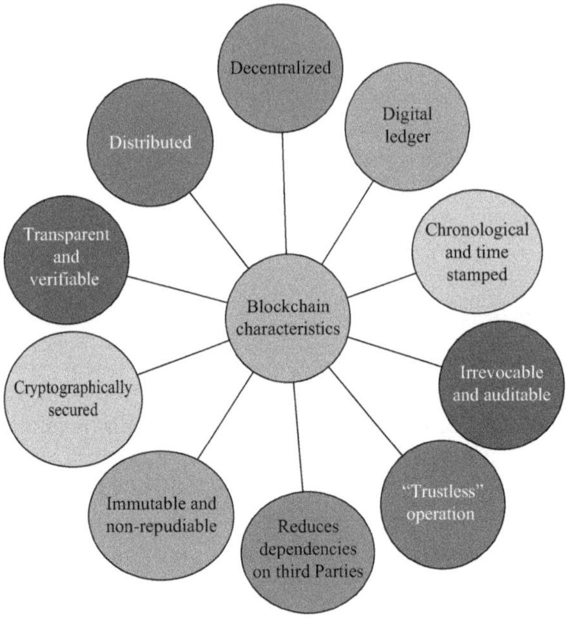

Figure 2.51 Blockchain characteristics

2.9.2 Blockchain – application

There are various applications to the blockchain. Mainly, the security it offers makes it suitable to be implemented in different domains. Figure 2.52 shows different applications of blockchain. It can be implemented in asset management where the security in managing and transferring the assets among different business networks is a priority. A software image or a physical entity like a computer or laptop can be an asset. It can be implemented from serialization until the step being deployed on the floor and it mainly focuses on the following issues: receiving and validating the assets, manufacturing serialization of assets, asset capitalization, installation of asset, and warranty activation.

Financial transactions can be a huge hassle when it comes to real estate. There are many entities from government databases, brokers, title companies, and so on. Blockchain will eliminate most of these entities, where every person and property can be digitally addressed and it may contain anything from financial, occupancy, legal, building performance, and physical attributes. Blockchain is mainly useful in financial transactions where the transfers of funds can be performed without the help of an intermediary. The blockchain implementation in financial transactions is not only cheap but also extremely fast. In case of a bank transaction, it takes 2–3 days where in the case of blockchain, it takes only about 5 min to complete a transaction.

Blockchain can also be used to improve the efficiency of IoT. There is a huge amount of data that is being collected in case of an IoT device. This data can be

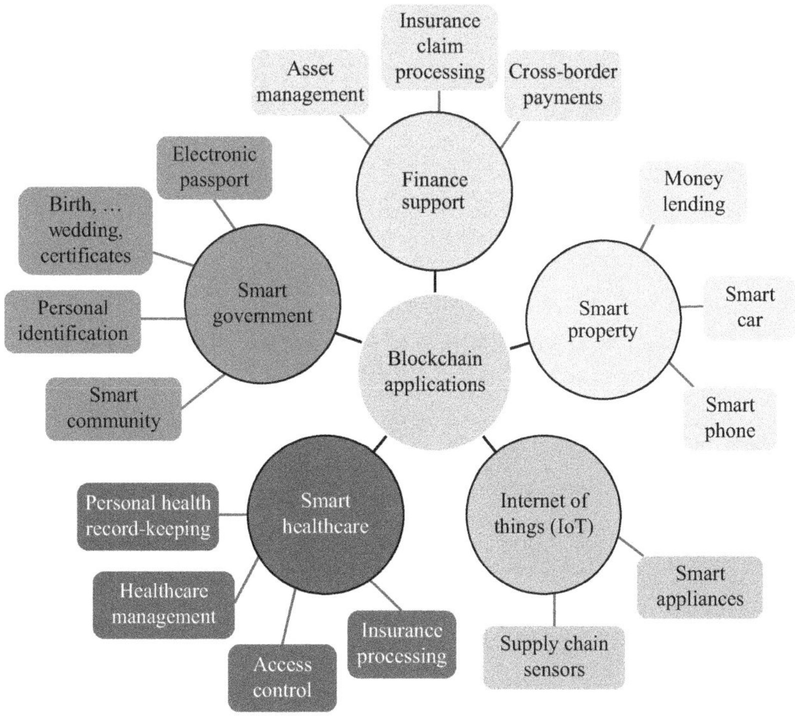

Figure 2.52 Blockchain applications

stored securely and trusted that no alterations are made to it. Blockchain-assisted weddings can also be made easy where searching for a bride and groom online then meeting them and made arrangements for wedding (Ngo, 2014). Healthcare is another domain where blockchain can be proven to be very useful. Patient's data should be stored securely and without compromise. This technology allows for such storage and also makes sure the data is protected from alterations.

2.9.3 Blockchain as a security framework

The consensus is a distributed and decentralized network where the user is the only entity that can determine the future updates of the state can occur. Figure 2.53 shows the process of adding a new value to the already existing chain and Figure 2.54 shows the consensus algorithm. In the blockchain, practical byzantine fault tolerance algorithm (PBFT) is responsible for establishing the consensus among different users or nodes. Each of the user will maintain a state which when receive a new message, is fed with the message data that is necessary for computing the value and reach a decision. Once a decision is reached, it is broadcasted to the network and when majority of the decisions match it leads to the next consensus. A consortium for businesses, blockchain technology is used and PBFT as its base is being developed by Hyperledger Project (https://www.hyperledger.org/).

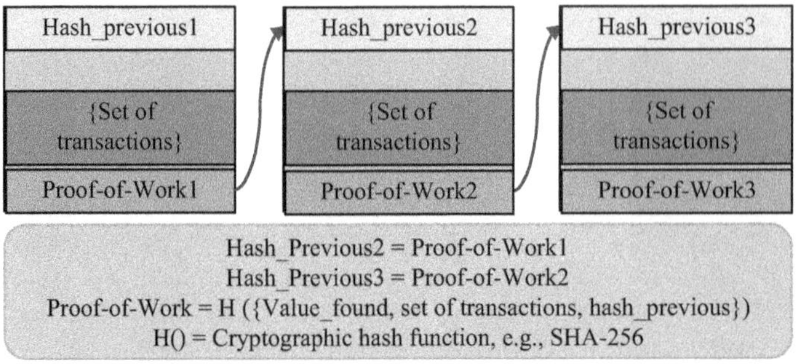

Figure 2.53 Blockchain process of adding a new value

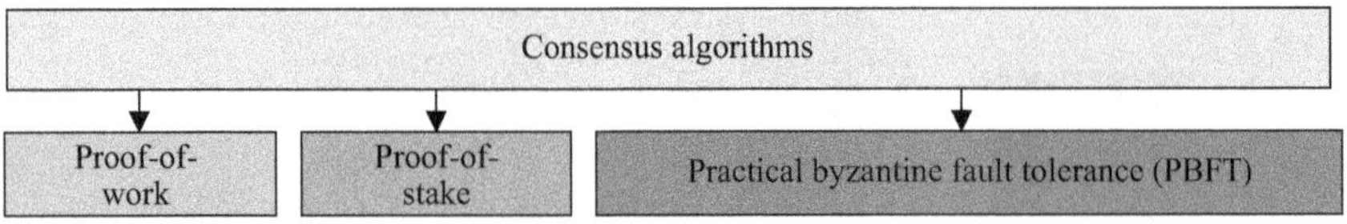

Figure 2.54 Blockchain consensus algorithm

Proof-of-work was initially the decentralized consensus protocol that was proposed by the developer of blockchain, Satoshi, for the consistency and security of Bitcoin. As the transactions of the bitcoin blockchain, they require a consensus for the validation of blocks. There are different nodes in the network and several of them compete with each other to calculate the value of the newly added block. Once a node calculates the value of hash, it waits for the other nodes to complete and verify the values together. Then, the generated hash is added to the blockchain. If an appropriate solution is found by multiple nodes, then there may be multiple blocks generated in the network which can cause a temporary branch. In such scenarios, all of them are acceptable and nodes closer to the miners accept the solution they receive and forward the same to other peers. At any time, the longest chain is accepted by the users in the network. Proof of stake has been developed to avoid the issues of high-power consumption by Proof Of Work in bitcoin. Etherium uses the Proof of Stake as the consensus. In case of a POS, the cryptocurrency is bought and then used in the network which reduces the high-power consumption in calculating the hash.

2.9.4 Blockchain – issues

As discussed in this section, there are various advantages to blockchain technology. But there are also some disadvantages when comes to this and the main being scalability. There is also need of the entire blockchain, i.e., the entire list of all the transactions that happened till date for the consensus and validation. This requires huge amounts of storage capacity. Restriction of block size can decrease the entire problem of scalability. But this also decreases the speed of transactions and increases the fees of transaction. But with a high block size, the entire transaction can take very long to execute. Bitcoin-NG was proposed which has a reduced block size and increased security. This can be achieved by dividing the block into two different parts.

When a Fork is created in the blockchain, a long time should be waited to confirm which chain is the longest and correct branch by all the nodes. This can cause much higher delay to the entire transaction. There is also a 50% attack problem in the blockchain where 50% of the nodes can fake the block and also reverse the transactions. In this case, the power of computation holds all the power so legitimate users might face a problem. Power consumption of bitcoin miners is very high. To mine a bitcoin needs very high amount of computational power which can also increase the power consumption drastically. This can lead to the problem where investment can be higher than the return.

2.10 Conclusions

This chapter discussed different security flaws of the IoT devices and modules. Appropriate solutions were also presented for the security issues for each of the module. Still there are many security issues that need addressing. There can also be some zero-day vulnerabilities which are still yet to be found. The attackers who find the security flaws can take the advantage of them and need not share the flaw with the world. There may be many security flaws and vulnerabilities that are present but not known to the security community which are not addressed in the solutions. There is a vast scope of future research in this domain where security can be increased, and robust design is implemented in an IoT environment.

2.11 Exercises

1. What is internet of things (IoT)?
2. Discuss different architectures of internet of things (IoT).
3. Discuss three applications of IoT.
4. What is a trusted consumer electronic system?
5. Discuss five challenges of internet of things?
6. What are the driving technologies of IoT? What is a THING? What are the attributes of a THING?
7. What is security? What is privacy?
8. Discuss the taxonomy of attacks in an internet of things environment.
9. What are the different attributes of trustworthy consumer electronic system?
10. Discuss the consumer electronic system security – aware design tradeoffs.
11. Discuss different aspects of security.
12. Discuss different aspects of privacy.

13. What is a memory? How many different types of memory are present in an embedded system?
14. Discuss two security attacks on a memory system and how to secure them?
15. What is RFID? How is it used in different IoT applications?
16. Discuss different RFID channel threats and system threats.
17. List different solutions to the RFID security issues and discuss three of them.
18. What is a Faraday cage? How does it work?
19. What is near-field communication? What are its advantages? What are the disadvantages of implementing it in an IoT environment?
20. Discuss four security issues in the NFC.
21. Discuss relay attack and eavesdropping attack on NFC modules.
22. What is Crypta tag? How does it help in securing NFC modules?
23. What are different parts of an autonomous car and which components are affected by the spoofing attack?
24. List out and discuss different attacks on a smart car?
25. What is decision chain? Discuss the decision chain of a smart car?
26. Discuss different solutions to the smart car security issues.
27. What is a UAV? What are its applications?
28. Discuss the big picture of UAV security.
29. What are the different effects of increased security in a UAV? How does it affect the flight?
30. Discuss how the spoofing attack is performed? How can it be detected and mitigated?
31. Discuss the big picture of smart healthcare security.
32. What is HIPAA and what are its functions?
33. Discuss three security issues of smart healthcare in detail.
34. Describe the insulin delivery system and how it can be attacked.
35. Discuss the security in insulin delivery system. How can an attack be detected and mitigated?
36. What is a Firmware? What is the difference between an operating system and a firmware?
37. Discuss the static and dynamic attacks on the firmware of the system.
38. Discuss the firmware security solutions.
39. What is blockchain? What led to the invention of blockchain technology?
40. Discuss how a new block is added to the current blockchain.
41. What are different applications of blockchain?
42. What are the problems that are stopping the broad adoption of blockchain technology?

References

Ahmadi M., Rajamani R., and Sezen, S., 'Transparent flexible active Faraday cage enables in vivo capacitance measurement in assembled microsensor,' *IEEE Sensors Letters,* 2017, vol. 1 (5), pp. 1–4.

Al-Kuwari M., Ramadan A., Ismael Y., Al-Sughair L., Gastli A., and Benammar M., 'Smart-home automation using IoT-based sensing and monitoring platform,' In: *2018 IEEE 12th International Conference on Compatibility, Power Electronics and Power Engineering* (CPE-POWERENG 2018), pp. 1–6 (2018).

Alamri A., 'Monitoring system for patients using multimedia for smart healthcare,' *IEEE Access,* 2018, vol. 6 (23), pp. 271–23, 276.

Almajali S., Salameh H. B., Ayyash M., and Elgala H., 'A framework for efficient and secured mobility of IoT devices in mobile edge computing,' In: *2018 Third International Conference on Fog and Mobile Edge Computing (FMEC)*, pp. 58–62 (2018).

Alrabady A. I. and Mahmud S. M., 'Analysis of attacks against the security of keyless-entry systems for vehicles and suggestions for improved designs,' *IEEE Transactions on Vehicular Technology,* 2005, vol. 54 (1), pp. 41–50.

Andrews C., 'Cyber car crime: Thieves turn to high tech [Transport car crime],' *Engineering Technology,* 2017, vol. 12 (2), pp. 32–35.

Asplund M. and Nadjm-Tehrani S., 'Attitudes and perceptions of IoT security in critical societal services,'. *IEEE Access,* 2016, vol. 4, pp. 2130–2138.

Barber S., Boyen X., Shi E., and Uzun E., 'Bitter to better – how to make bitcoin a better currency,' In: *International Conference on Financial Cryptography and Data Security*, Kralendijk, Bonaire, 2012, pp. 399–414.

Bhargava B., Farkas C., Lilien L., and Makedon F., 'Trust, privacy, and security summary of a workshop breakout session' at the National Science Foundation Information and Data Management (IDM) Workshop held in Seattle, Washington, September 14–16, 2003 (Version 2) (2018). http://www.cerias.purdue.edu/tools_and_resources/bibtex_archive/archive/2003-34.pdf

Bitcoin, Bitcoin: A Peer-to-Peer Electronic Cash System. https://bitcoin.org/en/bitcoin-paper, Last accessed on 1 May 2018.

Bolisetti S. K., Patwary M., Soliman A. H., and Abdel-Maguid M., 'RF sensing based target detector for smart sensing within internet of things in harsh sensing environments,' *IEEE Access,* 2017, vol. 5 (13), pp. 346–13, 363.

Carter B., Green S., Leeman R., and Chaulk N., 'SmartBay: Better information – Better decisions,'. In: OCEANS 2008, pp. 1–7 (2008).

Catarinucci L., de Donno D., Mainetti L., *et al.*, 'An IoT-aware architecture for smart healthcare systems,' *IEEE Internet of Things Journal,* 2015, vol. 2 (6), pp. 515–526.

Chang C. H., Zheng Y., and Zhang L., 'A retrospective and a look forward: Fifteen years of physical unclonable function advancement,' *IEEE Circuits and Systems Magazine,* 2017, vol. 17 (3), pp. 32–62.

Chaudhary R., Jindal A., Aujla G. S., Kumar N., Das A. K., and Saxena N., 'LSCSH: Lattice-based secure cryptosystem for smart healthcare in smart cities environment,' *IEEE Communications Magazine,* 2018, vol. 56 (4), pp. 24–32.

Choi B. C., Lee S. H., Na J. C., and Lee J. H., 'Secure firmware validation and update for consumer devices in home networking,' *IEEE Transactions on Consumer Electronics,* 2016, vol. 62 (1), pp. 39–44.

Dacus C. and Yannakogeorgos P. A., 'Designing cybersecurity into defense systems: An information economics approach,' *IEEE Security Privacy,* 2016, vol. 14 (3), pp. 44–51.

Danese A., Pravadelli G., and Bertacco V., 'Work-in-progress: DOVE: Pinpointing firmware security vulnerabilities via symbolic control flow assertion mining,' In: *International Conference on Hardware/Software Codesign and System Synthesis (CODES+ISSS)*, pp. 1–2 (2017).

Dupuis S., Flottes M. L., Natale G. D., and Rouzeyre B., 'Protection against hardware Trojans with logic testing: Proposed solutions and challenges ahead,' *IEEE Design Test,* 2018, vol. 35 (2), pp. 73–90.

Farooq M. J. and Zhu Q., 'On the secure and reconfigurable multi-layer network design for critical information dissemination in the internet of battlefield things (IoBT),' *IEEE Transactions on Wireless Communications,* 2018, vol. 17 (4), pp. 2618–2632.

Ferreira H. G. C., Canedo E. D., and de Sousa, R. T., 'IoT architecture to enable intercommunication through REST API and UPnP using IP, ZigBee and Arduino,' In: *IEEE 9th International Conference on Wireless and Mobile Computing, Networking and Communications (WiMob)*, 2013, pp. 53–60.

Freescale, http://www.nxp.com/assets/documents/data/en/supporting-information/DWF13_AMF_AUT_T0112_Detroit.pdf, Last accessed on 1 May 2018.

Ghoneim A., Muhammad G., Amin S. U., and Gupta B., 'Medical image forgery detection for smart healthcare,' *IEEE Communications Magazine,* 2018, vol. 56 (4), pp. 33–37.

Ghosh S., Majumder A., Goswami J., Kumar A., Mohanty S. P., and Bhattacharyya B. K., 'Swing-pay: One card meets all user payment and identity needs: A digital card module using NFC and biometric authentication for peer-to-peer payment,' *IEEE Consumer Electronics Magazine,* vol. 6 (1), pp. 82–93 (2017).

Greene J., 'Intel trusted execution technology,' *Tech. Rep., Intel,* 2016 (White paper).

Gubbi J., Buyya R., Marusic S., and Palaniswami M., 'Internet of things (IoT): A vision, architectural elements, and future directions,' *Elsevier Future Generation Computer Systems,* 2013, vol. 29 (7), pp. 1645–1660.

Guin U., Bhunia S., Forte D., and Tehranipoor M. M., 'SMA: A system-level mutual authentication for protecting electronic hardware and firmware,' *IEEE Transactions on Dependable and Secure Computing,* 2017, vol. 14 (3), pp. 265–278.

Halevi T., Li H., Ma D., Saxena N., Voris J., and Xiang, T., 'Context-aware defenses to RFID unauthorized reading and relay attacks,' *IEEE Transactions on Emerging Topics in Computing,* 2013, vol. 1 (2), pp. 307–318.

Harbison C., New Android NFC Attack Could Steal Money From Credit Cards Anytime Your Phone Is Near (2016). http://www.player.one/

Hartmann K. and Steup C., 'The vulnerability of UAVs to cyber attacks – An approach to the risk assessment,' In: *5th International Conference on Cyber Conflict (CYCON)*, 2013, pp. 1–23.

He D., Ye R., Chan S., Guizani M., and Xu Y., 'Privacy in the internet of things for smart healthcare,' *IEEE Communications Magazine,* 2018, vol. 56 (4), pp. 38–44.

He J., Zhao Y., Guo X., and Jin Y., 'Hardware Trojan detection through chip-free electromagnetic side-channel statistical analysis,' *IEEE Transactions on Very Large Scale Integration (VLSI) Systems,* 2017, vol. 25 (10), pp. 2939–2948.

Heartfield R., Loukas G., and Gan D., 'You are probably not the weakest link: Towards practical prediction of susceptibility to semantic social engineering attacks,' *IEEE Access,* 2016, vol. 4, pp. 6910–6928.

IEEE, Internet Initiative: Towards a Definition of the Internet of Things (IoT). IEEE Internet of Things, 2015.

Jalaian B., Gregory T., Suri N., Russell S., Sadler L., and Lee M., 'Evaluating LoRaWAN-based IoT devices for the tactical military environment,' In: *2018 IEEE 4th World Forum on Internet of Things (WF-IoT)*, 2018, pp. 124–128.

Jayakrishnan M. P., Vena A., Sorli B., and Perret E., 'Solid-state conductive-bridging reconfigurable RF-encoding particle for chipless RFID applications,' *IEEE Microwave and Wireless Components Letters,* 2018, vol. 28 (6), pp. 506–508.

Jin J., Gubbi J., Marusic S., and Palaniswami M., 'An information framework for creating a smart city through internet of things,' *IEEE Internet of Things Journal,* 2014, vol. 1 (2), pp. 112–121.

Kaspersky, BEWARE THE THINGBOT! (2014). https://usa.kaspersky.com/blog/beware-the-thingbot/3060/, Last accessed on 1 May 2018

Kim A., Wampler B., Goppert J., and Hwang I., 'Cyber attack vulnerabilities analysis for unmanned aerial vehicles,' In: *Infotech at Aerospace Conference*, California, 2012, pp. 1–30.

Kong H. K., Kim T. S., and Hong M. K., 'A security risk assessment framework for smart car,' In: *10th International Conference on Innovative Mobile and Internet Services in Ubiquitous Computing (IMIS)*, Fukuoka, 2016, pp. 102–108.

Kott A., Swami A., and West B. J., 'The internet of battle things,' *Computer,* 2016, vol. 49 (12), pp. 70–75.

Krishnamurthy R. K., Humble T., Cheung S. C., Lyke J., Mohanty S. P., and Casto M., Energy and Cybersecurity Constraints on Consumer Electronics. http://www.icce.org/expert-panels/ (January 13, 2018). Last visited on 20th Nov 2017.

Kundu S., Islam K. A., Jui T. T., Rail S., Hossain M. A., and Chowdhury I. H., 'Cyber crime trend in Bangladesh, an analysis and ways out to combat the threat,' In: *2018 20th International Conference on Advanced Communication Technology (ICACT)*, 2018, pp. 474–480.

Lee H. C. and Ke K. H., 'Monitoring of large-area IoT sensors using a LoRa wireless mesh network system: Design and evaluation,' *IEEE Transactions on Instrumentation and Measurement,* 2018, vol. 67 (9), pp. 2177–2187.

Lee Y. T., Hsiao W. H., Lin Y. S., and Chou S. C. T., 'Privacy-preserving data analytics in cloud-based smart home with community hierarchy,' *IEEE Transactions on Consumer Electronics,* 2017, vol. 63 (2), pp. 200–207.

Li C., Raghunathan A., and Jha N. K., 'Hijacking an insulin pump: Security attacks and defenses for a diabetes therapy system,' In: *2011 IEEE 13th International Conference on e-Health Networking, Applications and Services*, 2011, pp. 150–156.

Li H., Chen Y., and He Z., 'The survey of RFID attacks and defenses,' In: *8th International Conference on Wireless Communications, Networking and Mobile Computing*, 2012, pp. 1–4.

Li X., Jiang P., Chen T., Luo X., and Wen Q., 'A survey on the security of blockchain systems,' *Future Generation Computer Systems,* 2017 (submitted).

Lopez T., Ranasinghe D., Harrison M., and McFarlane D., 'Adding sense to the internet of things an architecture framework for smart objective systems,' *Pervasive Ubiquitous Computing,* 2012, vol. 16 (3), pp. 291–308.

Luo B., Wang Y., and Liu Y., 'Sensor fusion for vision-based indoor head pose tracking,' In: *2009 Fifth International Conference on Image and Graphics*, 2009, pp. 677–682.

Mazurczyk W., Holt T., and Szczypiorski K., 'Guest editors'; Introduction: Special issue on cyber crime,' *IEEE Transactions on Dependable and Secure Computing*, 2016, vol. 13 (2), pp. 146–147.

McGowen R., Poirier C. A., Bostak C., *et al.*, 'Power and temperature control on a 90-nm itanium family processor,' *IEEE Journal of Solid-State Circuits,* 2006, vol. 41 (1), pp. 229–237.

Mohanty S. P., Everything you Wanted to Know about Internet of Things (IoT). https://cesoc.ieee.org/images/files/pdf/Mohanty_IEEE-DL_IoT.PDF (16th November, 2017). Last visited on 20th Nov. 2017.

Mohanty S. P., Internet of Things (IoT) – Demystified. http://www.smohanty.org/Presentations/2017/Mohanty_ICIT-2017_Keynote_IoT.PDF (16th November, 2017). Last visited on 14 Feb 2018.

Mohanty S. P., Nanoelectronic Mixed-Signal System Design. 9780071825719. McGraw-Hill Education, New York, 2015.

Mohanty S. P., Choppali U., and Kougianos E., 'Everything you wanted to know about smart cities: The internet of things is the backbone,' *IEEE Consumer Electronics Magazine,* 2016, vol. 5 (3), pp. 60–70.

Mohanty S. P., Sengupta A., Guturu P., and Kougianos E., 'Everything you want to know about watermarking,' *IEEE Consumer Electronics Magazine,* 2017, vol. 6 (3), pp. 83–91.

Moon J.-G., Jang J. J., and Jung I. Y., 'Bot detection via IoT environment,' In: *2015 IEEE 17th International Conference on High Performance Computing and Communications*, 2015, pp. 1691–1692.

Nawir M., Amir A., Yaakob N., and Lynn O. B., 'Internet of things (IoT): Taxonomy of security attacks,' In: *2016 3rd International Conference on Electronic Design (ICED)*, 2016, pp. 321–326.

Ngo D., (2014). https://cointelegraph.com/news/david-and-joyces-wedding-demonstrates-how-easy-it-is-to-use-the-blockchain-technology-for-smart-contracts, Last accessed on 1 May 2018.

Nighswander T., Ledvina B., Diamond J., Brumley R., and Brumley D., 'GPS software attacks,' In: *ACM Conference on Computer and Communications Security*, 2012, pp. 450–461.

Nimgaonkar S., Gomathisankaran M., and Mohanty S. P., 'MEM-DnPï¿½A novel energy efficient approach for memory integrity detection and protection in embedded systems,' *Circuits, Systems and Signal Processing,* 2013, vol. 32 (6), pp. 2581–2604.

Notario M. A., *Privacidad vs Seguridad* (2016). https://blogs.deusto.es/master-informatica/privacidad-vs-seguridad/, Last accessed on 1 May 2018.

Obuchi Y., Yamasaki T., Aizawa K., Toriumi S., and Hayashi M., 'Measurement and evaluation of comfort levels of apartments using IoT sensors,' In: *2018 IEEE International Conference on Consumer Electronics (ICCE)*, 2018, pp. 1–6.

Oh D., Kim N.S., Chen C. C. P., Davoodi A., and Hu, Y. H., 'Runtime temperature-based power estimation for optimizing throughput of thermal-constrained multi-core processors,' In: *2010 15th Asia and South Pacific Design Automation Conference (ASP-DAC)*, 2010, pp. 593–599.

OWASP, OWASP Internet of Things (IoT) Project. https://www.owasp.org/index.php/OWASP_Internet_of_Things_Project, Last accessed on 1 May 2018.

Pacheco J., Satam S., Hariri S., Grijalva C., and Berkenbrock H., 'IoT security development framework for building trustworthy smart car services,' In: *2016 IEEE Conference on Intelligence and Security Informatics (ISI)*, 2016, pp. 237–242.

Petit J., Feiri M., and Kargl F., 'Revisiting attacker model for smart vehicles,' In: *IEEE 6th International Symposium on Wireless Vehicular Communications (WiVeC)*, 2014, pp. 1–5.

Petit J. and Shladover S. E., 'Potential cyberattacks on automated vehicles,' *IEEE Transactions on Intelligent Transportation Systems,* 2015, vol. 16 (2), pp. 546–556.

Plos T., Hutter M., Feldhofer M., Stiglic M., and Cavaliere F., 'Security-enabled near-field communication tag with flexible architecture supporting asymmetric cryptography,' *IEEE Transactions on Very Large Scale Integration (VLSI) Systems,* 2013, vol. 21 (11), pp. 1965–1974.

Hyperledger Project, https://www.hyperledger.org/

Prokofiev A. O., Smirnova Y. S., and Surov V. A., 'A method to detect internet of things botnets,' In: *2018 IEEE Conference of Russian Young Researchers in Electrical and Electronic Engineering (EIConRus)*, 2018, pp. 105–108.

Rauf A., Shaikh R. A., and Shah A., 'Security and privacy for IoT and fog computing paradigm,' In: *2018 15th Learning and Technology Conference (L T)*, 2018, pp. 96–101.

RS Components, 11 Internet of Things (IoT) Protocols You Need to Know About (2015). https://www.rs-online.com/designspark/eleven-internet-of-things-iot-protocols-you-need-to-know-about, Last accessed on 1 May 2018

Satam P., Pacheco J., Hariri S., and Horani M., 'Auto infotainment security development framework (ASDF) for smart cars,' In: *International Conference on Cloud and Autonomic Computing (ICCAC)*, 2017, pp. 153–159.

Schinianakis D., 'Alternative security options in the 5G and IoT era,' *IEEE Circuits and Systems Magazine,* 2017, vol. 17 (4), pp. 6–28.

Schneier B., 'IoT security: What's Plan B?,' *IEEE Security Privacy,* 2017, vol. 15 (5), pp. 96–96.

Sedjelmaci H., Senouci S. M., and Ansari N., 'A hierarchical detection and response system to enhance security against lethal cyber-attacks in UAV networks,' *IEEE Transactions on Systems, Man, and Cybernetics: Systems,* 2018, vol. 48 (9), pp. 1594–1606.

Sedjelmaci H., Senouci S. M., and Messous M. A., 'How to detect cyber-attacks in unmanned aerial vehicles network?,' In: *IEEE Global Communications Conference (GLOBECOM)*, 2016, pp. 1–6.

Sengupta A., 'Intellectual property cores: Protection designs for CE products,' *IEEE Consumer Electronics Magazine,* 2016, vol. 5 (1), pp. 83–88.

Sengupta A., 'Hardware vulnerabilities and their effects on ce devices: Design for security against trojans [hardware matters],' *IEEE Consumer Electronics Magazine,* 2017, vol. 6 (3), pp. 126–133.

Sengupta A. and Kundu S., 'Guest editorial securing iot hardware: Threat models and reliable, low-power design solutions,' *IEEE Transactions on Very Large Scale Integration (VLSI) Systems,* 2017, vol. 25 (12), pp. 3265–3267.

Sengupta A. and Roy D., 'Antipiracy-aware ip chipset design for CE devices: A robust watermarking approach [hardware matters],' *IEEE Consumer Electronics Magazine,* 2017, vol. 6 (2), pp. 118–124.

Sengupta A., Roy D., and Mohanty S. P., 'Triple-phase watermarking for reusable IP core protection during architecture synthesis,' *IEEE Transactions on Computer-Aided Design of Integrated Circuits and Systems,* 2018, vol. 37 (4), pp. 742–755.

Shen H., Tan H., Li H., Zhang F., and Li X., 'LMDet: A "Naturalness" statistical method for hardware Trojan detection,' *IEEE Transactions on Very Large Scale Integration (VLSI) Systems*, 2018, vol. 26 (4), pp. 720–732.

Shepard D. P., Bhatti J. A., Humphreys T. E., and Fansler A. A., 'Evaluation of smart grid and civilian UAV vulnerability to GPS spoofing attacks,' In: *ION GNSS Meeting*, 2012, pp. 1–15.

Shepard D. P. and Humphreys T. E., 'Characterization of receiver response to spoofing attacks,' In: *ION GNSS Meeting*, 2011, pp. 1–11.

Shila D. M., Geng P., and Lovett T., 'I can detect you: Using intrusion checkers to resist malicious firmware attacks,' In: *IEEE Symposium on Technologies for Homeland Security (HST)*, 2016, pp. 1–6.

Sivaraman V., Gharakheili H. H., Fernandes C., Clark N., and Karliychuk T., 'Smart IoT devices in the home: Security and privacy implications,' *IEEE Technology and Society Magazine,* 2018, vol. 37 (2), pp. 71–79.

Soja R., Automotive Security: From Standards to Implementation. https://www.nxp.com/docs/en/white-paper/AUTOSECURITYWP.pdf, Last accessed on 1 May 2018.

Soliman M., Abiodun T., Hamouda T., Zhou J., and Lung C. H., 'Smart home: Integrating internet of things with web services and cloud computing,' In: *IEEE 5th International Conference on Cloud Computing Technology and Science*, 2013, vol. 2, pp. 317–320.

Son S.-C., Kim N.-W., Lee B.-T., Cho C. H., and Chong J. W., 'A time synchronization technique for coap-based home automation systems,' *IEEE Transactions on Consumer Electronics,* 2016, vol. 62 (1), pp. 10–16.

Storm D., Black Hat Europe: It's Easy to Hack Self-Driving Car Sensors (2015). https://www.computerworld.com, Last accessed on 1 May 2018.

Sun D. Z. and Zhong J. D., 'A hash-based RFID security protocol for strong privacy protection,' *IEEE Transactions on Consumer Electronics,* 2012, vol. 58 (4), pp. 1246–1252.

Sundaravadivel P., Mohanty S. P., Kougianos E., and Albalawi U., 'An energy efficient sensor for thyroid monitoring through the IOT,' In: *Proceedings of the 17th International Conference on Thermal, Mechanical and Multi-Physics Simulation and Experiments in Microelectronics and Microsystems (EuroSimE)*, 2016, pp. 1–4.

Surendran S., Nassef A., and Beheshti B. D., 'A survey of cryptographic algorithms for IoT devices,' In: *2018 IEEE Long Island Systems, Applications and Technology Conference (LISAT)*, 2018, pp. 1–8.

Thompson D. R., Di J., and Daugherty M. K., 'Teaching RFID information systems security,' *IEEE Transactions on Education,* 2014, vol. 57 (1), pp. 42–47.

UAV, T.: https://www.theuav.com/, Last accessed on 1 May 2018.

Unwala I., Taqvi Z., and Lu J., 'IOT security: Z-wave and thread,' In: *2018 IEEE Green Technologies Conference (GreenTech)*, 2018, pp. 176–182.

Valea E., Silva M. D., Natale G. D., Flottes M. L., Dupuis S., and Rouzeyre B., 'SI ECCS: SECure Context Saving for IoT Devices,' In: *2018 13th International Conference on Design Technology of Integrated Systems In Nanoscale Era (DTIS)*, 2018, pp. 1–2.

Vattem S. and Anjali T., 'Complete RFID security solution for inventory management systems,' In: *International Conference on Advances in Computing, Communications and Informatics (ICACCI)*, 2017, pp. 179–183.

Wang J., Zhu R., and Liu S., 'A differentially private unscented Kalman filter for streaming data in IoT,' *IEEE Access,* 2018, vol. 6, pp. 6487–6495.

Wu H. T. and Horng G. J., 'Establishing an intelligent transportation system with a network security mechanism in an internet of vehicle environment,' *IEEE Access,* 2017, vol. 5, pp. 19, 239–19, 247.

Wu J. H., Scholvin J., del Alamo J. A., and Jenkins K. A., 'A Faraday cage isolation structure for substrate crosstalk suppression,' *IEEE Microwave and Wireless Components Letters,* 2001, vol. 11 (10), pp. 410–412.

Xiao F., Miao Q., Xie X., Sun L., and Wang R., 'Indoor anti-collision alarm system based on wearable internet of things for smart healthcare,' *IEEE Communications Magazine,* 2018, vol. 56 (4), pp. 53–59.

Xiao J. and Feng H., 'A low-cost extendable framework for embedded smart car security system,' In: *2009 International Conference on Networking, Sensing and Control*, 2009, pp. 829–833.

Xue H. and Ren S., 'Self-reference-based hardware Trojan detection,' *IEEE Transactions on Semiconductor Manufacturing,* 2018, vol. 31 (1), pp. 2–11.

Yanambaka V. P., Mohanty S. P., and Kougianos E., 'Novel FinFET based physical unclonable functions for efficient security in internet of things,' In: *Proceedings of the 2nd IEEE International Symposium on Nanoelectronic and Information Systems (iNIS)*, 2016a, pp. 172–177.

Yanambaka V. P., Mohanty S. P., and Kougianos E., 'Secure multi-key generation using ring oscillator based physical unclonable function,' In: *Proceedings of the 2nd IEEE International Symposium on Nanoelectronic and Information Systems (iNIS)*, 2016b, pp. 200–205.

Yanambaka V. P., Mohanty S. P., and Kougianos E., 'Making use of semiconductor manufacturing process variations: FinFET-based physical unclonable functions for efficient security integration in the IoT,' *IEEE Potentials,* 2017, vol. 93 (3), pp. 429–441.

Yanambaka V. P., Mohanty S. P., Kougianos E., and Ghai D., 'Nanoscale high-κ/metal gate CMOS and FinFET based logic libraries,' *Nano-CMOS and Post-CMOS Electronics: Devices and Modelling*, 2015, vol. 1, Chap. 6, pp. 169–211. Institute of Engineering and Technology.

Yang T., Kong L., Xin W., Hu J., and Chen Z., 'Resisting relay attacks on vehicular passive keyless entry and start systems,' In: *2012 9th International Conference on Fuzzy Systems and Knowledge Discovery*, 2012, pp. 2232–2236.

Chapter 3

Trojan security aware DSP IP core and integrated circuits

This chapter discusses different security approaches to designing digital signal processing (DSP) cores that have detection capability against functional-type hardware Trojans in a global supply chain. In the current design and fabrication supply chain, design houses, circuits and system core vendors, and manufacturing houses are globally scattered. It is quite possible that Trojans can be inserted in this design and manufacturing supply chain by anyone involved at any phase. Such Trojans can give backdoors to hackers and affect the operation of the system that uses the infected hardware. In a worst case, in critical applications, such as aircrafts and medical devices, functioning can be completely stopped, causing catastrophic consequences. The chapter is organized as follows: Section 3.2 discusses different features and threat models of hardware Trojans; Section 3.3 explains how a Trojan could be inserted in a 3PIP core and why it is difficult to detect; Section 3.4 discusses different hardware Trojan security approaches that are available in the literature with emphasis on DSP cores; Section 3.5 presents definition, goal, and design evaluation models for Trojan security aware DSP IP cores; Section 3.6 discusses the design process of Trojan secured DSP IP cores; Section 3.7 presents analysis and comparison of different Trojan security approaches for DSP IP cores; the chapter concludes with Section 3.8.

3.1 Introduction

The use of heterogeneous system-on-chip (SoC) architecture in modern consumer electronics (CE) devices such as smartphones, gaming consoles, tablets, and digital cameras has become a common practice in the semiconductor industry. The in-house SoC designer or the system integrator mostly imports intellectual property (IP) cores from third-party IP vendors to minimize design time and to meet strict product delivery deadlines. These third-party IP (3PIP) cores can be DSPs/media processors, such as JPEG image compression-decompression (CODEC), finite impulse response (FIR) filters, infinite impulse response (IIR) filters, fast Fourier transformation (FFT) algorithms, Haar wavelet transformation (HWT) algorithms, and discrete cosine transformation (DCT) algorithms, or they can be micro IP cores/blocks, such as adders, multipliers, and comparators. However, these 3PIP cores have severe trust issues, because an untrustworthy IP vendor may insert

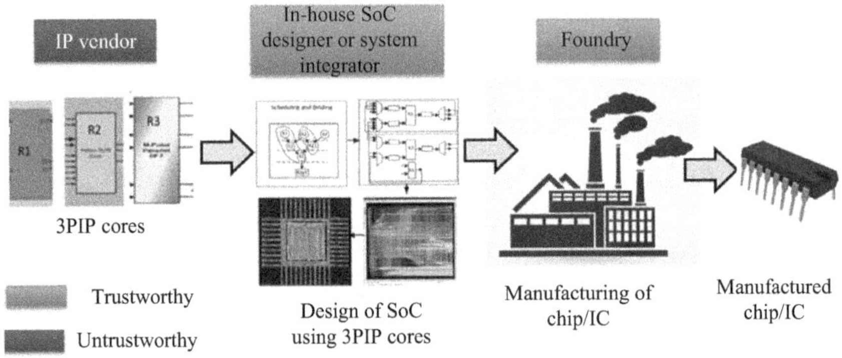

Figure 3.1 Design flow of a SoC

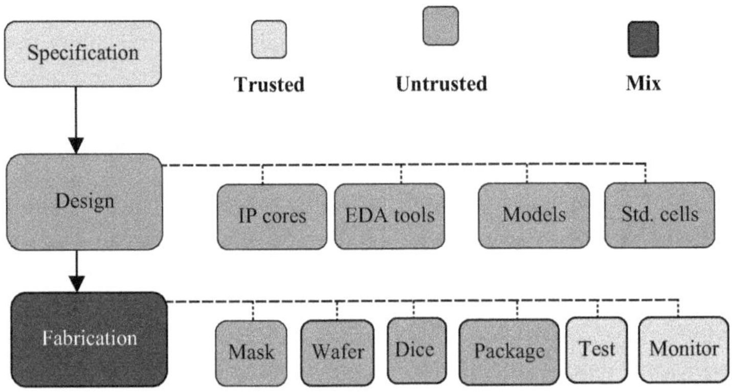

Figure 3.2 Life cycle of a SoC for a CE product and its possible vulnerabilities (Chakraborty et al., 2009)

malicious logic elements known as "hardware Trojans" into their IP cores to introduce malfunctioning/failure of the SoCs (Tehranipoor and Koushanfar, 2010; Liu *et al.*, 2014; Sengupta, 2016, 2017; Chakraborty *et al.*, 2009; Sengupta and Kundu, 2017) (a thematic representation of the aforesaid statement during an SoC life cycle is shown in Figure 3.1). Moreover, a Trojan-infected IP core may leak confidential information by creating a backdoor in the design or may degrade the performance of the processor (Bhunia *et al.*, 2013). The typical trusted and untrusted phases in the life cycle of a SoC design are shown in Figure 3.2.

Detection of a hardware Trojan is not always easy or straightforward, because an intelligent adversary always inserts malicious logic that will be activated under sporadic conditions (Wolff *et al.*, 2008) and/or exhibit its effect for a small duration to evade detection. Additionally, thorough analysis, test, and verification of a complete SoC is complex and time-consuming. The Trojan activation condition is termed the trigger condition and, once triggered, the effect of the Trojan is termed the payload

(Narasimhan *et al.*, 2010). Logic testing (Banga and Hsiao, 2009) and side-channel analysis (Narasimhan *et al.*, 2010; Koushanfar and Mirhoseini, 2011) are two commonly used techniques to detect hardware Trojans (Narasimhan *et al.*, 2010; Liu *et al.*, 2014). In the first approach, functional or structural testing is performed (Chakraborty *et al.*, 2009a; Wolff *et al.*, 2008; Banga and Hsiao, 2009) to generate rare conditions, so that the effect of hardware Trojan logic gets reflected in the output. This type of detection approach is effective for ultra-small Trojans (Narasimhan *et al.*, 2010). In the second approach, detection of hardware Trojans is performed by measuring physical "side-channel" parameters such as delay (Jin and Makris, 2008; Rai and Lach, 2009) or power signature (Agrawal *et al.*, 2007; Banga and Hsiao, 2008; Rad *et al.*, 2010). In this type of approach, detection of a hardware Trojan is possible without activating the Trojan logic and monitoring its effect on the output.

There are multiple classes of hardware Trojans which we discuss in Section 3.2.1; however, we mostly focus on how to provide security (detection) against a particular type of Trojan that changes the functionality or output of a DSP-based IP core. More specifically, we discuss the design process of a low-cost hardware Trojan secured datapath design for a DSP core through the dual modular redundant (DMR) schedule and vendor allocation processes (Rajendran *et al.*, 2013; Sengupta *et al.*, 2017b). To obtain a low-cost Trojan secured design, a particle swarm optimization (PSO)-based design space exploration (DSE) (Sengupta and Sedaghat, 2011, 2013) is performed for the DSP cores.

3.2 Types of hardware Trojans

This section discusses the different types of hardware Trojans, the advantages of handling this type of Trojan security for DSP cores at higher abstraction level, and details of the threat model of hardware Trojans.

3.2.1 Trojan features

We discuss here the major classification of Trojan-based hardware threats (shown in Figure 3.3) (Chakraborty *et al.*, 2009; Sinanoglu *et al.*, 2013; Sengupta, 2017). Digital hardware Trojans are classified into two types: trigger-based and payload-based. As mentioned earlier, triggering of a Trojan indicates activating it, which can be performed either through a combinational or a sequential mechanism. Similarly, the payload of a Trojan that indicates the effect of Trojan logic can change the functionality, degrade the performance, or leak secret information without changing the functionality.

Combinational-triggering-based hardware Trojan: This type of Trojan gets triggered under a very rare condition or event. Until that condition arises, the circuit works correctly. For example, as shown in Figure 3.4, the original output (Q) will be same as the Trojan-infected output (Q^*) for all input combinations of A and B except $A = B = 0$. In this particular case, the inserted Trojan logic triggers and the incorrect output in Q^* will be generated as payload. Therefore, in this circuit, $A = B = 0$ is the triggering condition.

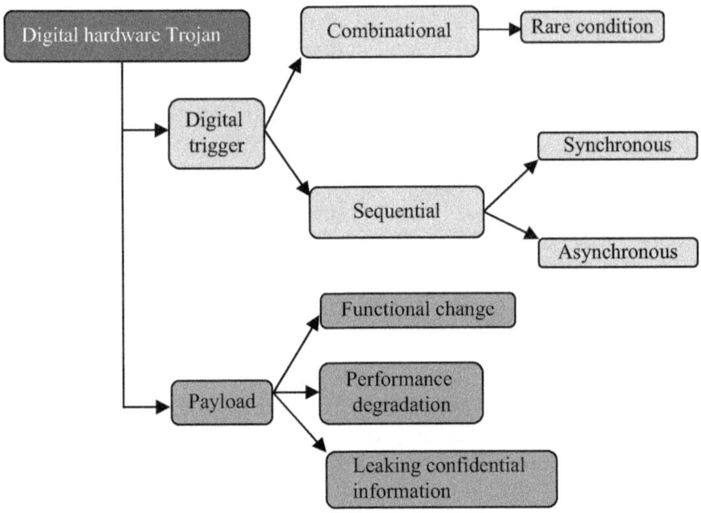

Figure 3.3 Broad classification of digital hardware threats

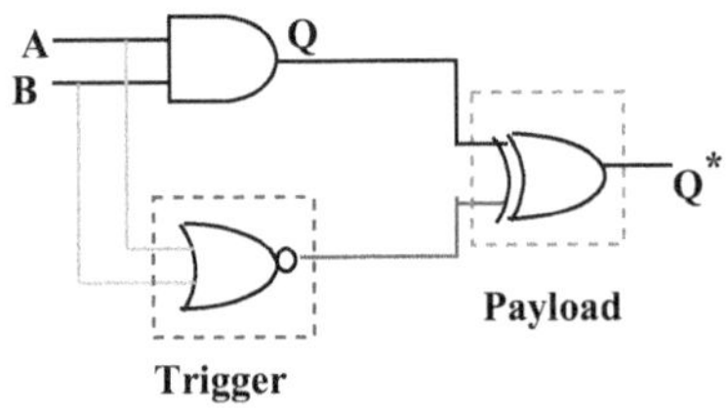

Figure 3.4 Combinational-triggering-based hardware Trojan under a rare event (Chakraborty et al., 2009)

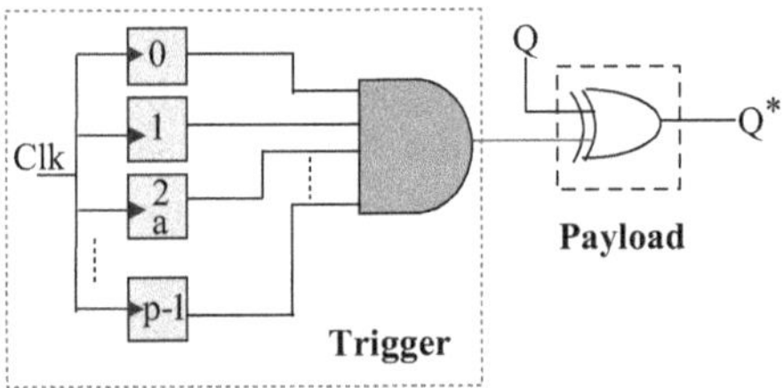

Figure 3.5 Synchronous counter sequential-triggered hardware Trojan

Sequential-triggering-based hardware Trojan: This type of Trojan gets triggered at a particular count of the counter. Once triggered, the circuit produces an incorrect output (Chakraborty *et al.*, 2009). For example, as shown in Figure 3.5, there are p counters and all the counters execute synchronously using a clock signal. Once the counter reaches $2^p - 1$ count, the Trojan logic gets triggered and the incorrect

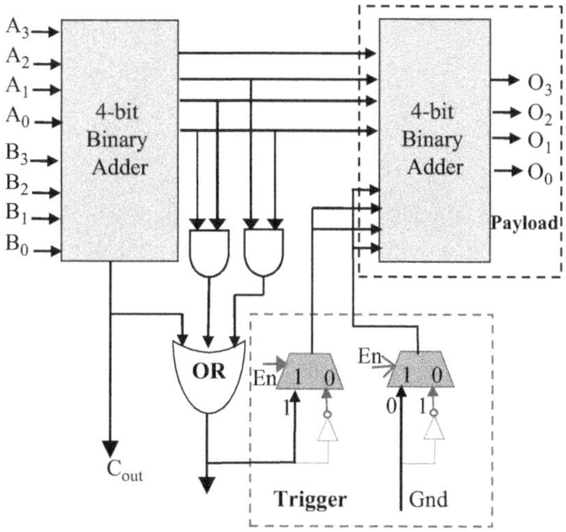

Figure 3.6 Trojan-infected BCD adder resulting in functionality change

output in Q^* will be generated as payload instead of Q. As this type of Trojan gets triggered in a certain time period, it is also known as a "time-bomb hardware Trojan."

Functionality changeable payload-based hardware Trojan: This type of Trojan affects the payload of a circuit by causing output malfunction. For example, as shown in Figure 3.6, the BCD adder has two primary inputs, A and B, each of 4-bit size (A_3-A_0 and B_3-B_0), one enable signal En of one bit and one primary output O of 4-bit size (O_3-O_0). The BCD adder will work correctly when the En signal is set to 1, but when the En signal is set to 0 to disable the component/module, the Trojan gets triggered and starts introducing malfunctioning. The system integrator would keep the component enabled (En = 1) during the testing phase, because he/she may assume that the En signal is for enabling and disabling the component (Sengupta, 2017); however, it is a covert triggering signal for the Trojan. Therefore, the rare condition test could evade and pass the functional simulation and test (explained later in greater detail).

Performance degradable payload-based hardware Trojan: This type of Trojan affects the payload of a circuit by degrading its performance. In other words, this type of hardware Trojan may increase execution delay or power consumption of a circuit without altering the correctness of the output. Therefore, this type of Trojan may go unnoticed during the functional verification phase. For example, as shown in Figure 3.7, for input values $A = 1$ and $B = 0$ or 1, the Trojan logic will not get triggered; therefore, there is no degradation of performance in terms of propagation delay of the 1-bit adder while calculating the sum (S). However, for input values $A = 0$ and $B = 0$ or 1, the Trojan logic gets triggered, therefore the propagation delay to calculate S will increase, thus degrading the performance of the circuit. However, in both the cases, the correct output will be generated, therefore, identification of this type of Trojan through functional/simulation verification

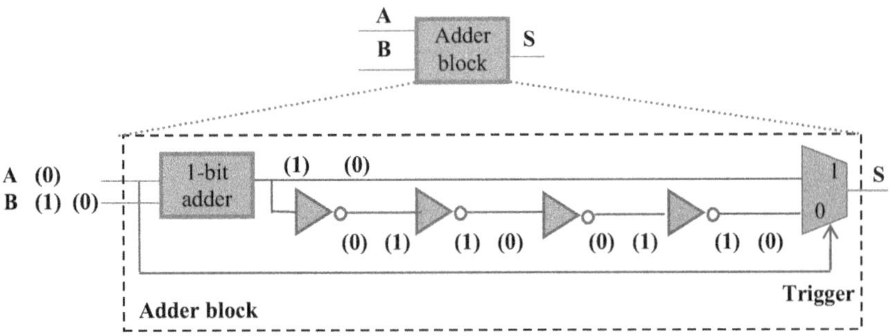

Figure 3.7 Hardware Trojan resulting in performance degradation

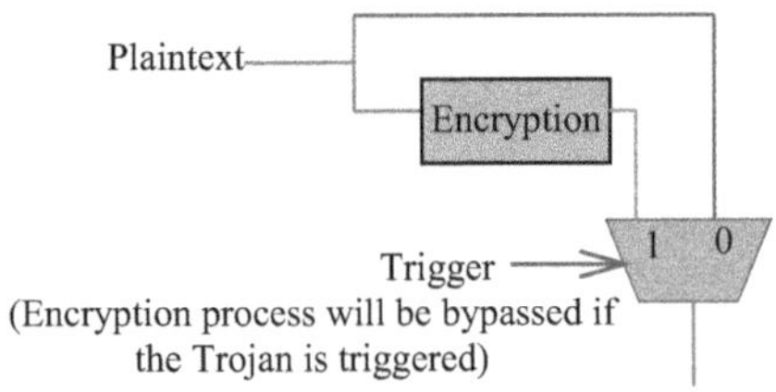

Figure 3.8 Hardware Trojan resulting in leakage of confidential information (Sinanoglu et al., 2013)

could be evaded. This type of Trojan may result in an occasional timing violation of the circuit (Sengupta, 2017).

Leaking confidential information and launching denial of service (DoS) payload-based hardware Trojan: This type of Trojan, which leaks confidential information, is designed in such a way, that it can bypass the encryption process applied on the confidential information. The Trojan could be a big threat for sensitive information. For example, as shown in Figure 3.8, once the Trojan is triggered, the plaintext may skip the encryption process, thus disclosing confidential information (Sinanoglu *et al.*, 2013).

The Trojan which launches DoS is designed in such a way that it forces the circuit to provide no output. For example, as shown in Figure 3.9, when the enable signal (En) is set to 0, it will trigger the Trojan. The effect of the payload is that data are not allowed to pass through the muxes, as for both the muxes, the second input is left uninitialized. Therefore, no output will be generated for the BCD adder.

3.2.2 Benefit of Trojan security at higher abstraction level

As mentioned earlier, most of the CE devices integrate IP cores from the DSP and multimedia domain in their SoC design process, such as JPEG CODEC (Sengupta *et al.*, 2018b), FIR, IIR, FFT, DCT, and HWT (Sengupta *et al.*, 2018a).These IP cores contain complex algorithms that are very challenging and time-consuming to design if the design process starts from the register transfer level or lower

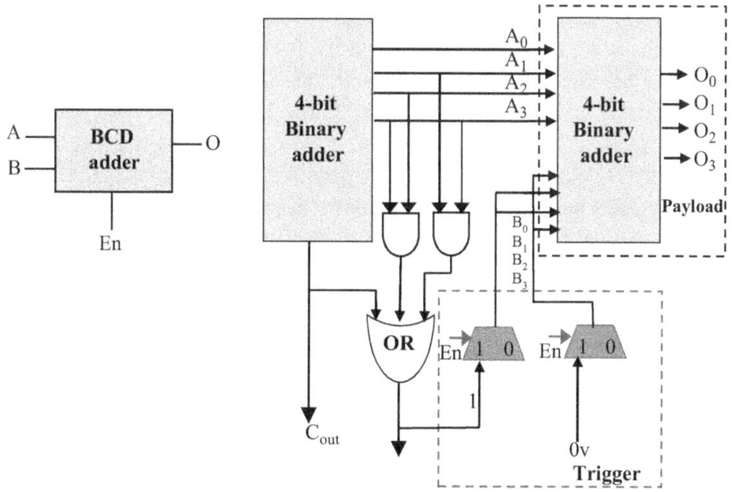

Figure 3.9 Trojan-infected BCD adder resulting in no output or denial of service (DoS). (The BCD adder available to a designer as a black box is shown on the right)

abstraction layer. Additionally, multiple design architectures (hardware variances) are possible for implementing these IP cores. Selecting arbitrary hardware design architecture may result in large overheads in terms of design implementation cost (area, delay, power etc.). A high-level synthesis (HLS) process integrated with a DSE engine is capable of exploring and designing a low-cost (in terms of power, performance, energy, and area) hardware architecture for DSP/multimedia IP cores within an acceptable runtime. Thus for designing of hardware for complex algorithms, within design constraints, while concurrently satisfying time-to-market pressure, HLS is the key.

While designing a DSP IP core, multiple 3PIP cores are usually available in the module library of an HLS tool. As these IP cores are imported from 3PIP vendors, which are untrustworthy, there is high possibility of their containing hardware Trojan (Rajendran *et al.*, 2013). In a typical case, the vendor of the HLS tool imports IP cores/blocks available in its library from 3PIP vendors. Again, as mentioned earlier, these IP cores may also contain hardware Trojan. As designing a DSP IP core is vulnerable to such Trojan, handling hardware Trojan for DSP cores is necessary (Rajendran *et al.*, 2013; Sengupta *et al.*, 2017b, 2017c). In Section 3.4, we discuss different Trojan secured DSP IP core design processes that are available in the literature.

3.2.3 Threat model

There are different variations of hardware Trojan as discussed in Section 3.2.1. However, in the context of DSP, since hardware Trojan threats affecting the output functionality is a well-studied topic, in this chapter we lay emphasis only on those type of Trojans which change the computational (functional) output of an IP core circuit. As mentioned earlier, this type of Trojan can be available in the library of an HLS tool inside an IP block/module. Insertion of Trojan logic into the IP block/module is possible in two ways: (i) through an adversary within the HLS design house team; or (ii) when the design team of the HLS tool vendor imports the IP block/module from a 3PIP vendor. As this type of Trojan is triggered under very rare conditions, its detection is not always possible (i.e., the chances of evasion are high). The literature has studied the Trojan threat model on DSP cores that consider 3PIP vendors as untrustworthy. Thus the in-house design integrator team and foundry are considered trustworthy and the 3PIP vendor is considered an untrustworthy entity in the SoC design cycle.

3.3 Hardware Trojan in a 3PIP module

This section describes examples of Trojan-infected 3PIP cores which change the functionality of a circuit and gives reasons why such Trojans could evade detection during regular validation. As already mentioned, in this chapter, we discuss only that type of Trojan which alters the functionality of an IP core but does not leak any secret information. Literature so far has studied the effect of these Trojans on DSP IP cores (Sengupta and Bhadauria, 2015).

3.3.1 Example of a hardware Trojan

As discussed earlier, owing to several reasons, the HLS framework has become integral in efficient DSP IP core design processes. There are multiple successful academic and commercial HLS tools available for designing DSP IP cores. Generally, these HLS frameworks shave some pre-defined modules or blocks available in their library. These pre-defined modules or blocks are also known as micro IPs. While designing a complete DSP IP core, the DSP IP designer directly uses these micro IPs to increase reusability and reduce design effort. If HLS tool vendor imports these micro IPs from another vendor who is non-trustworthy, then in such a scenario, these IP cores could be vulnerable to hardware Trojan. Thus there is no guaranteeing that a 3PIP core in an HLS library will be 100% Trojan-free.

Let us assume there is a BCD adder IP available in the module library of an HLS framework as a black box. It contains three inputs and one output: (i) the enable signal (En) which switches on/off the execution of the adder, (ii) the first primary input *A* of bit width 4 ($A_3A_2A_1$ A_0), (iii) the second primary input *B* of bit width 4 ($B_3B_2B_1$ B_0), and (d) one output *O* of bit width 4 ($O_3O_2O_1$ O_0). In a general scenario, during functional/simulation verification and test, the En signal is set (En = 1) to start the execution of the BCD adder IP. Once the verification process is completed, the En signal is reset (En = 0) to stop the execution of the BCD adder IP. Now, let us consider a BCD adder that is designed as shown in Figure 3.10. In this case, the BCD adder IP would behave correctly during its verification process as the design verifier will set the En signal "ON" during functional simulation. He/she may falsely believe that the En signal truly starts and stops the execution of the IP. However, an adversary could design an IP in such a way that the Trojan will

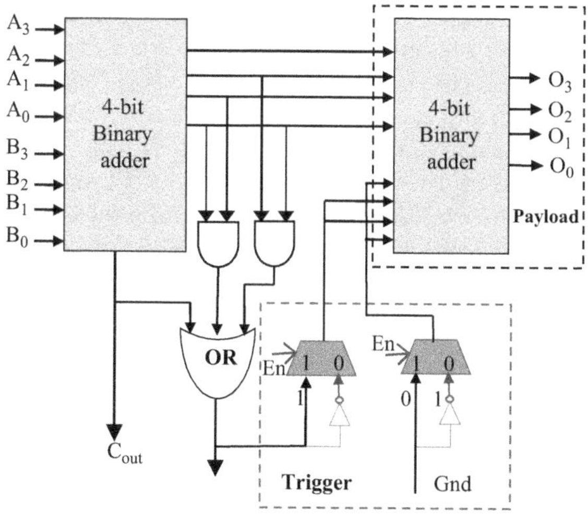

Figure 3.10 Trojan logic inserted by an adversary in a BCD Adder IP (inputs A and B). Note: When Enable (En) = 0 (rare event), the Trojan blocks get activated and produce incorrect computational output

be triggered when the En signal is reset. Therefore, whenever the system designer/integrator resets the En signal, it may maliciously generate the wrong output thereby impacting the output of the DSP IP core design. Moreover, other than the aforementioned case, triggering of the Trojan can also be achieved through the finite state machine (FSM) counter, external antenna, and so on. Moreover, there is no golden IP (i.e., with a guarantee that it is100% Trojan-free) available in the open market; thus detecting this type of Trojan is difficult during lower abstraction level. There are several other reasons as well, which are discussed in the following.

3.3.2 Trojan detectability in a 3PIP module at RTL/lower levels

Detection of hardware Trojan logic which corrupts the functional output of a 3PIP module is difficult during RTL validation or lower design abstraction level verification/tests for the following reasons:

1. A Trojan logic-infected 3PIP produces functionally correct output in normal condition, that is, until the Trojan logic is triggered.
2. The Trojan logic is triggered under very rare and specific time/condition (discussed earlier), therefore identifying such a condition or event is difficult during functional verification (Tehranipoor and Koushanfar, 2010).
3. During functional verification, in most cases, only the functionality of the entire SoC is verified; each and every IP/module used (imported) in the system (after system integration) is not checked.
4. Normally, the Trojan logic is set in an IP in such a way that it remains dormant during functional verification and triggered when the complete SoC is deployed in a real-time situation.

5. As there is no trustworthy golden IP (i.e., guarantee of a 100% Trojan-free IP) model available, the evasion of detection of hardware Trojan in a 3PIP during HLS design for DSP core is possible.
6. Physical inspection and reverse engineering, a very costly and complex procedure, does not guarantee the detection of such Trojans.
7. The testing process to detect manufacturing faults (i.e., bridging faults, delay faults, stuck-at faults, etc.) also does not guarantee the detection of such Trojans.
8. Due to evolution in technology, the physical feature size of a 3PIP decreases. The analysis of parametric signals to detect such Trojans is not considered very effective (Tehranipoor and Koushanfar, 2010).

3.4 Selected Trojan security approaches

In this section, we discuss Trojan security approaches for DSP IP cores and few selected Trojan security approaches for other hardware types such as combinational/sequential circuits. We first view multiple Trojan detection approaches for DSP cores and next, we discuss selected Trojan detection approaches for combinational/sequential circuits.

3.4.1 Trojan security approaches for DSP cores

The Trojan detection approaches for DSP cores discussed in the following is based on the threat model where hardware the Trojan is capable of producing functional (computational) changes in the DSP core output. In other words, the Trojan payload considered for these approaches is one that causes functional (computational) changes in the DSP core output.

3.4.1.1 Trojan security approach—Rajendran *et al.* (2013) and Liu *et al.* (2014)

Rajendran *et al.* (2013) and Liu *et al.* (2014) have devised duplication-based concurrent error detection (CED) techniques to achieve a Trojan secured design through HLS. The hardware Trojan that is available in a 3PIP core and affects its output or denies service is handled in this approach. In the duplication-based CED technique, replication of the same design is performed and both the designs are fed with the same set of inputs. The incorrect/dissimilar outputs due to activation of Trojan logic can be identified by comparing their outputs. However, generally an untrusted third party vendor makes a single Trojan infected IP core design and then exports multiple copies of it to the system integrator/SoC designer. Therefore, normal replication of the design cannot detect the Trojan (as both the designs will produce the same malicious outputs). In Rajendran *et al.* (2013), the source of the 3PIP core is diversified in such a way so that the same operation of both the design uses 3PIP cores coming from different sources. This process not only detects the malicious output of a 3PIP core caused by a hardware Trojan but also can prevent the activation of the Trojan by preventing collusion. Two types of collusion are

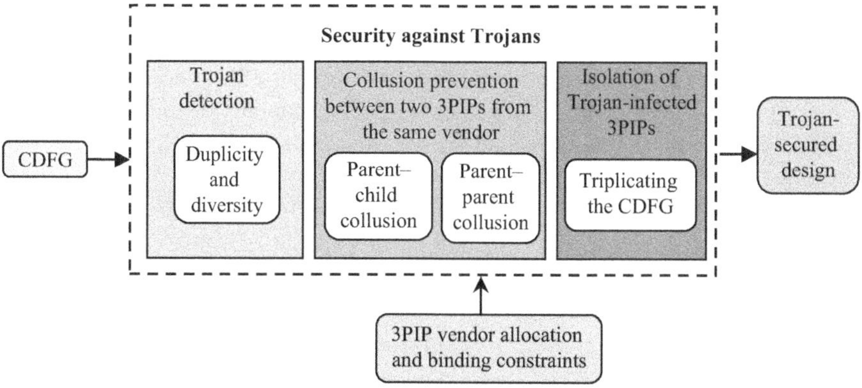

Figure 3.11 Top-level diagram of the Trojan security approach of Rajendran et al. (2013)

prevented in this approach, that is, parent-child collusion and parent-parent collusion. The binding rules to prevent collusion are as follows:

- A parent operation and its child operations in a design must be bound to the 3PIPs of different 3PIP vendors.
- At least one of the parent operations of a child operation in a design must be bound to a 3PIP from a different 3PIP vendor.

The generic overview of this approach is shown in Figure 3.11. Although this approach is capable of achieving Trojan detection and collusion prevention, it may incur some design overheads as no low-cost design architecture is explored during the scheduling phase of HLS. Moreover, distinctness during 3PIP vendor allocation can be achieved in multiple ways which have a direct impact on the overall design cost. In this approach, no optimal vendor allocation is explored to minimize the overall design security cost. Further, this approach does not handle loop-based control data flow graph digital signal processing (CDFG DSP) applications.

3.4.1.2 Trojan security approach—Sengupta *et al.* (2017a)

Sengupta *et al.* (2017a) devised a hardware Trojan security aware approach that is capable of detecting the Trojan for DSP cores. It accepts a non-loop-based DSP application in the form of a data flow graph (DFG), user constraints, bacterial foraging optimization algorithm (BFOA) control parameters (such as step size, # of chemotactic steps, # of elimination-dispersal steps, # of replication steps, terminating criteria, and population size), and a module library. It then obtains a low-cost Trojan secured design based on the input DSP core (fed as DFG).

To obtain low-cost Trojan secured DSP IP core, a design solution consisting of hardware architecture configuration and vendor allocation mode is explored iteratively through BFOA. This approach explores an optimal vendor allocation procedure using variable M_v in bacterial encoding. More explicitly, in this process, each bacterium is encoded with the number of resources of a particular resource type and the vendor allocation mode. The encoding of each bacterium can be represented as $B = (R_n, M_v)$, where R_n indicates the hardware configuration and M_v indicates the vendor allocation mode. Here, the vendor allocation mode includes two variances ($M_v = 0$ or 1). Before discussing the vendor allocation process, let us assume the scheduled design includes DMR logic consisting of the original unit (indicated as U^{OG}) and the duplicate unit (indicated as U^{DP}).

The rules for vendor allocation mode at $M_v = 0$ are as follows:

- In each control step for a particular unit (either U^{OG} or U^{DP}), alternate vendor allocation would be performed for similar types of operations. For example, if in control step 1, three addition operations (opn.1, opn.2 and opn.3) are available, then opn.1 and opn.3 would be allocated to hardware of same vendor and opn.2 would be allocated to hardware of another vendor.
- In the other unit, orthogonal vendor allocation is performed to maintain distinctness. In others words, the same operation of both the units must be allocated to different vendors. For example, if in U^{OG}, opn.1 and opn.3 are allocated to hardware of vendor type 1 (V1) and opn.2 is allocated to hardware of vendor type 2 (V2), then in U^{DP}, opn.1 and opn.3 must be allocated to hardware of V2 and opn.2 must be allocated to hardware of V1.

The rules for vendor allocation mode at $M_v = 1$ are as follows:

- All operations of a particular unit (either U^{OG} or U^{DP}) must be allocated to hardware of the same vendor. For example, all operations of U^{OG} must be allocated to hardware of V1.
- Similar operation of both the units must be allocated to distinct vendors. For example, if all operations of U^{OG} are allocated to hardware of V1 then all operations of U^{DP} must be allocated to hardware of V2 and vice versa.

The generic overview of this approach is shown in Figure 3.12. Although this approach is capable of handling hardware Trojan during HLS for non-loop based DSP applications, it does not provide any process to handle loop-based CDFG DSP cores.

3.4.1.3 Trojan security approach—Sengupta *et al.* (2017b)

Sengupta *et al.* (2017b) overcome the aforementioned shortcoming in their devised hardware Trojan security aware approach. This approach is capable of accepting single loop-based DSP applications in the form of a CDFG, user constraints, maximum loop iteration count (I), PSO control parameters (such as inertia weight, cognitive factor, and social factor), and module library. It finally produces a low-cost Trojan security aware DSP RTL design as output.

This approach obtains a low-cost design solution through a PSO-based DSE process within the user-specified design constraints (the details of the PSO-DSE approach is available in Section 3.6.3.1). The design solution in each iteration of the PSO-DSE contains not only hardware configuration and vendor allocation mode but also loop-unrolling factor, which is a crucial design parameter for loop-based CDFG. Moreover, in this approach, instead of searching among all the

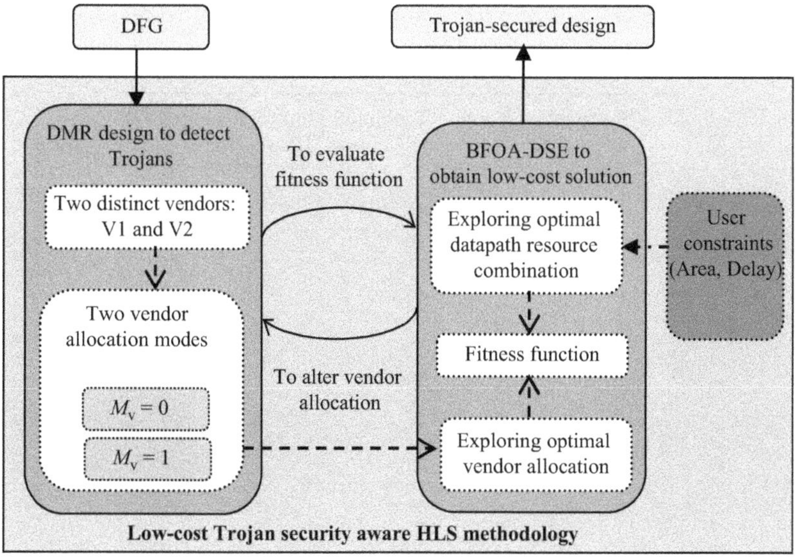

Figure 3.12 Top-level diagram of Trojan security approach of Sengupta et al. (2017a)

Preprocessing of unrolling factor

```
Input – value of 'I' (Total no. of loop iteration)
Output – screened set of unrolling factor (UF)
1 Begin
 //Screening of Unrolling Factor//
2. int k=0
3 For U =2 to I do
3.1        IF((I mod U < U/2 )&&  (U <= I/2))

           Then
//Add unrolling factor into the accepted list//
3.2        Accepted UF[k] = U
3.3        k++
3.4        End IF
3.5 End For
4 End
```

Figure 3.13(a) Pre-processing algorithm for screening unrolling factors

possible values of the unrolling factor, a list of selected unrolling factors is obtained through a pre-processing algorithm as shown in Figure 3.13(a). The input to the pre-processing algorithm is the maximum loop iteration count and the output is the screened list of unrolling factors. The algorithm works as follows: the unrolling

Table 3.1 Preprocessing of unrolling factors for FIR filter

U	Sequential Loop (I mod U)	Pipelined loop ($I - I$ mod U)	Accepted (I)
2	0	18	1
3	0	18	1
4	2	16	1
5	3	15	0
6	0	18	1
7	4	14	0
8	2	16	1
9	0	18	1
10	8	10	0
11	7	11	0
12	6	12	0
13	5	13	0
14	4	14	0
15	3	15	0
16	2	16	0
17	1	17	0
18	0	18	0

factors with the higher number of sequential execution and ones which increase code complexity without significantly improving the performance are not included in the screened list. This is because increase in sequential execution results in delay overhead, while increase in code complexity results in higher area/power. Further, large unrolling factors are not always monotonically increasing functions of performance. For example, the output of the pre-processing algorithm of an FIR filter is shown in Table 3.1. The C-code and CDFG of the FIR filter are described in Section 3.6.1.3 (Sengupta and Bhadauria, 2015).

In this approach, distinctness during vendor allocation is achieved in two ways which are similar to that of Sengupta *et al.* (2017a). The encoding of each particle can be represented as $P = (R_n, M_v, \mathrm{U})$, where U indicates the unrolling factor value; R_n and M_v have already been defined. The generic overview of this approach is shown in Figure 3.13(b). Although this approach is capable of handling hardware Trojan for single-loop CDFG-DSP cores, there are multiple nested loop-based CDFG applications available in DSP domain, which cannot be handled through this approach. Moreover, the authors here have presented only two-vendor allocation rules/modes for obtaining reduced design cost of Trojan secured architecture.

3.4.1.4 Trojan security approach—Sengupta *et al.* (2017c)

Sengupta *et al.* (2017c) overcome the aforementioned shortcoming in their devised hardware Trojan security aware approach. This approach is capable of accepting nested-loop DSP applications in the form of a CDFG, user constraints, maximum loop iteration count for all nested loops (I_1, I_2, … I_n), PSO control parameters

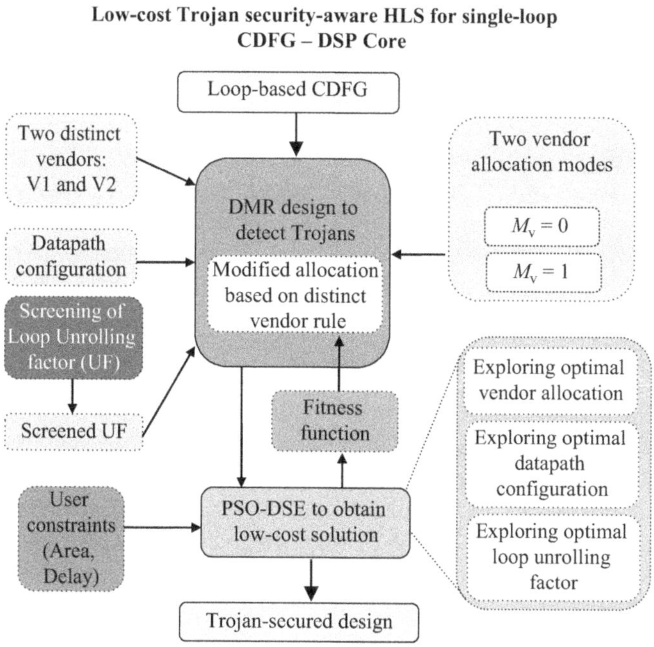

Figure 3.13(b) *Top-level diagram of Trojan security approach of Sengupta et al. (2017b)*

(such as inertia weight, cognitive factor, and social factor), and module library, finally producing a low-cost Trojan secured DSP RTL design as output.

As this approach handles nested loop-based applications, screening of unrolling factors for all the loops is performed to discard the unfit unrolling factors of each respective loop. The input to the pre-processing algorithm is the maximum loop iteration count of each loop and the output is the screened list of unrolling factors for that respective loop. Compared to Sengupta *et al.* (2017b), in this approach, the screening algorithm needs to be executed multiple times corresponding to multiple loops. Thus the implementation complexity of this approach is higher than that in Sengupta *et al.* (2017b). Further, in this approach, similar to that of Sengupta *et al.* (2017b), a low-cost design solution is obtained through a PSO-based DSE process within the user-specified design constraints. However, the encoding of each particle in this approach is significantly different from that of Sengupta *et al.* (2017b) and can be represented as $P = (R_n, M_v, \mathrm{UF}_1, \mathrm{UF}_2, \ldots \mathrm{UF}_n)$, where UF_1 indicates the unrolling factor value for the outermost nested loop and UF_n indicates the unrolling factor value for innermost nested loop. Moreover to achieve distinctness, this approach uses four vendor allocation modes to obtain low-cost Trojan secured design architecture for DSP cores. Here, the four different vendor allocation modes are $M_v = 00$, 01, 10, and 11, where $M_v = 00$ and 01 are the same as $M_v = 0$ and 1, respectively (discussed earlier). The remaining of the vendor allocation modes are defined in the following.

The rules for vendor allocation mode at $M_v = 10$ are as follows:

- In a particular unit (either U^{OG} or U^{DP}) all critical path operations must be allocated to the hardware of one vendor and all non-critical path operations must be allocated to the hardware of another vendor. For example, if all critical path operations of U^{OG} are allocated to the hardware of V1 then all non-critical path operations of U^{OG} must be allocated to the hardware of V2.
- Similar operation of both the units must be allocated to distinctive vendors. For example, if all critical path operations of U^{OG} are allocated to the hardware of V1 then all critical path operations of U^{DP} must be allocated to the hardware of V2 and vice versa.

The rules for vendor allocation technique at $M_v = 11$ are as follows:

- In a particular unit (either U^{OG} or U^{DP}) alternate vendor allocation is performed for the operations belonging to subsequent unrolled nested-loop iterations. For example, if all the operations of the first unrolled loop of U^{OG} are allocated to the hardware of V1 then the operations of the second unrolled loop of U^{OG} must be allocated to the hardware of V2, following which the operations of the third unrolled loop of U^{OG} must be allocated to hardware of V1, and so on.
- Similar operation of both the units must be allocated to distinctive vendors. For example, if all the operations of the first unrolled loop of U^{OG} are allocated to the hardware of V1 then all the operations of first unrolled loop of U^{DP} must be allocated to the hardware of V2 and vice versa.

The generic overview of this approach is shown in Figure 3.14. However, unlike Cui *et al.* (2014), this approach does not perform Trojan recovery during run-time. Nevertheless, Trojan correction through CED can be performed using TMR logic.

3.4.2 Trojan security approach for combinational/sequential circuits

Code-coverage analysis is one of the promising Trojan-detection approaches that are available in the literature. Hu (1997) has identified the list of suspicious signals from the RTL design. In this process, if a signal remains stable (does not change its state) during coverage analysis, it is identified as a suspicious signal. The reason for selecting these types of signal as suspicious is that a Trojan usually does not change its states until triggered. Thereafter, uncovered parts of the design are activated by adding test vectors. However, this approach suffers from huge verification time.

Equivalence analysis on only suspicious signals (instead of all the available signals of a design) can detect Trojan logic in an efficient way (Bushnell and Agrawal, 2013). However, this approach also incurs some runtime overhead. Additionally, all suspicious signals may not always be a Trojan logic. Moreover, the number of suspicious signals will increase with the complexity of the inserted Trojan. Thus, achieving 100% code-coverage for all IP cores through this approach is difficult. Salmani *et al.* (2009) devised a Trojan-detection mechanism by inserting dummy flip-flops to increase the Trojan activity. Dummy flip-flops are inserted in such a way that

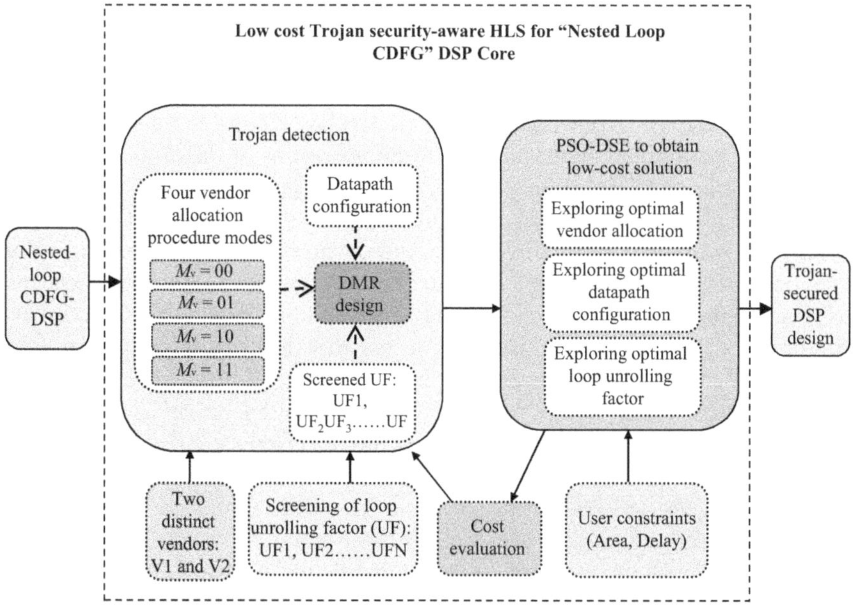

Figure 3.14 Top-level diagram of the Trojan security approach of Sengupta et al. (2017c)

they increase the transition probability of nets. Finally, they are capable of increasing Trojan activity while reducing Trojan activation time. Multiple Trojan secured methodologies for C-series benchmarks at lower design abstraction are also available in the literature (Agrawal *et al.*, 2007; Narasimhan *et al.*, 2010; Jin and Makris, 2008; Wang *et al.*, 2008). The Trojan detection process by Agrawal *et al.* (2007) is performed on the basis of comparison of characteristics of the manufactured chip, such as path delay and power consumption with the expected value. In detecting Trojans, Narasimhan *et al.* (2010) rely on the fact that the hardware Trojan inserted into a design will change the design parameters of the circuit. This approach is capable of detecting Trojans in the presence of large process variation-induced noise. Wang *et al.* (2008) devised a multiple-supply transient-current integration approach to detect hardware Trojan. This approach is capable of isolating the Trojan through a localized current analysis approach. However, process variations for detection and isolation of very small size Trojan circuits (less than few gates) can be negative.

3.5 Trojan security aware DSP IP core

In this section, we discuss the definition of "Trojan security aware DSP IP core design," objectives that must be satisfied while designing it, the design problem to achieve a low-cost Trojan secured DSP IP core, and the fitness models required to evaluate different design solutions (Sengupta *et al.*, 2017b; Rajendran *et al.*, 2013).

3.5.1 Definition

We discuss here one of the established methodologies to handle hardware Trojan (which corrupts the functional output of a DSP IP core) by generating a Trojan secured scheduling design. A Trojan secured schedule can be obtained in three consecutive steps: firstly, performing duplication of all the operations of the original CDFG to generate a duplicate CDFG, collectively known as a dual modular redundant (DMR) system; secondly, performing scheduling of operations of the DMR system based on the resource configuration (DSE-generated) to obtain the scheduled DMR; finally, achieving distinctness during hardware allocation of operations of the DMR schedule such that similar operations of the original CDFG and duplicate CDFG must have hardware of distinct vendors. Therefore, to achieve distinctness during hardware allocation, IP cores must be imported from a minimum of two distinct third-party vendors. As both the third-party vendors are not trustworthy, it may be possible that both the 3PIP cores are Trojan infected. However, having *exactly* the same Trojan payload in both the 3PIP cores is nearly impossible as both third-party vendors are mutually exclusive (unknown) to each other. Therefore, Trojan detection can be achieved by performing an equivalence check of the output of both the original CDFG and duplicate CDFG during the runtime. In other words, hardware Trojan detection can be achieved by assigning different sets of 3PIP vendors for similar operations in DMR design during the hardware-allocation step of HLS.

Trojan detection achieved through DMR for loop-based CDFG introduces duplication of all the operations, adding steering logic and internal buffering that results in incurring design overhead. Additionally, loop-based CDFGs can be implemented for different unrolling factor (UF) values, where different UF value have a different impact on the design latency, which may also incur design overhead (if not selected intelligently). Moreover, distinctness during vendor allocation can be achieved in several ways. However, each distinct vendor-allocation procedure may result in different area and execution delay of the Trojan secured design. The reason is that the same hardware imported from two different vendors may have different area and latency. Considering all these factors, to optimize the design overhead as well as to design a low-cost Trojan secured DSP IP core, an optimal design solution comprising of resource configuration, unrolling factor, and vendor allocation mode needs to be explored.

3.5.2 Goal

The goals of Trojan security aware design are as follows:

1. Generate a low-cost Trojan secured DMR schedule to detect hardware Trojan for DSP IP core.
2. The Trojan secured design should satisfy the user-specified design constraints (such as area, delay, and power).
3. For single- and nested-loop-based applications, an optimal unrolling factor should be additionally explored to optimize the Trojan secured DSP core design.
4. Perform distinct vendor allocation of the original and the duplicate unit of the DMR schedule.

5. Explore an optimal distinct vendor-allocation process out of several available possibilities.
6. Provide low-cost Trojan security aware solution.
7. Minimize the design overhead of Trojan secured DSP IP core.

3.5.3 Formulation

To generate a low-cost, Trojan secured DMR schedule for a DSP application (given in the form of a C-code or CDFG) within the user-specified constraints, which accepts module library and PSO control parameters as inputs, the formulation is as follows:

$$\text{Minimize: Trojan} - \text{secured design cost}(A_{\text{T}}^{\text{DMR}}, T_{\text{E}}^{\text{DMR}}),$$
$$\text{using optimal design solution}(Xi);$$
$$\text{Subjected to: } A_{\text{T}}^{\text{DMR}} \leq A_{\text{cons}}, T_{\text{E}}^{\text{DMR}} \leq T_{\text{cons}},$$

where $A_{\text{T}}^{\text{DMR}}$ and $T_{\text{E}}^{\text{DMR}}$ indicate the total area and total execution delay of the DMR design respectively; A_{cons} and T_{cons} indicate user-specified area and execution-delay constraints respectively.

On the basis of the DSP application type, the design solution X_i can be represented in different ways, such as the following:

$$X_i = \{N(R_1), N(R_2), \ldots N(R_D), M_v\}, \text{(for non-loop-based DFG)}$$
$$\{N(R_1), N(R_2), \ldots N(R_D), U, M_v\}, \text{ (for single-loop-based CDFG)}$$
$$X_i = \{N(R_1), N(R_2), \ldots N(R_D), U_1, U_2, \ldots U_n, M_v\},$$
$$\text{(for } n\text{-times nested-loop-based CDFG)}$$

where $N(R_D)$ indicates the number of resources of resource type R_D, U indicates the unrolling factor of single-loop-based CDFG, U_n indicates the unrolling factor of nth loop of n-times nested-loop-based CDFG and M_v indicates the vendor-allocation mode. The loop unrolling factor of both single and nested loops directly impacts the delay in a design, therefore the optimal design solution (X_i) depends on the unrolling factor. M_v is essential to provide distinctness in vendor allocation while designing the Trojan-secured DMR schedule. As the Trojan-secured DMR schedule designed using different vendor-allocation modes can have different design costs, an optimal design solution (X_i) also depends on M_v. There is a detailed discussion on different modes of vendor allocation in Section 3.6.3.2.

3.5.4 Models

On the basis of the hardware resource configuration, loop unrolling factor and vendor allocation mode, the total area and execution delay of a Trojan secured design is calculated. Thereafter, the implementation cost of the design is evaluated using the calculated area and delay of the design. (Note: area and delay have been considered as orthogonal design parameters for evaluating the design cost). In this section, we discuss the evaluation models of total design area, latency and design cost.

3.5.4.1 Design area evaluation model

Total design area consumed by a Trojan secured DSP IP core can be expressed by the following model (Sengupta *et al.*, 2017c):

$$A_T^{\mathrm{DMR}} = \sum_{j=1}^{n} \sum_{i=1}^{D} A(R_i^{v_j}) * N(R_i^{v_j}) \tag{3.1}$$

where $A(R_i^{v_j})$ indicates the area of resource type R_i corresponding to vendor type v_j, $N(R_i^{v_j})$ indicates the number of instances of resource type R_i corresponding to vendor type v_j used in the design. It is assumed there are 3PIP cores of n distinct vendors that are available in the module library, however 3PIP cores from a minimum of two different vendors are required to achieve distinctness. Here resource type R_i consists of functional resources (such as adder and multiplier), internal storage elements (latch, register), interconnecting units (multiplexer and de-multiplexer), and comparators (for error detection). Sengupta *et al.* (2017c) assessed the total design area with respect to a 90-nm CMOS-based technology node.

3.5.4.2 Design delay evaluation model

Design execution delay of a Trojan secured DSP IP core can be expressed in different models based on the DSP application type (Sengupta *et al.*, 2017b, 2017c):

Single-loop-based CDFG:

$$C_{\mathrm{total}}^{\mathrm{DMR}} = \left(C_{\mathrm{body}}^{\mathrm{DMR}} * \left\lfloor \frac{I}{UF} \right\rfloor \right) + (I \bmod U) * C_{\mathrm{first}}^{\mathrm{DMR}} \tag{3.2}$$

where $C_{\mathrm{body}}^{\mathrm{DMR}}$ is the number of control steps required to execute the loop body of the Trojan secured DMR CDFG; I is the total loop iteration (loop count); $C_{\mathrm{first}}^{\mathrm{DMR}}$ is the number of control steps required to execute the first iteration of the Trojan secured DMR CDFG. After calculating the total number of control steps required to execute the Trojan secured DMR CDFG ($C_{\mathrm{total}}^{\mathrm{DMR}}$), the execution delay can be calculated as follows:

$$T_E^{\mathrm{DMR}} = C_{\mathrm{total}}^{\mathrm{DMR}} * \Delta \tag{3.3}$$

where Δ indicates delay of a single control step.

DFG and nested-loop-based CDFG: The execution delay for nested-loop-based CDFG is calculated by completely unrolling the CDFG on the basis of the iteration counts ($I_1, I_2, \ldots I_n$).Therefore, the delay models for both nested loop and DFG-based applications are the same. The execution delay of DFG and nested-loop-based CDFG can be calculated as follows:

$$T_E^{\mathrm{DMR}} = \sum_{cs=1}^{cs(\max)} \sum_{j=1}^{n} \mathrm{Max}\left(D(R_i^{v_j}), \ldots D(R_k^{v_j}), \ldots D\left(R_i'^{v_j}\right), \ldots D\left(R_k'^{v_j}\right)\right)^{cs} \tag{3.4}$$

where $D(R_i^{v_j})$ and $D\left(R_i'^{v_j}\right)$ indicate the delay of resource type R_i corresponding to vendor type v_j of the original unit and resource type R_i' corresponding to vendor type v_j of the duplicate unit, respectively, cs indicates the control step number where the R_i is allocated.

3.5.4.3 Design cost evaluation model

The solution design cost of a Trojan secured DSP IP core is estimated using the following function (Sengupta *et al.*, 2017b):

$$C_f(X_i) = w_1 \frac{A_T^{\mathrm{DMR}} - A_{\mathrm{cons}}}{A_{\mathrm{max}}^{\mathrm{DMR}}} + w_2 \frac{T_E^{\mathrm{DMR}} - T_{\mathrm{cons}}}{T_{\mathrm{max}}^{\mathrm{DMR}}} \tag{3.5}$$

where $C_f(X_i)$ indicates the design cost of the Trojan secured DSP IP core with design solution X_i, $A_{\mathrm{max}}^{\mathrm{DMR}}$, and $T_{\mathrm{max}}^{\mathrm{DMR}}$ are the maximum area and execution in the design space, w_1 and w_2 are user-specified area and delay weightage factors (to provide equal preference, both are pre-set at 0.5). The remaining variables are defined in Section 3.6.3. As the objective of a Trojan-secured design is to satisfy the user-provided design constraints as well as minimizing the design cost, a higher negative value of $C_f(X_i)$ indicates a better design solution.

3.6 Design process of Trojan secured DSP IP core

In this section, we discuss the complete end-to-end design process of a Trojan secured DSP IP core. Each DSP IP core executes an algorithm or mathematical function during processing. These DSP IP cores could be realized as application-specific integrated circuits (ASICs) or SoC, which accepts appropriate inputs and, after executing the algorithm/function, produces the corresponding output (s). In the following, we discuss how to derive or generate a CDFG of a DSP IP core from its mathematical function. Next, we explain how to generate a DMR of a CDFG and finally we explain the process of generating a Trojan secured DSP IP core from the DMR CDFG through an illustrative example.

3.6.1 Deriving the CDFG of a DSP core

In this section, we discuss how to obtain the DFG/CDFG of some well-known transformation techniques used in DSP from the corresponding mathematical expressions (transfer functions) through examples. We demonstrate the process through three popular DSP core applications: the 8-point forward DCT function, the 8-point Decimation in Time (DIT) Discrete Fourier Transformation (DFT), and the FIR filter.

3.6.1.1 Deriving the CDFG of an 8-point DCT function

The generic equation of a forward DCT can be expressed as follows (Obukhov and Kharlamov, 2008):

$$X(m) = u(m)\sqrt{\frac{2}{N}}\sum_{i=0}^{N-1} x(i)\cos\left[\frac{(2i+1)m\pi}{2N}\right] \tag{3.6}$$

In the above expression,

$$u(m) = \begin{cases} \frac{1}{\sqrt{2}}; & \text{for } m = 0 \\ 1; & \text{for } m \neq 0 \end{cases} ; m = 0, 1, \ldots N-1,$$

where $x(i)$ is the input signal, $X(m)$ is the output signal and N indicates number of data points. Therefore, for $N=8$ the first output signal $X(0)$ can be expressed as follows:

$$\begin{aligned} X(0) &= \frac{1}{\sqrt{2}}\sqrt{\frac{2}{8}}\sum_{i=0}^{7} x(i)\cos[0] \\ &= \frac{1}{\sqrt{8}}x(0) + \frac{1}{\sqrt{8}}x(1) + \frac{1}{\sqrt{8}}x(2) + \frac{1}{\sqrt{8}}x(3) + \frac{1}{\sqrt{8}}x(4) \\ &\quad + \frac{1}{\sqrt{8}}x(5) + \frac{1}{\sqrt{8}}x(6) + \frac{1}{\sqrt{8}}x(7) \\ &= 0.3536\,x(0) + 0.3536\,x(1) + 0.3536\,x(2) + 0.3536\,x(3) \\ &\quad + 0.3536\,x(4) + 0.3536\,x(5) + 0.3536\,x(6) + 0.3536\,x(7) \end{aligned}$$

Similarly, for the same data point $X(1)$ and $X(2)$ can be expressed as follows:

$$\begin{aligned} X(1) &= 1\sqrt{\frac{2}{8}}\sum_{i=0}^{7} x(i)\cos\left[\frac{(2i+1)1\pi}{16}\right] \\ &= \frac{1}{2}x(0)\cos\left[\frac{\pi}{16}\right] + \frac{1}{2}x(1)\cos\left[\frac{3\pi}{16}\right] + \frac{1}{2}x(2)\cos\left[\frac{5\pi}{16}\right] + \frac{1}{2}x(3)\cos\left[\frac{7\pi}{16}\right] \\ &\quad + \frac{1}{2}x(4)\cos\left[\frac{9\pi}{16}\right] + \frac{1}{2}x(5)\cos\left[\frac{11\pi}{16}\right] + \frac{1}{2}x(6)\cos\left[\frac{13\pi}{16}\right] + \frac{1}{2}x(7)\cos\left[\frac{15\pi}{16}\right] \\ &= 0.4904x(0) + 0.4157x(1) + 0.2778x(2) + 0.0975x(3) + (-0.0975)x(4) \\ &\quad + (-0.2778)x(5) + (-0.4157)x(6) + (-0.4904)x(7) \end{aligned}$$

$$\begin{aligned} X(2) &= 1\sqrt{\frac{2}{8}}\sum_{i=0}^{7} x(i)\cos\left[\frac{(2i+1)2\pi}{16}\right] \\ &= \frac{1}{2}x(0)\cos\left[\frac{2\pi}{16}\right] + \frac{1}{2}x(1)\cos\left[\frac{6\pi}{16}\right] + \frac{1}{2}x(2)\cos\left[\frac{10\pi}{16}\right] + \frac{1}{2}x(3)\cos\left[\frac{14\pi}{16}\right] \\ &\quad + \frac{1}{2}x(4)\cos\left[\frac{18\pi}{16}\right] + \frac{1}{2}x(5)\cos\left[\frac{22\pi}{16}\right] + \frac{1}{2}x(6)\cos\left[\frac{26\pi}{16}\right] + \frac{1}{2}x(7)\cos\left[\frac{30\pi}{16}\right] \\ &= 0.4619x(0) + 0.1913x(1) + (-0.1913)x(2) + (-0.4619)x(3) \\ &\quad + (-0.4619)x(4) + (-0.1913)x(5) + 0.1913x(6) + 0.4619x(7) \end{aligned}$$

Similarly, $X(3)$ to $X(7)$ can be calculated using (3.6).

$$\begin{bmatrix} X(0) \\ X(1) \\ X(2) \\ X(3) \\ X(4) \\ X(5) \\ X(6) \\ X(7) \end{bmatrix} = \begin{bmatrix} 0.3536 & 0.3536 & 0.3536 & 0.3536 & 0.3536 & 0.3536 & 0.3536 & 0.3536 \\ 0.4904 & 0.4157 & 0.2778 & 0.0975 & -0.0975 & -0.2778 & -0.4157 & -0.4904 \\ 0.4619 & 0.1913 & -0.1913 & -0.4619 & -0.4619 & -0.1913 & 0.1913 & 0.4619 \\ 0.4157 & -0.0975 & -0.4904 & -0.2778 & 0.2778 & 0.4904 & 0.0975 & -0.4157 \\ 0.3536 & -0.3536 & -0.3536 & 0.3536 & 0.3536 & -0.3536 & -0.3536 & 0.3536 \\ 0.2778 & -0.4904 & 0.0975 & 0.4157 & -0.4157 & -0.0975 & 0.4904 & -0.2778 \\ 0.1913 & -0.4619 & 0.4619 & -0.1913 & -0.1913 & 0.4619 & -0.4619 & 0.1913 \\ 0.0975 & -0.2778 & 0.4157 & -0.4904 & 0.4904 & -0.4157 & 0.2778 & -0.0975 \end{bmatrix} \begin{bmatrix} x(0) \\ x(1) \\ x(2) \\ x(3) \\ x(4) \\ x(5) \\ x(6) \\ x(7) \end{bmatrix}$$

Figure 3.15 Output signal of 8-point DCT in the form of matrix multiplication

We can represent this calculation in the form of matrix multiplication as shown in Figure 3.15. Where the 8×1 output signal matrix can be derived by multiplying the 8-point DCT coefficient matrix with 8×1 input signal matrix. The 8-point DCT coefficient can also be represented as shown in Figure 3.16, where C_1 to C_7 indicate the DCT coefficients in ascending order (C_1 indicates the maximum positive DCT coefficient value and C_7 indicates the minimum positive DCT coefficient value). On the basis of matrix multiplication as shown in Figure 3.16, the output signals $X[0]$ to $X[7]$ can be represented as follows:

$$\begin{aligned}
X[0] &= C_4 * x[0] + C_4 * x[1] + C_4 * x[2] + C_4 * x[3] \\
&\quad + C_4 * x[4] + C_4 * x[5] + C_4 * x[6] + C_4 * x[7] \\
X[1] &= C_1 * x[0] + C_3 * x[1] + C_5 * x[2] + C_7 * x[3] \\
&\quad + (-C_7) * x[4] + (-C_5) * x[5] + (-C_3) * x[6] + (-C_1) * x[7] \\
X[2] &= C_2 * x[0] + C_6 * x[1] + (-C_6) * x[2] + (-C_2) * x[3] \\
&\quad + (-C_2) * x[4] + (-C_6) * x[5] + C_6 * x[6] + C_2 * x[7] \\
X[3] &= C_3 * x[0] + (-C_7) * x[1] + (-C_1) * x[2] + (-C_5) * x[3] + C_5 * x[4] \\
&\quad + C_1 * x[5] + C_7 * x[6] + (-C_3) * x[7] \\
X[4] &= C_4 * x[0] + (-C_4) * x[1] + (-C_4) * x[2] + C_4 * x[3] + C_4 * x[4] \\
&\quad + (-C_4) * x[5] + (-C_4) * x[6] + C_4 * x[7] \\
X[5] &= C_5 * x[0] + (-C_1) * x[1] + C_7 * x[2] + C_3 * x[3] + (-C_3) * x[4] \\
&\quad + (-C_7) * x[5] + C_1 * x[6] + (-C_5) * x[7] \\
X[6] &= C_6 * x[0] + (-C_2) * x[1] + C_2 * x[2] + (-C_6) * x[3] + (-C_6) * x[4] \\
&\quad + C_2 * x[5] + (-C_2) * x[6] + C_6 * x[7] \\
X[7] &= C_7 * x[0] + (-C_5) * x[1] + C_3 * x[2] + (-C_1) * x[3] + C_1 * x[4] \\
&\quad + (-C_3) * x[5] + C_5 * x[6] + (-C_7) * x[7]
\end{aligned}$$

$$\begin{bmatrix} X(0) \\ X(1) \\ X(2) \\ X(3) \\ X(4) \\ X(5) \\ X(6) \\ X(7) \end{bmatrix} = \begin{bmatrix} C_4 & C_4 & C_4 & C_4 & C_4 & C_4 & C_4 & C_4 \\ C_1 & C_3 & C_5 & C_7 & -C_7 & -C_5 & -C_3 & -C_1 \\ C_2 & C_6 & -C_6 & -C_2 & -C_2 & -C_6 & -C_6 & C_2 \\ C_3 & -C_7 & -C_1 & -C_5 & C_5 & C_1 & C_7 & -C_3 \\ C_4 & -C_4 & -C_4 & C_4 & C_4 & -C_4 & -C_4 & C_4 \\ C_5 & -C_1 & C_7 & C_3 & -C_3 & -C_7 & C_1 & -C_5 \\ C_6 & -C_2 & C_2 & -C_6 & -C_6 & C_2 & -C_2 & C_6 \\ C_7 & -C_5 & C_3 & -C_1 & C_1 & -C_3 & C_5 & -C_7 \end{bmatrix} \begin{bmatrix} x(0) \\ x(1) \\ x(2) \\ x(3) \\ x(4) \\ x(5) \\ x(6) \\ x(7) \end{bmatrix}$$

Figure 3.16 Another representation of output signal of 8-point DCT in the form of matrix multiplication

The generic equation of the 8-point DCT to compute the first sample is as follows:

$$\begin{aligned} X[0] &= k_1 * x[0] + k_2 * x[1] + k_3 * x[2] + k_4 * x[3] + k_5 * x[4] \\ &\quad + k_6 * x[5] + k_7 * x[6] + k_8 * x[7] \end{aligned} \tag{3.7}$$

The equation of an 8-point forward DCT consists of eight multiplication operations and seven addition operations. The equivalent DFG of an 8-point forward DCT is shown in Figure 3.17, where the primary inputs are of two types: (i) k_1 to k_8 indicate the DCT coefficients and (ii) $x[0]$ to $x[7]$ indicates the input signals' primary inputs, with light colored nodes indicating multiplication operations and deep colored nodes indicating addition operations, the corresponding operation/node numbers are mentioned using integer values (1 to 15) and finally the output is shown in a square box. It can be observed that all the output signals $X[0]$ to $X[7]$ can be computed by changing the k_1 to k_8 values.

3.6.1.2 Deriving the CDFG of an 8-point DFT

The generic equation of an N-point DFT can be expressed as follows (Salivahanan *et al.*, 2000):

$$X(k) = \sum_{n=0}^{N-1} x(n) W_N^{nk}, \quad 0 \le k \le N-1, \tag{3.8}$$

where $x(n)$ indicates a sequence of N values, where N can be represented as an integral power of 2, the complex-valued sequence $X(k)$ indicates the DFT of a sequence $x(n)$, and W_N indicates the complex-valued phase factor. The N-valued sequence as shown in Eqn. 3.8 can be decimated into two $N/2$ point sequences in the form of even- and odd-numbered values of $x(n)$. The decimated $x(n)$ obtained from Eqn. 3.8 can be expressed as follows:

$$X(k) = \sum_{n=0, n \text{ even}}^{N-1} x(n) W_N^{nk} + \sum_{n=0, n \text{ odd}}^{N-1} x(n) W_N^{nk} \tag{3.9}$$

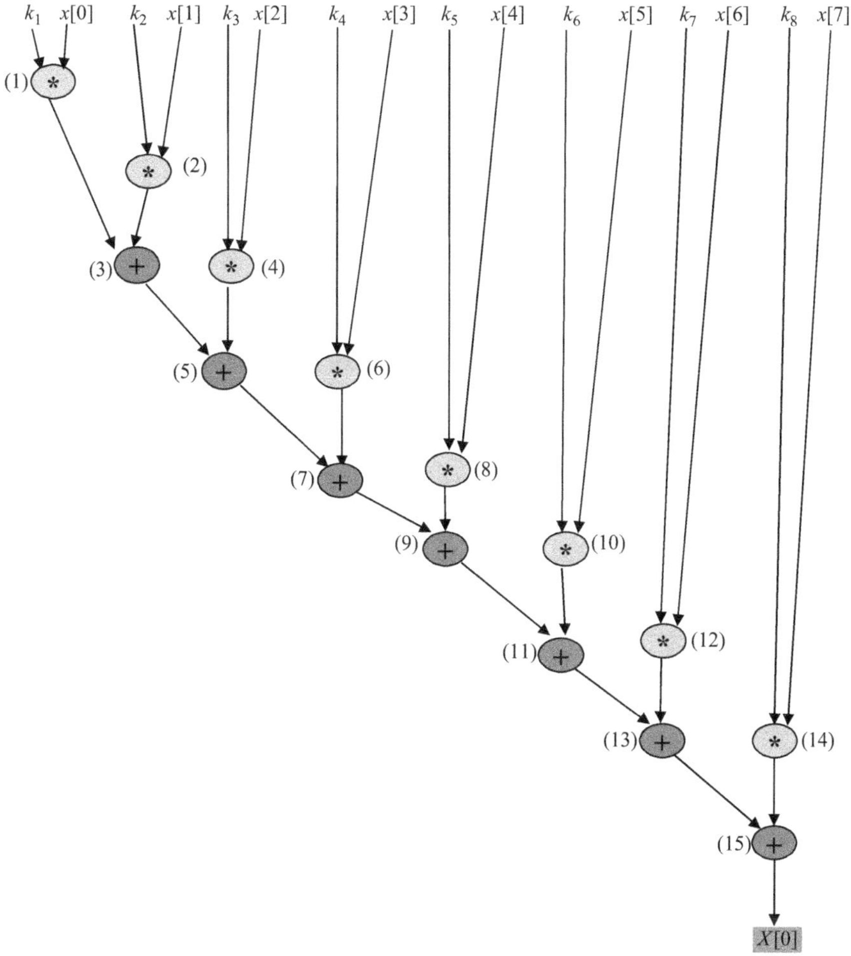

Figure 3.17 Data flow graph of an 8-point DCT

If we substitute $n = 2r$ when n is even and $n = 2r + 1$ when n is odd in Eqn. 3.9, we get,

$$\begin{aligned} X(k) &= \sum_{r=0}^{\left(\frac{N}{2}-1\right)} x(2r)W_N^{2rk} + \sum_{r=0}^{\left(\frac{N}{2}-1\right)} x(2r+1)W_N^{(2r+1)k} \\ &= \sum_{r=0}^{\left(\frac{N}{2}-1\right)} x(2r)\left(W_N^2\right)^{rk} + W_N^k \sum_{r=0}^{\left(\frac{N}{2}-1\right)} x(2r+1)\left(W_N^2\right)^{rk} \end{aligned} \tag{3.10}$$

Here,

$$W_N^2 = \left[e^{-j\left(\frac{2\pi}{N}\right)}\right]^2 = e^{-j\frac{\left(\frac{2\pi}{N}\right)}{2}} = W_{N/2}$$

Therefore, (3.10) can be written as follows:

$$X(k) = \sum_{r=0}^{\left(\frac{N}{2}-1\right)} x(2r)W_{N/2}^{rk} + W_N^k \sum_{r=0}^{\left(\frac{N}{2}-1\right)} x(2r+1)W_{N/2}^{rk}$$
$$X(k) = G(k) + W_N^k H(k), k = 0, 1, \ldots \frac{N}{2} - 1 \tag{3.11}$$

In the above expression, $G(k)$ and $H(k)$ are the two $N/2$-point DFTs of even and odd sequences respectively, therefore k values for both $G(k)$ and $H(k)$ lie between 0 and $(N/2)-1$, as shown in 3.11. The corresponding signal flow graph for $N=8$ is shown in Figure 3.18. It can be observed from Figure 3.18 that $X(0)$ can be obtained by multiplying $H(0)$ by $W^0{}_8$ then adding the product to $G(0)$. Similarly, the values $X(1)$ to $X(3)$ can be obtained; owing to the periodicity property of $G(k)$ and $H(k)$ over k, $G(0)=G(4)$ and $H(0)=H(4)$. Therefore, $X(4)$ can be obtained by multiplying $H(0)$ by $W^4{}_8$ then adding the product to $G(0)$. All the equations for the outputs $X(0)$ to $X(7)$ are expressed as follows:

$$X[0] = H[0] * W_8^0 + G[0]$$
$$X[1] = H[1] * W_8^1 + G[1]$$
$$X[2] = H[2] * W_8^2 + G[2]$$
$$X[3] = H[3] * W_8^3 + G[3]$$
$$X[4] = H[0] * W_8^4 + G[0]$$
$$X[5] = H[1] * W_8^5 + G[1]$$
$$X[6] = H[2] * W_8^6 + G[2]$$
$$X[7] = H[3] * W_8^7 + G[3]$$

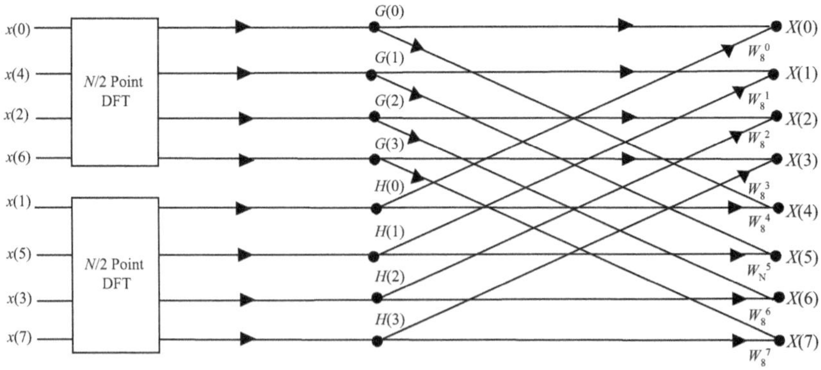

Figure 3.18 Signal flow graph of the first stage DIT DFT for N = 8

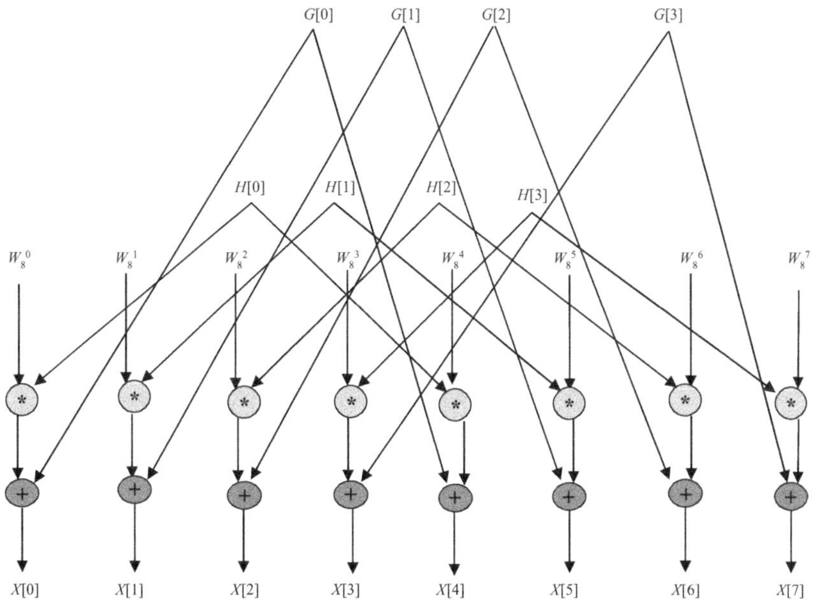

Figure 3.19 Equivalent data flow graph of the first-stage DIT DFT for $N = 8$

The equivalent DFG of the first-stage DIT of DFT for $N = 8$ is shown in Figure 3.19.

Similarly, each of the $N/2$-point DFTs, that is, $G(x)$ and $H(x)$, can be decimated again into two $N/4$ point DFTs. Therefore, each $G(x)$ and $H(x)$ in Eqn. 3.11 can be expressed as follows:

$$
\begin{aligned}
G(k) &= \sum_{r=0}^{\left(\frac{N}{2}\right)-1} g(r) W_{N/2}^{rk} \\
&= \sum_{l=0}^{\left(\frac{N}{4}\right)-1} g(2l) W_{N/2}^{2lk} + \sum_{l=0}^{\left(\frac{N}{4}\right)-1} g(2l+1) W_{N/2}^{(2l+1)k} \\
&= \sum_{l=0}^{\left(\frac{N}{4}\right)-1} g(2l) W_{N/4}^{lk} + W_{N/2}^{k} \sum_{l=0}^{\left(\frac{N}{4}\right)-1} g(2l+1) W_{N/4}^{lk} \\
G(k) &= A(k) + W_{N/2}^{k} B(k)
\end{aligned}
\tag{3.12}
$$

In the above expression, $A(k)$ and $B(k)$ are the two $N/4$-point DFTs of even and odd sequences respectively, therefore the value k for both $A(k)$ and $B(k)$ lies between 0 and $(N/4)-1$.

Similarly,

$$H(k) = \sum_{l=0}^{\left(\frac{N}{4}\right)-1} h(2l) W_{N/4}^{lk} + W_{N/2}^{k} \sum_{l=0}^{\left(\frac{N}{4}\right)-1} h(2l+1) W_{N/4}^{lk}$$
$$H(k) = C(k) + W_{N/2}^{k} D(k) \tag{3.13}$$

The corresponding signal flow graph for $N = 8$ is shown in Figure 3.20. It can be observed from Figure 3.20 that, $G(0)$ can be obtained by multiplying $B(0)$ and $W^0{}_8$, then adding the product to $A(0)$. Similarly, $H(0)$ can be obtained by multiplying $D(0)$ and $\mathrm{W}^0{}_8$, then adding the product to $C(0)$. Owing to periodicity property of $A(k)$, $B(k)$, $C(k)$, and $D(k)$ over k, $A(0) = A(2)$, $B(0) = B(2)$, and so on. All the equations for outputs $G(0)$ to $G(3)$ and $H(0)$ to $H(3)$ are shown below:

$$G[0] = B[0] * W_8^0 + A[0]$$
$$G[1] = B[1] * W_8^2 + A[1]$$
$$G[2] = B[0] * W_8^4 + A[0]$$
$$G[3] = B[1] * W_8^6 + A[1]$$
$$H[0] = D[0] * W_8^0 + C[0]$$
$$H[1] = D[1] * W_8^2 + C[1]$$
$$H[2] = D[0] * W_8^4 + C[0]$$
$$H[3] = D[1] * W_8^6 + C[1]$$

The equivalent DFG of the second-stage DIT DFT for N = 8 is shown in Figure 3.21.

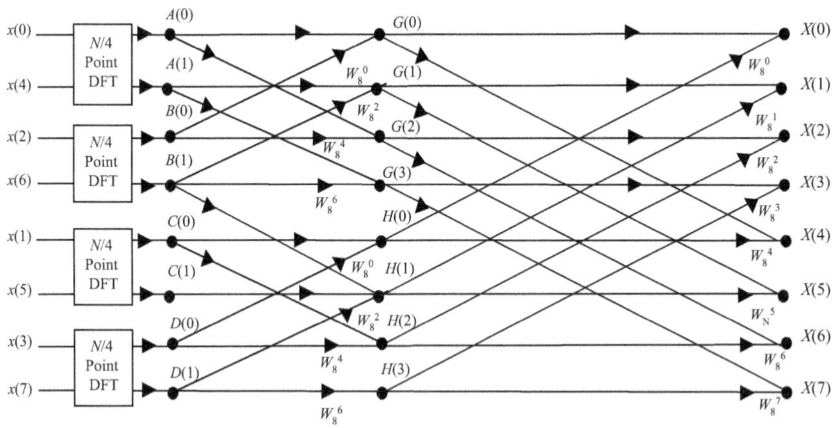

Figure 3.20 Signal flow graph of the second-stage DIT DFT for N = 8

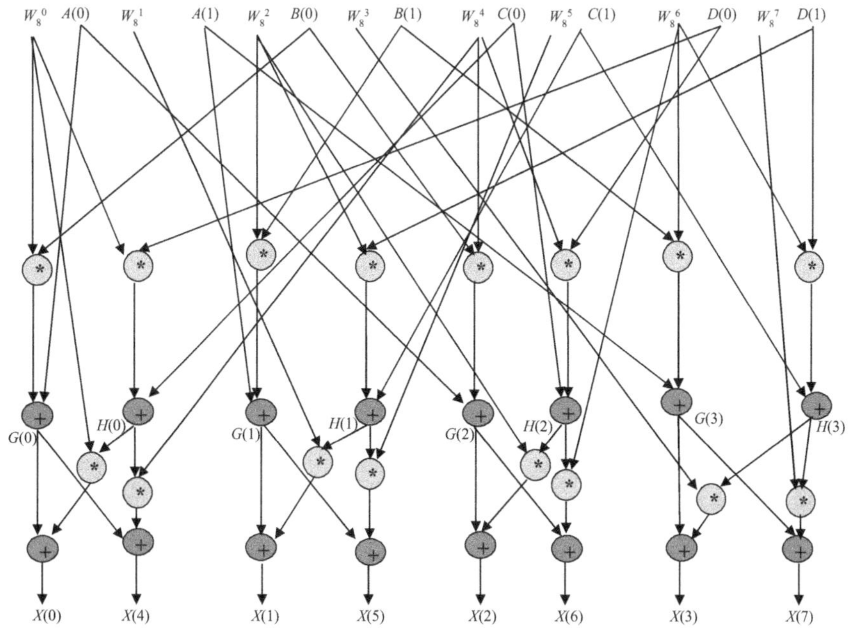

Figure 3.21 Equivalent data flow graph of the second-stage DIT DFT for N = 8

3.6.1.3 Deriving the CDFG of the FIR filter

For simplicity and easier understanding of readers, we use the FIR filter to demonstrate a complete Trojan secured DSP IP core design procedure using Sengupta *et al.* (2017c). We use this approach for explanation because it provides comprehensive details of optimizing Trojan secured DSP architecture based on a multi-vendor procedure and PSO-DSE as well as its end-to-end design process. Now, we continue our explanation using the FIR filter example in each step. However, readers may choose any DSP application to follow these steps. In the previous two examples, we have demonstrated how to obtain the DFG of other DSP applications (DCT and DFT) from their corresponding transfer functions. Now, we also demonstrate how to obtain the CDFG alternatively from a single loop-based C-code. We have used the FIR application to demonstrate the Trojan secured design process. The generic equation of the FIR filter can be expressed as follows:

$$y(n) = \sum_{i=0}^{M} h_i * x(n-i), \tag{3.14}$$

where M is the filter order of the FIR, h_i the FIR coefficient, $x(n)$ the input impulse and $y(n)$ the output impulse. Using the above equation, the FIR filter for the unrolling factor can be represented as,

$$y(n) = h_0 * x(n) + h_1 * x(n-1) \tag{3.15}$$

```
for(i=0; i<M; i=(i+2)) // M = order of FIR filter
{
        y(n)+=h_i*x(n-i);
        y(n)+=h_{i+1}*x(n-(i+1));
}
```

Figure 3.22 C-code of the FIR benchmark for the unrolling factor 2

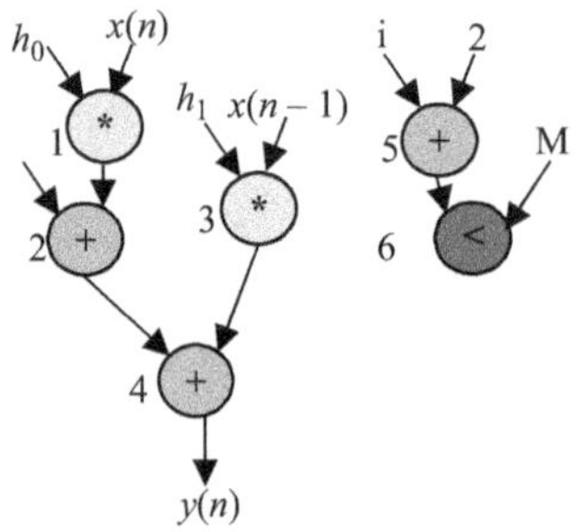

Figure 3.23 CDFG of the FIR benchmark for unrolling factor 2

The corresponding C-code of the FIR application is shown in Figure 3.22. For the unrolling factor 2, the corresponding CDFG of the FIR is shown in Figure 3.23. The light colored node indicates multiplier, the deep colored node the adder, and the very deep colored node the comparator.

3.6.2 Generating the DMR of the CDFG

For designing a Trojan secured DSP IP core, we already discussed how to obtain the CDFG of a DSP core from its mathematical equation/transfer function or C-code. The next step is to construct the DMR of the application. In order to obtain the DMR CDFG, complete duplication of all the unrolled operations is performed. The unrolling factor is obtained from the candidate design solution X_i (explained earlier in Section 3.5.3). For example, the DMR of the FIR filter for unrolling factor 2 is shown in Figure 3.24, where U^{OG} indicates operations of original unit and U^{DP} indicates operations of the duplicate unit.

3.6.3 Trojan secured scheduling of DMR CDFG

After obtaining the DMR CDFG, the next step is to schedule all the operations of the original and duplication units simultaneously on the basis of the resource configuration. The resource configuration is obtained from the candidate design solution X_i. Once the scheduling of the DMR is performed, the next step is to allocate 3PIP cores imported from different vendors. In this context, four distinct vendor allocation rules, devised in Sengupta *et al.* (2017c) are used (discussed earlier in Section 3.4.1.2). As the same functional resources (3PIP cores) imported from different vendor have different area, delay and power, each vendor allocation mode

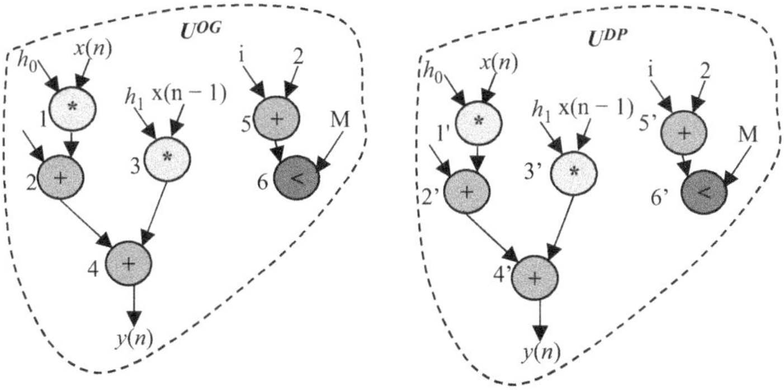

Figure 3.24 DMR CDFG of the FIR benchmark for unrolling factor 2

has an impact on the final Trojan secured design. Hence, to obtain a low-cost Trojan secured DSP core, optimal hardware resource configuration, unrolling factor(s), and vendor allocation mode must be explored. In Sengupta *et al.* (2017c), this is obtained through a PSO-DSE approach (explained in the next paragraph).

3.6.3.1 Determining resource configuration

Particle initialization

To explore a low-cost Trojan secured design for a loop-based DSP core (CDFG), each particle of the PSO process is encoded through hardware resource configuration, unrolling factor(s), and vendor allocation mode. For example, the *i*th candidate design solution (X_i) in the particle population can be represented as:

$$X_i = \{N(R_1), N(R_2), ..N(R_D), U_1, U_2, ..U_n, M_v\}$$

where $N(R_D)$ indicates the number of resource of *D*th dimension of the particle, and $U_{1..n}$ and M_v are defined before. To uniformly distribute the particles over the design space based on the aforementioned design solution template, the first three particles (X_1, X_2 and X_3) are ws:

$$\begin{aligned}
X_1 &= \{\min(R_1),\ \min(R_2), \ldots,\ \min(R_D),\ \min(U_{1\ldots n}), 00\},\\
X_2 &= \{\max(R_1), \max(R_2), \ldots, \max(R_D), \max(U_{1\ldots n}), 11\},\\
X_3 &= \{(\min(R_1)) + (\max(R_1))/2, (\min(R_2)) + (\max(R_2))/2, \ldots,\\
&\quad (\min(R_D)) + (\max(R_D))/2, (\min(U_{1\ldots n})) + (\max(U_{1\ldots n}))/2, 01\}.
\end{aligned}$$

The rest of the particles are initialized as follows:

$$\begin{aligned}
X_j = [&\{(\min(R_1)) + (\max(R_1))/2 \pm \alpha\}, (\min(R_2))\\
&+ (\max(R_2))/2 \pm \alpha\}, \ldots, \{(\min(R_D)) + (\max(R_D))/2 \pm \alpha\},\\
&\{(\min(U_{1\ldots n})) + (\max(U_{1\ldots n}))/2 \pm \alpha\}, \text{random}(M_v)]
\end{aligned}$$

In the above expressions, α indicates a random value within the range of the maximum and minimum possible values.

Particle movement

The fitness of each particle is calculated using (3.5).The particle position which has the highest fitness (lower cost value indicates higher fitness) among its positions so far is known as local best particle. Further, among all the local best particles the particle having highest fitness is known as global best particle. On the basis of the local and global best, the velocity of each particle is calculated. Each dimension (d) of a particle velocity (V_{di}) is updated using the following expression:

$$V_{d_i}^{+} = \omega V_{d_i} + b_1 r_1 \left[R_{d_{lb_i}} - R_{d_i}\right] + b_2 r_2 \left[R_{d_{gb}} - R_{d_i}\right], \tag{3.16}$$

where V^{+}_{di} and V_{di} are new and current velocity of the ith particle in the dth, dimension respectively; R_{di} is the resource value of the ith particle in the dth dimension; R_{dlbi} is the local best position of the ith particle in the dth dimension; and R_{dgb} is the global best of the dth dimension; ω is the inertia weight, b_1 and b_2 are social and cognitive factors; r_1 and r_2 are random numbers between 0 and 1. On the basis of the updated velocity of a particle calculated in Eqn. 3.16, the updated particle position is calculated. The updated particle position can be determined as follows:

$$R_{d_i}{}^{+} = R_{d_i} + V_{d_i}{}^{+}, \tag{3.17}$$

where R^{+}_{di} is the updated resource value of ith particle in dth dimension.

Velocity clamping

To control unwanted exploration drift of a particle, velocity clamping is required. Velocity clamping is required when the new velocity (V^{+}_{di}) of a particle crosses its maximum ($+V^{\max}_{\mathrm{di}}$) and minimum ($-V^{\max}_{\mathrm{di}}$) limits. Velocity clamping can be performed as follows:

$$V_{d_i}^{+} = \begin{cases} +V_{d_i}^{\max}, \text{ if } V_{d_i}^{+} > +V_{d_i}^{\max} \\ -V_{d_i}^{\max}, \text{ if } V_{d_i}^{+} < -V_{d_i}^{\max} \\ V_{d_i}^{+}, \text{else} \end{cases} \tag{3.18}$$

In the above expression,

$$\pm V_{d_i}^{\max} = \pm\left(\frac{\max(N(R_d)) - \min(N(R_d))}{2}\right)$$

Terminating criteria

PSO searches for a low-cost hardware architecture design solution for DSP cores iteratively. In each iteration, it explores an equal or better solution in terms of design cost. However, this iterative searching process terminates on the basis of the following two conditions: (i) the number of iteration exceeds the maximum count ($I_{\max}$); (ii) the global best remains unchanged for the maximum consecutive iteration count (δ). In Sengupta *et al.* (2017b), the used value of $I_{\max}$ and δ during the experimental setup is 100 and 10, respectively.

3.6.3.2 Vendor allocation procedure

As mentioned earlier, scheduling and hardware allocation are performed for the DMR CDFG on the basis of resource configuration and vendor allocation mode. For example for the DMR of the FIR filter, the hardware resources required to schedule are the adder, multiplier and comparator (shown in Figure 3.24) . Now to implement distinctness during hardware allocation, the same 3PIP cores (adders, multipliers and comparators) are imported from two different vendors. Let us assume the two different vendors are V1 and V2; the multiplier and adder imported from V1 have area = 2468 au and 2034 au, latency = 10,000 ns and 265 ns, respectively; while the multiplier and adder imported from V2 have area = 2464 au and 2032 au, latency = 11,000 ns and 270 ns, respectively (*Note*: the values are used based on 90 nm technology scale; assuming 1 au = 1 transistor; ns = nanosecond).

Now consider the Trojan secured DMR CDFG of the FIR that is designed on the basis of the candidate solution $X_i = 2(+), 2(*), 1(<), U = 2, M_v = 00$ (illustrated in Figure 3.25(a)). This indicates that the DMR CDFG of the FIR filter is unrolled two times, scheduled based on two adders, two multipliers, and one comparator, and vendor hardware allocation is performed using mode '00'. The corresponding execution delay and area of the design is: $T_E^{DMR} = 45{,}080$ ns and $A_T^{DMR} = 13{,}064$ au. As shown in Figure 3.25(a), based on the vendor allocation rule in mode 00 (all modes are explained in Section 3.4.1.2), operations of U^{OG} are allocated to the vendor in an alternative way in a control step (CS). For example, opn.1 and opn.3 of U^{OG} are allocated alternatively to hardware of V1 and V2 in control step 1. Further, opn.1′ and opn.3′ of U^{DP} are allocated to hardware of V2 and V1, respectively.

If the same DMR CDFG of the FIR filter is designed based on the candidate solution $X_i = 2(+), 2(*), 1(<), U = 2, M_v = 01$ (illustrated in Figure 3.25(b)), the corresponding execution delay and area of the design are $T_E^{DMR} = 43{,}080$ ns and

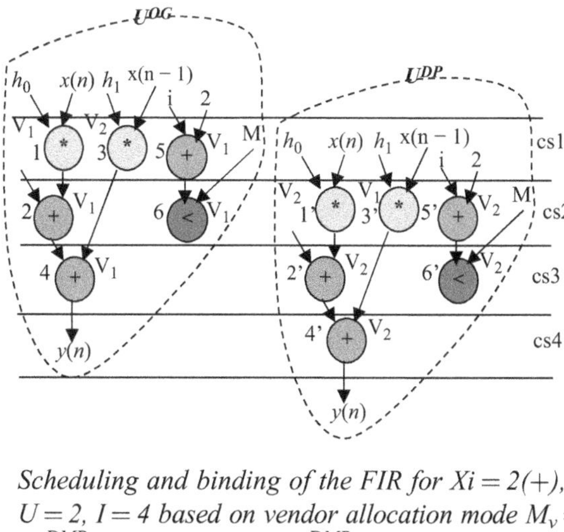

Figure 3.25(a) Scheduling and binding of the FIR for Xi = 2(+), 2(*), 2(<), U = 2, I = 4 based on vendor allocation mode $M_v = 00$; $T_E^{DMR} = 45{,}080$ ns and $A_T^{DMR} = 13{,}064$ au

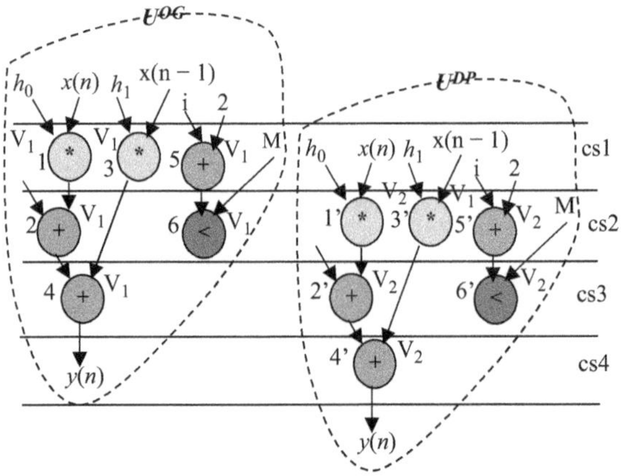

Figure 3.25(b) Scheduling and binding of the FIR for Xi = 2(+), 2(*), 2(<), U = 2, I = 4 based on vendor allocation mode $A_v = 00$; $T_E^{DMR} = 45{,}080$ ns and $A_T^{DMR} = 17{,}996$ au

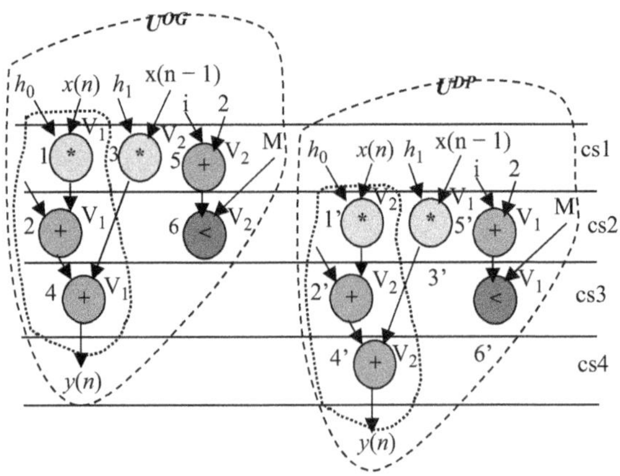

Figure 3.25(c) Scheduling and binding of the FIR for Xi = 2(+), 2(*), 2(<), U = 2, I = 4 based on vendor allocation mode $A_v = 00$; $T_E^{DMR} = 45{,}080$ ns and $A_T^{DMR} = 13{,}064$ au and opn.4 of U^{OG} are allocated to the hardware of V2

$A_T^{\mathrm{DMR}} = 17{,}996$ au. As shown in Figure 3.25(b), based on the vendor allocation rule in mode 01, all operations of U^{OG} are allocated to the hardware of V1 and all operations of U^{DP} are allocated to the hardware of V2.

If the same DMR CDFG of the FIR filter is designed based on candidate solution $X_i = 2(+), 2(*), 1(<), U = 2, M_v = 10$ (illustrated in Figure 3.25(c)), the corresponding execution delay and area of the design is: $T_E^{\mathrm{DMR}} =$ '45,080 ns and $A_T^{\mathrm{DMR}} =$ '13,064 au. As shown in Figure 3.25(c) based on vendor allocation rule in mode 10, all the critical path operations, that is, opn.1, opn.2, and opn.4 of U^{OG}

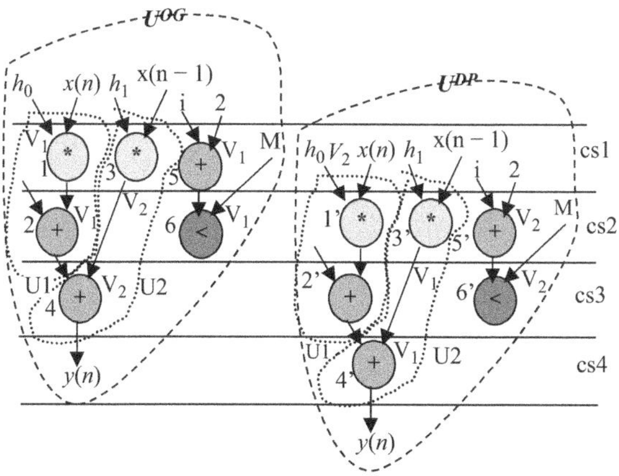

Figure 3.25(d) *Scheduling and binding of the FIR for Xi = 2(+), 2(*), 2(<), U = 2, I = 4 based on vendor allocation mode $A_v = 10$; $T_E^{DMR} = 45{,}070$ ns and $A_T^{DMR} = 15{,}096$ au. Hardware of distinct vendor is allocated in U^{DP}*

(marked in a dotted circle) are allocated to the hardware of V1 and all non-critical path operations, that is, opn.3, opn.5, and opn.6 of U^{OG} are allocated to the hardware of V2. The hardware of the distinct vendor is allocated in U^{DP}.

If the same DMR CDFG of the FIR filter is designed based on candidate solution $X_i = 2(+)$, 2(*), 1(<), $U = 2$, $M_v = 11$ (illustrated in Figure 3.25(d)), the corresponding execution delay and area of the design are $T_E^{DMR} = 45{,}070$ ns and $A_T^{DMR} = 15{,}096$ au. As shown in Figure 3.25(d) based on vendor allocation rule in mode 11, alternate vendor allocation is performed to the operations belonging to subsequent unrolled loop iterations, for example all operations of the first unrolled loop, that is, opn.1, opn.2, opn.5, and opn.6 of U^{OG} are allocated to the hardware of V1 and all operations of the second unrolled loop, that is, opn.3, and opn.4 of U^{OG} are allocated to the hardware of V2. Hardware of the distinct vendor is allocated in U^{DP}.

3.7 Analysis of case studies/test cases

In this section, we analyze and compare four different Trojan security approaches for DSP cores that are available in the literature in terms of their security and design cost. This section is divided into four sub-sections: (i) DSP core applications and system setup, (ii) security analysis of Sengupta *et al.* (2017b), (iii) design cost analysis of Sengupta *et al.* (2017a), Rajendran *et al.* (2013) and Sengupta *et al.* (2017c) (iv) comparative study among them.

3.7.1 DSP applications and system setup for the case studies

Results of approaches HLS-ST (Rajendran *et al.*, 2013), LC-HLS (Sengupta *et al.*, 2017a), TL-HLS (Sengupta *et al.*, 2017b) and NL-TS (Sengupta *et al.*, 2017c) reported in the literature were based on the implementation carried out in Java 8 and executed on Windows platform with Intel Core i53210M CPU, 4GB DDR3 primary memory, and processor frequency of 3.20 GHz. The design parameters were evaluated with respect to 90 nm CMOS-based technology node. The design cost of Trojan secured DSP core is calculated based on (3.5) where w_1 and $w_2 = 0.5$. The design area is calculated based on (3.1), design latency for single-loop CDFG is calculated based on Eqn. 3.3 and nested-loop CDFG and non-loop CDFG is calculated based on Eqn. 3.4. During the PSO-DSE approach, each design solution is run 15 times to obtain low-cost design solution. Further, the optimal PSO control parameters are set as follows: $\omega =$ linearly decreasing between 0.9 and 0.1; $b1$ and $b2 = 2$; $r1$ and $r2 = 1$; terminating criterion $I_{max} = 100$ and $\delta = 10$ (Sengupta *et al.*, 2017b). (*Note*: all the variables are defined in Section 3.6.3).

3.7.2 Security analysis

Authors on TL-HLS (Sengupta *et al.*, 2017b) presented security analysis to show the robustness of their approach. The number of vulnerabilities detected by this approach is proportional to its robustness. In the design flow of an SoC, each 3PIP core used is considered as potentially untrustworthy. The number of 3PIP cores used in a design is directly proportional to the number of operations in the DMR design. In Sengupta *et al.* (2017b) such vulnerable 3PIP cores are detected using their devised concept. Figure 3.26 shows the numbers of vulnerable 3PIP cores that are detected through this approach for different non-loop CDFG, single-loop CDFG, and nested-loop CDFG benchmarks. As it can be observed from Figure 3.26, the maximum numbers of vulnerable 3PIP cores are detected for the DHMC nested-loop CDFG.

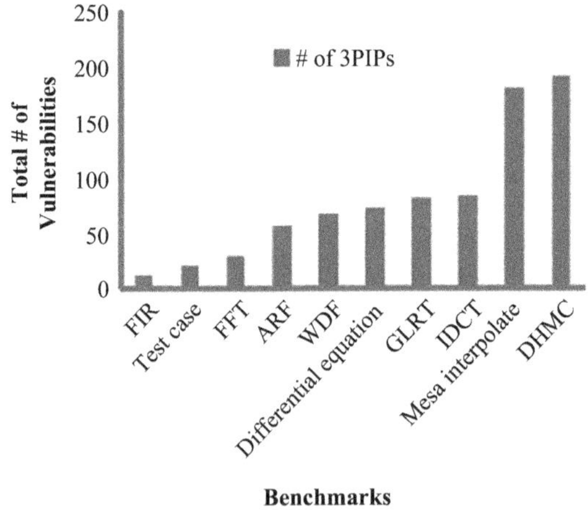

Figure 3.26 Total number of vulnerable 3PIPs detected through TL-HLS (Sengupta et al., 2017b)

3.7.3 Design cost analysis

3.7.3.1 Non-loop CDFG DSP core

The design cost of a non-loop Trojan secured DSP core for the LC-HLS (Sengupta *et al.*, 2017a) and HLS-ST (Rajendran *et al.*, 2013) approaches is reported in Table 3.2. In HLS-ST (Rajendran *et al.*, 2013), no low-cost design solution is achieved while designing the Trojan secured DSP core. Moreover, no optimal vendor allocation is performed during this approach. On the other hand, in LC-HLS (Sengupta *et al.*, 2017a), a low-cost design solution is explored through a BFOA-based DSE approach. Moreover, out of the two devised vendor allocation modes ($M_v = 0$ or 1), the optimal vendor allocation mode is also explored. For this reason, Sengupta *et al.* (2017a) achieve lower design cost than Rajendran *et al.* (2013). The design cost of less than zero indicates it satisfies the user-provided design parameter constraints such as area, delay, and power.

3.7.3.2 Single- and nested-loop CDFG DSP cores

The design cost of single- and nested-loop Trojan secured DSP cores for NL-TS (Sengupta *et al.*, 2017c) and HLS-ST (Rajendran *et al.*, 2013) approaches is reported in Table 3.3, where FIR, FFT, and differential equation applications are single-loop CDFGs and auto-correlation filter and DHMC applications are nested-loop CDFGs. As is evident from the table, because of lack of optimization framework, the design cost for Trojan secured DSP core of HLS-ST approach is higher than that of Sengupta *et al.* (2017c). In the NL-TS approach, low-cost design solution, the optimal loop unrolling factor, and optimal vendor allocation mode (out of the four devised security rules) are explored through PSO-DSE. Owing to the PSO-driven multi-dimensional optimization framework, NL-TS (Sengupta *et al.*, 2017c) achieves lower design cost than HLS-ST (Rajendran *et al.*, 2013). In Table 3.3, UF_1 indicates the optimal unrolling factor for the outermost loop

Table 3.2 Design cost analysis of LC-HLS (Sengupta et al., 2017a) and HLS-ST (Rajendran et al., 2013) for non-loop CDFG

Benchmark	Design architecture		Design cost	
	LC-HLS (Sengupta *et al.*, 2017a)	**HLS-ST (Rajendran *et al.*, 2013)**	**LC-HLS (Sengupta *et al.*, 2017a)**	**HLS-ST (Rajendran *et al.*, 2013)**
IIR Butterworth (second order)	2(+), 5(*), $M_v = 0$	2(+), 3(*), $M_v = 1$	−0.15	−0.04
FIR(6-tap)	6(+), 4(*), $M_v = 0$	5(+), 5(*), $M_v = 1$	−0.03	0.00
ARF	2(+), 4(*), $M_v = 0$	5(+), 3(*), $M_v = 1$	−0.23	−0.06
DCT	4(+), 2(*), $M_v = 0$	5(+), 3(*), $M_v = 1$	−0.21	−0.12
MPEG(motion vectors)	2(+), 10(*), $M_v = 0$	3(+), 8(*), $M_v = 1$	−0.24	−0.22

Table 3.3 Design cost analysis of NL-TS (Sengupta et al., 2017c) and HLS-ST (Rajendran et al., 2013) for loop-based DSP cores (CDFGs)

Benchmark	**Design architecture**		**Design cost**	
	NL-TS (Sengupta *et al.*, 2017c)	**HLS-ST (Rajendran *et al.*, 2013)**	**NL-TS (Sengupta *et al.*, 2017c)**	**HLS-ST (Rajendran *et al.*, 2013)**
Autocorrelation filter	2(+), 4(*), 2(<), $I_1 = 2$, $I_2=3$, $M_v = 00$	3(+), 5(*), 2(<), $UF_1 = 2$, $UF_2=3$, $A_v=01$	−0.134	0.096
DHMC	3(+), 4(*), 2(<), $I_1=3$, $I_2=3$, $M_v = 00$	3(+), 5(*), 2(<), $UF_1=3$, $UF_2=3$, $A_v=01$	−0.131	0.173
FIR (7th order)	4(+), 8(*), 2(<), $U=6$, $M_v=00$	3(+), 5(*), 2(<), U = 6, $A_v=01$	−0.175	−0.074
Differential equation	2(+), 2(−), 5(*), 2(<),$U=4$, $M_v = 00$	2(+), 2(−), 5(*), 2 (<), U = 4, $A_v=01$	−0.232	−0.193
FFT	4(+), 2(−), 4(*), 1(<),$U=2$, $M_v = 00$	5(+), 2(−), 5(*), 2 (<), U = 4, $A_v=01$	−0.185	−0.088

and UF_2 indicates the optimal unrolling factor for the innermost loop for auto-correlation filter and DHMC applications, while *U* indicates the explored optimal loop unrolling factor for FIR, FFT, and differential equation applications.

3.7.4 Comparative perspectives

We now highlight some state-of-the-art approaches by providing a comparative perspective on them. To present a comparative perspective, design latency parameter has been used by the authors. The comparative analysis between LC-HLS and HLS-ST is presented for non-loop DSP cores (DFGs) and highlighted in Figure 3.27. It was observed that for almost all the reported DSP/multimedia applications, LC-HLS achieves lesser design latency than HLS-ST does, owing to optimal hardware configuration and vendor allocation mode. The comparative perspective between LC-HLS and HLS-ST in terms of design latency for single- and nested-loop DSP cores (CDFGs) is also highlighted in Figure 3.28. It was observed by authors that, for all the reported benchmarks, the NL-TS approach achieves lesser design latency than HLS-ST does, except for the auto-correlation filter and DHMC applications. The underlying reason was found out to be that for both the benchmarks, the hardware resource configuration used in HLS-ST is higher than that in NL-TS. However, this results in higher design area of HLS-ST than that of NL-TS. Thus overall, the final design cost of both these benchmarks is better in NL-TS than in HLS-ST as reported in Table 3.3.

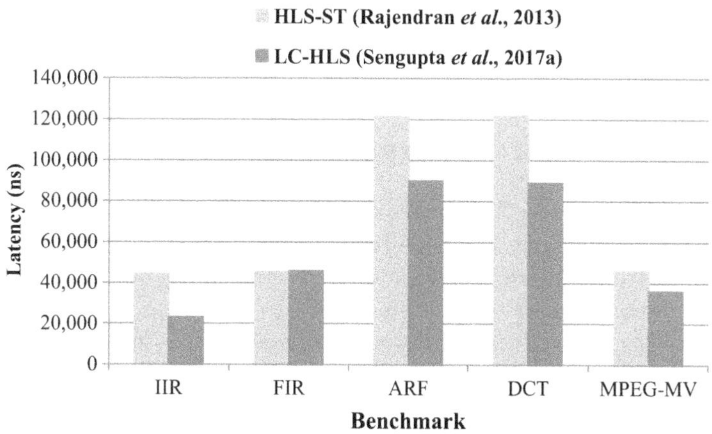

Figure 3.27 Comparison between HLS-ST and LC-HLS in terms of design latency

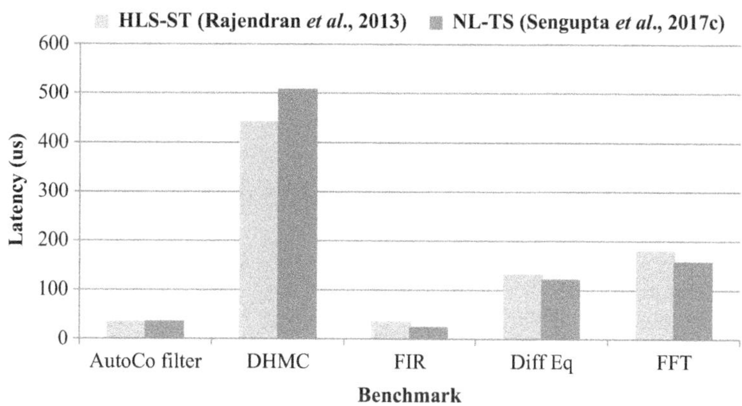

Figure 3.28 Comparison between HLS-ST and NL-TS in terms of design latency

3.8 Conclusion

This chapter presents multiple Trojan security approaches, with emphasis on DSP IP cores. The mechanism mostly used to design a Trojan secured DSP IP core is the CED- or DMR-based approach. Authors have mentioned different types of vendor allocation techniques to achieve that. Rajendran *et al.* (2013) devised collusion prevention in addition to CED, while Sengupta *et al.* (2017a) devised a BFOA-driven low-cost Trojan secured design for non-loop DSP cores. Further, Sengupta *et al.* (2017c) devised a Trojan secured design for nested-loop DSP cores and multiple vendor allocation rules to obtain a PSO-driven low-cost DSP IP core. Finally, it can be observed that both Sengupta *et al.* (2017a) and Sengupta *et al.* (2017c) obtained better design solution in terms of design cost.

3.9 Exercises

1. What is a hardware Trojan?
2. What are the types of Trojan?
3. What is the difference between payload and triggering?
4. What is a time bomb Trojan?
5. What is a rare event Trojan?
6. What is DMR in the context of Trojan?
7. How is Trojan security achieved for DSP cores?
8. What is the vendor allocation procedure?
9. How many types of vendor allocation procedures are available? How are they useful?
10. How do you evaluate the latency and area of a DMR design? Write out the functions and explain them.
11. How does loop unrolling play a role in area and latency evaluation of a Trojan secured design?
12. What is the particle encoding of Trojan secured solutions?
13. How does PSO ensure convergence?
14. What is the Trojan detection block in the case of DSP cores composed of?
15. What does PSO-DSE optimize? Name all factors and explain how the optimization is achieved.
16. What is a 3PIP?
17. Design Trojan infected modules in the HLS tool library for the BCD adder, decoder, and comparator.
18. Why are the tests used for detecting manufacturing faults not suitable for Trojan detection?
19. Derive the function for evaluating the delay of Trojan secured CDFGs for the following:
    (a) when the unrolling factor divides the iteration count evenly;
    (b) when the unrolling factor does not divide the iteration count evenly;
    (c) when there is no unrolling.
20. What is the function of velocity clamping?
21. Explain the pre-processing algorithm of loop unrolling factors.
22. How do you find the number of potential untrustworthy 3PIPs as vulnerabilities?
23. Calculate the delay of Trojan secured differential equation solver CDFG for Pv = 01. Given: 2(+), 4(*), 2(−), 2(<), $U = 3$.
24. Calculate the delay of Trojan secured differential equation solver CDFG for Pv = 10. Given: 2(+), 4(*), 2(−), 2(<), $U = 3$.
25. Calculate the delay of Trojan secured differential equation solver CDFG for Pv = 11. Given: 2(+), 4(*), 2(−), 2(<), $U = 3$.
26. Explain exploration-exploitation balance in case of PSO-DSE in Trojan security of DSP cores?
27. How do you evaluate the delay overhead for Trojan secured designs?

References

Agrawal D., Baktir S., Karakoyunlu D., Rohatgi P., and Sunar B. 'Trojan detection using IC fingerprinting', *IEEE Symposium on Security and Privacy*, 2007, pp. 296–310.

Banga M. and Hsiao M. S., 'A novel sustained vector technique for the detection of hardware Trojans,' in *Proceedings of 22nd International Conference on VLSI Design*, 2009, pp. 327–332.

Banga M. and Hsiao M. S., 'A region based approach for the identification of hardware Trojans', *IEEE International Workshop on Hardware-Oriented Security and Trust*, 2008, pp. 40–47.

Bhunia S., Abramovici M., Agrawal D., *et al.*, 'Protection against hardware Trojan attacks: Towards a comprehensive solution,' in *IEEE Design & Test*, 2013, vol. 30 (3), pp. 6–17.

Bushnell M. and Agrawal V., 'Essentials of electronic testing for digital, memory and mixed-signal VLSI circuits', 2013, New York, NY, USA: Springer.

Chakraborty R.S., Narasimhan S., and Bhunia S., 'Hardware Trojan: Threats and emerging solutions', *Proceedings of IEEE International High Level Design Validation Test Workshop*, 2009, pp. 166–171.

Chakraborty R. S., Wolff F., Paul S., Papachristou C., and Bhunia S., 'MERO: A statistical approach for hardware Trojan detection', *Proceedings of Workshop Cryptograph. Hardware Embedded Syst.*, 2009a, pp. 396–410.

Cui X., Ma K., Shi L., and Wu K., 'High-level synthesis for run-time hardware Trojan detection and recovery,' *51st ACM/EDAC/IEEE Design Automation Conference (DAC),* San Francisco, CA, 2014, pp. 1–6.

Hu A.J., 'Formal hardware verification with BDDs: An introduction,' in *Proceedings of IEEE Pacific Rim Conference on Communications Computing and Signal Processing*. 10 Years PACRIM 1987–1997 Netw. Pac. Rim, 1997, vol. 2. Victoria, BC, Canada, pp. 677–682.

Jin Y. and Makris Y., 'Hardware Trojan detection using path delay fingerprint', *IEEE International Workshop on Hardware-Oriented Security and Trust*, 2008, pp. 51–57.

Koushanfar F. and Mirhoseini A., 'A unified framework for multimodal submodular integrated circuits Trojan detection,' *IEEE Trans. Inf. Forensics Security*, 2011, vol. 6 (1), pp. 162–174.

Liu C., Rajendran J., Yang C., and Karri R., 'Shielding heterogeneous MPSoCs from untrustworthy 3PIPs through security-driven task scheduling,' *IEEE Transactions on Emerging Topics in Computing*, 2014, vol. 2 (4), pp. 461–472.

Narasimhan S., Du D., Chakraborty R.S., *et al.*, 'Multiple-parameter side-channel analysis: A noninvasive hardware Trojan detection approach,' in *IEEE International Workshop on Hardware-Oriented Security and Trust*, 2010, pp. 13–18.

Obukhov A. and Kharlamov A., 'Discrete cosine transform for 8×8 blocks with CUDA' Nvidia whitepaper document, 2008.

Rad R., Plusquellic J., and Tehranipoor M., 'A sensitivity analysis of power signal methods for detecting hardware Trojans under real process and environmental conditions,' *IEEE Transactions on Very Large Scale Integration (VLSI) Systems*, 2010, vol. 18 (12), pp. 1735–1744.

Rai D. and Lach J., 'Performance of delay-based Trojan detection techniques under parameter variations,' *IEEE International Workshop on Hardware-Oriented Security and Trust*, 2009, pp. 58–65.

Rajendran J., Zhang H., Sinanoglu O., and Karri R., 'High-level synthesis for security and trust,' *IEEE 19th International On-Line Testing Symposium (IOLTS)*, 2013, pp. 232–233.

Salivahanan S., Vallavaraj A., and Gnanapriya C., 'Digital signal processing' Tata McGraw-Hill, New Delhi, ISBN 007463996X , 2000, vol. xiii, pp. 808.

Salmani H., Tehranipoor M., and Plusquellic J., 'New design strategy for improving hardware Trojan detection and reducing Trojan activation time,' in *Proceedings of IEEE International Workshop Hardware Oriented Security Trust (HOST)*, 2009, pp. 66–73.

Sengupta A., 'Intellectual property cores: Protection designs for CE products,' *IEEE Consumer Electronics Magazine*, 2016, vol. 5 (1), pp. 83–88.

Sengupta A., 'Hardware vulnerabilities and their effects on CE devices: Design for security against Trojans,' in *IEEE Consumer Electronics Magazine*, 2017, vol. 6 (3), pp. 126–133.

Sengupta A. and Bhadauria S., 'Untrusted third party digital IP cores: Power-delay trade-off driven exploration of hardware Trojan secured datapath during high level synthesis' *In Proceedings of the 25th edition on Great Lakes Symposium on VLSI (GLSVLSI'15)*. ACM, New York, NY, USA, 2015, pp. 167–172.

Sengupta A. and Kundu S., 'Securing IoT hardware: Threat models and reliable, low-power design solutions' in *IEEE Transactions on VLSI Systems*, 2017, vol. 25 (12), pp. 3265–3267.

Sengupta A. and Sedaghat R., 'Integrated scheduling, allocation and binding in high level synthesis using multi structure genetic algorithm based design space exploration,' *International Symposium on. Quality Electronic Design*, 2011, pp. 1–9.

Sengupta A. and Sedaghat R., 'System and methodology for development of a system architecture using optimization parameters', *US Patent by United Sates Patent and Trademark Office (USPTO)*, 2013, Patent no. US 8,397, 204.

Sengupta A., Bhadauria S., and Mohanty S. P., 'Low-cost security aware HLS methodology,' in *IET Computers & Digital Techniques*, 2017a, vol. 11 (2), pp. 68–79.

Sengupta A., Bhadauria S., and Mohanty S. P., 'TL-HLS: Methodology for low cost hardware Trojan security aware scheduling with optimal loop unrolling factor during high level synthesis,' *IEEE Transactions on CAD of Integrated Circuits and Systems*, 2017b, vol. 36 (4), pp. 655–668.

Sengupta A., Roy D., and Bhadauria S., 'Low cost optimized Trojan secured schedule at behavioral level for single & nested loop control data flow graphs', *Integration, the VLSI Journal*, 2017c, vol. 58, pp. 378–389.

Sengupta A., Roy D., Mohanty S. P., and Corcoran P., 'A framework for hardware efficient reusable IP core for grayscale image CODEC,' in *IEEE Access*, 2018a, vol. 6, pp. 871–882.

Sengupta A., Roy D., Mohanty S. P., and Corcoran P., 'Low-cost obfuscated JPEG CODEC IP core for secure CE hardware,' in *IEEE Transactions on Consumer Electronics*, 2018b, vol. 64 (3), pp. 365–374.

Sinanoglu O., Karimi N., Rajendran J., *et al.*, 'Reconciling the IC test and security dichotomy', *18th IEEE European Test Symposium (ETS)*, 2013, pp. 176–181.

Tehranipoor M. and Koushanfar F., 'A survey of hardware Trojan taxonomy and detection', *IEEE Design & Test of Computers*, 2010, vol. 27 (1), pp. 10–25.

Wang X., Salmani H., Tehranipoor M., and Plusquellic J., 'Hardware Trojan detection and isolation using current integration and localized current analysis,' in *Proceedings of IEEE International Symposium on Defect Fault Tolerance VLSI Syst. (DFTVS)*, 2008, pp. 87–95.

Wolff F., Papachristou C., Bhunia S., and Chakraborty R. S., 'Towards Trojan-free trusted ICs: Problem analysis and detection scheme', Design, Automation and Test in Europe, 2008, pp. 1362–1365.

Chapter 4

IP core and integrated circuit protection using robust watermarking

This chapter discusses robust watermarking approaches for the ownership protection of hardware cores [a.k.a. intellectual property (IP) cores]. Watermarking ensures some additional attributes inserted in the hardware core such a way that it can be used to verify ownership of the hardware core when required. Robust watermarking approaches have been discussed as these are resilient to various attacks that happen in global supply chain for various reasons. The remaining chapter is structured as follows: Section 4.2 provides an overview on selected hardware/IP core watermarking approaches; Section 4.3 discusses design process of watermarked IP core; Section 4.4 analyses the hardware watermarking approaches on case studies/test cases; and finally, Section 4.5 draws the conclusion.

4.1 Introduction

A watermark is a secret mark implanted into an entity such as official documents, currency notes, postal stamps, audio/video files, etc., to protect ownership right of an owner. Usually, a watermark embedded into an entity is expected to preserve its quality and functionality. However, for the domains in which watermarking is applied such as audio/video, documents, etc., slight degradation of the quality is usually noticed. Given the advantages of a watermark, it may be considered an extremely useful tool for protecting hardware [e.g., digital signal processor (DSP) core] ownership as well. However, the prime difference from watermarking in other domains is zero tolerance toward modification of quality/functionality of the hardware design. This makes hardware watermarking challenging but extremely useful for protecting legal rights of an IP vendor/owner. It is widely acknowledged that complex consumer electronics (CE) system design relies heavily on DSP IP cores realized as system-on-chip (SoC). Thus, hardware watermarking plays an integral role in IP core protection of CE devices (Mohanty *et al.*, 2017; Voyatzis and Pitas, 1999; Cox *et al.*, 2006; Roy *et al.*, 2013; Sengupta, 2017; Sengupta and Sedaghat, 2011; Sengupta *et al.*, 2013).

In the modern era of CE, use of DSP IP cores in global supply chains become an inexorable part of complex SoC design process. IP cores not only speedup the design productivity massively but also decreases the design period immensely.

Importing these IP cores from third-party IP vendors by the system integrator or SoC designer has become a common industry practice. Previously, IP vendors have mainly focused on IP performance and IP functionality but have neglected IP security. As evidence it can be observed that in a typical IP design flow, measurement of performance and functionality is only included in the IP specification. However, prevailing usage of IP cores in SoC design process increases the rivalry between multiple IP core vendors, which invites security threats like IP core forgery, dishonest claim of ownership, etc., which can be tackled by hardware watermarking. Thus, the threat model can be defined as:

Threat model: The targeted threat models for watermark-based IP core protection mechanism are: (i) false claim of ownership, (ii) ownership conflict, and (iii) IP piracy/forgery in IP cores.

The designing process of an IP core involves many man-hours of research, investment, and substantiation; therefore, this effort must be protected. As IP core can be copied easily, IP piracy and ownership abuse is a rising security concern now. Hence, protecting the ownership of an original IP owner is very crucial. For example, after importing an IP core from a third-party IP vendor, a SoC integrator may claim himself as the original owner of the IP or may reuse the IP core without acknowledging the original IP owner. Thus, nullifying false claim of ownership or detecting the illegal misuse of an IP core is essential and complex for an original IP owner. One popular way to safeguard an IP core against aforementioned potential threats is embedding vendor signature in the IP core during its design process in the vendor house itself (Alkabani and Koushanfar, 2008; Koushanfar *et al.*, 2005; Sengupta, 2016; Guajardo *et al.*, 2008; Sengupta *et al.*, 2017; Sengupta and Roy, 2017).

4.2 Selected watermarking approaches

This section discusses few selected watermarking mechanisms for IP core protection. Ziener and Teich (2008) devised an IP core watermarking methodology for field programmable gate array (FPGA) architecture. The IP owner signature as a watermark is embedded in the netlist and the bit-stream file of the IP core. The watermarking algorithm embeds the signature into the functional part of the IP core in such a way that it becomes challenging to remove the watermark without destroying the IP core. The signature is identified from the power supply pins of the watermarked FPGA. More specifically, the authors have devised a signature detection algorithm that decodes the watermark through voltage trace with high reliability. Additionally, two robustness enhancement algorithms are also devised so that the signature can be decoded correctly in the presence of significant noise. The overview of Ziener and Teich (2008) approach is shown in Figure 4.1. Two in-synthesis IP core watermarking methodology: a low-cost watermark and a costless watermark are possible (Le Gal and Bossuet, 2012). The vendor signature in the form of a watermark is embedded based on the 'temporally free' IP output slot. In low-cost watermarking approach, the watermark is characterized by a set of arbitrarily selected internal computation values of the IP which get transferred through a 'temporally free'

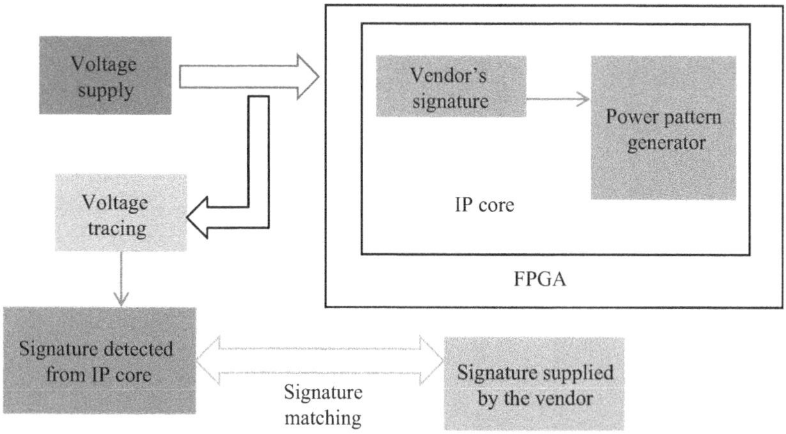

Figure 4.1 Top level diagram of watermarking process of Ziener and Teich (2008)

output slot. In costless watermarking approach, the watermark uses dynamic transient outputs unit to drive selected data to the output ports when the slots are temporarily free which can be achieved by modifying the IP control. As high level synthesis (HLS) tools design a control unit using a finite state machine, therefore, the costless watermarking approach only influences the finite state machine of an IP core. The desired mathematic relations used to generate watermarked IP core can be validated using equality-checking tests of the 'temporally free' IP output slots. A watermark can be represented as (Le Gal and Bossuet, 2012):

$$\text{Watermark} = \{(M_1(n), C_1), \ldots, (M_p(n), C_p)\}, \tag{4.1}$$

where $M_i(n)$ indicates ith sub-mark or mathematical function and C_i indicates the clock cycle number during which the sub-mark is sent to a 'temporally free' IP output port. Different DSP applications are used as the benchmark to embed this watermarking approach and the result is compared with the approach (Koushanfar *et al.*, 2005) in terms of area and timing overhead. The overview of Le Gal and Bossuet (2012) approach is shown in Figure 4.2.

Koushanfar *et al.* (2005) devised a dynamic watermarking process to shield a DSP IP core (Coussy *et al.*, 2009). A set of design and timing constraints in the form of vendor watermark is embedded in the IP core. In this approach, a vendor's signature is transformed into its equivalent binary form where each digit (0 or 1) indicates a watermarking constraint. More specifically, these watermarking constraints are embedded in register allocation and binding phase. This is achieved by adding additional edges in the equivalent colored interval graph (CIG) of the IP design. In a CIG, each node indicates a storage variable. The lifetime of a storage variable lies between the time it is created (value written into it) and the last use (value read from it). An edge between two nodes indicates there is an overlap in

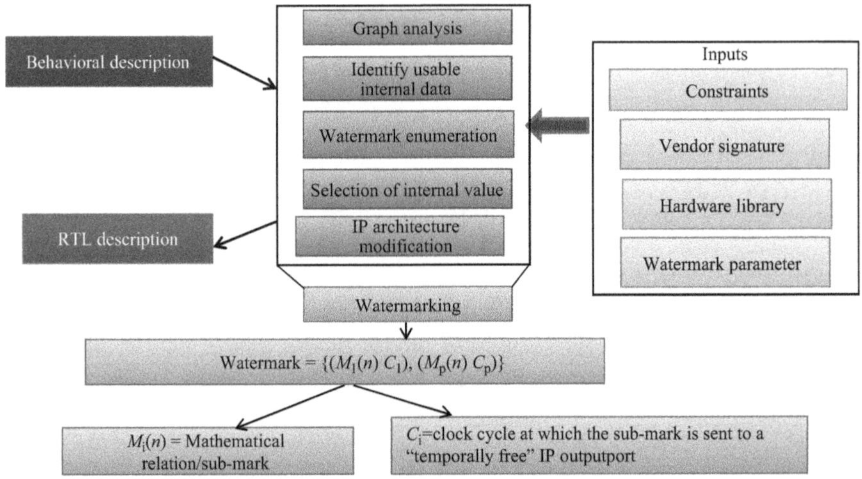

Figure 4.2 Top level diagram of watermarking process of Le Gal and Bossuet (2012)

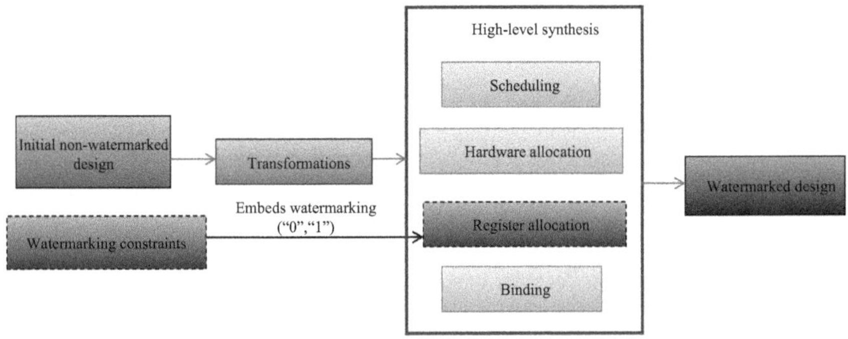

Figure 4.3 Top level diagram of watermarking process of Koushanfar et al. (2005)

lifetime between those storage variables. The color of the node indicates the allocated register of that corresponding storage variable. Therefore, two storage variables whose lifetimes overlap cannot be stored in the same register. In other words, two nodes having a common edge between them cannot have same node color. Thus, adding additional edges in the CIG as watermarking constraints will force to execute two storage variables in two different registers. The overview of Koushanfar *et al.* (2005) approach is shown in Figure 4.3.

Unlike dual variable (0 and 1) signature encoding rule in Koushanfar *et al.* (2005), Sengupta and Bhadauria (2016) devised a single phase multi-variable (combination of four) watermarking methodology to protect a reusable IP core. Each variable which carries a unique encoding rule (unknown to an attacker) adds

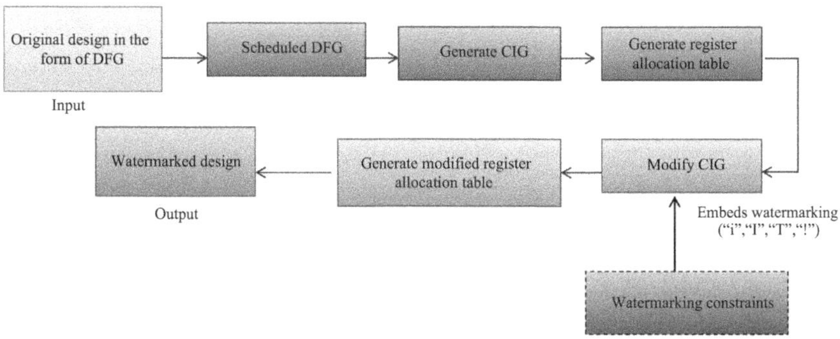

Figure 4.4 Top level diagram of watermarking process of Sengupta et al. (2016)

an additional edge between two node pair in the CIG. Thus, Sengupta *et al.* (2016) embed a more robust and tamper-proof watermark compared with Koushanfar *et al.* (2005). Due to insertion of watermarking constraints in both Sengupta *et al.* (2016) and Koushanfar *et al.* (2005) area overhead may occur. Therefore, to minimize the design cost, a low-cost design solution is explored through particle swarm optimization in Sengupta *et al.* (2016). The embedded signature can be detected by inspecting the datapath hardware description language (HDL) file of the watermarked IP. As the encoding rules are hidden to an attacker, it is almost impossible to identify the watermarking constraints from its datapath. The overview of Sengupta *et al.* (2016) approach is shown in Figure 4.4.

Sengupta *et al.* (2018) devised a triple-phase watermarking process to secure a DSP IP core (Sengupta and Sedaghat, 2011). In this approach, vendor signature consists of 7-variables. The IP owner signature as watermarking constraints is embedded in three dissimilar (but interconnected) phases of HLS, i.e., scheduling phase, hardware allocation phase, and register allocation phase. Due to insertion of watermarking constraints in three different phases, vendor signature is distributed throughout the IP core design. Thus, complete removal or malicious alteration of the watermark is extremely difficult. Besides, embedding 7-variable encoding makes this approach tremendously strong. Additionally, the watermark in scheduling phase is independent of the watermark in hardware allocation and register allocation phase. Similarly, the watermark in register allocation phase is independent of both scheduling and hardware allocation phase watermarks. Due to these benefits, the triple-phase watermarking approach is tremendously robust as well as tamper resistant. The embedded signature can be detected by inspecting the datapath and controller HDL file of the watermarked IP. As the watermarking constraints are well dispersed in the IP design and the encoding rules are hidden to an attacker, it is extremely difficult to identify the watermarking constraints from its datapath and controller by an adversary. The overview of Sengupta *et al.* (2018) approach is shown in Figure 4.5.

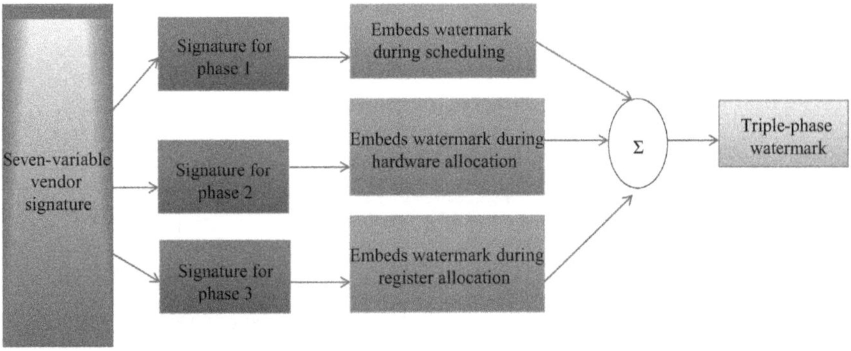

Figure 4.5 Top level diagram of watermarking process of Sengupta et al. (2018)

4.3 Design process of watermarked IP core/hardware

As mentioned earlier, IP/hardware watermarking is a process of embedding owner's signature covertly in the design of an IP/hardware to proof his/her ownership in case of any conflict (Sengupta *et al.*, 2017). In this section, we discuss problem formulation and design process of two (i.e., single-phase and triple-phase) IP core watermarking processes in detail. Both the approaches are demonstrated through a motivational example.

4.3.1 Problem formulation

Inputs: (i) DSP application, (ii) resource configuration, (iii) module library, and (iv) vendor signature.

Output: Watermarked IP core design.

Sengupta *et al.* (2016, 2018) accepted DSP application in the form of data flow graph (DFG), module library in the form of NanGate 15-nm technology (NanGate Open Cell Library, 2018) or 90-nm library/other library, and user-specified resource configuration (R). Here, R is represented as: $N(X_1)$, $N(X_2)$, …, $N(X_D)$, and $N(X_D)$ indicating the number of resources of type X_D. Sengupta *et al.* (2016) proposed a vendor signature that contains four encoding variables used for generating watermarking constraints. These watermarking constraints are embedded in a single HLS phase of a DSP IP core design process. On the other hand, Sengupta *et al.* (2018) proposed a vendor signature that contains seven encoding variables. These watermarking constraints are embedded in three different HLS design phases of a DSP IP core design process. Figure 4.6 depicts the inputs and output of an IP core watermarking process.

Target platform: The targeted platform for Sengupta *et al.* (2016, 2018) watermarking approach is any modern computer-aided design (CAD) tool. Both watermarking approaches generate the final watermarked DSP IP core design in

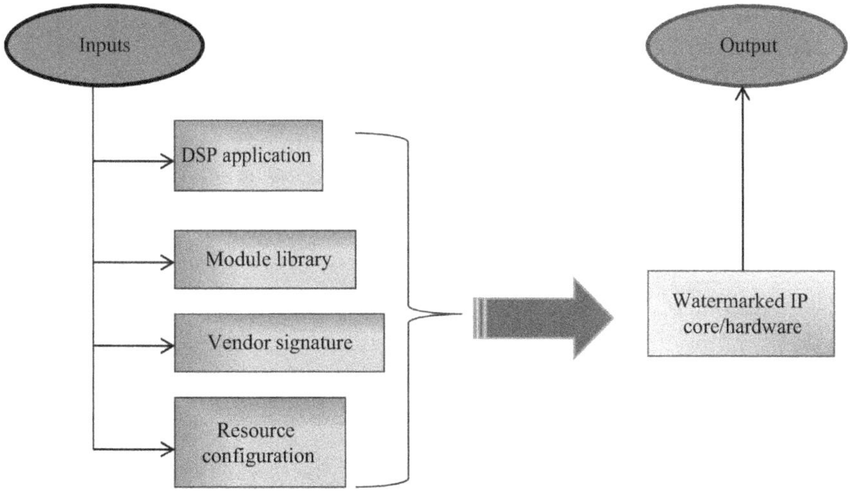

Figure 4.6 Inputs and output for watermarked IP core

HDL; therefore, any HDL-based CAD tool used for generating IP core can easily adopt these watermarking algorithms.

4.3.2 Design process of single-phase watermarked IP core/hardware

In this section, we describe the design of a watermarked DSP IP core/hardware (especially for DSP/multimedia applications).

4.3.2.1 Process of generating DFG from a DSP function

In this section, as an example, we discuss how to obtain DFG of a digital signal processing filter from its corresponding transfer function. We demonstrate the process through a popular DSP application, i.e., finite impulse response (FIR) filter. The generic equation of FIR filter can be expressed as follows:

$$y(n) = \sum_{i=0}^{M} h_i \times x(n-i), \tag{4.2}$$

where M is the filter order of a digital FIR, h_i is the FIR coefficient, $x(n)$ is input impulse, and $y(n)$ is output impulse.

Using the above expression, a 7th order (8-tap) FIR filter can be represented as:

$$y(7) = h_0 \times x(7) + h_1 \times x(6) + h_2 \times x(5) + h_3 \times x(4) + h_4 \times x(3) + h_5 \times x(2) + h_6 \times x(1) + h_7 \times x(0). \tag{4.3}$$

The equation of 7th order FIR filter consists of eight multiplication operations and seven addition operations. The corresponding block diagram is shown in Figure 4.7. The equivalent DFG of 7th order FIR filter is shown in Figure 4.8, where primary inputs are shown in gray boxes, light nodes indicate multiplication operation, and dark nodes indicate addition operations, corresponding operation/node number are mentioned using integer value (1–15) and finally the output is shown in a square box.

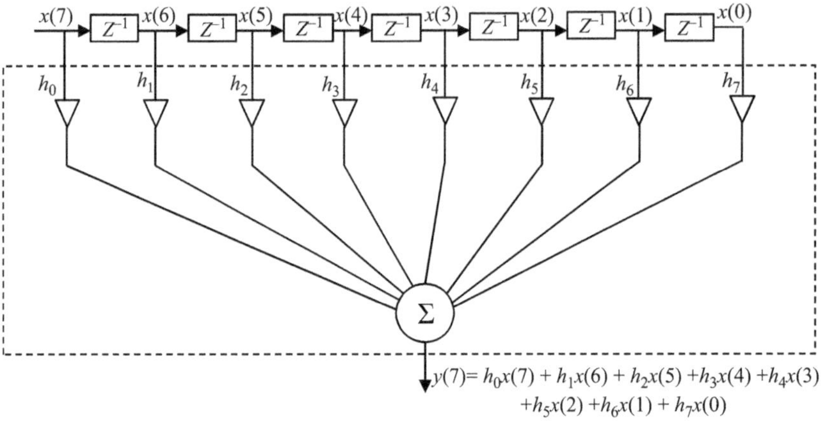

Figure 4.7 Block diagram of a 7th order FIR filter

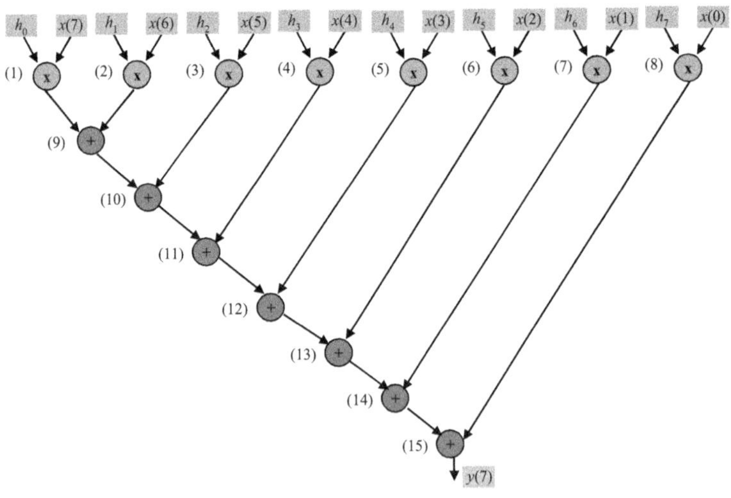

Figure 4.8 Data flow graph of a 7th order FIR filter

Now, we discuss how to obtain a DFG of a digital infinite impulse response (IIR) Butterworth filter that satisfies predefined constraints. Let us assume the given problem is as follow (Salivahanan *et al.*, 2000):

Problem: Design a digital IIR Butterworth filter that satisfies the following constraints using bilinear transformation. Assume $T = 1$ s.

$$\begin{aligned} &0.9 < |H(e^{j\omega})| \le 1, && \text{for } 0 \le \omega \le \pi/2 \\ &|H(e^{j\omega})| \le 0.2, && \text{for } 3\pi/4 \le \omega \le \pi \end{aligned}, \tag{4.4}$$

Solution: Given $\delta_1 = 0.9$, $\delta_2 = 0.2$, $\omega_1 = \pi/2$, and $\omega_2 = 3\pi/4$, where δ_1 and δ_2 are the parameters specifying *allowable pass-band* and *stop-band*, respectively.

Step 1: Determination of *edge frequencies* of analog filter:

$$\Omega_1 = 2/T\,\tan(\omega_1/2) = 2/T\,\tan(\pi/4) = 2;$$
$$\Omega_2 = 2/T\,\tan(\omega_2/2) = 2/T\,\tan(3\pi/8) = 4.828.$$

Therefore, $\Omega_2/\Omega_1 = 2.414$.

Step 2: Determination of order of the filter:

Order of a filter (N) can be determined using the following equation (Salivahanan *et al.*, 2000):

$$N \ge \frac{1}{2}\frac{\log((1/\delta_2^2) - 1)/(1/\delta_1^2) - 1))}{\log(\Omega_2/\Omega_1)}. \tag{4.5}$$

Using the above expression, we can obtain the following:

$$N \ge ½\log(24/0.2346)/\log(2.414) = 2.626.$$

Thus, order $N = 3$.

Step 3: Determination of –3 dB cut-off frequency:

Cut-off frequency (Ω_c) can be calculated using following equation (Salivahanan *et al.*, 2000):

$$\Omega_c = \frac{\Omega_1}{\left[(1/\delta_1^2 - 1)\right]^{1/2N}}. \tag{4.6}$$

Using the above expression, we can obtain:

$$\Omega_c = 2/\left[(1/0.9^2) - 1\right]^{1/6} = 2.5467.$$

Step 4: Determination of transfer function ($H(s)$):

As N is odd, transfer function of IIR Butterworth can be derived using following equation (Salivahanan *et al.*, 2000):

$$H(s) = \frac{B_0\Omega_c}{s + c_0\Omega_c} \prod_{k=1}^{(N-1)/2} \frac{B_k\Omega_c^2}{s^2 + b_k\Omega_c s + c_k\Omega_c^2} \tag{4.7}$$

Therefore, the *transfer function* of the current filter is:

$$H(s) = \left(\frac{B_0\Omega_c}{s + c_0\Omega_c}\right)\left(\frac{B_1\Omega_c^2}{s^2 + b_1\Omega_c s + c_1\Omega_c^2}\right), \tag{4.8}$$

where $b_1 = 2\sin(\pi/6) = 1$, $c_0 = 1$ and $c_1 = 1$, $B_0B_1 = 1$, therefore, $B_0 = B_1 = 1$.

Therefore, $H(s) = \left(\frac{2.5467}{s+2.5467}\right)\left(\frac{6.4857}{s^2+2.5467\,s+6.4857}\right)$

Step 5: Determination of equivalent *system function* ($H(z)$) using bilinear transformation:

For bilinear transformation, $s = \frac{2}{T}\left(\frac{z-1}{z+1}\right)$.

Therefore, $H(z) = \left(\frac{2.5467}{s((z-1)/(z+1))+2.5467}\right)\left(\frac{6.4857}{[2((z-1)/(z+1))]+5.0934((z-1)(z+1))+6.4857}\right)$,

$$H(z) = \left(\frac{16.5171(z+1)^3}{70.83z^3 + 31.1205z^2 + 27.2351z + 2.948}\right)$$

$$= \left(\frac{16.5171z^3 + 49.5513z^2 + 49.5513z + 16.5171}{70.83z^3 + 31.1205z^2 + 27.2351z + 2.948}\right),$$

$$H(z) = \frac{Y(z)}{X(z)}\left(\frac{16.5171z^3 + 49.5513z^2 + 49.5513z + 16.5171}{70.83z^3 + 31.1205z^2 + 27.2351z + 2.948}\right)$$

$$= \left(\frac{0.2332 + 0.4664z^{-1} + 0.4664z^{-2} + 0.2332z^{-3}}{1 + 0.4394z^{-1} + 0.3845z^{-2} + 0.0416z^{-3}}\right).$$

$$\begin{aligned} y(n) = {} & 0.2332x(n) + 0.4664x(n-1) + 0.4664x(n-2) + 0.2332x(n-3) \\ & - 0.4394y(n-1) - 0.3845y(n-2) - 0.0416y(n-3). \end{aligned} \tag{4.9}$$

In the above expression, $H(z)$ represents the transfer function of the filter in frequency domain. We can convert the transfer function from frequency domain to time domain. Therefore, the final transfer function of the filter in time domain is shown in Eqn. (4.9). The equation of the filter consists of seven multiplication operations, three addition operations, and three subtraction operations; $x(n)$, $x(n-1)$, $x(n-2)$, and $x(n-3)$ represent the current and part input impulse of the filter in time domain; $y(n)$, $y(n-1)$, $y(n-2)$, and $y(n-3)$ represent the current and past output

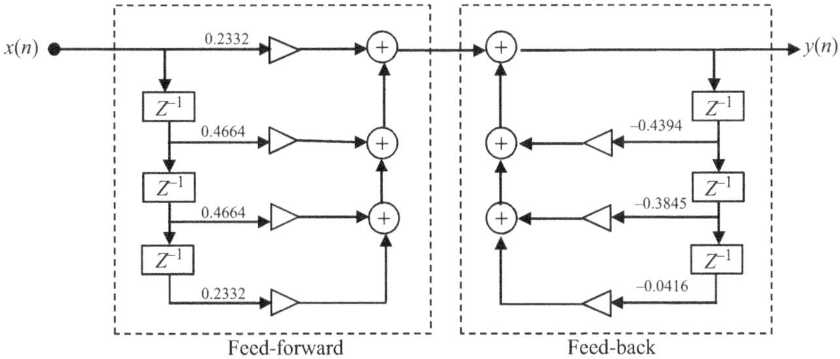

Figure 4.9 *Block diagram of a 3rd order IIR digital Butterworth filter*

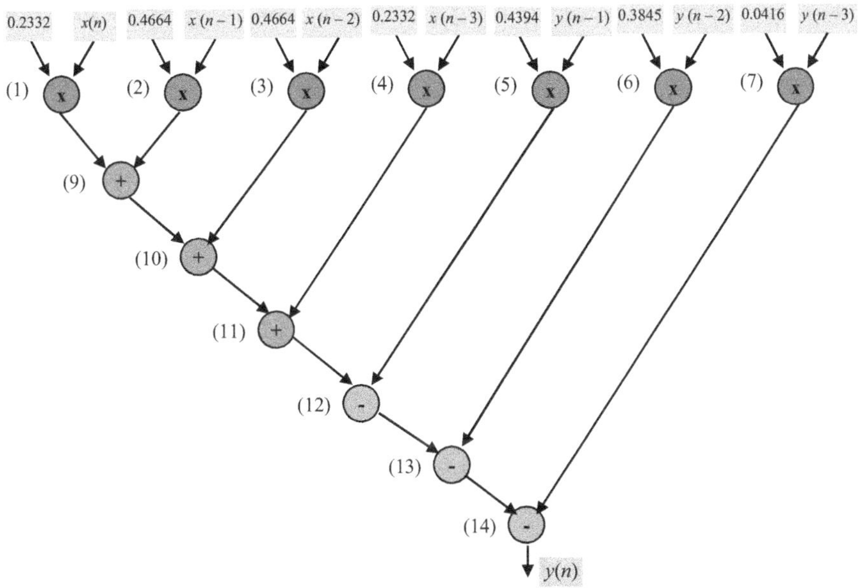

Figure 4.10 *Data flow graph of an IIR Butterworth filter*

impulse of the filter in time domain. The equivalent block diagram and its corresponding DFG are shown in Figures 4.9 and 4.10, respectively.

4.3.2.2 Watermarking process for DSP IP core

As an example, we have used single-phase watermarking approach presented in Sengupta *et al.* (2016) to demonstrate the design process. Watermark in DSP IP core can be embedded as additional design constraints in one of the HLS design phases. Sengupta *et al.* (2016) embed watermarking constraints in the register allocation phase of HLS. To embed additional constraints, a register allocation table is generated using the aid of CIG. In CIG, the nodes indicates the storage variables of the scheduled design, the edges indicates existence of overlapping of lifetime between two storage variables, and the color of the node indicates allocated register of that corresponding node. Therefore, if there is an edge between two storage variables (nodes) then the same register cannot be assigned between them. In other words, if there is an edge between two nodes, then the color of them must be different.

In Sengupta *et al.* (2016) and Koushanfar *et al.* (2005), watermarking constraints are added as supplementary edges between the nodes of a CIG. The color of the nodes are adjusted after embedding additional edges and the register allocation table of an IP core is generated from the final modified CIG. More specifically, inserting each edge as watermarking constraints, force two storage variables of a CIG to execute through distinct registers. Thus, embedding additional edges in a CIG affect the allocation of the register of a design. However, these additional edges should be derived from a vendor's hidden signature. This mandates designer to design a strong signature encoding process. Sengupta *et al.* (2016) devised a multi-variable signature encoding scheme to embeds a robust watermark. Compared with the authors in Koushanfar *et al.* (2005), who have used two variables signature encoding scheme, a more robust and stronger signature encoding scheme is devised in Sengupta *et al.* (2016). The signature encoding scheme of Sengupta *et al.* (2016) is discussed in the rest of this section.

4.3.2.3 Signature encoding scheme

A four variables signature encoding scheme is devised in Sengupta *et al.* (2016) where each variable indicates an encoded watermarking constraint. More specifically, each decoded signature digit inserts an additional edge between two nodes of a CIG as a watermarking constraint. An IP core owner can select his desired signature using any random combination of four variables, i.e., 'i', 'I', 'T', and '!'. The encoded rule of each variable is as follows:

- 'i' = Add an additional edge between (prime, prime) node pair of CIG.
- 'I' = Add an additional edge between (even, even) node pair of CIG.
- 'T' = Add an additional edge between (odd, even) node pair of CIG.
- '!' = Add an additional edge between (0, any integer) node pair of CIG.

In Sengupta *et al.* (2016), as IP sellers signature is embedded based on multi-variable signature encoding scheme and distributed throughout the DSP design, it is harder for an adversary to detect and remove the complete watermark. Thus, provides significant robustness and stronger encoding.

4.3.2.4 Watermark embedding process for DSP IP core

In the previous sub-section, we have discussed the signature encoding rule using four variables. Here, we discuss how to embed an IP seller's signature into an original (non-watermarked) IP design. As mentioned earlier, all the aforesaid encoding variables embed watermarking constraints in the register allocation phase of HLS; the phases of HLS appearing before register allocation phase, is described as the pre-processing steps of watermark. Therefore, watermark embedding process is divided into two parts, viz., pre-processing steps and watermark embedding steps. Detailed steps of the complete watermarking process (Sengupta *et al.*, 2016) are as follows:

A. **Pre-processing steps**
   1. Accept an original/non-watermarked DSP application in the form of CDFG with resource configuration as input.
   2. Schedule the CDFG of DSP application based on the resource configuration (constraint).
   3. Assign storage variable for each operation of the scheduled CDFG.
   4. Construct the CIG to identify the minimal registers essential for register allocation.
   5. Generate the register allocation table from the constructed CIG.
   6. Sort storage variables in ascending order according to their number.
   7. Accept an *n*-digit vendor's signature as a combination of 'i', 'I', 'T', and '!' variables.

B. **Watermark embedding steps**
   1. Extract the decoded meaning of each signature digit based on encoded rule of each variable.
   2. Construct a list of additional edge from the decoded value of vendor's signature by traversing the sorted nodes.
   3. Insert those additional edges in the CIG (constructed in pre-processing step 4) one by one as watermarking constraints and adjust the CIG if required.
   4. After embedding all the additional edges in the CIG, modify the register allocation table (constructed in pre-processing step 5) based on the updated CIG.

After following the above steps, IP owner's watermark gets embedded into the DSP IP core. The watermarking constraints are embedded into the register allocation phase of HLS. Each constraint forces two storage variables to be executed through a distinctive register. Thus, manipulates the register sharing which creates an impact on the final datapath of the IP core and its corresponding gate-level structure (netlist). Therefore, embedding this watermark retains the correctness and functionality of a DSP IP core.

4.3.2.5 Didactic example of single-phase watermarking process

In this sub-section, we demonstrate end-to-end single-phase watermark embedding process through a representative example. The CDFG of FIR-2 filter adopted from DSP Benchmark Suite (2018) shown in Figure 4.12.

Figure 4.11 is taken as input (according to step A.1). The accepted CDFG is then scheduled based on two adders and two multipliers (according to step A.2). Storage variables (V0–V30) are assigned to each operation of the scheduled DFG (according to step A.3). A CIG is constructed (according to step A.4) with

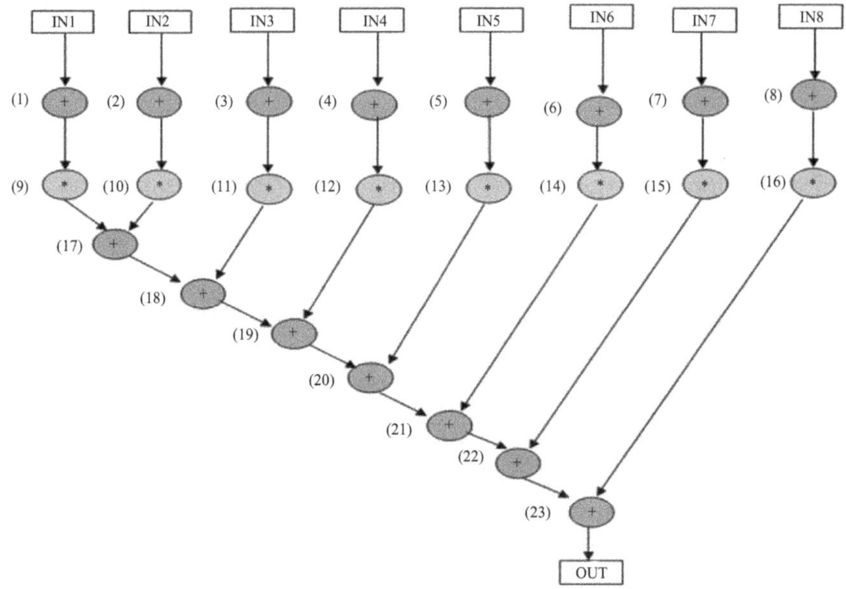

Figure 4.11 Control data flow graph of FIR-2 benchmark

31 nodes, i.e., V0–V30 and eight registers, i.e., register purple (Reg_P), register dark_blue (Reg_DB), register light_blue (Reg_LB), register green (Reg_G), register yellow (Reg_Y), register red (Reg_R), register orange (Reg_O), and register black (Reg_B). The corresponding CIG is shown in Figure 4.12. The scheduled CDFG after performing storage variable assignment and register allocation is shown in Figure 4.13. The number of control steps (cs) required to schedule the FIR-2 CDFG is 9. The register allocation table corresponding to the constructed CIG (shown in Figure 4.12) is reported in Table 4.1 (according to step A.5). All the storage variables (V0–V30) are sorted in ascending order according to variable number and a list is generated (according to step A.6). As an example, a 14-digits vendor signature: **'i I I T I TT I IT ! TTT'** is selected which is a combination of four aforementioned signature variables (according to step A.7). All the pre-processing steps to embed single-phase watermark is performed now. The next step is to decode the 14-digits signature, based on aforementioned encoding rules (according to step B.1). The vendor's signatures and its corresponding decoded meaning are shown in Table 4.2. According to the decoded signature shown in Table 4.2, the list of additional edges are between nodes (V2 and V3), (V2 and V4), (V2 and V6), (V1 and V2), (V2 and V8), (V1 and V4), (V1 and V6), (V2 and V10), (V2 and V12), (V1 and V8), (V0 and V1), (V1 and V10), (V1 and V12), and (V1 and V14) (according to step B.2). Although according to the vendor's signature, there are 14 additional edges to be added in the CIG, however, by default eight edges (i.e., edge between nodes (V2 and V3), (V2 and V4), (V2 and V6), (V1 and V2), (V2 and V8), (V1 and V4), (V1 and V6), and (V0 and V1)) are already present

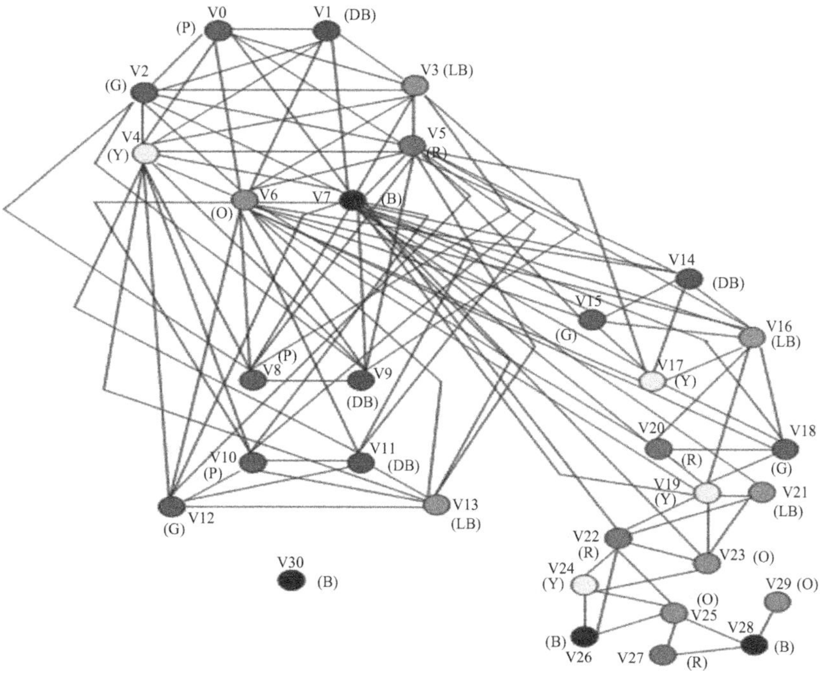

Figure 4.12 Colored interval graph of FIR-2 before embedding watermarking constraints

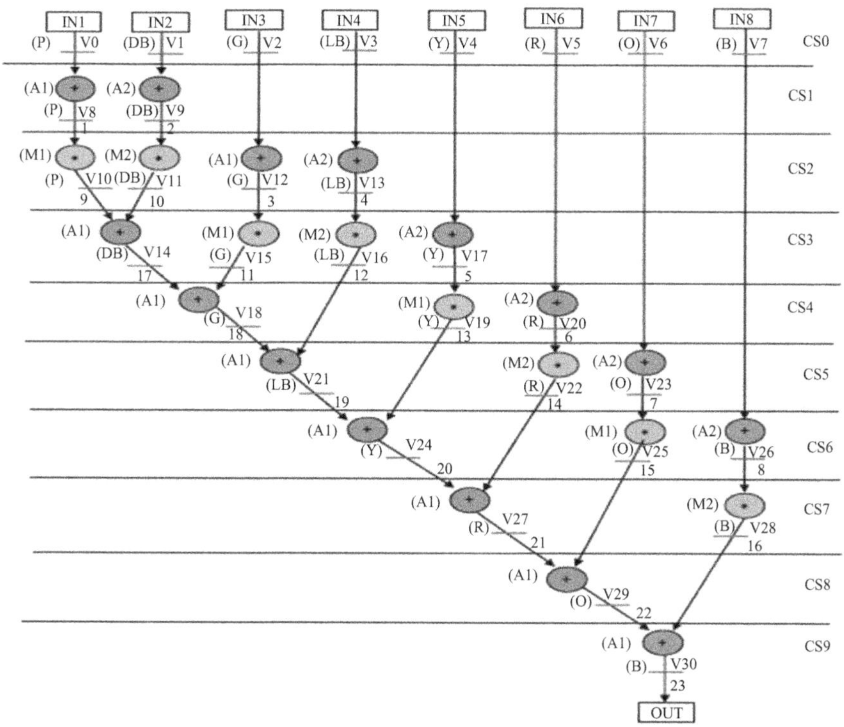

Figure 4.13 Scheduled CDFG of FIR-2 using two adders and two multipliers

Table 4.1 Register allocation table of FIR-2 before embedding watermarking constraints

Control step	Reg_P	Reg_DB	Reg_G	Reg_LB	Reg_Y	Reg_R	Reg_O	Reg_B
0	V0	V1	V2	V3	V4	V5	V6	V7
1	V8	V9	V2	V3	V4	V5	V6	V7
2	V10	V11	V12	V13	V4	V5	V6	V7
3	–	V14	V15	V16	V17	V5	V6	V7
4	–	–	V18	V16	V19	V20	V6	V7
5	–	–	–	V21	V19	V22	V23	V7
6	–	–	–	–	V24	V22	V25	V26
7	–	–	–	–	–	V27	V25	V28
8	–	–	–	–	–	–	V29	V28
9	–	–	–	–	–	–	–	V30

Table 4.2 Vendor signature and its decoded meaning

Desired Signature	Corresponding additional edges
i	2.3
I	2.4
I	2.6
T	1.2
I	2.8
T	1.4
T	1.6
I	2.10
I	2.12
T	1.8
!	0.1
T	1.10
T	1.12
T	1.14

in the graph. Hence, the constraints for these eight constraints are satisfied by default. Therefore, the objective is to satisfy the other six remaining constraints (i.e., edge between nodes (V2 and V10), (V2 and V12), (V1 and V8), (V1 and V10), (V1 and V12), and (V1 and V14)) in the IP core design. It can be observed that edge between node (V2 and V12) forces storage variable V2 and V12 to be executed through distinct register, therefore node color of V2 and V12 must be different which is green (G) now (see Figure 4.12). Hence to accommodate this constraint, the allocated register for storage variable V2 (Reg_G) needs to swap with other storage variable of same control step. Scanning the storage variables from left to right, we can observe that swapping V2 with V0 is violating a constraint (due to edge between nodes V2 and V8), similarly, swapping V2 with V1 is violating a constraint (due to edge between nodes V1 and V12). However, swapping V2 with V3 does not violate any constraints; therefore, storage variable V2 is allocated to Reg_LB and storage variable V3 is allocated to Reg_G. Similarly, edge between nodes V1 and V14 forces storage variable V1 and V14 to be executed through distinct register which need swapping. Swapping V14 with V15 does not violate any constraints, therefore storage variable V14 is allocated to Reg_G (previously allocated to Reg_DB) and storage variable V15 is allocated to Reg_DB (previously allocated to Reg_G). (*Note: If no storage other variable is found to swap with the conflicting storage variable, a new register should be added in the CIG and assigned to that conflicting storage variable.*) After embedding all these additional edges in the CIG, the final modified graph (according to step B.3) is shown in Figure 4.14. The register allocation table corresponding to the modified CIG (shown in Figure 4.14) is reported in Table 4.3 (according to step B.3). The scheduled CDFG after modifying storage variable assignment and register allocation is shown in Figure 4.15. The corresponding multiplexing schemes of 'Reg_G', 'Reg_DB', and 'Reg_LB' before and after embedding watermark are shown in Tables 4.4(a), 4.4(b) and 4.4(c), respectively. The multiplexers and demultiplexers

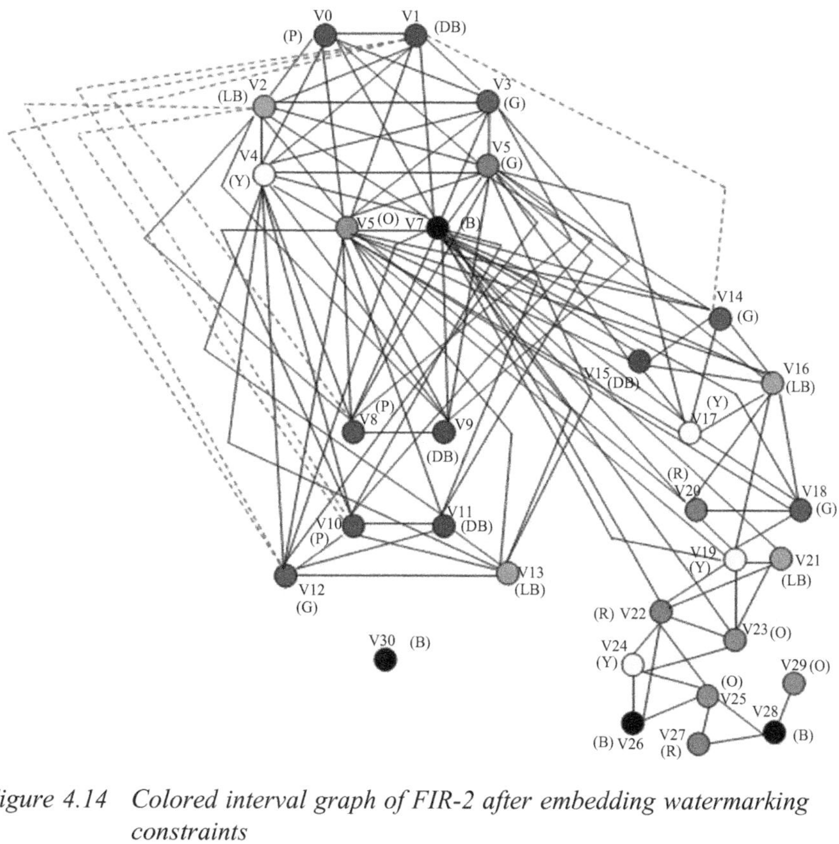

Figure 4.14 Colored interval graph of FIR-2 after embedding watermarking constraints

Table 4.3 Register allocation table of FIR-2 after embedding watermarking constraints

Control Step	Reg_P	Reg_DB	Reg_G	Reg_LB	Reg_Y	Reg_R	Reg_O	Reg_B
0	V0	V1	V3	V2	V4	V5	V6	V7
1	V8	V9	V3	V2	V4	V5	V6	V7
2	V10	V11	V12	V13	V4	V5	V6	V7
3	–	V15	V14	V16	V17	V5	V6	V7
4	–	–	V18	V16	V19	V20	V6	V7
5	–	–	–	V21	V19	V22	V23	V7
6	–	–	–	–	V24	V22	V25	V26
7	–	–	–	–	–	V27	V25	V28
8	–	–	–	–	–	–	V29	V28
9	–	–	–	–	–	–	–	V30

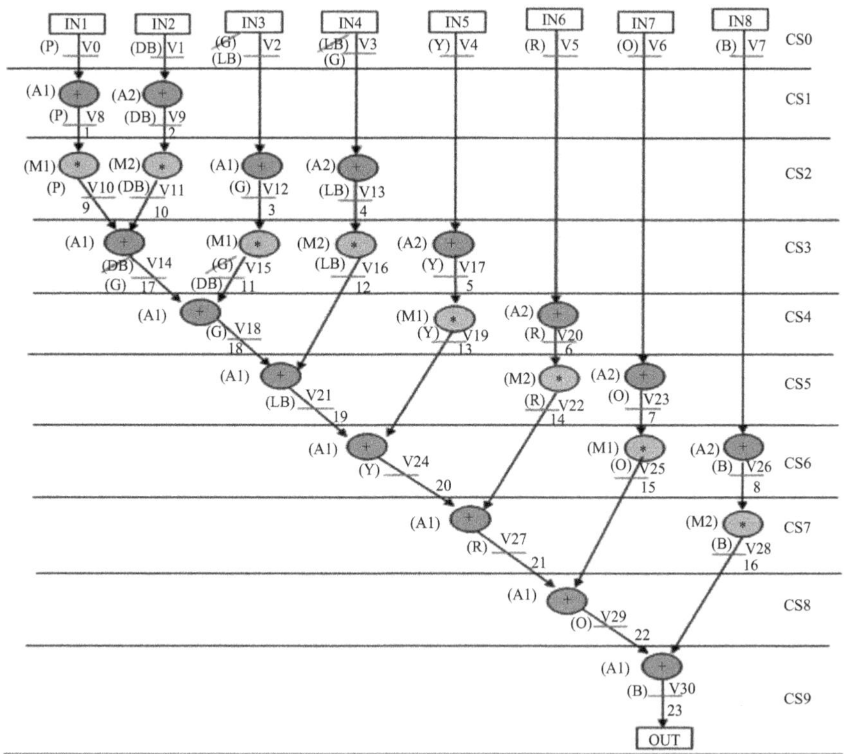

Figure 4.15 Scheduled CDFG of FIR-2 using two adders and two multipliers after embedding watermark

Table 4.4(a) Multiplexing scheme of register green (Reg_G) before and after embedding watermark

Multiplexing scheme before watermark			Multiplexing scheme after watermark		
CS	**Input**	**Output**	**CS**	**Input**	**Output**
0	IN3	A1_I(1)	0	IN4	A2_I(1)
2	A1_O(1)	M1_I(1)	2	A1_O(1)	M1_I(1)
3	M1_O(1)	A1_I(3)	3	A1_O(2)	A1_I(3)
4	A1_O(3)	A1_I(4)	4	A1_O(3)	A1_I(4)

(*Notes*: Different inputs of A1 are indicated as A1_I(0), A1_I(1)… A1_I(8); different outputs of A1 are indicated as A1_O(0), A1_O(1)… A1_O(8); different input of M1 are indicated as M1_I(0), M1_I(1)… M1_I(3); different outputs of M1 are indicated as M1_O(0), M1_O(1)… M1_O(3)).

Table 4.4(b) Multiplexing scheme of register dark blue (Reg_DB) before and after embedding watermark

Multiplexing scheme before watermark			Multiplexing scheme after watermark		
CS	**Input**	**Output**	**CS**	**Input**	**Output**
0	IN2	A2_I(0)	0	IN2	A2_I(0)
1	A2_O(0)	M2_I(0)	1	A2_O(0)	M2_I(0)
2	M2_O(0)	A1_I(2)	2	M2_O(0)	A1_I(2)
3	A1_O(2)	A1_I(3)	3	M1_O(1)	A1_I(3)

Table 4.4(c) Multiplexing scheme of register light blue (Reg_LB) before and after embedding watermark

Multiplexing scheme before watermark			Multiplexing scheme after watermark		
CS	**Input**	**Output**	**CS**	**Input**	**Output**
0	IN4	A2_I(1)	0	IN3	A2_I(1)
2	A2_O(1)	M2_I(1)	2	A2_O(1)	M2_I(1)
3	M2_O(1)	A1_I(4)	3	M2_O(1)	A1_I(4)
5	A1_O(4)	A1_I(5)	5	A1_O(4)	A1_I(5)

of 'Reg_G', 'Reg_DB', and 'Reg_LB' before embedding watermark are shown in Figures 4.16(a), 4.17(a), and 4.18(a), respectively. The corresponding multiplexer and demultiplexer of 'Reg_G', 'Reg_DB', and 'Reg_LB' after embedding watermark are shown in Figures 4.16(b), 4.17(b), and 4.18(b), respectively. The change in datapath after embedding watermark is marked as dotted lines to the mux inputs. For example, due to the swapping of storage variable V2 and V3 (due to the

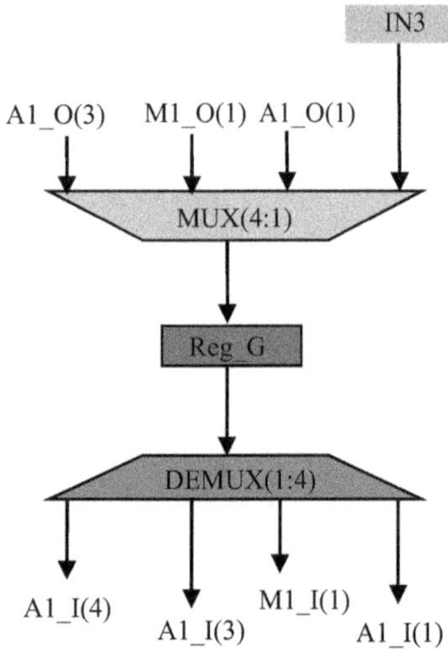

Figure 4.16(a) Illustration of the segment of the datapath without watermark in register G (A1_I(3) is the 2nd input of 4th run of Adder1 through 2nd 16:1 MUX)

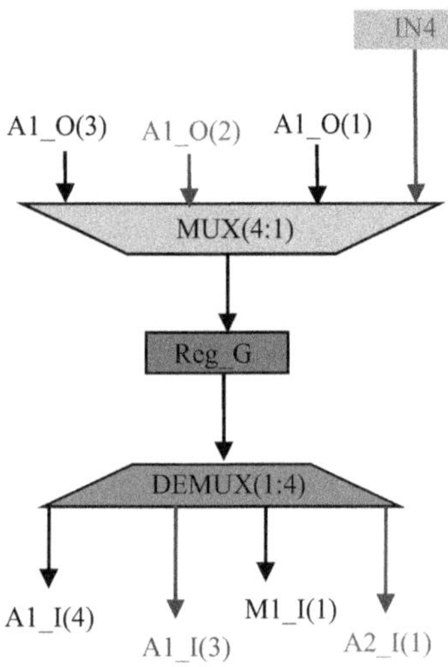

Figure 4.16(b) Illustration of the segment of the datapath with watermark in register G (A1_I(3) is the 1st Input of 4th run of Adder 1 through 1st 16:1 MUX)

insertion of edge between nodes V1 and V12), input IN3 of Reg_G moved Reg_LB, similarly input IN4 of Reg_LB moved to Reg_G. The reflection of this alteration is also visible in the gate structure of the design. The partial gate structure of Mux of Reg_G before embedding watermark is shown in Figures 4.19(a) and (c).

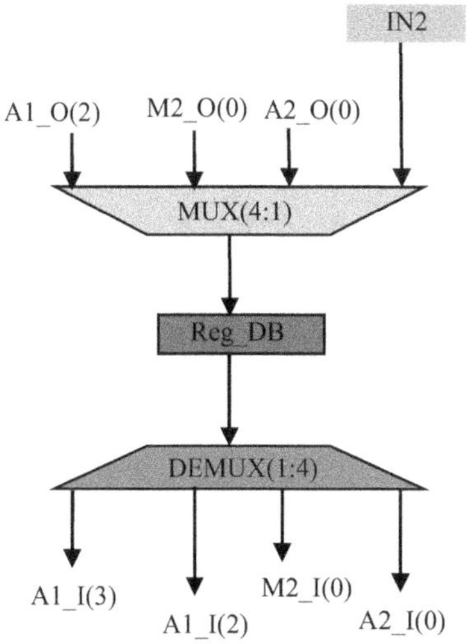

Figure 4.17(a) *Illustration of the segment of the datapath without watermark in register DB (A1_I(3) is the 1st Input of 4th run of Adder 1 through 1st 16:1 MUX)*

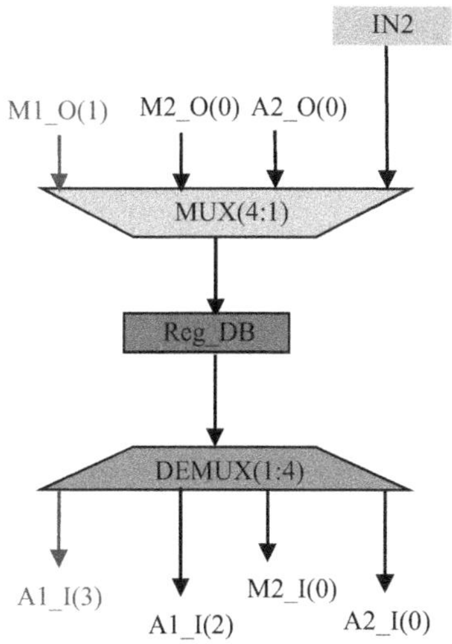

Figure 4.17(b) *(b) Illustration of the segment of the datapath without watermark in register DB (A1_I(3) is the 2nd Input of 4th run of Adder 1 through 2nd 16:1 MUX)*

The corresponding gate structure of the same part of Mux of Reg_G after embedding watermark is shown in Figures 4.19(b) and (d).

The change is gate structure can be observed between Figure 4.19(a) vs. Figure 4.19(b) and Figure 4.19(c) vs. Figure 4.19(d). The final watermarked FIR-2

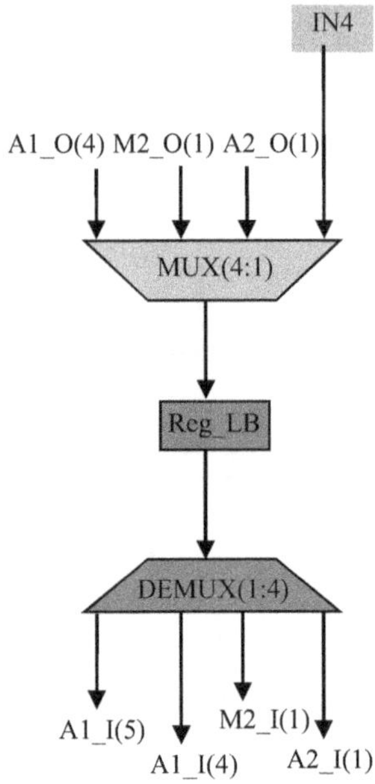

Figure 4.18(a) Illustration of the segment of the datapath without watermark in register LB

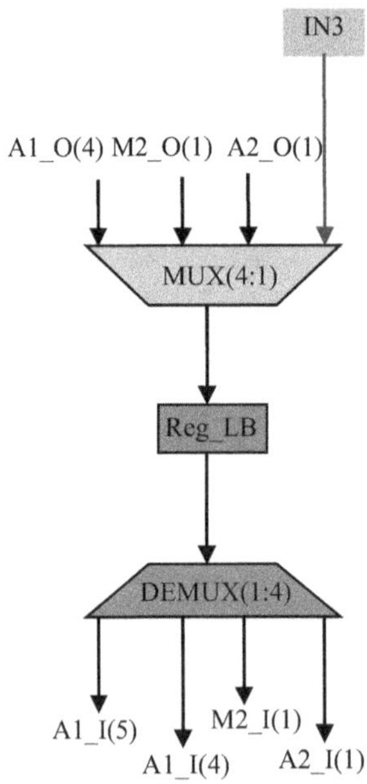

Figure 4.18(b) Illustration of the segment of the datapath without watermark in register LB

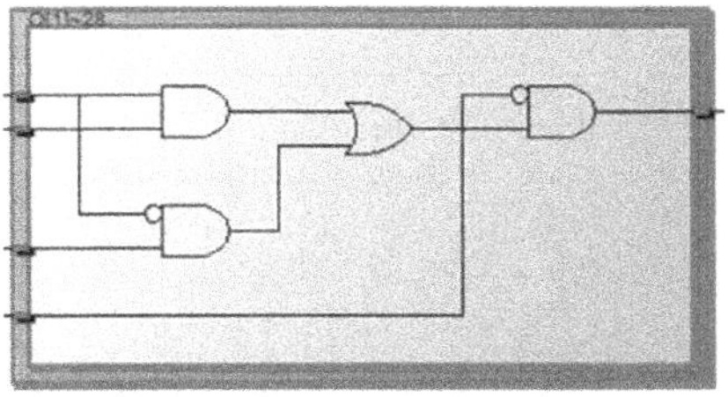

Figure 4.19(a) Partial gate representation of Mux of Reg_G before embedding watermark

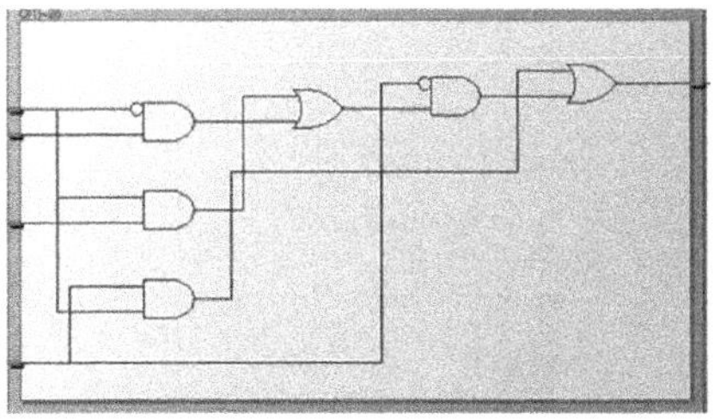

Figure 4.19(b) Partial gate representation of Mux of Reg_G after embedding watermark

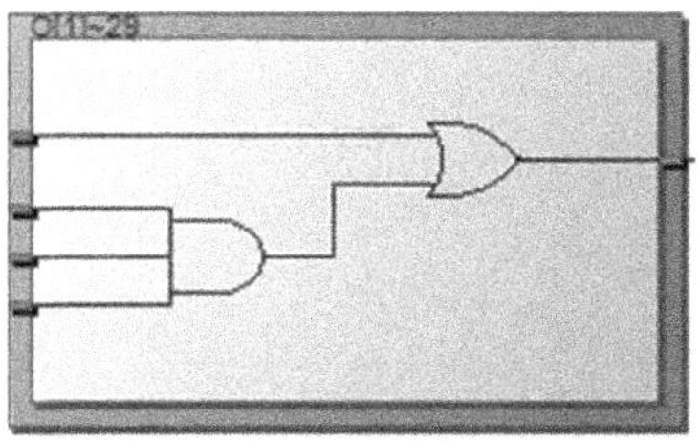

Figure 4.19(c) Partial gate representation of Mux of Reg_G before embedding watermark

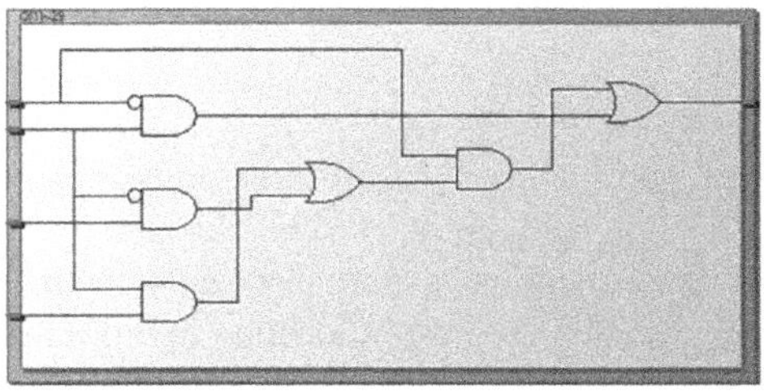

Figure 4.19(d) Partial gate representation of Mux of Reg_G after embedding watermark

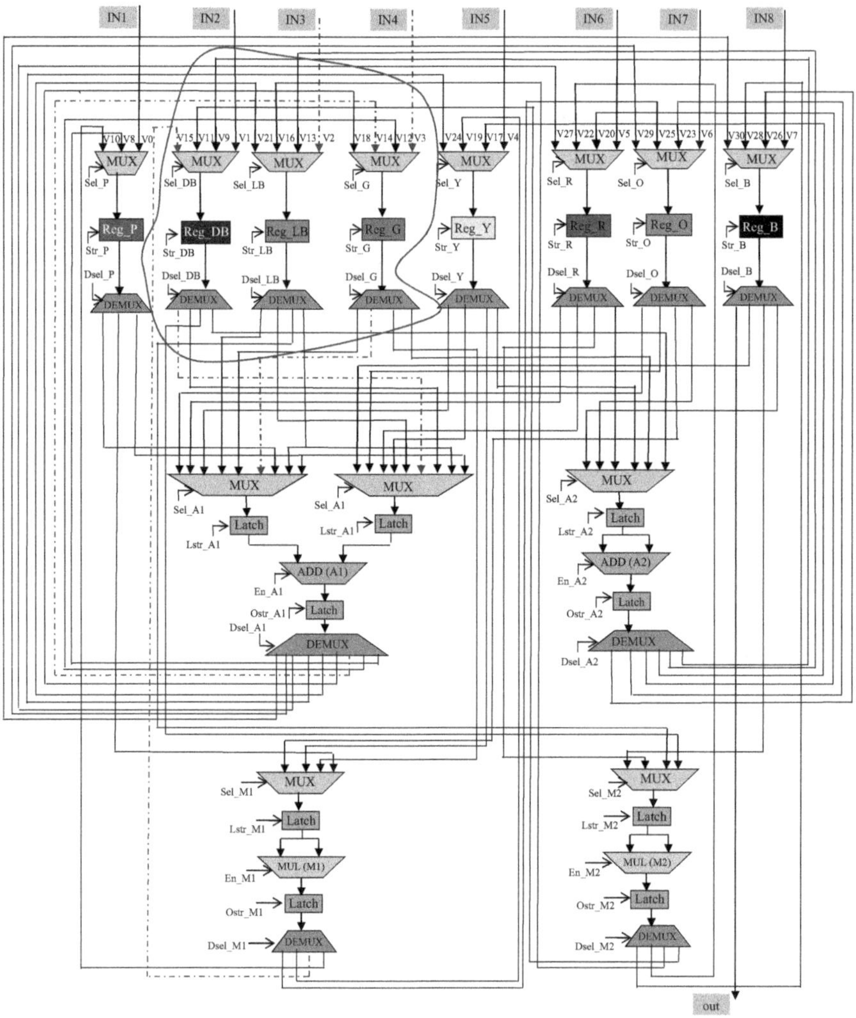

Figure 4.20 Datapath of single-phase watermarked FIR-2 (change in register allocation is marked using encircled lines)

datapath is shown in Figure 4.20. The corresponding non-watermarked (original) FIR-2 benchmark is shown in Figure 4.21. The changes in datapath connectivity due to insertion of watermarking constraints are marked in *dotted* line in Figure 4.20. Similarly, change in register allocation in the watermarked datapath is highlighted using *boundary curve* in the same figure. Both watermarked and non-watermarked design of FIR-2 benchmark are implemented using Altera Quartus (2018) 13.0 v and the targeted FPGA is cyclone II (Cyclone II FPGA, 2018). The schematic representation of implemented watermarked and non-watermarked

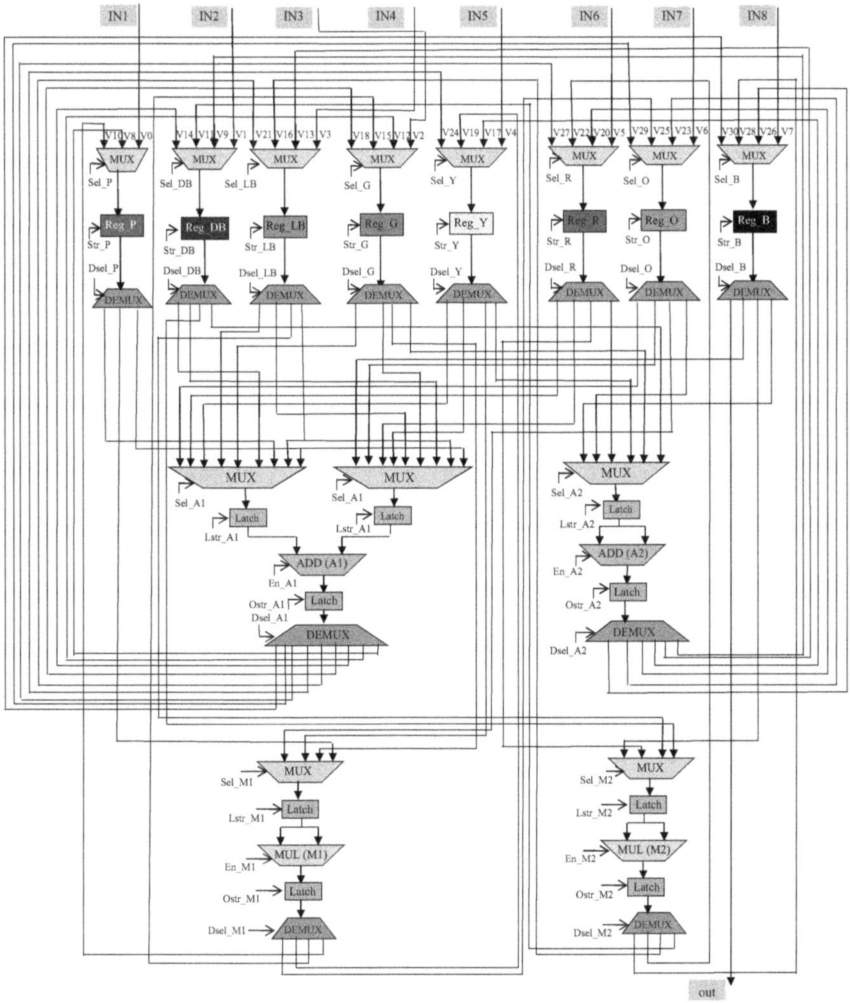

Figure 4.21 Datapath of non-watermarked FIR-2

design is shown in Figures 4.22 and 4.23, respectively. As embedding single-phase watermarking constraints in a design only changes its datapath structure but does not change the controller, therefore both watermarked and non-watermarked should have same controller structure. The controller structure of FIR-2 benchmark corresponding to two adders and two multipliers resources is shown in Figure 4.24. Simulation result of non-watermarked and watermarked FIR-2 is shown in Figures 4.25(a) and (b). It can be observed that for same set of inputs both non-watermarked and watermarked design produce same output. Thus, single-phase watermark preserves the functionality and correctness of a watermarked design. The comparison of resource utilization between non-watermarked and water-marked FIR-2 design is shown in Table 4.5. It can be observed that due to

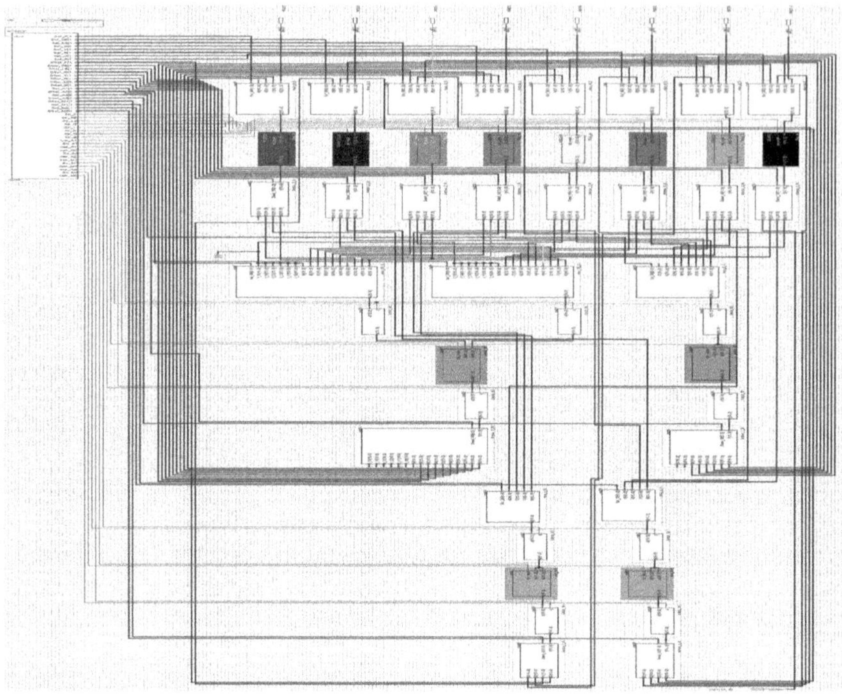

Figure 4.22 Schematic representation of single-phase watermarked FIR-2

insertion of 14-digit watermark the overhead in terms of total logic elements are almost negligible.

4.3.3 Design process of triple-phase watermarked IP core/ hardware

In this section, we describe the design process of another watermarked DSP IP core/hardware. We describe a more robust triple-phase watermarking approach presented in Sengupta *et al.* (2018) to design a DSP IP core. This approach does not protect non-HDL IP cores such as standard/pre-defined blocks or symbols that are present in the in-built library of a logic CAD tool. The triple-phase watermarking methodology protects the ownership of an IP core (especially for DSP/multimedia applications). In the process of embedding triple-phase watermark, it is presumed that same type of micro-IPs are available from at least two different third-party vendors to design a macro-IP core (which is a common practice in the industry).

Unlike single-phase watermarking approach (Sengupta *et al.*, 2016), triple-phase watermarking approach (Sengupta *et al.*, 2018) for DSP core embeds watermarking constraints in multiple phases of HLS. More specifically, triple-phase watermarking approach embeds watermarking constraints in the scheduling, hardware allocation, and register allocation phase of HLS. In addition to implanting

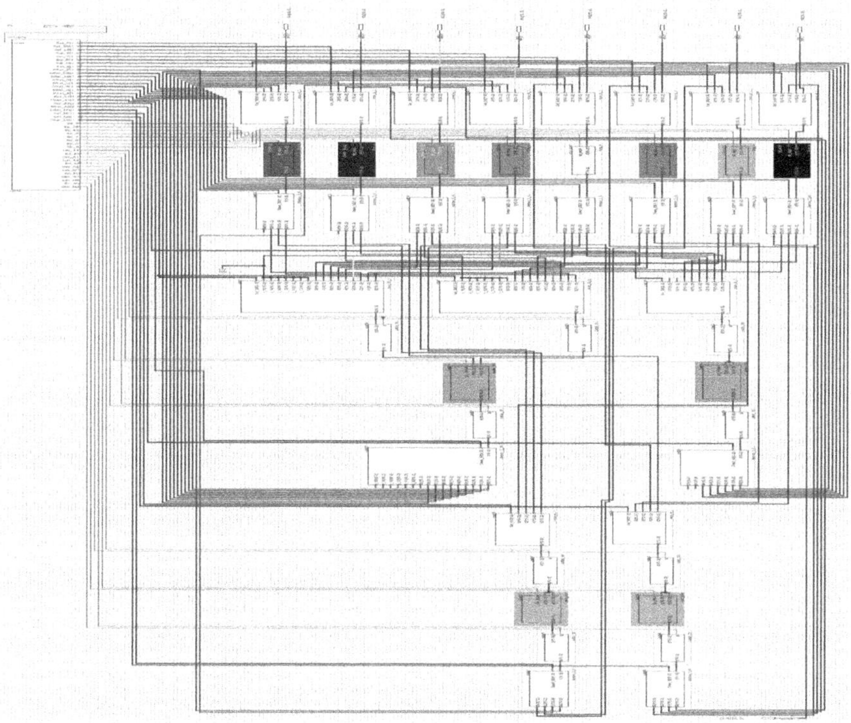

Figure 4.23 Schematic representation of non-watermarked FIR-2

watermarking constraints in three different phases of HLS, the signature encoding scheme consists of seven different variables, where each variable carries a unique encoding rule. Thus, triple-phase watermarking approach enhances the robustness and minimizes the chance of malicious alteration or removal of the signature. Furthermore, for an adversary, it is also difficult to recognize the synthesis phase and in what way watermark constraints are implanted in the design.

In Sengupta *et al.* (2018), watermarking constraints are embedded during scheduling, hardware allocation, and register allocation phases by modifying three information: (a) 'Non-critical operations ($\mu_m > 0$)' timing table (μ_m = the mobility of an operation), (b) 'Functional unit allocation' table, and (c) 'Register allocation' table (adopted from Sengupta *et al.*, 2016). As mentioned earlier, this watermarking approach has seven different variables, viz. 'α', 'β', 'γ', 'i', 'I', 'T', and '!'. Variable 'γ' modifies the timing table of 'non-critical operations ($\mu_m > 0$)' thus embedding watermarking constraints during scheduling (first phase). Variables 'α' and 'β' modifies the 'functional unit allocation' table thus embedding watermarking constraints during hardware allocation (second phase). Finally, variables 'i', 'I', 'T', '!' modify the 'register allocation' table thus embedding watermarking constraints during register allocation (third phase). The signature encoding process of each variable is described below (Sengupta *et al.*, 2018).

Control Signals	clk0	1	2	3	4	5	6	7	8	9	10	11	12	13	14	15	16	17	18	19	20	21	22	23	24	25	26	27	28	29	30	31	32	33	34	35	36	37	38
Sel_P	00					01				10																													
Sel_DB	00					01				10				11																									
Sel_LB	00									01				10								11																	
Sel_G	00									01				10				11																					
Sel_Y	00													01				10								11													
Sel_R	00																	01				10								11									
Sel_O	00																					01				10								11					
Sel_B	00																									01				10								11	
Str_P	0	1																																					
Str_DB	0	1																																					
Str_LB	0	1																																					
Str_G	0	1																																					
Str_Y	0	1																																					
Str_R	0	1																																					
Str_O	0	1																																					
Str_B	0	1																																					
Dsel_P	00						01				10																												
Dsel_DB	00						01				10				11																								
Dsel_LB	00										01				10								11																
Dsel_G	00										01				10				11																				
Dsel_Y	00														01				10								11												
Dsel_R	00																		01				10								11								
Dsel_O	00																						01				10								11				
Dsel_B	00																										01				10								11
Sel_A1	0000						0001				0010				0011				0100				0101				0110				0111				1000				
Lstr_A1	0		1	0			1	0			1	0			1	0			1	0			1	0			1	0			1	0			1	0			
En_A1	0			1																																			
Ostr_A1	0				1	0			1	0			1	0			1	0			1	0			1	0			1	0			1	0			1	0	
Dsel_A1	0000								0001				0010				0011				0100				0101				0110				0111				1000		
Sel_A2	000						001				010				011				100				101																
Lstr_A2	0		1	0			1	0			1	0			1	0			1	0			1	0															
En_A2	0			1																																			
Ostr_A2	0				1	0			1	0			1	0			1	0			1	0			1	0													
Dsel_A2	000								001				010				011				100				101														
Sel_M1	00										01				10								11																
Lstr_M1	0						1	0			1	0			1	0							1	0															
En_M1	0							1																															
Ostr_M1	0								1	0			1	0			1	0							1	0													
Dsel_M1	00												01				10								11														
Sel_M2	00										01								10								11												
Lstr_M2	0						1	0			1	0							1	0							1	0											
En_M2	0							1																															
Ostr_M2	0								1	0			1	0							1	0							1	0									
Dsel_M2	00												01								10								11										

Figure 4.24 Controller of FIR-2 benchmark corresponding to two adders and two multipliers

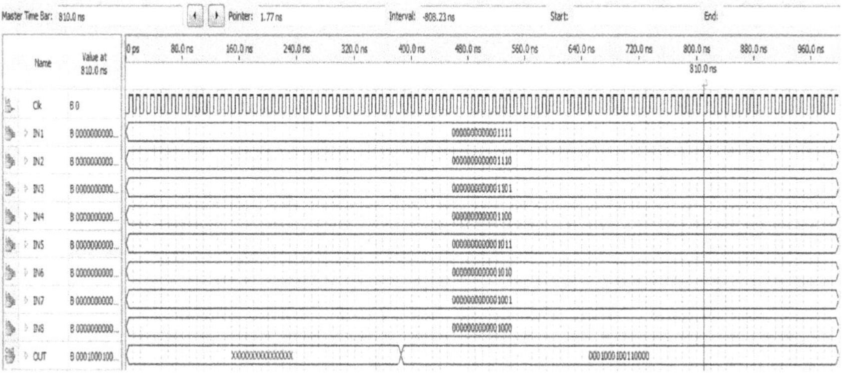

Figure 4.25(a) Simulation result of non-watermarked FIR-2

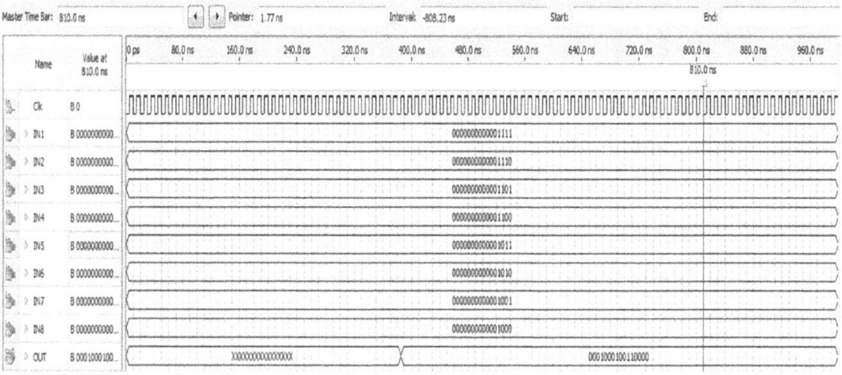

Figure 4.25(b) Simulation result of single-phase watermarked FIR-2

Table 4.5 Comparison of resource utilization between non-watermarked and single-phase watermarked FIR-2

Resource utilization summary	Non-watermarked FIR-2	Watermarked FIR-2
Family	Cyclone II	Cyclone II
Device	EP2C70F896C6	EP2C70F896C6
Total logic elements	2,034/68,416 (3%)	2,036/68,416 (3%)
Total combinational functions	2,029/68,416 (3%)	2,028/68,416 (3%)
Dedicated logic registers	69/68,416 (<1%)	69/68,416 (<1%)
Total registers	69	69
Total pins	145/622 (23%)	145/622 (23%)

4.3.3.1 Signature encoding scheme

A seven variables signature encoding scheme is devised in Sengupta *et al.* (2018) where each variable embeds decoded watermarking constraint in any of the three phases (scheduling, hardware allocation, and register allocation) of synthesis process. More specifically, each decoded signature digit inserts a watermarking constraint by modifying any of the aforementioned tables. An IP owner can select his desired signature using any random combination of these seven variables, i.e., 'α', 'β', 'γ', 'i', 'I', 'T' and '!'. The encoded rule of each variable is as follows:

- 'γ' = Move/shift a non-critical path's operation with the highest mobility into immediate next control step (cs).
- 'α' = *In odd control steps only*: Even operation must be allocated to the hardware of vendor type 2 (U2) and odd operation must be allocated to the hardware of vendor type 1 (U1).
- 'β' = *In even control steps only*: Even operation must be allocated to the hardware of vendor type 1 (U1) and odd operation must be allocated to the hardware of vendor type 2 (U2).
- 'i' = Add an additional edge between (prime, prime) node pair of CIG.
- 'I' = Add an additional edge between (even, even) node pair of CIG.
- 'T' = Add an additional edge between (odd, even) node pair of CIG.
- '!' = Add an additional edge between (0, any integer) node pair of CIG.

In Sengupta *et al.* (2018), 3rd phase watermarking variables, i.e., 'i', 'I', 'T' and '!' are independent of both 1st phase watermarking variable, i.e., 'γ' and 2nd phase watermarking variables, i.e., 'α' and 'β'. Similarly, watermarking variable of 1st phase is independent of both watermarking variables of 2nd and 3rd phase. These advantages make the triple-phase watermarking approach exceptionally robust and tamper tolerant for an adversary. However, watermarking variable of 2nd phase is dependent on watermarking variables of 1st phase. Therefore, watermarking constraints of 2nd phase may be affected due to alteration of 1st phase watermark which is a limitation of this approach. Nevertheless, as mentioned earlier, it is tremendously tough for an adversary to detect in which step of synthesis, the 1st phase watermark is implanted, and how. Moreover, if 1st phase watermark is tampered somehow by an attacker, then still 3rd phase watermarking constraints (which is independent of 1st- and 2nd-phase watermark) are capable of resolving the ownership conflict and identifying the real IP core owner. Thus, triple-phase watermark provides a strong tamper tolerant watermarked IP core design.

4.3.3.2 Watermark embedding process

In the previous sub-section, we have discussed encoding rule of seven variables of the triple-phase watermark. In this sub-section, we will discuss about how to embed a triple-phase watermark in a non-watermarked IP design. The triple-phase watermark embedding process is divided into four parts, i.e., pre-processing steps, 1st phase watermark embedding steps, 2nd phase watermark embedding steps, and 3rd phase watermark embedding steps. Detailed steps of this watermarking process are as follows (Sengupta *et al.*, 2018):

A. **Pre-processing steps**
   1. Accept a non-watermarked DSP application as CDFG and schedule it based on user resource configuration.

2. Perform functional unit allocation and register allocation.
3. Construct the timing table for 'non-critical operations ($\mu_m > 0$),' 'functional unit allocation' table, and 'register allocation' table from the non-watermarked IP core design.
4. Sort the operations control stepwise based on their number in increasing order.
5. Accept an n-digit vendor's signature as a combination of 'α', 'β', 'γ', 'i', 'I', 'T' and '!' variables.

B. **1st phase watermark embedding steps**

1. For each occurrence of 'γ', shift an operation of the non-critical path into its very next control step while checking from control step 1 onward (without re-occurrence), if both of the following conditions satisfied:
   (i) Respective operation does not have any child operation in very next control step.
   (ii) Shifting does not violate the hardware resource constraints. (*Note: In case more than one operation is eligible to be shifted in the next control step, the operation with the highest mobility will be given the priority*).
2. Adjust the timing table of 'non-critical operations ($\mu_m > 0$)' after embedding all 'γ' variables.

C. **2nd phase watermark embedding steps**

1. Accept the 1st phase watermark embedded scheduled design as input to embed 2nd phase watermark.
2. For each occurrence of 'α' and/or 'β' variable in the vendor's signature, re-allocation of functional unit needs to be performed according to the encoding rule of 'α' and 'β'.
3. Modify the 'functional unit allocation' table after embedding all 'α' and 'β' variables.

D. **3rd phase watermark embedding steps**

1. Accept the dual phase (1st and 2nd phase) watermark embedded design as input to embed 3rd phase watermark.
2. Rest of the steps is similar to single-phase watermark embedding steps B.1–B.4 (refer Section 4.3.B).

4.3.3.3 Signature detection process

The signature detection process is a crucial and compulsory step to resolve ownership conflict. The inputs required to detect triple-phase watermark are: (a) decoded vendor signature, (b) application, and (c) watermarked IP core in the form of controller and datapath HDL. The triple-phase watermark can be achieved in three steps and each step verifies the presence of watermarking constraints of each phase. In the signature detection and verification process of phase 1 watermark, the control steps of modified operations are extracted from the decoded signature which is then compared with the controller HDL of the watermarked IP. To detect and verify the phase 2 watermark, the allocated hardware of the watermarked operations are extracted from the decoded signature which is then identified with the help of

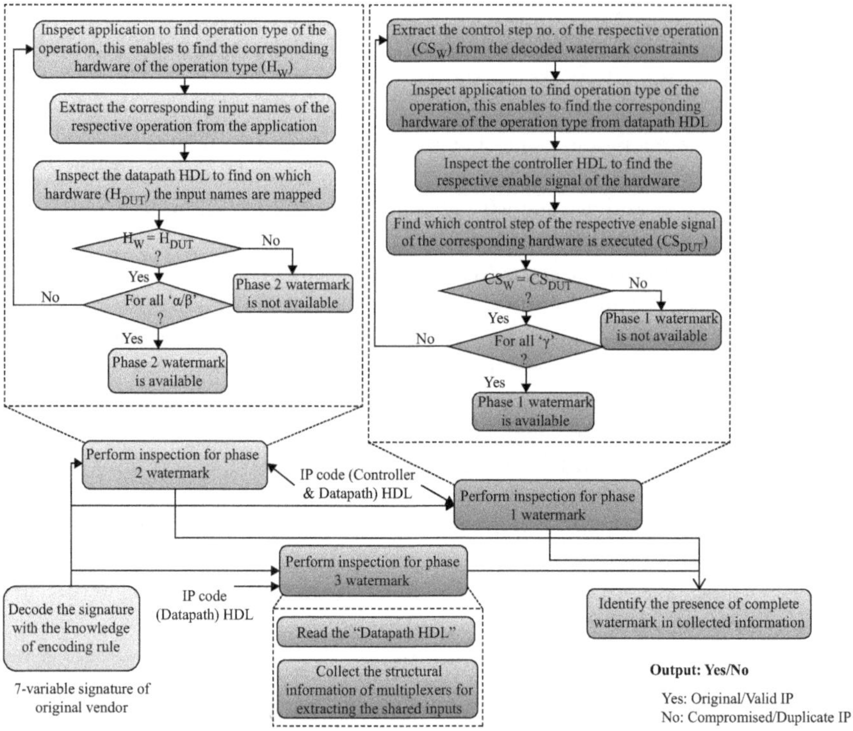

Figure 4.26 Signature detection process of triple phase watermark

original application and its corresponding inputs. Then, the watermarked hardware information fetched from the decoded watermark is compared with the datapath HDL of the watermarked IP. Similarly, to detect and verify the phase 3 watermark, the shared inputs of each multiplexer should be fetched from the datapath HDL of the watermarked IP. It is then compared with the 3rd phase watermarking constraints extracted from the decoded signature. If all the constraints of decode signature are available in the watermarked IP core then the vendor who provided the decoded signature is awarded as the original owner of that IP core. The complete signature detection process of triple phase watermark is shown in Figure 4.26.

4.3.3.4 Didactic example of triple-phase watermarking process

In this sub-section, we demonstrate an end-to-end triple-phase IP core watermark embedding process through a representative example. The DFG of discrete wavelet transformation (DWT) benchmark shown in Figure 4.27 is taken as input and then scheduled using user given hardware configuration, i.e., three adders and three multipliers (according to step A.1). After that, the functional unit allocation of each operation and the register allocation of each storage variable (V0–V21) are performed (according to step A.2). In the triple-phase watermarking approach, two distinctive IP core vendors are employed to achieve better security. For example, in

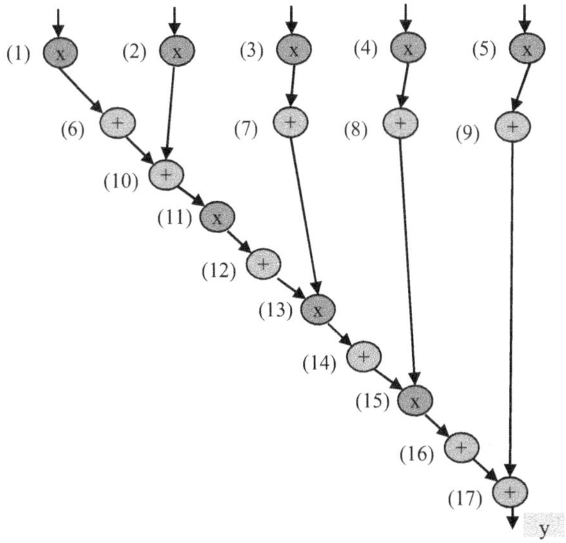

Figure 4.27 Data flow graph of a DWT

this case, out of three adders, two are acquired from vendor type 1 (A1) and one is acquired from vendor type 2 (A2), similarly, out of three multipliers, two are obtained from vendor type 1 (M1) and one is obtained from vendor type 2 (M2). The scheduled DFG after performing functional unit re-allocation and register re-allocation (before watermarking) is shown in Figure 4.28. The IP core design before embedding watermarking constraints uses list scheduling algorithm and random functional unit allocation. Register allocation for the same design (before watermark) is performed based on CIG (shown in Figure 4.29). A CIG is constructed with 22 nodes, i.e., V0–V21 and five registers, i.e., register red (R), register green (G), register blue (B), register yellow (Y), and register purple (P). The timing table for 'non-critical operations,' 'functional unit allocation' table, and 'register allocation' table of the non-watermarked design are generated as shown in Tables 4.6, 4.7, and 4.8, respectively (according to step A.3). All the operations are sorted control step wise in ascending order (according to step A.4). As an example, a 12-digits vendor signature: 'ɣ ɣ ɑ β ɑ β β **i I IT !**' is selected which is a combination of seven aforementioned signature variables (according to step A.5). All the pre-processing steps to embed single-phase watermark is complete now. The vendor's signatures and its corresponding decoded meaning are shown in Table 4.9. In the 1st phase of the triple-phase watermark, all the 'γ' digits should be embedded. According to the encoding rule of 'γ', the first 'γ' digit of vendor's signature moves operation (opn.) two from control step (cs) 1 to cs 2. For next 'γ' digit, the next operation which satisfies all the rules of B.1 is opn.9, because operations 3, 4, and 5 do not satisfy the rule mentioned in B.1(i), and opn.7 does not satisfy the rule mentioned in B.1(ii). Because between operations 8 and 9, opn.9 has higher mobility value, hence it is moved from cs 3 to cs 4. The modified timing table for

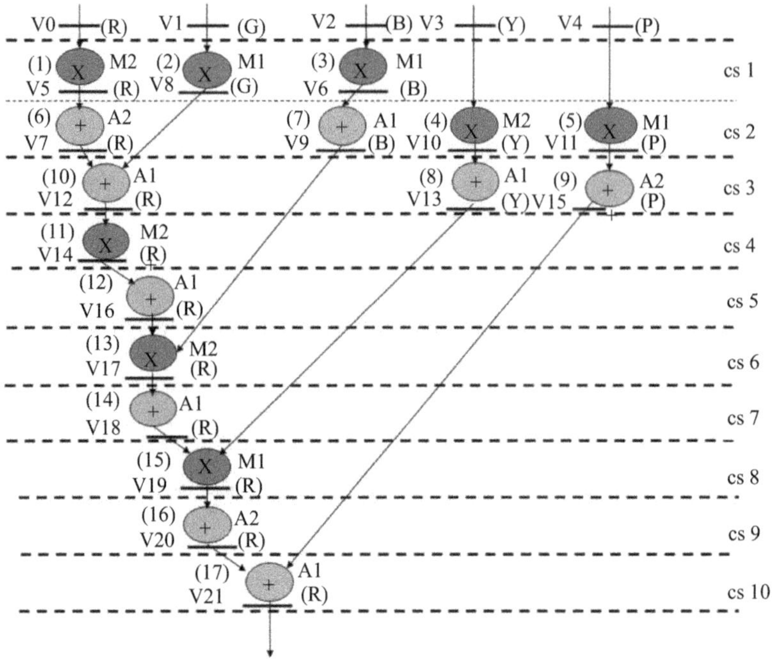

Figure 4.28 Scheduled DFG of DWT based on three adders and three multipliers before embedding watermark

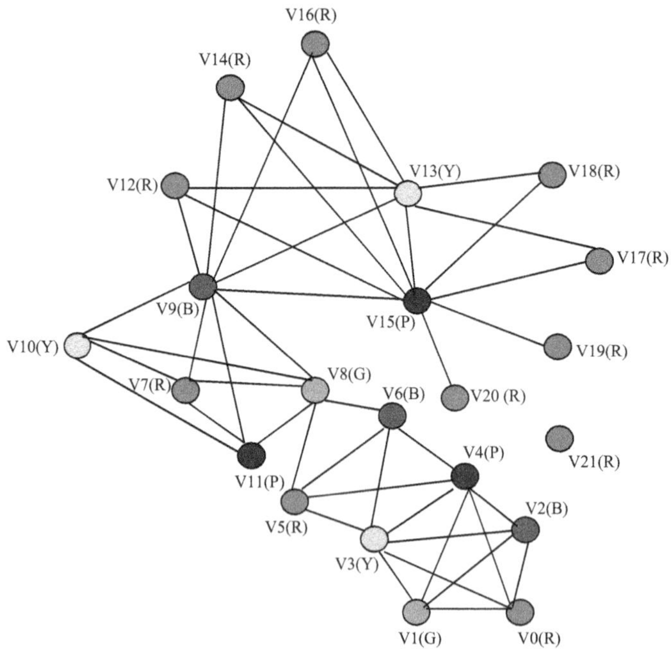

Figure 4.29 Colored interval graph before implanting watermark

Table 4.6 'Timing table for non-critical operations' (before embedding 1st phase watermark)

Operation no.	2	3	4	5	7	8	9
Control step	1			2		3	

Table 4.7 'Functional unit allocation' table (before embedding 2nd phase watermark)

Odd control step	Operation no.	1	2	3	8	9	10	12	14	16
	Allocated functional unit	M2	M1	M1	A1	A2	A1	A1	A1	A2
Even control step	Operation no.	4	5	6	7	11	13	15	17	–
	Allocated functional unit	M2	M1	A2	A1	M2	M2	M1	A1	–

Table 4.8 'Register allocation' table (before embedding 3rd phase watermark)

Control step	**R**	**G**	**B**	**Y**	**P**
0	V0	V1	**V2**	V3	V4
1	V5	V8	**V6**	V3	V4
2	V7	V8	V9	V10	V11
3	V12	–	V9	V13	V15
4	V14	–	V9	V13	V15
5	V16	–	V9	V13	V15
6	V17	–	–	V13	V15
7	V18	–	–	V13	V15
8	V19	–	–	–	V15
9	V20	–	–	–	V15
10	V21				

'non-critical operations' and its corresponding scheduled DFG is shown in Table 4.10 and Figure 4.30 (according to step B.2). In the 2nd phase of the triple-phase watermark, all the 'α' and 'β' digits should be embedded on the top of 1st phase watermarked design (according to step C.1). According to the encoding rule of 'α', the third digit of the signature modifies the hardware assignment of opn.1 from M2 (multiplier obtained from vendor type 2) to M1 (multiplier obtained from vendor type 1). Similarly, according to the encoding rule of 'β', the sixth digit of the signature modifies the hardware assignment of opn.4 from M2 to M1. (*Note: no modification of hardware assignment is required for the fourth and fifth digit of the signature.*) Similarly, the seventh digit of the signature modifies the hardware assignment of opn.5 from M1 to M2. The modified 'functional unit allocation'

Table 4.9 Vendor signature and its decoded meaning

Desired signature (12-digits)	Corresponding operation to shift (Phase 1)	Allocate FU type (Phase 2)	Additional edges to insert between nodes in the colored interval graph (Phase 3)	Observations
ɣ	opn.2 from cs 1 to 2	–	–	cs shift to be done
ɣ	opn.9 from cs 3 to 4	–	–	cs shift to be done
ɑ	–	opn.1 with vendor 1	–	FU reallocation to be done
б	–	opn.2 with vendor 1	–	No change occurred
ɑ	–	opn.3 with vendor 1	–	No change occurred
б	–	opn.4 with vendor 1	–	FU reallocation to be done
б	–	opn.5 with vendor 2	–	FU reallocation to be done
I	–	–	(v2, v3)	Exists by default
I	–	–	(v2, v4)	Exists by default
I	–	–	(v2, v6)	New edge to be added
T	–	–	(v1, v2)	Exists by default
!	–	–	(v0, v1)	Exists by default

Table 4.10 'Timing table for non-critical operations' (after embedding 1st phase watermark)

Operation No.	3	2	4	5	7	8	9
Control step	1			2		3	4

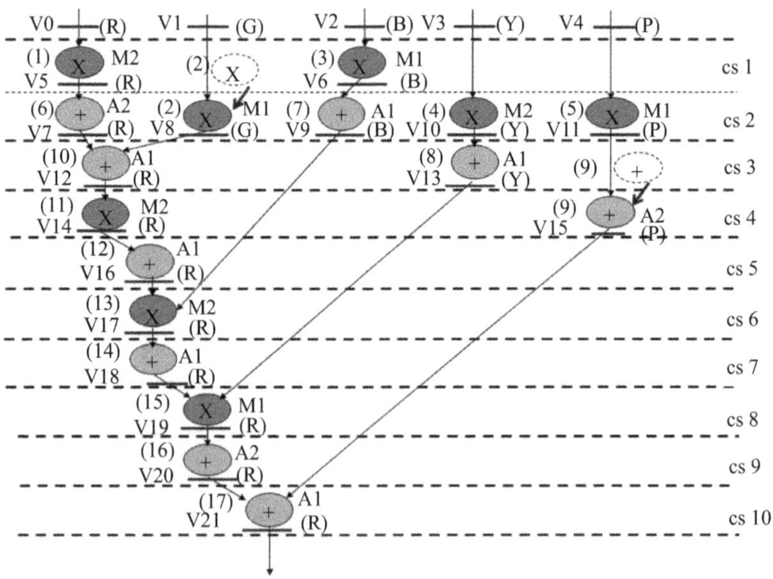

Figure 4.30 Modified scheduled DFG after implanting 1st phase watermark

table and its corresponding scheduled DFG is shown in Table 4.11 and Figure 4.31 (according to step C.3). In the 3rd phase of the triple-phase watermark, hall the 'i', 'I', 'T', and '!' digits should be embedded on the top of 1st and 2nd phase watermarked design (according to step D.1). According to the decoded signature shown in Table 4.9, the list of additional edges includes edges between nodes (V2 and V3), (V2 and V4), (V2 and V6), (V1 and V2), and (V0 and V1). Among these five edges except the edge between nodes (V2 and V6), all are already present in the constructed CIG. It can be observed that edge between node (V2 and V6) forces storage variable V2 and V6 to be executed through distinct register, therefore node color of V2 and V6 must be different which is blue (B) now (see Figure 4.29 and Table 4.8 marked in bold). Hence to accommodate this constraint, the allocated register for storage variable V6 (Reg_B) needs to be swapped with another storage variable in same control step. Swapping V6 with V5 does not violate any constraints; therefore, storage variable V6 is now allocated to Reg_R (previously

Table 4.11 'Functional unit allocation' table (after embedding 2nd phase watermark)

Odd CS	Operation no.	1	3	8	10	12	14	16	–	–	–
	Allocated functional unit	M1	M1	A1	A1	A1	A1	A2	–	–	–
Even CS	Operation no.	4	2	5	6	7	9	11	13	15	17
	Allocated functional unit	M1	M1	M2	A2	A1	A2	M2	M2	M1	A1

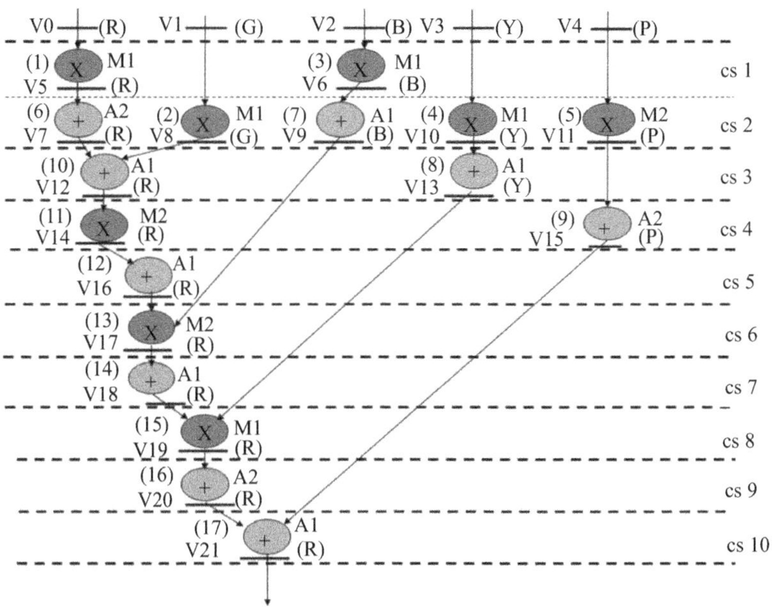

Figure 4.31 Altered scheduled DFG after implanting 1st- and 2nd-phase watermark

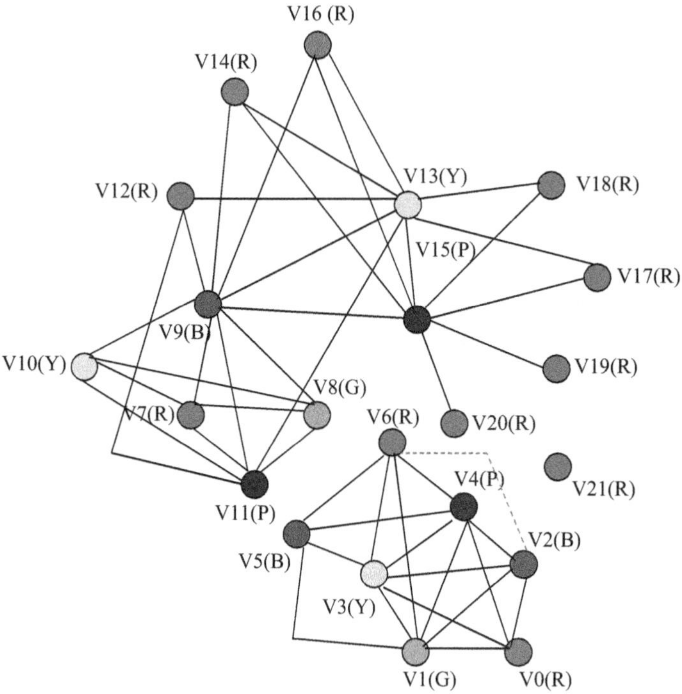

Figure 4.32 Colored interval graph before implanting watermark

Table 4.12 'Register allocation' table (after embedding 3rd phase watermark)

Control step	R	G	B	Y	P
0	V0	V1	V2	V3	V4
1	V6	V1	V5	V3	V4
2	V7	V8	V9	V10	V11
3	V12	–	V9	V13	V11
4	V14	–	V9	V13	V15
5	V16	–	V9	V13	V15
6	V17	–	–	V13	V15
7	V18	–	–	V13	V15
8	V19	–	–	–	V15
9	V20	–	–	–	V15
10	V21				

allocated to Reg_B) and storage variable V5 is now allocated to Reg_B (previously allocated to Reg_R). After embedding the additional edges in the CIG, the final modified graph is shown in Figure 4.32. The modified register allocation table and its corresponding modified scheduled DFG containing watermark information is shown in Table 4.12 and Figure 4.33.

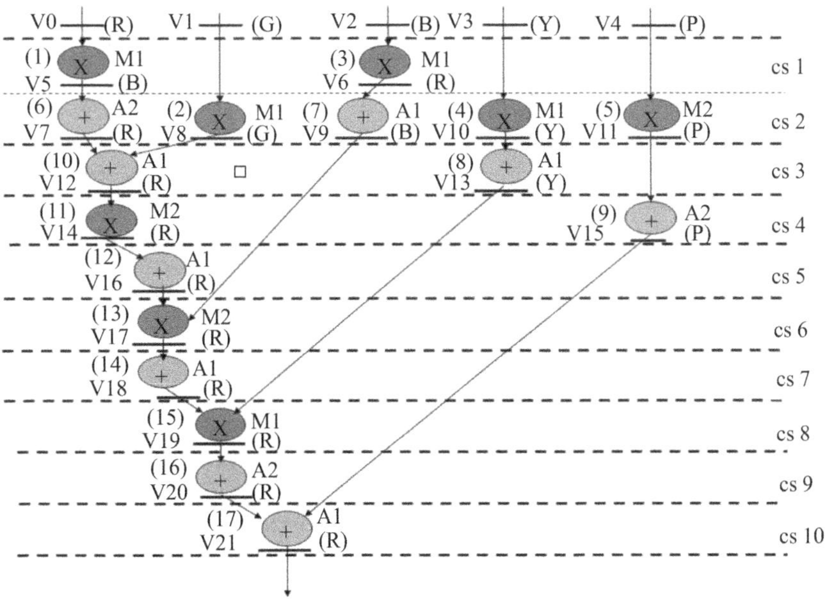

Figure 4.33 Modified scheduled DFG after embedding 1st-, 2nd-, and 3rd-phase watermark

4.3.4 Desired properties of IP core watermark

Every ideal IP core watermark should satisfy following properties:

(i) **Low-design overhead**: Embedding additional design constraints in the form of watermark may incur design overhead. However, an ideal IP core watermarking process should incur minimal design overhead as possible. The design overhead of both single- and triple-phase watermark discussed in this chapter in terms of design area and design latency is significantly low.

(ii) **Strong resiliency**: An ideal IP core watermark should be resilient against removal attack. Therefore, the watermarking constraints should be implanted as a core part of an IP core design. Both single- and triple-phase watermarks contain multi-variable encoding rules where the watermarking constraints are dispersed all over the design (especially in triple-phase watermarking process). Therefore, complete removal of watermarking constraints from the entire IP core design is enormously difficult for an attacker. Thus, substantial resiliency against removal attack is achieved in both watermarking processes.

(iii) **Strong tamper tolerance**: An ideal IP core watermark should have strong tamper tolerance ability, i.e., partial removal should not impact the proof of ownership of an IP core. The watermarking constraints of single- and triple-phase presented in this chapter are distributed using multiple encoding variables. Moreover, in triple-phase watermarking process, watermark constraints are embedded in three different phases of HLS. Additionally, as mentioned earlier, 1st and 3rd phase of triple-phase watermark is independent of other two phases. Thus, strong tamper tolerance ability is achieved in both watermarking process.

(iv) **Low-signature embedding and detection time**: A robust signature may incur complex signature decoding and high-embedding time; therefore, the total runtime to generate a watermarked IP core may increase massively. An ideal IP core watermark should be embedded with lower embedding time. Additionally, signature detection process for a genuine vendor (who has the thorough information of encoding rule and the signature) should be quick. Due to unobvious encoding rules and robust signature embedding process, the embedding time of both single- and triple-phase watermark phase presented in this chapter is low. Furthermore, signature detection time for a genuine entity is also straightforward and speedy.

(v) **Correct functionality**: In the process of embedding watermarking constraints, the correct functionality of an IP core must be preserved. Both single- and triple-phase watermarking processes phase presented in this chapter do not perform any alteration of IP core functionality while embedding watermarking constraints. Thus, functionality of an IP core does not get affected at all.

4.3.5 Possible cases of dishonest claim of IP core/hardware ownership and its resolution

This section presents the possible threat scenarios of the false claim of IP core ownership and its resolution using watermarking process presented in this chapter. Let us assume 'Alice' is an IP vendor, designed a watermarked IP core with her signature and sold it to 'Bob.' In that case, Bob can claim his IP core ownership using any three possible scenarios as discussed below (Koushanfar *et al.*, 2005):

(i) **Claiming any arbitrary design information as the signature**: After purchasing the watermarked IP core, Bob can reverse engineer the IP core and extract the complete information of the IP core design. Then, he may claim any random design information as his watermarking constraints, thus may claim his ownership of that IP core. Bob's claim may valid for a single design, but will invalid for other watermarked design. Thus, Bob's claim is not strong enough to prove his ownership. On the other hand, Alice's watermarking constraints will be more meaningful and consistent for all watermarked design. In such scenario, the entity having more meaningful and consistent watermark will be awarded to the original owner of the IP core. The presented watermarking in this chapter is capable to resolve this threat.

(ii) **Inserting illegal signature**: Bob can embed his own signature on the top of the watermarked design purchased from Alice and then claim for his IP ownership. In such scenario, Alice can prove her ownership easily because the watermarked design coming from Alice contents only Alice's watermark; however, the watermarked design coming from Bob contains both Alice's and Bob's watermark. The presented watermarking in this chapter is also capable to resolve this threat.

(iii) **Tampering/partial removing of original watermark**: Though it is recommended that watermarking constraints should be an integral part of the design, still sometime it may be possible to identify some design information as the watermarking constraints of the design. In such case, Bob may perform some modification/alteration in the design to remove those watermarks. In such scenario, if the watermarking constraints are independent of each other and distributed throughout the design removal of the complete watermark is extremely difficult. Therefore, ownership of Alice can be proved. Additionally, sometimes tampering of embedded watermarking constraints may force to perform all the pre-synthesis steps from the beginning, which is a costly and time-consuming process. The presented watermarking in this chapter is also capable to resolve this threat.

4.4 Analysis on case studies

In this section, we discuss the triple-phase watermarking process (Sengupta *et al.*, 2018) on case studies/test cases and compare it with a multiple-variable single-phase watermarking process (Sengupta *et al.*, 2016) and a dual-variable single-phase watermarking process (Koushanfar *et al.*, 2005). This section is divided into three sub-sections: (i) system setup, (ii) security analysis of triple-phase watermark, and (iii) design cost analysis of triple-phase watermark.

The baseline (original non-watermarked) design, dual-variable single-phase watermark design (Koushanfar *et al.*, 2005), multi-variable single-phase watermark design (Sengupta *et al.*, 2016), and triple-phase watermark design (Sengupta *et al.*, 2018) were all implemented in Java 8 and executed on Windows platform with 4 GB DDR3 primary memory with processor frequency of 3.20 GHz 15-nm NanGate based technology scales (NanGate Open Cell Library, 2018) is employed to assess both watermarked design area and design latency. While calculating the design cost, equal weightage is provided to both design area and design latency.

4.4.1 Security analysis of triple-phase watermark for DSP IP cores

The possibility of generating the same solution before and after embedding watermark is calculated based on the probability of coincidence (P_c) metric. Lower P_c value indicates a higher strength of watermark. The metric partially adopted from Koushanfar *et al.* (2005) can be expressed below:

$$P_c = \left(1 - \frac{1}{c \times \prod_{i=1}^{D} N(R_i)}\right)^w, \tag{4.10}$$

where, 'w' is the watermark strength/size, $N(R_i)$ is the number of functional unit of type i, 'D' indicates the type of functional unit, and 'c' is the number of colors used

Table 4.13 Comparison of probability of coincidence between, triple-phase (Sengupta et al., 2016), dual-variable (Koushanfar et al., 2005), and multi-variable (Sengupta et al., 2016) for W = 80

Benchmarks	# of register before watermark	P_c			# of times lower P_c of triple-phase approach compared with dual and multi variable
		Triple-phase	Dual-variable	Multi-variable	
ARF	8	3.3×10^{-27}	2.2×10^{-5}	2.2×10^{-5}	6.9×10^{21}
JPEG	12	1.9×10^{-23}	9.4×10^{-4}	9.4×10^{-4}	5.0×10^{19}
DWT	5	8.3×10^{-35}	1.7×10^{-8}	1.7×10^{-8}	2.1×10^{26}
MPEG	14	3.8×10^{-31}	2.6×10^{-3}	2.6×10^{-3}	6.9×10^{27}
DCT	8	3.7×10^{-21}	2.2×10^{-5}	2.2×10^{-5}	6.1×10^{15}

in CIG before embedding watermark. Table 4.13 reports the comparative study of the probability of coincidence between triple-phase watermarking (Sengupta *et al.*, 2016, 2018; Koushanfar *et al.*, 2005) for signature size 80. It can be observed from Table 4.13 that the triple-phase approach achieves 1.4×10^{27} times lower P_c value in average compared with Sengupta *et al.* (2016) and Koushanfar *et al.* (2005) which indicates the higher robustness of watermark. Low P_c value is desirable which is achievable by incrementing the watermark size. However, too bigger size of watermarking constraints may result in register overhead, thereby increasing the design cost. According to Sengupta *et al.* (2016), the desirable maximum range of signature size should be 80.

A watermark has higher tamper tolerance (T_t) ability if finding the correct signature from a watermarked design is highly complex and time-consuming. If the total number of possible signature combination is large then identifying the exact watermark for an attacker through brute-force search is time consuming and cost expending process. The total number of possible signature combinations for '*v*' variable encoding and '*w*' digits watermark can be calculated by the following expression:

$$T_t = v^W, \tag{4.11}$$

Table 4.14 reports the comparison of the total number of possible signature combinations for different watermark sizes for triple-phase watermarking (Sengupta *et al.*, 2016, 2018; Koushanfar *et al.*, 2005). It can be observed from Table 4.14 that the triple-phase approach achieves higher T_t value for all watermark size compared with Sengupta *et al.* (2016) and Koushanfar *et al.* (2005) which thus indicates higher tamper tolerance ability.

4.4.2 Design cost analysis of triple-phase watermark for DSP IP cores

Design cost of the triple-phase watermark is calculated based on area and latency of the watermarked IP core design, where watermarked design cost is evaluated based

Table 4.14 Comparison of total number of possible signature combinations between triple-phase (Sengupta et al., 2016), dual-variable (Koushanfar et al., 2005), and multi-variable (Sengupta et al., 2016)

Signature size (w)	**# of possible signature combination (T_t)**			**# of times higher tamper-tolerance of proposed approach compared with dual and multi-variable**	
	Triple-phase	**Dual-variable**	**Multi-variable**	**Dual-variable**	**Multi-variable**
15	4.8×10^{12}	32,768	10.7×10^{8}	14.5×10^{7}	4421
30	2.3×10^{25}	1.1×10^{9}	1.2×10^{18}	2.1×10^{16}	19.5×10^{6}
45	1.1×10^{38}	3.5×10^{13}	1.2×10^{27}	3.0×10^{24}	8.6×10^{10}
60	5.1×10^{50}	1.2×10^{18}	1.3×10^{36}	4.4×10^{32}	3.8×10^{14}
80	4.1×10^{67}	1.2×10^{24}	1.5×10^{48}	3.4×10^{43}	2.8×10^{19}

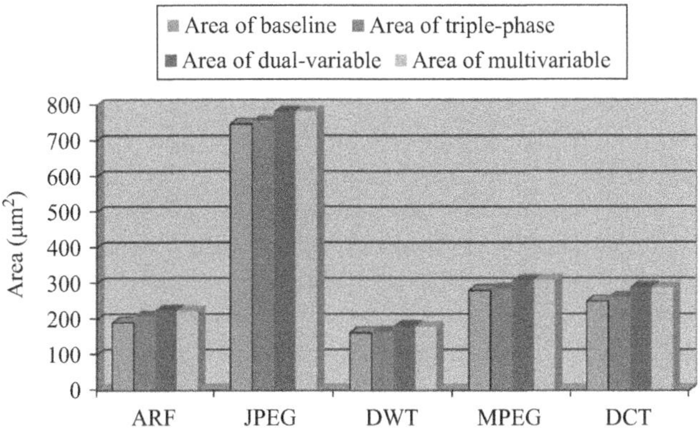

Figure 4.34 Comparison of design area of baseline, triple-phase (Sengupta et al., 2016), dual-variable (Koushanfar et al., 2005), and multi-variable (Sengupta et al., 2016)

on the following expression:

$$C_f(X_i) = w_1 \frac{L_{\text{total}}}{L_{\max}} + w_2 \frac{A_{\text{total}}}{A_{\max}}, \tag{4.12}$$

where, $C_f\,(X_i)$ is the watermarked design cost for user provided resource configuration X_i, L_{total}, and A_{total} are total watermarked design latency and area, respectively. The total area and latency of design are calculated based on the models proposed in Sengupta *et al.* (2016). $L_{\max}$ and $A_{\max}$ indicate maximum design latency and area based on the available resources in the module library, w_1 and w_2 are weightage parameters for latency and area, respectively. Figure 4.34 shows the comparison of design area between baseline and triple-phase (Sengupta *et al.*,

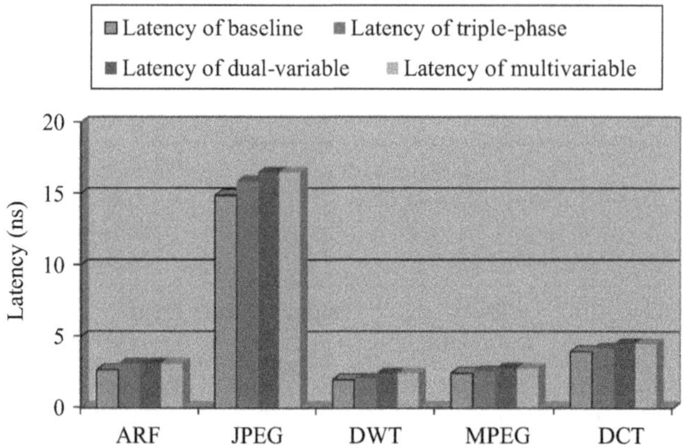

Figure 4.35 Comparison of design latency of baseline, triple-phase (Sengupta et al., 2016), dual-variable (Koushanfar et al., 2005), and multivariable (Sengupta et al., 2016)

2016, 2018; Koushanfar *et al.*, 2005). Triple-phase watermark achieves lesser design area compared with Sengupta *et al.* (2016) and Koushanfar *et al.* (2005) and minimal area overhead compared with baseline design. Figure 4.35 shows the comparison of design latency between baseline and triple phase (Sengupta *et al.*, 2016, 2018; Koushanfar *et al.*, 2005). Triple-phase watermark achieves lesser or equal design latency compared with Sengupta *et al.* (2016) and Koushanfar *et al.* (2005) and minimal latency overhead compared with baseline design. The comparison of design cost between baseline and triple phase (Sengupta *et al.*, 2016, 2018; Koushanfar *et al.*, 2005) is reported in Table 4.15. The design cost overhead of triple-phase watermark compared with the baseline design and the reduction of design cost compared with Sengupta *et al.* (2016) and Koushanfar *et al.* (2005) are also reported in Table 4.15. It can be observed that the triple-phase watermarking produces on avg. 6.3% design cost overhead compared with the baseline design. However, it achieves on avg. 7.8% and 6.8% reduction of watermarked design cost overhead compared with Koushanfar *et al.* (2005) and Sengupta *et al.* (2016), respectively.

4.5 Conclusion

This chapter discussed in details about hardware watermarking of DSP application based IP cores used in CE devices. Several selected state-of-the-art watermarking approaches at different abstraction levels were reviewed. Two watermarking approaches used for DSP cores have been presented with greater details. These approaches implant watermark signature digits before RTL synthesis (during high-level synthesis). Embedding watermarking before RTL synthesis enables watermarking constraints to propagate to the lower levels of design. For example, how gate level design of a DSP core reflects the watermarking constraints that is

Table 4.15 Comparison of design cost between baseline, triple-phase (Sengupta et al., 2016), dual-variable (Koushanfar et al., 2005), and multi-variable (Sengupta et al., 2016)

Benchmarks	**Resource configuration**	**Design cost**				**Cost overhead %**	**Cost reduction %**	**Cost reduction %**
		Baseline	**Triple-phase**	**dual-variable**	**multi-variable**	**Triple-phase approach with respect to baseline**	**Triple-phase approach with respect to dual-variable**	**Triple-phase approach with respect to multi-variable**
ARF	5(+), 3(*)	0.77	0.87	0.92	0.90	12.98	5.44	3.33
JPEG	5(+), 5(*)	0.72	0.76	0.79	0.79	5.56	3.80	3.80
DWT	2(+), 4(*)	0.78	0.81	0.93	0.92	3.85	12.90	11.96
MPEG	3(+), 8(*)	0.73	0.76	0.81	0.81	4.11	6.17	6.17
DCT	6(+), 3(*)	0.80	0.84	0.94	0.92	5.00	10.64	8.70

embedded at its higher design level is explained and demonstrated with examples in this chapter. This chapter also lucidly explains the design flow of designing a watermarked DSP core with some industry standard examples of FIR filter, IIR low-pass Butterworth filter, etc. Details on security metrics were also presented and analyzed. We also demonstrate why hardware watermarking is an extremely useful tool for combating false claim of ownership, IP piracy, forgery, and other attacks.

4.6 Exercises

1. What is watermark?
2. What is H/W watermark?
3. What are the desired properties of hardware watermark?
   (a) What are the steps in HLS where watermark can be embedded?
   (b) Explain advantages and disadvantages of each?
4. What is fault tolerance property of watermark?
5. What threat does V protect against?
   (a) What are the IP threats which watermark cannot protect?
6. What is probability of coincidence?
7. What is signature detection process? Explain.
8. What is the trade-off between signature size and design overhead?
9. Where are the effects of single- and triple-phase watermark reflected in a watermarked design (RT level and gate level)?
10. What are the ways to increase the tamper tolerance ability of a watermarked design?
11. What is the effect of mobility on the γ digit of watermark encoding?
12. Implement a watermarked DSP core of an IIR filter for the following 16-digits signature '$\gamma\ \gamma\ \alpha\ \beta\ \beta\ \alpha\ \gamma$ iII!TIiTI.'
13. How to reduce the impact of watermark overhead on design latency and area?
14. What are the inputs to a triple-phase watermarking in HLS of DSP cores?
15. Explain the concept of mobility in scheduling?
16. What is the role of mobility in triple-phase watermarking of DSP cores?
17. Explain the pre-watermark embedding steps?
18. What is the role of colored interval graph in case of watermarking?
19. What is tamper tolerance in watermarking? How is it evaluated?
20. How do you estimate robustness of your watermark design?
21. Explain the different types of threat scenarios in case of false claim of ownership.
22. How does increase in watermark strength affect probability of coincidence?
23. What are the other alternative watermarking methods for DSP cores?
24. How cost overhead is reduced in watermarking?

References

Alkabani Y. and Koushanfar F., 'Active control and digital rights management of integrated circuit IP cores,' in *Proceedings of International Conference on Compilers, Architecture. and Synthesis of Embedded Systems (CASES)*, 2008, pp. 227–234.

Altera Quartus, Available: https://dl.altera.com/13.0sp1, Last accessed on 1 February 2018.

Coussy P., Gajski D. D., Meredith M., and Takach A., 'An introduction to high-level synthesis,' in *IEEE Design & Test of Computers*, 2009, vol. 26, no. 4, pp. 8–17.

Cox J., Doërr G., and Furon T., 'Watermarking is not cryptography,' in *Proceedings of International Workshop on Digital Watermarking*, 2006, pp. 1–15.

Cyclone II FPGA, Available: https://www.altera.com/products/fpga/cyclone-series/cyclone-ii/overview.html, last accessed on 2018, Last accessed on 1 February 2018.

DSP Benchmark Suite, Available: http://www.ece.ucsb.edu/EXPRESS/benchmark/, Last accessed on 1 April 2018.

Guajardo J., Kumar S. S., Schrijen G. J., and Tuyls P., 'Brand and IP protection with physical unclonable functions,' in *Proceedings of IEEE International Symposium on Circuits and Systems*, 2008, pp. 3186–3189.

Koushanfar F., Hong I., and Potkonjak M., 'Behavioral synthesis techniques for intellectual property protection', *ACM Transactions on Design Automation of Electronic Systems*, 2005, vol. 10 (3), pp. 523–545.

Le Gal B. and Bossuet L., 'Automatic low-cost IP watermarking technique based on output mark insertions,' *Design Automation Embedded Systems*, 2012, vol. 16 (2), pp. 71–92.

Mohanty S. P., Nanoelectronic Mixed-Signal System Design. McGraw-Hill Education, New York, 2015, no. 0071825711.

Mohanty S. P., Sengupta A., Guturu P., and Kougianos E., 'Everything you want to know about watermarking: From paper marks to hardware protection: From paper marks to hardware protection,' *IEEE Consumer Electronics Magazine*, 2017, vol. 6 (3), pp. 83–91.

NanGate, 15 nm open cell library. Available: http: //www.nangate.com/?pageid=2328, last accessed on 2018.

Roy S. D., Li X., Shoshan Y., Fish A., and Yadid-Pecht O., 'Hardware implementation of a digital watermarking system for video authentication,' *IEEE Transactions on Circuits and Systems for Video Technology*, 2013, vol. 23 (2), pp. 289–301.

Salivahanan S., Vallavaraj A., and Gnanapriya C., 'Digital signal processing' Tata McGraw-Hill, ISBN 007463996X, vol. xiii, 2000, pp. 808.

Sengupta A., 'Intellectual property cores: Protection designs for CE products,' *IEEE Consumer Electronics Magazine*, 2016, vol. 5 (1), pp. 83–88.

Sengupta A., 'Hardware security of CE devices,' *IEEE Consumer Electronics Magazine*, 2017, vol. 6 (1), pp. 130–133.

Sengupta A. and Bhadauria S., 'Exploring low cost optimal watermark for reusable IP cores during high level synthesis,' *IEEE Access*, 2016, vol. 4, pp. 2198–2215.

Sengupta A. and Roy D., 'Antipiracy-aware IP chipset design for CE devices: A robust watermarking approach', in IEEE Consumer Electronics Magazine, *2017*, vol. 6 (2), pp. 118–124.

Sengupta A. and Sedaghat R., 'Integrated scheduling, allocation and binding in High Level Synthesis using multi structure genetic algorithm based design space exploration,' *International Symposium on Quality Electronic Design*, California, 2011, pp. 1–9.

Sengupta A. and Sedaghat R., 'System and methodology for development of a system architecture using optimization parameters', *US Patent by United Sates Patent and Trademark Office* (USPTO), 2013, Patent no. US 8,397,204.

Sengupta A., Bhadauria S., and Mohanty S. P., 'TL-HLS: Methodology for low cost hardware Trojan security aware scheduling with optimal loop unrolling factor during high level synthesis,' *IEEE Transactions on CAD of Integrated Circuits and Systems*, 2017, vol. 36 (4), pp. 655–668.

Sengupta A., Roy D., and Mohanty S. P., 'Triple-phase watermarking for reusable IP core protection during architecture synthesis,' in *IEEE Transactions on CAD of Integrated Circuits and Systems*, 2018, vol. 37 (4), pp. 742–755.

Voyatzis G. and Pitas I., 'The use of watermarks in the protection of digital multimedia products,' *Proceedings of IEEE*, 1999, vol. 87(7), pp. 1197–1207.

Ziener D. and Teich J., 'Power signature watermarking of IP cores for FPGAs,' *Journal of Signal Processing Systems*, 1999, vol. 51 (1), pp. 123–136.

Chapter 5

Symmetrical protection of DSP IP core and integrated circuits using fingerprinting and watermarking

This chapter discusses the use of watermarking and fingerprinting for symmetrical protection of DSP IP cores and integrated circuits. Symmetrical IP core protection is a mechanism in which both seller and buyers of an IP can have signature for double-proof of ownership and whereas significantly reducing false ownership claims. The chapter is structured as follows: Section 5.2 discusses the fundamentals of IP core protection with emphasis on symmetrical IP core protection techniques for DSP cores; Section 5.3 discusses the low-cost DSP core IP core protection; Section 5.4 explained that methodology with a motivational example. Section 5.5 discusses the results of that approach. The conclusion of this chapter is provided in Section 5.6.

5.1 Introduction

The designs of integrated circuits have evolved into greater sophistication than ever before with the revolution of implementation and application technology in consumer electronics industry. The usage of digital signal processing (DSP) intellectual property (IP) cores (Castillo *et al.*, 2007) generated using architectural synthesis not only maintains a proper balance between time-to-market pressure and design productivity but additionally help in design cost reduction. High-level synthesis (HLS) (Sengupta and Sedaghat, 2011, 2013) is an automated design process that transforms an algorithmic/behavioral description of a digital circuit or IP into its corresponding register transfer level (RTL) digital hardware through numerous sub-processes like scheduling, hardware and register allocation, and binding (Mohanty, 2015). In the process of manufacturing an IP core, two entities are involved, viz. seller and buyer. An IP seller also known as IP vendor is the creator of an IP, whereas an IP buyer also known as IP user is the purchaser of an IP. However, to maintain a viable application of IP cores in composite SoC-based designs, protection of both the entities against threats is extremely crucial. Let us now discuss on the protection aspect of IP buyer and IP seller. In an IP core, an IP buyer may claim buyer privilege as a buyer so that the same IP copy should not be usable/accessible to his competitors in the market. This is possible when customized specifications of an IP core are obtained by an IP seller from an IP buyer, thus creating an exclusive one-to-one mapping between both parties. Embedding a unique buyer's signature (known as buyer fingerprint) into an IP core design facilitates detection of unlawfully redistributed/resold duplicates of an IP core by a deceitful seller (Gang and Potkonjak, 2007). Likewise, an IP seller must protect his design from piracy and false claim of ownership before selling it to an IP buyer. Embedding a unique seller's signature (known as seller watermark) into an IP core design protects an IP core from ownership abuse (Alkabani *et al.*, 2007; Sengupta and Kundu, 2017; Sengupta *et al.*, 2017, 2018; Kougianos *et al.*, 2009; Maltoni *et al.*, 2009; Roy *et al.*, 2008; Alkabani and Koushanfar, 2008; Sengupta, 2016, 2017; Sengupta and Roy, 2017).

Therefore, symmetrical protection of DSP IP cores provides user right as well as nullifies IP ownership abuse. A thematic summary is highlighted in Figure 5.1. Significant investments are required in terms of man-power, time and cost to design, and verify an IP core; however, it can be cloned, modified, or resold easily by a rough entity. Therefore, both seller and buyer of IP designs want to nullify the possibilities of false claiming and illegal reselling of an IP core. Symmetrical protection can be achieved by implanting dual entity signature scheme in a DSP IP core during any of the design abstraction level. Standard IP protection methods (Yuan *et al.*, 2006) such as patents, copyrights, trademarks, etc., are distributed, restricted in capacities, and insufficient in securing IP cores from buyer and seller viewpoint. Additionally, several signature embedding schemes in different multimedia applications, like image, text, video, and audio are also not appropriate to protect an IP core as it does not preserve the functionality and precision of a design in signature insertion. Thus, from the perspective of IP core protection, necessary

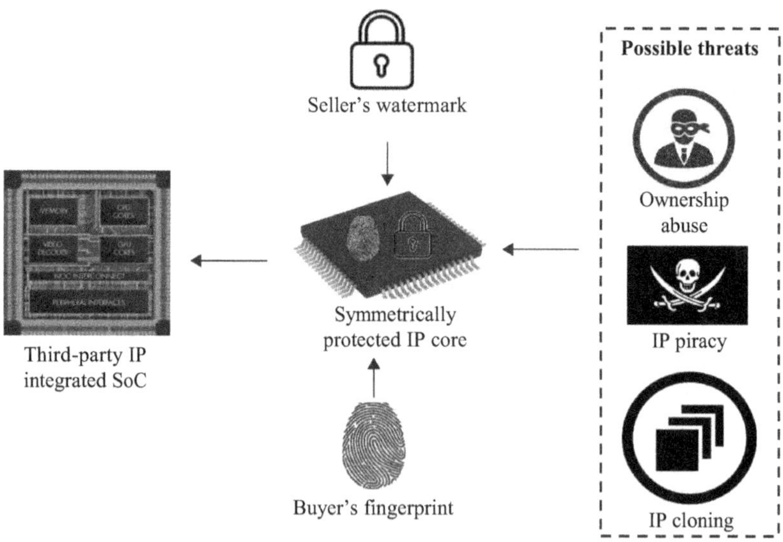

Figure 5.1 Thematic overview of symmetrical IP core protection

action needs be taken to preserve the functionality and correctness of the design while embedding any signature.

5.1.1 Background on watermark and fingerprint

Watermark and fingerprint represents the signature of two distinct entities of a DSP reusable IP core to protect their corresponding ownership right. While fingerprint conveys the signature of an IP buyer, watermark is intended for an IP seller. An ideal signature for both cases must meet the below features:

- It must be implanted as vital part of an IP design so that it must not be perceptible through traditional scanning.
- It must be inimitable in nature so that the proof of ownership can be established easily.
- For manifold signature implanting mechanism, each signature must be self-governing of each other.
- It should be tamper tolerant so that partial removal of signature should not impact on the proof on ownership.
- It should be resilient against attacks like reverse engineering.
- Low embedding runtime.
- Nominal latency and hardware overhead after implanting signatures.
- Seamless signature detection process for an authentic owner with comprehensive cognizance of encoding rules (Koushanfar *et al.*, 2005; Cox *et al.*, 2007).
- Must uphold logic correctness of an IP core post-implanting signatures.

5.1.2 Threat model

The ultimate goals of IP core security are:

1. ***Buyer's standpoint:*** detect unlawfully redistributed/resold copies in the market by a deceitful IP seller (Gang and Potkonjak, 2007).
2. ***Seller's standpoint:*** security against ownership exploitation (Yu and Zhu, 2011).

5.1.3 Benefits of protection at higher abstraction

Although DSP IP core protection can be provided in any of the design abstraction level, however, embedding buyer fingerprint and seller watermark architecture level (Mohanty *et al.*, 2002, 2014) are beneficial as the signature constraints embedded in higher design abstraction level will automatically propagate in the lower design abstraction level; thus, it will protect the higher level as well as the lower level of the design. Additionally, symmetrical protection in higher level can be achieved in lesser signature embedding complexity while incurring lower design overhead. Roy and Sengupta (2017) presented a multi-variable fingerprinting and watermarking-based symmetrical IP core protection mechanism during multiple sub-steps of architectural synthesis/HLS process. Additionally, the approach incurs minimal design overhead while incorporating both signature constraints.

5.2 Fundamentals of IP core protection

This chapter discusses fundamentals of different reusable IP core protection mechanisms (shown in Figure 5.2) with emphasis on higher design abstraction level-based IP core protections that are available in the literature and how they are different when compared with symmetrical IP core protection. Reusable IP core protection mechanism can be classified into two parts: (a) non-symmetric approach and (b) symmetric approach.

5.2.1 Overview on non-symmetric IP core protection techniques

According to the literature, non-symmetric IP core protection techniques mainly provide protection from a seller's perspective. It can be performed in three different techniques: (i) hardware obfuscation, (ii) computational forensic engineering (CFE), and (iii) watermark.

1. ***Hardware obfuscation:*** Obfuscation is the process of transforming an original application or design into its functionally equivalent form to make the reverse engineering process significantly more challenging (Chakraborty and Bhunia, 2009). It can protect an IP core either by obscuring the IP design/code, known as structural obfuscation (Hada, 2000; Sengupta *et al.*, 2017) or by generating a locked netlist (Chakraborty and Bhunia, 2009). A person who has the proper knowledge of the design of the correct key combination can only use it. In other words, structural obfuscation hides the functionality of an IP core whereas functional obfuscation hides the implementation of an IP core.

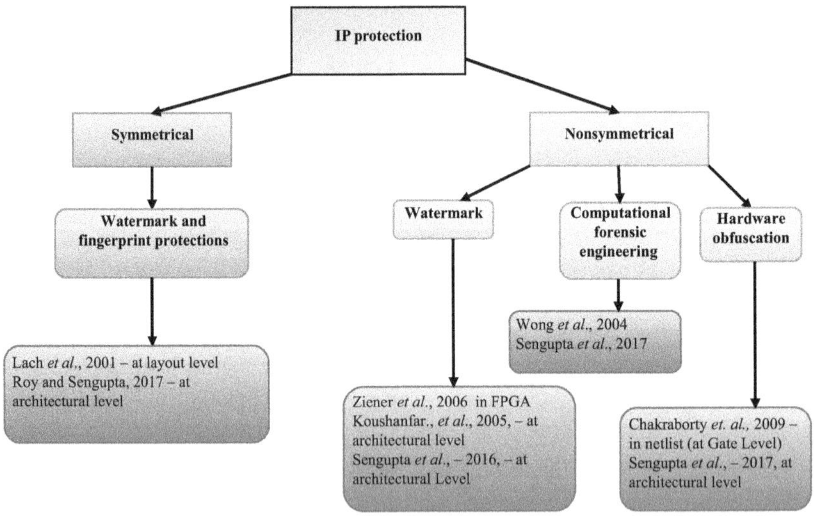

Figure 5.2 Classification and corresponding works on IP core protection

2. ***Computational forensic engineering:*** Computational forensic engineering (CFE) is a signature-free driven process which can be performed in three steps, feature collection, feature extraction, and IP validation (Sengupta and Kachave, 2018). It tries to identify a list of features that can be used to identify an IP core uniquely (Wong *et al.*, 2004). In case of ownership conflict, these features can be used to recognize the real owner of the IP core.
3. ***Watermark:*** IP protection of a single entity (for IP seller only) can be achieved by embedding a seller watermark in to the IP core design. Watermark can be inserted either at lower design abstraction level (Ziener and Teich, 2006, 2008; Nie *et al.*, 2013) or at higher design abstraction level (Sengupta and Bhadauria, 2016; Koushanfar *et al.*, 2005; Hong and Potkonjak, 1999; Le Gal and Bossuet, 2012; Sengupta *et al.*, 2016). Watermark at lower design level can be detected from the power supply pins of field programmable gate array (FPGA) (Ziener and Teich, 2006), or from the bit-stream of the IP core design (Nie *et al.*, 2013), or by tracing the voltage (Ziener and Teich, 2008), etc. On the other hand, watermark at higher design level be identified from the datapath and/or the control path of the IP design.

5.2.2 Overview on symmetric IP core protection techniques

The main goal of symmetrical IP core protection is to generating unique one to one mapped IP core between IP user and IP seller. From the seller side, it will not only protect the ownership of the IP seller but also he can identify the traitors among the list of IP buyers. If all the buyers get identical IP from the seller and one of them may unlawfully redistribute it identifying the actual dishonest buyer will become insurmountable (Gang and Potkonjak, 2007). There are only two approaches in the literature which have presented symmetrical IP core protection techniques. In the approach presented in Lach *et al.* (2001), a hidden encrypted mark is embedded into the physical layout of a digital circuit when it is placed and routed onto the FPGA. This mark uniquely identifies both the source of the circuit and original recipient of the circuit, which is also difficult to remove. Thus, the approach provides protection for both the entities (IP buyer and IP seller) but in lower design abstraction level, i.e. layout level, not in higher design abstraction level. On the other hand, the approach presented in Roy and Sengupta (2017) provides symmetrical protection during HLS. This approach embeds hidden buyer's and seller's signature constraints as multi-variable fingerprint and watermark during register allocation and scheduling stages of architectural synthesis. Signature embedding process is not time consuming and less complex; additionally, overhead generated due to additional constraints is significantly low. Furthermore, detail explanation of this approach is discussed in the next section.

5.3 Symmetrical IP core protection for DSP core

The symmetrical protection approach presented in Roy and Sengupta (2017) covertly inserts fingerprinting and watermarking constraints to protect the IP buyer as well as the IP seller during HLS. Based on the signature constraints provided by a buyer, an IP seller inserts buyer fingerprint with his own watermark; thus, safeguards the privileges of seller and buyer against popular IP threats like IP piracy, IP cloning, false claim of ownership, etc. (discussed in Section 5.1). Both buyer fingerprint and seller watermark taken as input from the respective entities are decoded to obtain the corresponding signature constraints. Firstly, the fingerprinting constraints are embedded in scheduling and register allocation phase within HLS (shown in Figure 5.3). Once the fingerprint implanted design is obtained; then, lastly the watermarking constraints are implanted in register allocation phase. Thus, a DSP IP core design has both fingerprinted and watermarked constraints, in addition to its normal design constraints (i.e. module library, resource configurations, etc.). The multiplexing process for each resource is generated to identify the resource interconnectivity, multiplexer size, and inputs/outputs of the watermarked and fingerprinted design. Using generated multiplexing scheme of each resource, the complete datapath and timing description of the control unit of the DSP IP core are developed. Thus, the final output of a symmetrically protected RTL design of an IP core is created as datapath and controller where each component is designed using hardware description language (HDL).

Thereafter, these HDL files are fed into computer-aided design (CAD) software for RTL synthesis. It produces optimal gate level description of the respective design. This is then subsequently verified and transformed to corresponding bitstreams for the validation on a programmable hardware. After being validated, it is then fed into CAD tools for physical designing. Lastly, the GDS design file generated is sent to the fab.

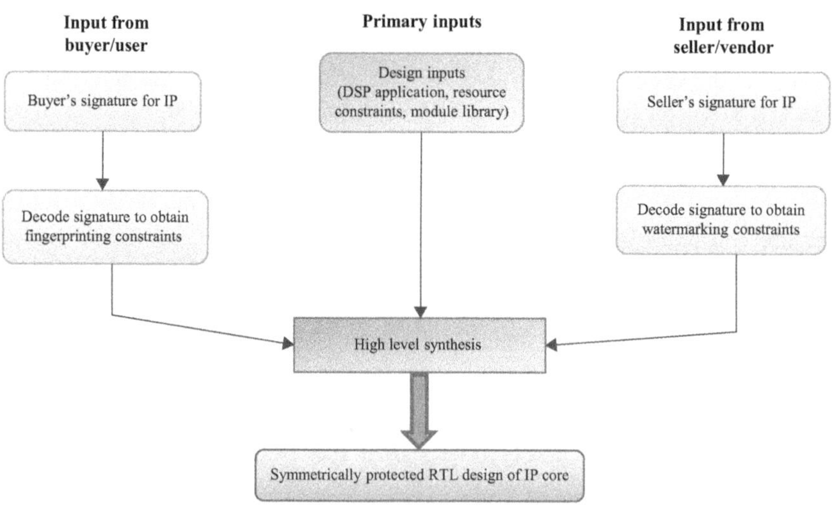

Figure 5.3 Symmetrical IP core protection mechanism for DSP cores

5.3.1 Problem formulation

To determine an optimal symmetrically protected design solution for a particular data flow graph (DFG) within user specified area-delay constraints. It may be expressed as:

Minimize: Embedding cost (A_{total}, L_{total}) for an optimal design solution (X_i), where $(X_i) = N(R_1), N(R_2), \ldots, N(R_D)$; and $N(R_D)$ is the number of resource type R_D.

Subject to: $A_{\text{total}} \le A_{\text{cons}}$ and $L_{\text{total}} \le L_{\text{cons}}$ to symmetrically protect an IP core, where A_{total} and L_{total} are total area and latency of the symmetrically protected design and A_{cons} and L_{cons} are user provided area and latency constraints respectively.

5.3.2 Symmetrically protected design—area evaluation model

Total design area (A_{total}) consumed by a symmetrically protected reusable IP core is formulated by the following model (Roy and Sengupta, 2017):

$$A_{\text{total}} = \sum_{i=1}^{D} A(R_i) * R_i + A(\text{mux}) * N(\text{mux}) + A(\text{buffer}) * N(\text{buffer}) \qquad (5.1)$$

Total design area is evaluated by adding total area of hardware resource units (like adder, subtractor, multiplier, etc.), interconnecting units (such as multiplexer and de-multiplexer), and storage units (such as register). The total design area is evaluated for module library of 15-nm NanGate-based technology (NanGate, 2017).

5.3.3 Symmetrically protected design—delay evaluation model

Total design latency (L_{total}) of symmetrically protected reusable IP core can be expressed by the metric below (Roy and Sengupta, 2017):

$$L_{\text{total}} = \text{total}_{\text{cs}} * \text{delay}_{\text{cs}} \qquad (5.2)$$

where total_{cs} and delay_{cs} indicate the total number of control step (CS) of the design and execution delay of each CS, respectively. The total design latency is evaluated with respect to module library of 15-nm NanGate-based technology (NanGate, 2017) with 1, GHz clock frequency.

5.3.4 Symmetrically protected design—cost evaluation function

The design cost is assessed by following metric (Roy and Sengupta, 2017):

$$C_f(X_i) = w_1 \frac{A_{\text{total}}}{A_{\max}} + w_2 \frac{L_{\text{total}}}{L_{\max}} \qquad (5.3)$$

where, $C_f(X_i)$ is the design cost of the symmetrically protected reusable IP core with resource configuration X_i, $A_{\max}$, and $L_{\max}$ are the maximal area and latency in the design space, $w1$ and $w2$ are user provided area and latency weightage factor, to provide equal preference both is kept as 0.5.

5.3.5 *Encoding rules of buyer fingerprint and seller watermark for DSP IP cores*

Signatures for both parties can be embedded into the design during HLS as encoded additional constraints. Roy and Sengupta (2017) encoded fingerprinting and watermarking constraints are implanted covertly in scheduling and register allocation phases of DSP core design methodology (using HLS as backbone). More specifically, buyer fingerprint constraints in scheduling step are performed by forcing particular operations in a particular CS while resolving scheduling conflict. It is a very intelligent process of embedding buyer fingerprint, as it will incur zero hardware and may incur nominal latency overhead. Additionally, during register allocation, the fingerprinting and watermarking constraints are inserted with the notion of colored interval graph (CIG), where each node of the graph indicates storage variables of the design and the connecting edges indicates execution-time overlapping between conforming storage variables.

More explicitly, if there is execution-time overlap between two storage variables, then there will be an edge between those corresponding nodes. Therefore, injecting extra edges as encoded signature constraints will force storage variables of a CIG to implement using distinctive registers. Corresponding illustration of the approach devised in Roy and Sengupta (2017) is shown is Figure 5.4.

Fingerprint encoding rules: The fingerprint encoding process devised in Roy and Sengupta (2017) embeds additional constraints in both scheduling and register allocation phases of DSP core design to enhance the signature strength. The definition of fingerprint encoding rule for each variable devised in Roy and Sengupta (2017) is as follows:

- A = While resolving scheduling conflict force even operation in odd CS in scheduling phase.
- B = While resolving scheduling conflict force odd operation in even CS in scheduling phase.
- C = Add additional edge between node pair number (odd, odd) in the CIG.

The operations need to be sorted initially based on their operation number to apply the encoded fingerprinting constraints for ‘A’ and ‘B.’ Similarly, the storage variables need to be sorted before applying encoded fingerprinting constraints for ‘C’ (note: this encoded fingerprint embedding approach is demonstrated using a working application in the upcoming section). The robustness of buyer fingerprint is directly proportional with the signature strength. Roy and Sengupta (2017) are capable to accept and embed any big size fingerprint. As seen, this fingerprint embedding process includes of three different variables and each variable carries a distinctive encoding rule which un-invites significant overhead to the design with respect to area and latency while preserving robustness.

It is recommended that while selecting buyer fingerprint, the signature should have higher number of ‘A’ and ‘B’ digits than ‘C’ digits. The reason is that, the encoding of ‘A’ and ‘B’ digit are formulated in a way that it incurs nil hardware area and trivial latency overhead. This is conceivable as both the digits

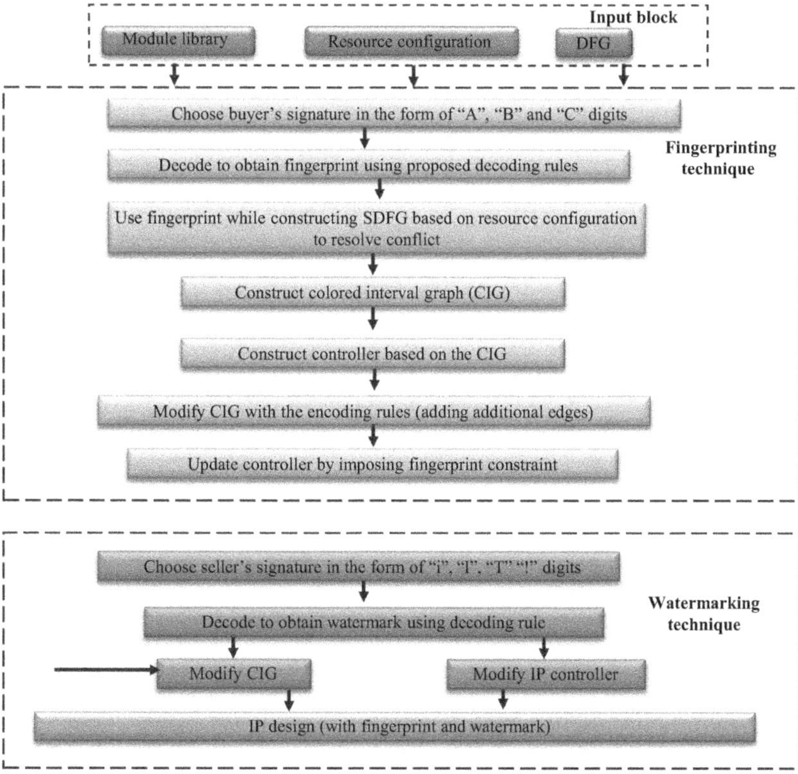

Figure 5.4 Design flow of symmetrical IP core protection mechanism for DSP cores using fingerprint and watermark

are injected while resolving operation scheduling conflict, i.e. fingerprint encoding rule acts as conflict resolver. On the contrary, too many 'C' digits in the signature may yield hardware overhead. Nonetheless, it is a key digit as it augments additional security layer by implanting the fingerprint constraint in another phase of HLS.

Watermark encoding rules: The watermark encoding mechanism used in Roy and Sengupta (2017) is adopted from (Sengupta and Bhadauria 2016). In (Sengupta and Bhadauria 2016), supplementary constraints for watermark are implanted in the register allocation by injecting the extra edges between the nodes of a CIG. Inserting extra edges mandates storage variables of a CIG to be implemented in distinctive registers. This methodology is flexible for injecting any number of extra edges in the register allocation step. The robustness of the watermark surges with the number of extra edges (i.e. signature magnitude) implanted. Although strength upsurges by increment in size of watermark signature, a bulky signature invites major design overhead. Consequently, it becomes obligatory to use a strong watermarking that delivers extensive protection with marginal design cost burden. As mentioned earlier, (Roy and Sengupta 2017) use a multi-variable watermark process from (Sengupta and Bhadauria 2016). The watermark generation mechanism in Sengupta and Bhadauria (2016) consist of four different variables. The seller can generate unique watermark through an arbitrary blend of these four variables. Each variable depicts a unique encoding rule injects an extra edge between two specific nodes. More explicitly, choice of node pair from CIG to insert an extra edge is well explained in the encoding scheme. The explanation of the encoding scheme has been described below:

- i = Add additional edge between node pair number (prime, prime) in the CIG.
- I = Add additional edge between node pair number (even, even) in the CIG.
- T = Add additional edge between node pair number (odd, even) in the CIG.
- ! = Add additional edge between node pair number (0, any integer) in the CIG.

Although ‘C’ digit of fingerprint and ‘i’, ‘I’, ‘T’, and ‘!’ digits of watermark, implant signature constraints in register allocation phase through CIG, nevertheless as their encoding rules are distinctive; therefore, the ensuing constraints will be totally dissimilar.

5.3.6 Multi-variable signature embedding process

The steps to achieve symmetrical IP protection by implanting fingerprint and watermark process are described in the rest of the section.

Process of implanting fingerprint

Broad steps:

1. Choose buyer signature.
2. Decode the signature to obtain equivalent fingerprinting constraints.
3. Resolve operation conflict during scheduling to embed fingerprinting constraints.
4. Construct the CIG based on assigning storage variables into register.
5. Perform re-allocation of register if necessary while inserting additional edges in CIG based on fingerprinting constraints.

Detailed steps:

1. Receive a preferred signature as random mixture of: ‘A,’ ‘B,’ and ‘C’ digits from buyer.
2. Construct a sorted list of operations of the DFG in increasing order based on their operation number.
3. Decode fingerprinting constraint corresponding to digits ‘A’ and ‘B’ that will help to resolve operation conflict during scheduling phase.
4. Initiate DFG scheduling based on the given resource configuration.
5. In case of operation conflict during scheduling, apply decoded fingerprint constraints (in step 3) to resolve the conflict. *Note*: *When there are no*

fingerprinting constraints available, conflict is broken based on sorted ordering.

6. Allocate storage variables in the partially fingerprinting constraints embedded scheduling and construct a CIG to identify the least number of registers compulsory for assignment.
7. Sort the allocated storage variables in increasing order based on their number.
8. Construct a timing table for register allocation based on the generated CIG.
9. Generate a list of extra edge pairs conforming to 'C' digit(s) by traversing the sorted storage variables in step 7.
10. Inject the extra edges in the CIG based on decoded fingerprint constraints and perform re-allocation of register if necessary.

After executing the aforesaid steps, a buyer fingerprint implanted IP design is produced. This buyer fingerprint implanted IP design acts as an input for implanting the seller watermark next.

Process of implanting watermark

Broad steps:

1. Choose seller signature.
2. Decode the signature to obtain equivalent watermarking constraints.
3. Construct the CIG from the fingerprint constraints embedded scheduling.
4. Perform re-allocation of register if necessary while inserting extra edges in CIG based on watermarking constraints.

Detailed steps:

1. Accept fingerprint constraints implanted IP design as an input.
2. Sort the allocated storage variables in increasing order based on their number.
3. Admit a chosen signature as random blend of 'i', 'I', 'T', and '!' from seller where each character carries a unique value of additional edges to be inserted.
4. Create a list of additional edge pairs corresponding to its encoded values by scanning the sorted nodes.
5. Inject the extra edges in the CIG of fingerprint implanted design as watermark constraints.
6. Alter the timing table design of register allocation based on generated watermark.

Post-execution of the aforesaid steps, an IP design with seller watermark is also produced. Therefore, the IP design contains signature of the dual entities.

5.3.7 Signature detection process

This process is obligatory for any encoded signature embedding scheme. As evident, in this phase, the implanted signature constraints are detected from an IP core.

It is a crucial process to resolve ownership conflict. Signature detection presented in Roy and Sengupta (2017) is a two-step process:

- *Design inspection*: Design inspection step collects the pertinent evidence in terms of specification, structural feature, etc., from a signature implanted IP core design. The objective is to rebuild the controller and the datapath.
- *Signature verification*: Signature verification step performs validation of the signatures in the received IP core. The existences of signature constraints are corroborated in the controller and datapath design to determine the seller's and buyer's right.

5.3.8 Desirable properties of signature

The signatures embedded into the symmetrically protected IP core design contain the properties below:

- *Lower embedding cost*: An ideal signature should incur marginal design overhead with respect to hardware area and latency. In Roy and Sengupta (2017), both fingerprint and watermark imposes low design overhead.
- *Resiliency against attacks*: An ideal signature must be so strong and robust that it cannot be removed or tampered easily by an attacker. In Roy and Sengupta (2017), both fingerprint and watermark are equipped with strong protection. In the absence of comprehensive information of both encoding rules, it is extremely challenging to eliminate or meddle the fingerprint and watermark. Additionally, it is also difficult to identify in which HLS step the signatures are embedded. Besides, the signatures' constraints are distributed all over the IP design which renders it more resistant against attacks.
- *Fault tolerance ability*: An ideal signature must be distributed enough throughout the design so that partial removal of the signature constraints does not prevent the proof of ownership. As in Roy and Sengupta (2017), the signature scheme dispenses the constraints all over the design; therefore, ownership remains intact even after fractional elimination of the signature constraints.
- *Adaptability to modern CAD tool*: An ideal signature must be easily adaptable into any modern CAD tool. In the approach (Roy and Sengupta, 2017), both the signature inserting procedure is well-suited to any CAD software. It can be incorporated effortlessly.
- *Lower time*: An ideal signature creation process must not be very composite to generate a signature. In Roy and Sengupta (2017), the signature creation time for both fingerprint and watermark are significantly low. Additionally, for an authentic entity (who has the comprehensive cognizance of encoding rules), the signature detection methodology is straightforward.

5.4 Case study of symmetrical IP core protection

This section is divided into two subsections: the first subsection demonstrates the fingerprint formation and embedding process while next portion demonstrates the watermark formation and embedding scheme of Roy and Sengupta (2017). As stated earlier, fingerprint constraints of 'A' and 'B' have been implanted during scheduling phase by allocating particular operations to particular CS, during schedule conflict resolution mechanism. On the other hand, variable 'C' injects extra edges in register allocation phase by forcing specific storage variables to implement in distinctive registers. On the contrary, watermarking constraints (as a mix of 'i', 'I', 'T' and '!') are implanted in the register allocation step by forcing specific storage variables to implement in diverse registers.

5.4.1 Demonstration of fingerprinting constraints embedding process

DFG of MESA benchmark is taken for demonstration purpose. Additionally, two adders and three multipliers are provided as resource configuration. The scheduled DFG of MESA benchmark using aforesaid resource configuration is shown in Figure 5.5. No signature constraint is embedded in that design. Corresponding tabular representation of the same scheduling is shown in Table 5.1. Now let us assume that a 7-digit buyer signature is provided as: **'AAABACC'**. The signature provided by the buyer is then deciphered. The signature and its respective deciphered constraints are depicted in Table 5.2. Deciphered constraints are attained using the fingerprint encoding rules described in Section 5.3.E. Now, the DFG of MESA benchmark needs is subjected to scheduling based on the resource configuration with fingerprinting constraints. In cs 1, four multiplication operations, viz.,

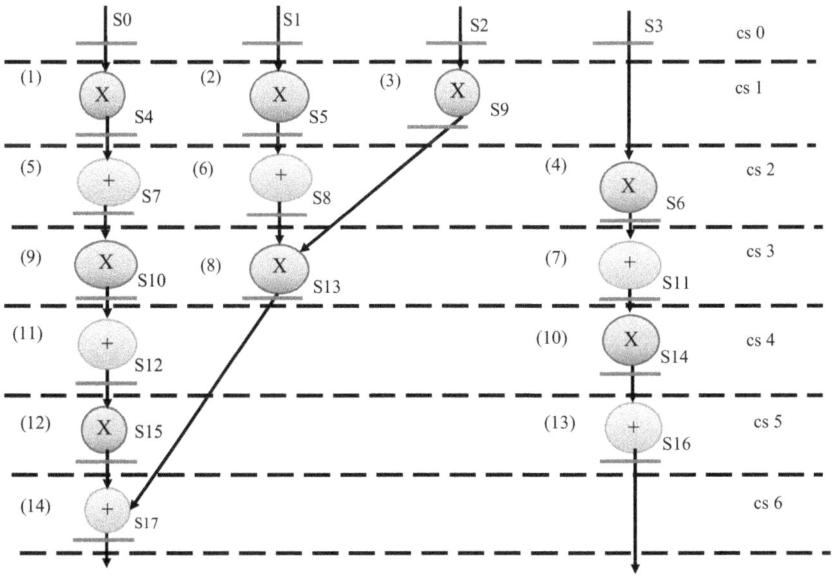

Figure 5.5 Scheduled DFG of MESA benchmark using two adders and three multipliers without embedding any constraints

Table 5.1 Tabular representation of scheduled MESA benchmark using two adders and three multipliers without embedding any constraints

CS	Operation number				
	Multiplier 1	**Multiplier 2**	**Multiplier 3**	**Adder 1**	**Adder 2**
cs 1	opn. 1	opn. 2	opn. 3	–	–
cs 2	opn. 4	–	–	opn. 5	opn. 6
cs 3	opn. 9	opn. 8	–	opn. 7	–
cs 4	opn. 10	–	–	opn. 11	–
cs 5	opn. 12	–	–	opn. 13	–
cs 6	–	–	–	opn. 14	–

Table 5.2 Fingerprint and watermark and its corresponding decoded constraints

Fingerprint			Watermark	
Desired signature (7-digit)	**Allocation of operation in corresponding cs based on constraints**	**Additional edge to add between nodes in CIG based on constraints**	**Desired signature (7-digit)**	**Additional edge to add between nodes in CIG based on constraints**
A	Allocate operation 2 in cs 1	–	i	(S2, S3)
A	Allocate operation 4 in cs 1	–	i	(S2, S5)
A	–	–	i	(S2, S7)
B	Allocate operation 5 in cs 2	–	I	(S2, S4)
A	Allocate operation 6 in cs 3	–	T	(S1, S2)
C	–	(S1, S3)	T	(S1, S4)
C	–	(S1, S5)	!	(S0, S1)

opn 1, opn 2, opn 3, and opn 4, are ready to be scheduled but only three multiplier units are present. This creates schedule operation conflict situation. This conflict can be resolved by inserting decoded fingerprint constraints in the scheduling phase. For instance, first digit 'A' forces even multiplication operation, i.e. opn 2 into odd CS, i.e. cs 1 while resolving scheduling conflict. Likewise, the next digit 'A' forces multiplication opn 4 in odd cs 1, while resolving the scheduling conflict. In case of the third digit 'A', no even operation is available to schedule; therefore, next multiplication is scheduled in cs 1 by merely picking the first operation from the sorted list. Hence, opn 1 is chosen and scheduled in cs 1. Similarly, for fourth digit 'B' odd operation, i.e. opn 5 is scheduled in even cs 2, while resolving scheduling conflict. Further even operation, i.e. opn 6 is scheduled in odd cs 3 for

Table 5.3 Tabular representation of scheduled MESA benchmark using two adders and three multipliers post implanting fingerprint constraints

CS	**Operation number**				
	Multiplier 1	**Multiplier 2**	**Multiplier 3**	**Adder 1**	**Adder 2**
cs 1	opn. 2	opn. 4	opn. 1	–	–
cs 2	opn. 3	–	–	opn. 7	opn. 5
cs 3	opn. 9	opn. 10	–	opn. 6	–
cs 4	opn. 8	–	–	opn. 13	opn. 11
cs 5	opn. 12	–	–	–	–
cs 6	–	–	–	opn. 14	–

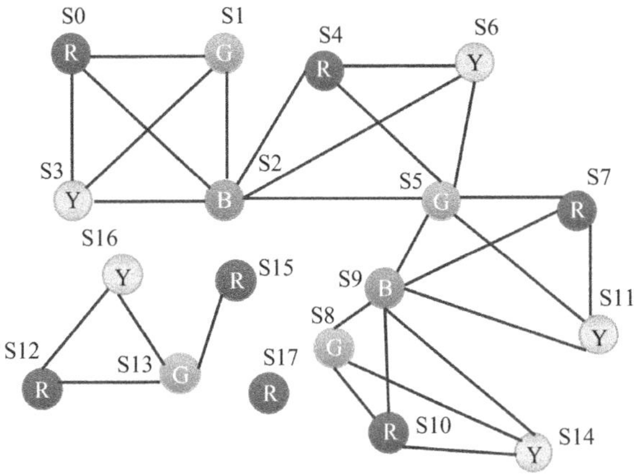

Figure 5.6 Colored interval graph before embedding any constraints as additional edge

fingerprint encoding of digit 'A'. This fingerprint constraint implanting mechanism in scheduling will operate up until all operation conflicts resolutions are done. The final schedule after embedding buyer fingerprint is shown in Table 5.3. It can be observed from Table 5.1 (no constraints embedded) and Table 5.3 (with fingerprint constraints) that both the scheduled design has same number of CS and same resource configuration. Consequently, the buyer fingerprint is implanted in the design with completely nil area and nil latency overhead, without changing the functionality.

Subsequently, the storage variables (S1–S17) are allocated in the fingerprint embedded scheduling and its equivalent CIG is generated (shown in Figure 5.6). In the CIG of Figures 5.6 and 5.7, red node is indicated by R, yellow node by Y, blue node by B, and green node by G. Similar convention is applicable for Tables 5.4 and 5.5. The timing table of register allocation of the same CIG is shown in Table 5.4. Corresponding to two 'C' digits, two additional edges need to be inserted in the generated CIG as fingerprint constraints. Although two extra edges have to

Table 5.4 Timing table for register allocation pre-implanting 'C' digits

cs	Register_Red	Register_Green	Register_Blue	Register_Yellow
cs 0	S0	S1	S2	S3
cs 1	S4	S5	S2	S6
cs 2	S7	S5	S9	S11
cs 3	S10	S8	S9	S14
cs 4	S12	S13	–	S16
cs 5	S15	S13	–	–
cs 6	S17	–	–	–

Table 5.5 Timing table for register allocation post implanting 'C' digits

cs	Register_Red	Register_Green	Register_Blue	Register_Yellow
cs 0	S0	S1	S2	S3
cs 1	S5	S4	S2	S6
cs 2	S5	S7	S9	S11
cs 3	S10	S8	S9	S14
cs 4	S12	S13	–	S16
cs 5	S15	S13	–	–
cs 6	S17	–	–	–

be inserted, but fortuitously an edge between S1, S3 is previously present in the CIG. Therefore, only one new edge between S1, S5 needs to be added. This indicates storage variable S1 and S5 must be executed through distinct register. According to Table 5.4, both S1 and S5 are executed through same register (Green), which indicates that modification in register allocation process is required. The revised timing table of register allocation is evident in Table 5.5, where S1 is executed through register green and S5 is executed through register red.

5.4.2 Demonstration of watermarking constraints embedding process

After implanting buyer fingerprinting constraints in the design, the fingerprint embedded design is now fed into the block for watermark insertion. For demo purpose, a 7-digit seller signature is chosen as: **'i i i I I T !'**. The signature and its equivalent deciphered constraints as extra edge between nodes in CIG are shown in Table 5.2. It can be observed using the CIG shown in Figure 5.6, only two extra edges ((S2, S7) and (S1, S4)) as constraints requires implanting in the CIG, because rest of the five edges are previously existing. Because of an extra edge injected between S1 and S4, both are forced to execute in distinctive registers. S2 and S7 are by default implemented using distinctive registers as apparent in Table 5.5. The final revised CIG is shown in Figure 5.7, and Table 5.6 depicts its corresponding

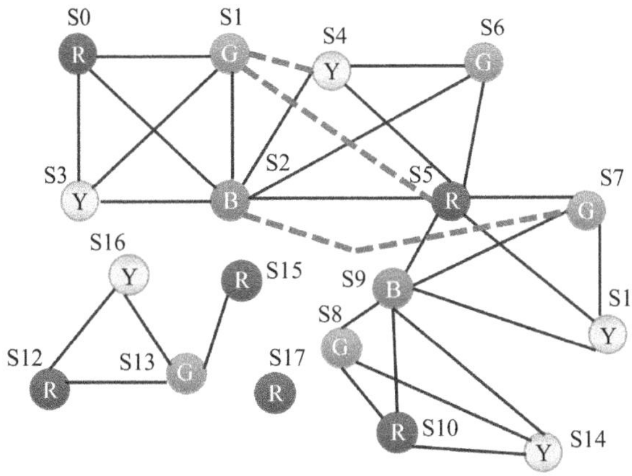

Figure 5.7 *Modified colored interval graph after embedding fingerprint and watermark constraints as extra edges*

Table 5.6 *Timing table for register allocation post implanting fingerprint and watermark constraints*

CS	Register_Red	Register_Green	Register_Blue	Register_Yellow
cs 0	S0	S1	S2	S3
cs 1	S5	S6	S2	S4
cs 2	S5	S7	S9	S11
cs 3	S10	S8	S9	S14
cs 4	S12	S13	–	S16
cs 5	S15	S13	–	–
cs 6	S17	–	–	–

timing table for register allocation. In Figure 5.7, dotted edge between S1 and S5 indicates fingerprint constraints and dotted edge between S2 and S7 indicates watermark constraints. It is evident that nil area and latency overhead are obtained after implanting watermark. Therefore, in case of MESA application, implanting both fingerprint and watermark attains completely nil area and latency design.

5.5 Analysis of case studies for DSP cores

Symmetrical IP core protection mechanism in Roy and Sengupta (2017) generates a low overhead strongly secured DSP IP core design. To validate the robustness and low overhead, following metrics are used by Roy and Sengupta (2017):

- Implanting cost for both signatures in a design indicating design quality.
- Area overhead with respect to resources and extra register needed while implanting fingerprint and watermark.

- Time consumed to generate signature.
- Strength of signature in terms of probability of coincidence.
- Influence of large size fingerprint and watermark on design area, design delay, and implanting cost.
- Scalability to deal with bigger size applications.

Therefore, the analysis/case study section is divided as follows: (i) result of symmetrical IP core protection approach (Roy *et al.*, 2008) in context of design latency, design area, embedding cost, and strength of signature, and (ii) comparative study between Roy *et al.* (2008) and unsymmetrical IP core protection approach (Koushanfar *et al.*, 2005). (Note: as no approach for symmetrical IP core protection exists for DSP core.)

5.5.1 Analysis of embedding cost, security metric on DSP Cores symmetrical protection

The analysis of design latency, design area, and implanting cost of DSP cores for both fingerprint strength (*F*) and watermark strength (*W*) equal to 30 is shown in Table 5.7. For instance, for JPEG, the resultant latency, area, and embedding cost are 14.24 ps, 724.30 μm^2, and 0.7349, respectively. (Note: module library includes resources corresponding to 15-nm technology scale extracted from NanGate (2017).) The cost of each solution is calculated based on Eq. (3).

The strength of watermark signature is indicated through probability of coincidence (P_c) metric. It calculates the possibility of redeveloping the same design solution with the watermark signature; thus, designates the evidence of authorship. The P_c which is derived from Koushanfar *et al.* (2005) can be defined as follows:

$$P_c = \left(1 - \frac{1}{c}\right)^W, \qquad (5.4)$$

where c is the number of color required and W is the size of watermarking constraints. It can be observed from Table 5.8 that with the increment in signature size,

Table 5.7 Results of symmetrical IP core protection design (Roy et al., 2008) with F = 30 and W = 30

Benchmarks (DSP benchmark suite, 2017)	**# of nodes**	**Resource configuration for symmetrical IP core protection**	**Total design latency (ns)**	**Total design area (μm^2)**	**Design cost**
JPEG	33	4(+), 4(*)	14.24	724.30	0.7349
FIR	23	4(+), 4(*)	1.86	180.09	0.7526
ARF	28	2(+), 4(*)	2.59	196.61	0.8393
MPEG	28	3(+), 5(*)	2.38	224.13	0.6645
IDCT	42	4(+), 2(*)	3.73	224.13	0.7349

Table 5.8 Measuring strength of watermark using probability of coincidence (P_c) (Roy et al., 2008)

Benchmarks	**# of register before fingerprint**	**P_c**		
		# of watermark constraints (*W*)		
		$W = 10$	**$W = 20$**	**$W = 30$**
JPEG	12	0.41890	0.17548	0.07351
FIR	8	0.26308	0.06921	0.01821
ARF	8	0.26308	0.06921	0.01821
MPEG	14	0.47660	0.22715	0.10826
IDCT	9	0.30795	0.09483	0.0292

Table 5.9 Change of latency, area and cost with the change of fingerprint constraints (Roy et al., 2008)

Benchmarks	**# of fingerprint constraints (*F*)**								
	$F = 10$			**$F = 20$**			**$F = 30$**		
	Latency (ns)	**Area (μm^2)**	**Cost**	**Latency (ns)**	**Area (μm^2)**	**Cost**	**Latency (ns)**	**Area (μm^2)**	**Cost**
JPEG	14.24	724.30	0.7349	14.24	724.30	0.7349	14.24	724.30	0.7349
FIR	1.81	179.31	0.7408	1.86	179.31	0.7521	1.86	179.31	0.7521
ARF	2.47	195.82	0.8211	2.52	195.82	0.8288	2.59	195.82	0.8391
MPEG	2.36	224.13	0.6639	2.36	224.13	0.6639	2.38	224.13	0.6645
IDCT	3.72	222.56	0.8246	3.73	222.56	0.8265	3.73	223.35	0.8267

P_c decreases. A lower P_c value indicates stronger proof of ownership. Therefore, to achieve a sturdier proof of ownership, a bigger size signature should be selected.

The effect of increasing fingerprint strength on the design latency, hardware area, and implanting cost is shown in Table 5.9. Three different fingerprint sizes is selected, viz. 10, 20, and 30 to observe the comparative outcome on various design metrics. It can be observed from Table 5.9 that for most of the benchmarks, the increment of design area is zero with the increment of fingerprint size. The reason is that out of three different fingerprint encoding variables only '*C*' variable embeds fingerprinting constraints during register allocation phase, which may incur area overhead. Therefore, with the increase in fingerprint size, the possibility of incurring design area overhead is very minimal. On the other hand, minimal increment in design latency can be observed with the increment of fingerprint size from Table 5.9. The reason is that adding fingerprinting constraints during scheduling phase while resolving operation conflict. However, for both design area and design latency, the increment is minimal. Therefore, nominal increment in cost with increment in fingerprint size can be observed for some benchmarks.

The effect of increasing watermark strength (after implanting fingerprint) on the design latency, hardware area, and implanting cost is shown in Table 5.10.

Table 5.10 Change of latency, area and cost with the augmentation of watermark constraints post implanting fingerprint (Roy and Sengupta, 2017)

Benchmarks	# of watermark constraints (*W*) after implanting fingerprint								
	F = 30, *W* = 10			*F* = 30, *W* = 20			*F* = 30, *W* = 30		
	Latency (ns)	Area (μm^2)	Cost	Latency (ns)	Area (μm^2)	Cost	Latency (ns)	Area (μm^2)	Cost
JPEG	14.24	724.30	0.7349	14.24	724.30	0.7349	14.24	724.30	0.7349
FIR	1.86	179.31	0.7521	1.86	180.09	0.7526	1.86	180.09	0.7526
ARF	2.59	195.82	0.8391	2.59	196.61	0.8393	2.59	196.61	0.8393
MPEG	2.38	224.13	0.6645	2.38	224.13	0.6645	2.38	224.13	0.6645
IDCT	3.73	223.35	0.8267	3.73	223.35	0.8267	3.73	223.35	0.8267

As mentioned earlier, watermark is embedded after embedding fingerprint; therefore, diverse watermark sizes chosen are shown by both *W* and *F*. In Table 5.10, the seller watermark strength is altered (by retaining the buyer fingerprint strength to maximal strength, i.e. $F = 30$) to observe its comparative consequence on aforementioned design objectives. Three different watermark sizes are selected, viz. $W = 10$ and $F = 30$, $W = 20$ and $F = 30$, and $W = 30$ and $F = 30$. It can be observed from Table 5.10 that for all the benchmarks the increment of design latency is zero with the augmentation of watermark size. The reason is watermarking constraints are not implanted during scheduling (which thus does not incur latency overhead). Therefore, increment in latency is zero with the increment in watermark size. On the other hand, minimal increment in design area can be observed by increasing the watermark size (in Table 5.10). The reason is that augmenting more watermarking constraints during register allocation phase may result into register overhead. Therefore, marginal upsurge in cost by increasing watermark size can be witnessed for some benchmarks.

5.5.2 Comparative study between symmetrical and non-symmetrical technique

The DSP IP designs produced through Roy *et al.* (2008) are compared with baseline IP core design, where baseline design indicates an unprotected design (i.e. a design without fingerprint and watermark) in terms of design latency, design area, and implanting cost. As shown in Table 5.11, symmetrical IP protection can be achieved at minimal latency overhead (avg. 1.87%), at minimal area overhead (avg. 0.24%), and at very minimal design cost overhead (less than 1%) in comparison to original baseline.

As stated previously, except Roy *et al.* (2008), there is no symmetrical IP core protection approach for DSP cores. Therefore, symmetrical protection approach (Roy *et al.*, 2008) is compared with Koushanfar *et al.* (2005) (inserts seller watermark only) in context of design latency, design area, and implanting cost. It can be observed from Table 5.12, symmetrical IP protection can be achieved at nil area overhead (0%), trivial latency overhead (avg. 1.87%), and at very low design cost overhead (lesser than 1%) relative to Koushanfar *et al.* (2005).

Table 5.11 Comparative study between baseline design and symmetrically protected IP core design with respect to latency, area, and cost (Roy and Sengupta, 2017)

Benchmarks	Latency (ns)			Area (μm^2)			Cost		
	Baseline	Symmetrical	Overhead (%)	Baseline	Symmetrical	Overhead (%)	Baseline	Symmetrical	Overhead (%)
JPEG	14.24	14.24	0	724.30	724.30	0	0.7349	0.7349	0
FIR	1.80	1.86	3.23	179.31	180.09	0.43	0.7370	0.7526	2.07
ARF	2.46	2.59	5.02	195.82	196.61	0.40	0.8185	0.8393	2.48
MPEG	2.36	2.38	0.84	224.13	224.13	0	0.6639	0.6645	0.09
IDCT	3.72	3.73	0.27	222.56	223.35	0.35	0.8246	0.8267	0.25

Table 5.12 Comparative study between a non-symmetrical design (Koushanfar et al.*, 2005) and symmetrically protected IP core design with respect to latency, area, and cost (Roy et al., 2008)*

Benchmarks	Latency (ns)			Area (μm^2)			Cost		
	Unsym-IP (Koushanfar *et al.*, 2005)	Symmetrical IP (Roy *et al.*, 2008)	Ov. (%)	Unsym-IP (Koushanfar *et al.*, 2005)	Symmetrical IP (Roy *et al.*, 2008)	Ov. (%)	Unsym-IP (Koushanfar *et al.*, 2005)	Symmetrical IP (Roy *et al.*, 2008)	Ov. (%)
JPEG	14.24	14.24	0	724.30	724.30	0	0.7349	0.7349	0
FIR	1.80	1.86	3.23	180.09	180.09	0	0.7370	0.7526	2.07
ARF	2.46	2.59	5.02	196.61	196.61	0	0.8185	0.8393	2.48
MPEG	2.36	2.38	0.84	224.13	224.13	0	0.6639	0.6645	0.09
IDCT	3.72	3.73	0.27	223.35	223.35	0	0.8246	0.8267	0.25

Table 5.13 Comparative study between a non-symmetrical design (Koushanfar et al., 2005) and symmetrically protected IP core design in terms of number of register used and signature creation time (Roy et al., 2008)

Benchmarks	**# of register used**			**Register overhead % of symmetrical**		**Signature generation time (ms)**	
	Baseline	**Unsym-IP (Koushanfar *et al.*, 2005)**	**Symmetrical IP (Roy *et al.*, 2008)**	**w.r.t. Baseline**	**w.r.t. unsym-IP (Koushanfar *et al.*, 2005)**	**Unsym-IP (Koushanfar *et al.*, 2005)**	**Symmetrical IP (Roy *et al.*, 2008)**
JPEG	12	12	12	0	0	61	70
FIR	8	9	9	0.11	0	31	37
ARF	8	9	9	0.11	0	17	19
MPEG	14	14	14	0	0	14	16
IDCT	8	9	9	0.11	0	14	17

The comparison of Roy *et al.* (2008) with Koushanfar *et al.* (2005) and baseline design in terms of number of register used and corresponding register overhead is shown in Table 5.13. It can be observed from the table that the register overhead of Roy *et al.* (2008) is zero compared with Koushanfar *et al.* (2005), where only seller watermark is present. Furthermore, the register overhead of the symmetrical protection approach (Roy *et al.*, 2008) is almost negligible compared with baseline design (less than 0.07%). Additionally, Table 5.13 also shows comparison of Roy *et al.* (2008) with Koushanfar *et al.* (2005) in terms of signature generation time. For Roy *et al.* (2008), signature generation time contains creation time of buyer fingerprint and seller watermark; on the contrary, for Koushanfar *et al.* (2005) signature, generation time contains creation time of seller watermark only. It can be observed from the table that Roy *et al.* (2008), while embedding both multi-variable fingerprint and watermark marginally increments the runtime overhead relative to Koushanfar *et al.* (2005) (which only inserts two-variable based signature). This pattern can be seen for all size application; thus, confirming the scalability of Roy and Sengupta (2017).

Design cost overhead for different sizes tested DSP cores is shown in Figure 5.8. It can be observed from the graph shown in Figure 5.8 that increment in benchmark size (i.e. amount of parallel operations) causes design cost overhead percentage to fall. Cost overhead of Roy and Sengupta (2017) is maximum for ARF benchmark (which includes least possible number of parallel operations), while it is minimal for JPEG benchmark (which includes maximum number of parallel operations). The reason of the following: Roy and Sengupta (2017) embed fingerprint constraints, while resolving operation scheduling conflict which happens frequently for larger size benchmarks (having huge amount of parallel operations) operating within a given/explored resource constraint. Thus, Roy and Sengupta (2017) attain almost insignificant overhead for large size benchmarks, thus satisfying the scalability property.

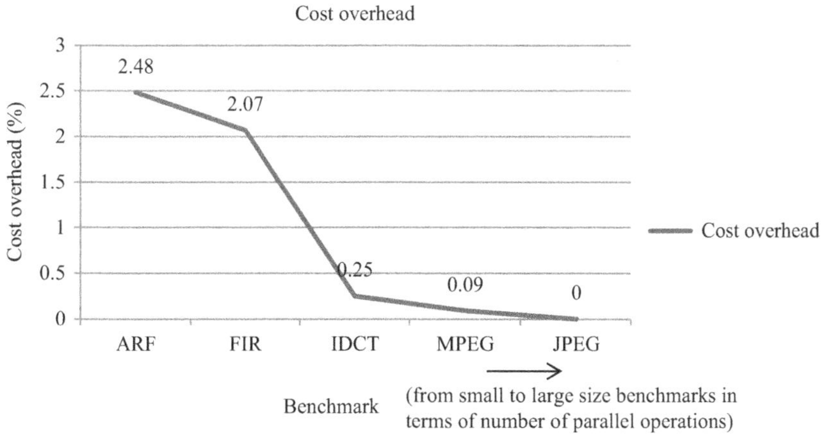

Figure 5.8 Change in cost overhead % for different size benchmarks

5.6 Conclusion

This chapter presents a symmetrical IP core protection mechanism for DSP cores. This approach incurs very low overhead in the context of design latency, design area, and embedding cost compared with baseline design. Furthermore, the fingerprint and watermark encoding mechanisms discussed in this chapter are very strong in terms of security. In addition to producing marginal overhead and providing strong robustness, this approach has fault tolerance ability, lower signature embedding time, and adaptability to any CAD tool. The robustness of the symmetrical protection discussed in chapter can be improved by embedding signature constraints in other steps of HLS like resource binding, hardware allocation, etc. This would allow executing a relative investigative study between distinctive HLS phases to explore the well-matched step for signature implanting. Additionally, robustness can also be improved by adding more variables in the signature encoding process.

5.7 Exercises

1. Why symmetrical IP core protection necessary?
2. Why embedding symmetrical protection scheme is necessary?
3. Which HLS step enables insertion of fingerprint?
4. Write the difference between watermark and fingerprint?
5. In symmetrical protection of DSP core which entity protection is performed earlier?
6. What are the steps to obtain CIG?
7. Where are the storage variables inserted and how?
8. How is resource const. scheduling performed using list scheduling?
9. What does additional edge in CIG represents?
10. What is the impact of FP size on the number of additional edges?
11. Implement a 6th-order FIR filter that incorporate eight digit fingerprint and watermark.
12. How can a design reduce design overhead during symmetrical protect of IP core without compromising security?
13. Explain the threat models of buyer entity and seller entity?
14. Explain some of the existing IP protection schemes?
15. Why symmetrical IP core protection is performed at architecture level?
16. Explain some of the desirable properties of a fingerprint?
17. What is the role of a colored interval graph in fingerprinting? How is it different from its role in watermarking?
18. Explain the detailed steps of fingerprinting of DSP cores?
19. What is fault tolerance in case of symmetrical IP core protection?
20. How do you ascertain the design cost overhead of a symmetrically protected IP core? What is the trend? Which DSP cores have higher cost overhead and why?

References

Alkabani Y. and Koushanfar F., 'Active control and digital rights management of integrated circuit IP cores', in *Proceedings of International Conference on Compilers, Architecture. and Synthesis of Embedded Systems (CASES)*, 2008, pp. 227–234.

Alkabani Y., Koushanfar F., and Potkonjak M., 'Remote activation of ICs for piracy prevention and digital right management', in *Proceedings of IEEE/ ACM International Conference on Computer-Aided Design*, Nov. 2007, pp. 674–677.

Castillo E., Meyer-Baese U., Garca A., Parrilla L., and Lloris A., 'IPP@HDL: Efficient intellectual property protection scheme for IP cores', *IEEE Transactions on Very large Scale Integrated (VLSI) Systems*, 2007, vol. 15 (5), pp. 578–591.

Chakraborty R. S. and Bhunia S., 'HARPOON: An obfuscation based SoC design methodology for hardware protection', *IEEE Transactions on Computer-Aided Design of Integrated Circuits Systems*, 2009, vol. 28 (10) pp. 1493–1502.

Cox I. J., Miller M. L., Bloom J. A., Fridrich J., and Kalker T., 'Digital Watermarking and Steganography', *Morgan Kaufmann*, San Mateo, CA, USA, 2007.

DSP benchmark suite. [Online]. Available: http://www.ece.ucsb.edu/ EXPRESS/ benchmark/, last accessed on June 2017.

Gang Q. and Potkonjak M., '*Intellectual Property Protection in VLSI Designs: Theory and Practice*', Springer Science & Business Media, New York, USA, 2007.

Hada S., 'Zero-knowledge and code obfuscation', in *Advances in Cryp-Tology*, Springer, Berlin, Germany, 2000, pp. 443–457.

Hong I. and Potkonjak M., 'Behavioral synthesis techniques for intellectual property protection', in *Proceedings of 36th Annual ACM/IEEE Design Automated Conference*, Jun. 1999, pp. 849–854.

Kougianos E., Mohanty S. P., and Mahapatra R. N., 'Hardware assisted watermarking for multimedia', *Computers & Electrical Engineering*, 2009, vol. 35 (2) pp. 339–358.

Koushanfar F., Hong I., and Potkonjak M., 'Behavioral synthesis techniques for intellectual property protection', *ACM Transactions on Design Automation of Electronic Systems*, 2005, vol. 10 (3), pp. 523–545.

Lach J., Mangione-Smith W. H., and Potkonjak M., 'Fingerprinting techniques for field-programmable gate array intellectual property protection', *IEEE Transactions on Computer-Aided Design of Integrated Circuits Systems*, 2001, vol. 20 (10), pp. 1253–1261.

Le Gal B. and Bossuet L., 'Automatic low-cost IP watermarking technique based on output mark insertions', *Design Automation of Electronic Systems*, 2012, vol. 16 (2), pp. 71–92.

Maltoni D., Maio D., Jain A., and Prabhakar S., '*Handbook of Fingerprint Recognition*', Springer, New York, NY, USA, 2009.

Mohanty S. P., '*Nanoelectronic Mixed-Signal System Design*', McGraw- Hill Education, New York, USA, 2015.

Mohanty S. P., Ranganathan N., and Krishna V., 'Datapath scheduling using dynamic frequency clocking', in *Proceedings of of the IEEE Computer Society Annual Symposium on VLSI*, 2002, pp. 58–63.

Mohanty S. P., Gomathisankaran M., and Kougianos E., 'Variability-aware architecture level optimization techniques for robust nanoscale chip design', *Computers & Electrical Engineering*, 2014, vol. 40 (1), pp. 168–193.

NanGate 15 nm open cell library. [Online]. Available: http: //www.nangate.com/?pageid=2328, last accessed on June 2017.

Nie T., Zhou L., and Li Y., 'Hierarchical watermarking method for FPGA IP protection', *IETE Technical Review*, 2013, vol. 30 (5), pp. 367–374.

Roy D. and Sengupta A., 'Low overhead symmetrical protection of reusable IP core using robust fingerprinting and watermarking during high level synthesis', *Future Generation Computer Systems*, 2017, vol. 71, pp. 89–101.

Roy J.A., Koushanfar F., and Markov I.L., 'EPIC: Ending piracy of integrated circuits', in *Proceedings of Design, Automation Test Europe (DATE)*, Mar. 2008, pp. 1069–1074.

Sengupta A., 'Intellectual property cores: Protection designs for CE products,' *IEEE Consumer Electronics Magazine*, 2016, vol. 5 (1), pp. 83– 88.

Sengupta A., 'Hardware security of CE devices,' *IEEE Consumer Electronics Magazine*, 2017, vol. 6 (1), pp. 130–133.

Sengupta A. and Bhadauria S., 'Exploring low cost optimal watermark for reusable IP cores during high level synthesis', *IEEE Access Journal*, 2016, vol. 4 (99), pp. 2198–2215.

Sengupta A. and Kachave D., 'Forensic engineering for resolving ownership problem of reusable IP core generated during high level synthesis', *Future Generation Computer Systems*, 2018, vol. 80, pp. 29–46.

Sengupta A. and Kundu S., 'Securing IoT hardware: Threat models and reliable, low-power design solutions' *IEEE Transactions on VLSI Systems*, 2017, vol. 25 (12), pp. 3265–3267.

Sengupta A. and Roy D., 'Antipiracy-aware IP chipset design for CE devices: A robust watermarking approach', *IEEE Consumer Electronics Magazine*, 2017, vol. 6 (2), pp. 118–124.

Sengupta A. and Sedaghat R., 'Integrated scheduling, allocation and binding in high level synthesis using multi structure genetic algorithm based design space exploration,' *International Symposium on Quality Electronic Design*, California, USA, 2011, pp. 1–9.

Sengupta A. and Sedaghat R., 'System and methodology for development of a system architecture using optimization parameters', Google Patents, 2013, https://www.google.com/patents/US8397204.

Sengupta A., Bhadauria S., and Mohanty S. P., 'TL-HLS: Methodology for low cost hardware Trojan security aware scheduling with optimal loop unrolling factor during high level synthesis,' *IEEE Transactions on CAD of Integrated Circuits and Systems*, 2017, vol. 36 (4), pp. 655–668.

Sengupta A., Roy D., Mohanty S. P., and Corcoran P., 'DSP design protection in CE through algorithmic transformation based structural obfuscation', *IEEE Transactions on Consumer Electronics* , 2017, vol. 63 (4), pp. 467–476.

Sengupta A., Roy D., and Mohanty S. P., 'Triple-phase watermarking for reusable IP core protection during architecture synthesis,' *IEEE Transactions on CAD of Integrated Circuits and Systems*, 2018, vol. 37 (4), pp. 742–755.

Sengupta A., Bhadauria S., and Mohanty S. P., 'Embedding low cost optimal watermark during high level synthesis for reusable IP core protection', in *Proceedings of 48th IEEE International Symposium on Circuits & Systems (ISCAS),* Montreal, May 2016, pp. 974–977.

Wong J.L., Kirovski D., and Potkonjak M., 'Computational forensic techniques for intellectual property protection', *IEEE Transactions on Computer-Aided Design of Integrated Circuits Systems*, 2004, vol. 23 (6), pp. 987–994.

Yu T. and Zhu Y., 'A new watermarking method for soft IP protection', *Proceedings of International Conference Consumer Electronics, Communications and Networks (CECNet)*, Apr. 2011, pp. 3839–3842.

Yuan L., Qu G., Ghouti L., and Bouridane A., 'VLSI design IP protection: Solutions, new challenges, and opportunities', in *Proceedings of 1st NASA/ESA Conference Adapting Hardware Systems (AHS)*, 2006, pp. 469–476.

Ziener D. and Teich J., 'FPGA core watermarking based on power signature analysis', in *Proceedings of IEEE International Conference Field Programmable Technology (FPT)*, Dec. 2006, pp. 205–212.

Ziener D. and Teich J., 'Power signature watermarking of IP cores for FPGAs', *Journal of Signal Processing Systems*, 2008, vol. 51 (1), pp. 123–136.

Chapter 6

Computational forensic engineering for resolving ownership conflict of DSP IP core

Some of previous chapters focused on watermarking and fingerprinting for ownership protection. This chapter focuses on another technique called forensic engineering (FE) for ownership protection. FE extracts the features of IP cores and matches to statistically suggest the original ownership. It also discusses various steps involved in FE of a hard IP core and presents specific details with case study examples.

6.1 Introduction

The rapid proliferation of electronic/SoC industry along with the fierce competition should be considered to remain competent. To beat the competition, companies seek to reduce time-to-market and design cost. These goals can be easily met through utilization of reusable IP core(s) (Sengupta, 2016; Kachave and Sengupta, 2016). This is because reusable IP core(s) minimize design time by reducing the manpower required to reproduce an IP. Therefore, reusable IP core has become a mandatory component of generic SoC/IC design flow. Moreover, a design process at higher level of design abstraction such as HLS is always crucial for complex SoCs such as DSP/multimedia cores (Sengupta and Sedaghat, 2010, 2011). This is because a higher level of design abstraction reduces design complexity and identifies optimal (low-cost) design architecture based on several orthogonal design objectives. Therefore, reusable IP core(s), such as DSP/multimedia core generated through HLS, is an essential component for consumer electronics devices. However, an IP core is vulnerable to various threats such as IP piracy, Trojan insertion, IP overbuilding, and false claim of ownership (Sengupta, 2017; Yasin *et al.*, 2016; Sengupta and Kundu, 2017). An estimate based on the report presented in the International Chamber of Commerce (2011) shows that electronics industry lose roughly minimum 1.5 trillion USD annually due to piracy and counterfeiting (Zhang, 2016). Therefore, it becomes mandatory to devise methodologies that can safeguard an IP core from these aforementioned threats. There are several IP protection mechanisms to overcome these threats, as shown in Figure 6.1.

Another technique that targets protection of IP core(s) is hardware obfuscation. Obfuscation targets protection against IP piracy and Trojan insertion (Yasin *et al.*, 2016). It aims to obfuscate a design through either structural (Sengupta *et al.*, 2017; Li and Zhou, 2013) or functional obfuscation (Lao and Parhi, 2014). The primary aim is to increase the efforts of an attacker to identify correct functionality of an IP

IP protection mechanism	Technique of protection	Overhead	Quality of the design	Nature of protection	Complexity	Security	Possibility of threat
Watermarking	Inserts authors signature by imposing constraints	Hardware overhead may exist resulting from storage registers	Degradation in quality or performance may occur, if not taken care	Nullifying/ resolving false claim of ownership	Complex as it involves two layered data security, viz., signature encoding and encryption	Robustly secured when two-layered data security is provided with complex signature	Higher due to the existence of reverse engineering step in signature detection
CFE	Identifies the source of the IP creator by feature extraction.	No overhead	No degradation in quality can occur as no signature/ID is inserted	For identifying IP creator	Complex as watermarking as it involves three-stage process.	Equally fail-proof as watermarking in identifying IP creator	Threat due to reverse engineering does not exist

Figure 6.1 CFE vs. watermarking for IP core protection

(Yasin *et al.*, 2016), thus making it hard for an attacker to secretly insert Trojan that can go undetectable during testing phase of an IC design. Further, this also makes it difficult to resell or utilize an IP core because its correct functionality is unknown to an adversary. Obfuscation does not aim to provide protection against ownership conflict (since it does not insert a unique ID or signature). However, functional obfuscation can be used to protect against reverse engineering of an IP core. Functional obfuscation implements a locking technique that can be used to safeguard an IP core against reverse engineering of its netlist. However, an adversary can reverse engineer an IP core, identify the correct key through various attacks, and misuse the IP core.

Yet another technique named "IP metering" aims to provide protection against IP piracy and IP overbuilding. This methodology seeks to distinguish one IP from another made through same fabrication process (Koushanfar and Qu, 2001). Therefore, when the IP core passes through untrustworthy design phases (Yasin *et al.*, 2016) of a generic IC design flow, those which are pirated or not reported by the foundry (overbuild) can be detected. Since each IC contains a unique ID, they can detect piracy or overbuilding whenever required.

Another mechanism that primarily aims to provide protection against false claim of ownership is watermarking (Sengupta *et al.*, 2016, 2018; Sengupta and Bhadauria, 2016). Watermarking embeds a signature in an IP during design phases and detects it to resolve false claim of ownership (see Chapter 3). However, it is vulnerable to reverse engineering-based attacks such as signature tampering. Further, signature insertion may incur hardware overhead (Sengupta *et al.*, 2018; Sengupta and Roy, 2017). To overcome some of the limitations of hardware watermarking, another alternative method suggested is computational forensics engineering-based IP core protection that can resolve ownership conflicts with 0% design overhead and no chances of signature tampering.

6.1.1 *Overview of forensic engineering*

FE has always been an impactful technique for resolving conflicts during criminal cases. Although it has been used intensively in fields such as image tampering and failure analysis, there has been little to no effort made in the literature to utilize such an impactful technique for intellectual property core protection. A good description of computational forensics can be found in Franke and Srihari (2008). A generic computational forensic engineering problem can be formulated as follows: given a solution S to a problem P, having a finite set of algorithms/tools A_n ($n = 1, \ldots, I$), applicable to problem P that can produce solution S, CFE targets automated identification with a very high degree of confidence that the algorithm/tool A_i has been applied, to produce solution S. The approaches that utilize FE for IP protection are Kachave and Sengupta (2016), Sengupta and Kachave (2018), and Wong *et al.* (2004). However, method presented by Wong *et al.* (2004) requires multiple instances of an IP core (obtained through distinct implementation of an algorithm) for providing IP protection, which is not always feasible in real-world scenarios. Further, the technique ("feature-set") presented in that approach is limited/only applicable to graph coloring and Boolean satisfiability. Additionally, the approach presented in Wong *et al.* (2004) is not applicable to DSP IP core.

A generic CFE approach comprises four steps, namely feature and statistical data collection, feature extraction, entity clustering, and validation (as shown in Figure 6.2). The purpose of these steps is explained as follows:

- Feature and statistical data collection: In this step, the goal is to observe the solution(s) to problem P obtained by algorithm A_i ($i = 1, \ldots, n$) and deciding various features that could help in distinguishing one IP core solution from another. These unique features are typically processed through some statistical techniques. Therefore, this step is termed as feature and statistical data collection.
- *Feature extraction*: Once the features are decided, one IP core solution from another solution can be distinguished, the next aim is to extract all these features from each IP core solution through some feature extraction rules.
- *Entity clustering*: In this step, our aim is to divide "s" number of solutions into "k" number of clusters such that each cluster contains solutions that are closely matched. Through this step, the aim is to reduce complexity of problem. Thus, instead of considering "s" solutions separately, it is grouped based on the similarities, and then "k" groups are considered instead of "s" solutions. Further, a generic assumption is that solutions produced through different

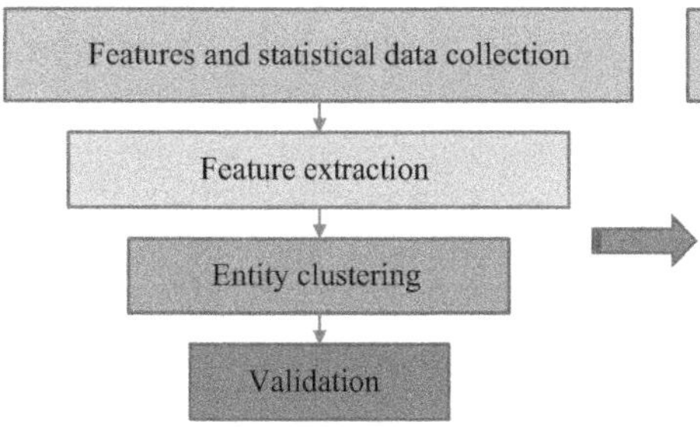

Figure 6.2 Generic CFE

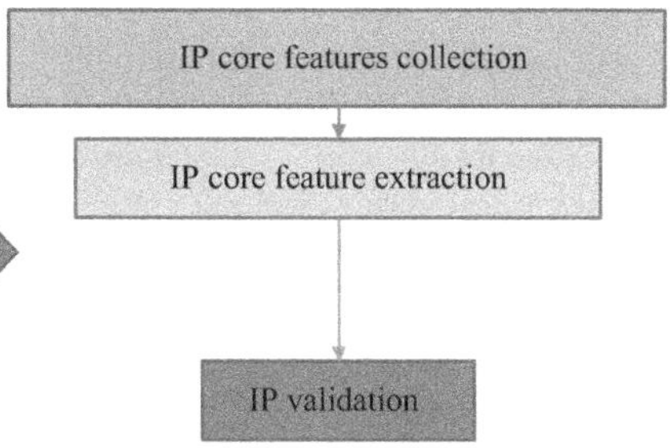

Figure 6.3 Customized CFE

implementations of same algorithm A_*i* will have closely matched features. Thus, a cluster should contain solutions generated from different implementations of same algorithm.

- *Validation*: In this step, the features of IP core solution "S" obtained through algorithm A_*i* is matched with the features of the IP core under test (see problem statement).

6.2 Computational FE technology

A generic CFE-based methodology comprises four steps as discussed earlier. However, customized CFE approach (Sengupta and Kachave, 2018) adopts only three steps (as shown in Figure 6.3) and customizes them for resolving IP core ownership conflict. The entity-clustering step is not adopted, because in a typical IP core ownership conflict scenario, the number of entities claiming ownership of an IP is not huge. Therefore, entity-clustering step is not mandatory in customized CFE approach. Figure 6.4 (Sengupta and Kachave, 2018) shows the flow of customized CFE approach. The customized CFE approach takes input in the form of DSP IP cores generated through HLS tool of vendors competing for ownership of an IP core (termed as IP_{Vn}, where "n" is the number of vendors claiming ownership of IP core) as well as IP core whose ownership is to be awarded/IP core under test (termed as $\mathrm{IP}_{\mathrm{ID}}$). Note that only executable version of competing vendor's tool is required. Thus, competing vendors are not concerned about their source code or trade secrets being revealed through the executable version of the HLS tool. Further, if a competing vendor refuses to provide executable version of his/her HLS tool, he/she is discarded from the ownership-awarding process. This is because the rightful owner of an IP core will never hesitate to provide executable version of his/her HLS tool (since ownership of his/her IP is at stake). The initial step of customized CFE approach (Sengupta and Kachave, 2018) comprises four substeps as shown in Figure 6.4. In the initial substep, studies of HLS tools as well as the sample DSP IP cores generated through corresponding HLS tools of each competing vendor are made. Based on this study, various features are decided that can distinguish one IP core solution from the other. The decided features are clubbed as "feature set." These features can further be classified as generic features and unique features. Generic features are those steps that are either integral part of HLS process (such as scheduling, allocation, and binding) or features ubiquitously present in every DSP IP core generated through HLS tool. The identified generic IP core features for CFE are *scheduling algorithm, resource type*, and *bus width support*. A naïve question may be how generic features can distinguish one IP core from another, if the feature is present in all other competing IP cores. The reason is, for example, in case of scheduling algorithm feature there are several algorithms to implement scheduling which will not be same for every HLS tool. Thus, though property (scheduling) is integral to a HLS methodology, feature (scheduling algorithm) can be different for each HLS tool. Similar reasoning exists for other generic

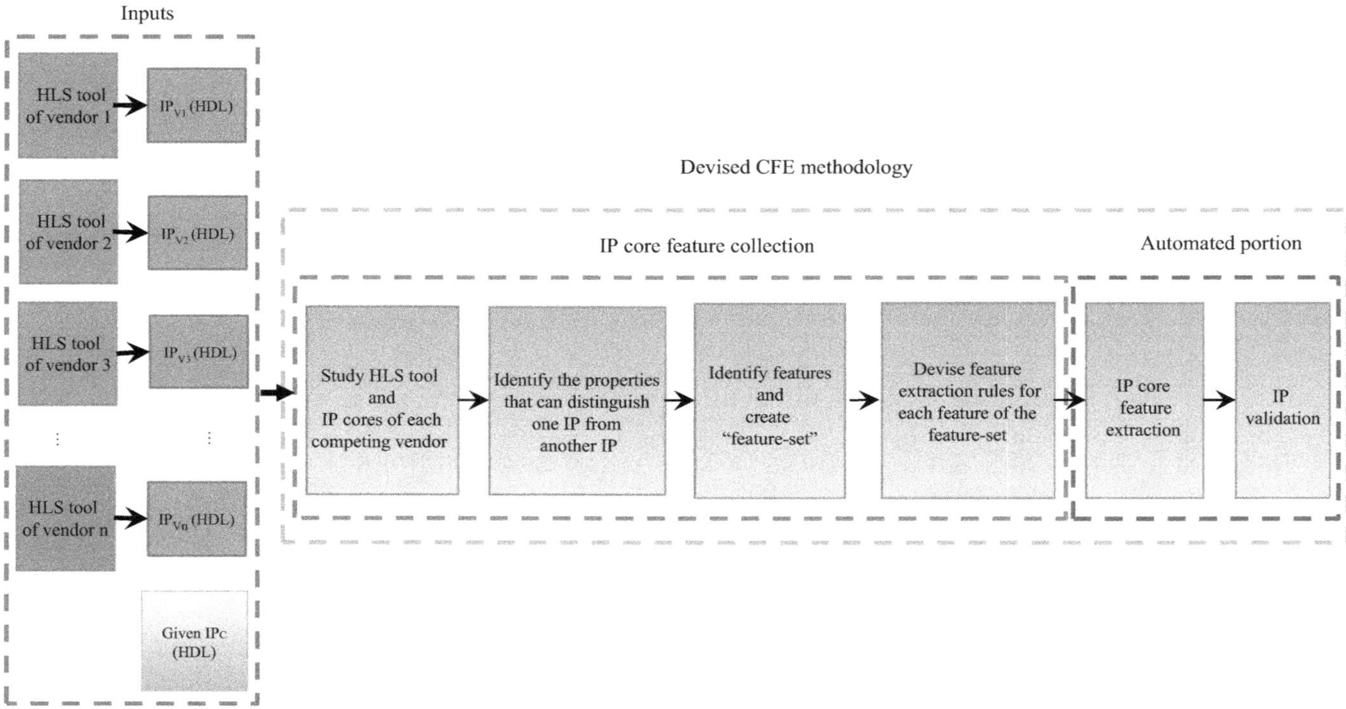

Figure 6.4 Customized CFE approach

features such as *resource type* and *bus width support.* The property is integral for all HLS tools; however, the feature may be different.

Unique features are the ones that are not present in each DSP IP core generated through HLS. However, these features are present in IP core of at least one of the competing HLS tools. Thus, unique features will be able to distinguish at least one IP core from all remaining IPs. Once the feature-set is identified, feature extraction rules are devised for extracting these features from each IP core available in the form of HDL. *Note: these feature extraction rules are automated in the form of a computer program.* Further, in the next step of customized CFE approach, automated feature extraction of IP core corresponding to each competing HLS tool as well as IP_{ID} core is executed. Finally, ownership of IP_{ID} is awarded to vendor whose "feature-set" matches exactly with the "feature-set" of IP_{ID}. This final step is termed as IP core validation. The customized CFE approach can be understood more effectively through following case study.

6.3 IP core feature extraction algorithms

In a scenario where there is a DSP IP core ownership conflict among tools vendors of Harish Ram *et al.* (2011), Inoue *et al.* (2011), Sengupta and Sedaghat (2015), Koushanfar *et al.* (2005), Rajendran *et al.* (2013), Mishra and Sengupta (2014), and Sengupta and Bhadauria (2015), the IP core extraction rules can be explained through the aid of a demonstration. Further, it is assumed that executable versions of HLS tools corresponding to each competing vendor are also available.

IP cores are generated and properties are decided that can distinguish at least one IP from another. Based on these properties, features are devised and a "feature-set" is created. The created "feature-set" (Sengupta and Kachave, 2018) may comprise generic features and unique features as shown below:

$$\begin{aligned}
&\text{Feature} - \text{set} &&= \{\text{generic features, unique features}\};\ \text{where}\\
&\text{Generic feature} &&= \{\text{“scheduling algorithm”, “bus width support”, “resource type”}\};\ \text{and}\\
&\text{Unique feature} &&= \{\text{“pipelining”, “chaining”, “multi-cycling”, “reliability”,}\\
& && \quad \text{“loop support”, “design objective”, “Trojan security”}\};
\end{aligned}$$

Once feature-set is determined, feature extraction rules are devised for each feature of the feature-set. Section 6.3.1 discusses feature extraction rules in detail.

6.3.1 Feature extraction rules

Scheduling algorithm: Scheduling is an integral step in DSP IP core design (Paulin and Knight, 1989; McFarland, 1988). The three most commonly used scheduling algorithms are As Soon As Possible (ASAP), As Late As Possible (ALAP), and List algorithm. There are few other useful scheduling algorithms, but this case study is limited to these three most commonly used scheduling algorithms. The flow graph to identify which scheduling algorithm is utilized to create a DSP IP core is shown in Figure 6.5. The feature extraction methodology initially checks for the presence

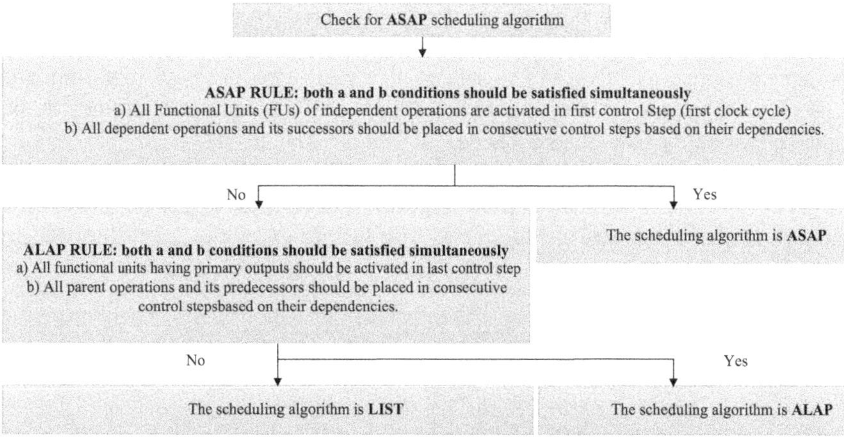

Figure 6.5 Feature extraction methodology for scheduling algorithm

```
component Mul_4x4
  port ( enable_R1 : in  std_logic;
      Data_out7  : in  std_logic_vector  (3 downto  0);
      Data_out8  : in  std_logic_vector  (3 downto  0);
      Data_in9   : out  std_logic_vector  (7 downto  0));
 end component;
```

Figure 6.6 Partial HDL code of an IP core

of ASAP scheduling in controller HDL of IP core. If both the rules to identify ASAP scheduling are satisfied, then the scheduling algorithm utilized to create an IP is detected as ASAP. However, if any one of the rules is not satisfied, then the possibility of employing ALAP scheduling in creating an IP core is checked. If both the conditions of ALAP scheduling are satisfied, the scheduling algorithm in IP core is detected as ALAP. Otherwise, the scheduling algorithm detected is LIST.

Resource type: To identify the presence of resource type feature, the top-level entity datapath HDL of IP core is checked and name of the components are identified. Further, a set of names resembling that of a functional unit (FU) are created as indicators as "type of resources" used to generate a DSP IP core. As shown in Figure 6.6, the name of the component is "Mul_4×4" which indicates a 4×4 multiplier. Similarly, all other component's names are checked.

Bus width support: The feature extraction process, devised to identify bus width in an IP core, reads the top-level entity datapath HDL of the respective IP core as input. Subsequently, a list of component names is prepared. Next, the input/output size for each component is extracted as the bus width in that IP core. Figure 6.7 shows partial HDL code of a representative IP core. The variable size of the register component named as “Reg_r1” is size of r1 = size of reg_r1 = 16 bits. Similarly, sizes of other components are determined and variable size is determined as bus width supported by that IP core design.

Chaining: As explained in Sengupta and Kachave (2018) and Kachave and Sengupta (2017), the flow diagram of algorithm devised to detect the presence of chaining feature is shown in Figure 6.8. The feature extraction algorithm accepts

```
component Reg_r1
    port ( r1   : in   std_logic_vector (15 downto 0 );
         reg_r1: out  std_logic_vector (15 downto 0 );
         strobe: in   std_logic);
  end component;
```

Figure 6.7 Partial HDL code of an IP core

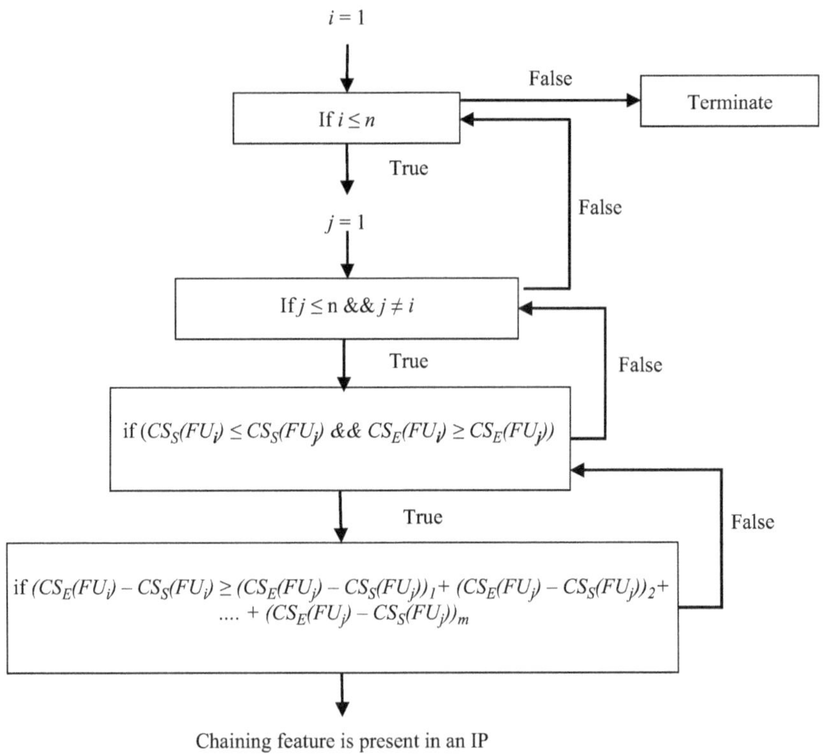

Figure 6.8 Algorithm to extract chaining feature

controller HDL of an IP core as input and produces output as whether chaining feature is detected or not. Detection of chaining is ascertained as follows: if during execution time of *i*th FU, multiple ("m") instances of *j*th FU are executed then chaining feature is present in an IP core design. The algorithm is executed for $i =$ 1–*n*, where "*n*" is the total number of FUs present in a particular IP core. The second "if" statement (in Figure 6.8) ensures that algorithm runs for all *j*th FU, other than the *i*th FU. Subsequently, the third "if" statement allows only those instances of *j*th FU to be compared with *i*th FU that started (represented as $CS_S(FU_i) \leq CS_S(FU_j)$) and ended (represented as $CS_E(FU_i) \geq CS_E(FU_j)$) between a single instance of *i*th FU, where $CS_S(FU_i)$ and $CS_S(FU_j)$ represent starting control step of *i*th and *j*th FU, respectively. Similarly, $CS_E(FU_i)$ and $CS_E(FU_j)$ symbolize ending control step of *i*th and *j*th FU, respectively. Finally, the last "if" statement checks whether during execution time ($CS_E(FU_i) - CS_S(FU_i)$) of FU of *i*th type, multiple instances of FU of *j*th type are executed or not. Thus, if the last *if* condition is satisfied, then chaining feature is detected in a DSP IP core. Figure 6.9 shows the partial HDL code of controller of an IP core. As shown in controller's HDL, the multiplier FU is enabled in control step 1 represented by statement MUL_En_1<= "1". Adder FU is enabled in the first CS and disabled in the second CS. Further, second instance of Adder FU is activated in third CS and deactivated in fourth CS. Last, multiplier FU execution continues until it is disabled in CS = 40. Thus, during execution of single multiplier FU, multiple instances of adder FU are executed. This indicates that chaining feature is present in this IP core design.

1. *Multicycling*: To detect multicycling feature in an IP core, the feature extraction rule reads the controller HDL code of the respective IP core. It checks to detect if the execution time of any FU consumes more than one CS. If the condition is true then it is inferred that multicycling feature is present in that IP core design. The condition to detect the presence of multicycling in the controller HDL code of an IP core is given as: $CS_E(FU_i) > CS_S(FU_i)$, where $CS_S(FU_i)$ and $CS_E(FU_i)$ represent starting and ending control steps of *i*th FU, respectively.
2. *Data pipelining*: Another optimization technique termed as "Data Pipelining" targets the latency reduction of overall schedule delay of an IP core. This is achieved through performing some operations of the consecutive data sets in parallel as shown in Figure 6.10. The devised feature extraction algorithm accepts controller HDL of an IP core as input. Subsequently, if the following condition stated in (6.1) is satisfied, the time overlap between operations of consecutive datasets is detected in an IP core controller design.

$$(CS_E(N)_1 - CS_S(N)_1) > (CS_E(N)_2 - CS_E(N)_1) \tag{6.1}$$

where $CS_S(N)_1$ represents the starting control step of data set 1, $CS_E(N)_1$ and $CS_E(N)_2$ represent the ending control step of data set 1 and 2, respectively. As shown in Figure 6.11(a) and 6.11(b), the starting control step is identified as the control step at which the first FU (such as Multiplier FU in this example) is

```
--------------------------------CONTROL STEP 1--------------------------
if CS=1 then
                                        MUL_EN_1<='1';
//start of multiplication operation//

                    ADD_EN_1<='1';

                    ADD_EN_1<='1';

                    :
                    :



                                                              CS
<= CS+1;
end if;

--------------------------------CONTROL STEP 2--------------------------
if CS=2 then
                                        ADD_EN_1<='0';
                                        ADD_EN_1<='1';

                    :
                    :

                                        CS <= CS+1;

end if;
--------------------------------CONTROL STEP 3--------------------------
if CS=3 then
                                        ADD_EN_1<='0';
                               :
CS <= CS+1;

end if;
--------------------------------CONTROL STEP 4--------------------------
if CS=4 then


                               :
                               :
                               :
                        CS <= CS+1;
end if;
--------------------------------CONTROL STEP 40-------------------------
if CS=40 then



                    :
                    :
                    :


                    MUL_EN_1 <= '0'; //end of
multiplication operation//


                                                              CS
<= CS+1;
end if;
```

Figure 6.9 Partial controller HDL code of an IP

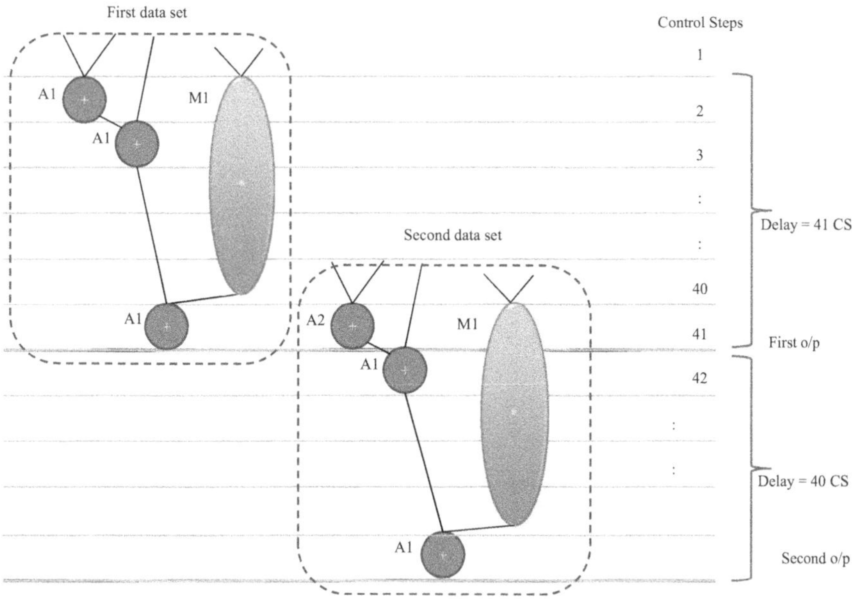

Figure 6.10 Schedule diagram representing data pipelining between two data sets of a representative IP core

activated. Further, ending control step of data set 1 is the CS at which the output register (Reg Y) of a data set is enabled for the first time. Similarly, the ending control step of data set 2 is the CS at which the output register (Reg Y) is enabled for the second time. As shown in Figure 6.11(a) and 6.11(b), register Y is activated for first time at CS 42 and is again activated at CS 82 for the second time. Thus, the condition stated in (6.1) is satisfied. Therefore, data pipelining is detected in an IP core, as demonstrated through this representative example.

3. *Design objective*: A generic HLS tool used to design an IP core typically has at least one design objective. The design objectives are typically optimization based on power-delay trade-off or area-delay tradeoff, etc. Based on the design objectives supported, one tool used for IP core creation can be distinguished from another. The feature extraction algorithm for extracting design objective feature takes user interface (graphical or nongraphical) of executable version of HLS tool as input. Subsequently, the extraction algorithm reads the interface and determines the design objectives available for selection in a tool's user interface. Subsequently, the interface's design objectives are matched and validated with the design objective details mentioned in the datasheet of the IP core under test.
4. *Reliability*: Reliability can take multiple forms such as reliability (detectability/security) against transient fault, reliability (detectability/security) against permanent faults (Inoue *et al.*, 2011; Nomura *et al.*, 2011). A typical

```
--------------------------------CONTROL STEP 1--------------------------
                    if CS=1 then
                        if count=first_count

                      MUL_EN_1<='1'; //start of
            multiplication operation//

                    ..............

                                        ..............
                              ..............

                                        ..............

                                        ..............

                                        ..............
                                        ..............
                    CS <= CS+1;
                      end if;

--------------------------------CONTROL STEP 2--------------------------
                    if CS=2 then
                                        ..............

                                        ..............

                                        ..............;
                        ..............

                                        ..............

                                        ..............

                                        ..............
                                        ..............
                    CS <= CS+1;
                      end if;
--------------------------------CONTROL STEP 3--------------------------
                              :
                              :
                              :
                              :
--------------------------------CONTROL STEP 41-------------------------
                    if CS=41 then
                                        ..............

                    ..............
                                        ..............

                                        ..............
                                 MUL_EN_1 <= '1';

                                REG_Y <='0';

                                          :
                    :

                                                              ;

                    CS <= CS+1;
                      end if;
```

Figure 6.11(a) Partial controller HDL code of an IP

approach that aims to design a reliable system utilizes modular redundancy (Inoue *et al.*, 2011; Yu and McCluskey, 2001; D'Angelo *et al.*, 1999). Redundancy may be dual modular redundancy (DMR), triple modular redundancy, or multimodular redundancy. However, so far in the literature, such as Inoue *et al.* (2011), DSP IP core vendors have only implemented reliability against transient fault using DMR. Therefore, feature extraction rule is discussed only for this feature. The rule aims to detect the presence of DMR logic in DSP IP core design.

Reliability feature extraction process accepts datapath HDL of an IP core as input and aims to detect the presence of a DMR system. As shown in

```
---------------------------CONTROL STEP 42---------------------------
          if CS=42 then
                        ..............
          REG_Y <='1';//output of dataset
          1 available//
                        ..............
              ..............     end if;
                        :
                        :
                        ..............
   ..............
                        ..............
                     CS<=CS+1;
              end if;
---------------------------CONTROL STEP 43---------------------------
                  :
                  :
---------------------------CONTROL STEP 81---------------------------
          if CS=81 then
                        ..............
                        ..............
              MUL_EN_1 <= '1';
                REG_Y <='0';
                        :
                        :
                        ..............
                        ..............
                        ..............
                              CS <= CS+1;
              end if;
-------------------------------CONTROL STEP 82----------------------
            if CS=82 then
                        ..............
          REG_Y <='1';//output of dataset 2 available//
                        ..............
                        ..............
                        :
                        :
                        ..............
                        ..............
                     CS<=CS+1;
              end if;
```

Figure 6.11(b) Partial controller HDL code of an IP

Figure 6.12, in a typical DMR system, the outputs of each module of a DMR system are compared using a comparator component to determine whether the fault is detected or not. This comparator component accepts one input from output register of the first module (original) and another from the second module (duplicate). If a port mapping similar to following port map statement in HDL: *port map (output register signal 1, output register signal 2, comparator output signal)* is detected, the reliability feature is present in the respective IP core.

5. *Loop support feature*: The loop support feature aims to detect the type of application that can be fed as an input to an HLS tool used in the IP core design. In other words, this feature extracts whether input format of a tool can

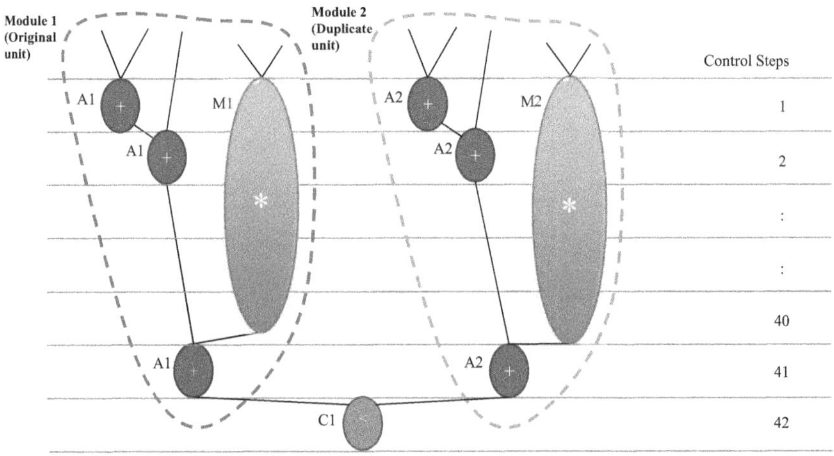

Figure 6.12 Schedule diagram of a DMR system

support loop-based application in the form of control data flow graph (CDFG) or non-loop-based application in the form of DFG (Gupta, 2003; Kurra and Panda, 2007; Sengupta *et al.*, 2017). The devised rule to identify loop support feature requires input application in C-code. This C-code is fed into executable version of HLS tool. If HLS tool runs correctly without any error and produce comprehensible RTL code, then it supports the respective input format.

Note: this case study only considers executable versions of HLS tools that do not require manual intervention.

6. *Trojan security feature*: In hardware domain, Trojan can be understood as a malicious modification of a hardware design that aims to achieve malicious intention of an attacker (Rajendran *et al.*, 2013; Bhunia *et al.*, 2013; Zhang and Tehranipoor, 2011). An attacker may intend to leak information, generate incorrect output, trigger denial of service attacks, etc. There can be several techniques to provide security against Trojan based on their intent. However, so far in the literature for DSP IP cores, protection against Trojan, which intends to harm correct functionality of an IP by generating incorrect output, has been considered. Thus, during the feature extraction rule, only this special Trojan for DSP cores is accounted for. Typically, Trojan security for DSP cores requires modules from at least two different IP vendors. Further, Trojan detection (security) is typically performed using modules from at least two distinct vendors and using them in DMR system as shown in Figure 6.13. Therefore, the feature extraction algorithm accepts datapath HDL of an IP core as well as module library of HLS tool as inputs. The presence of DMR is detected (as discussed in reliability feature extraction process) by identifying the presence of a comparator component having port map similar to following port mapping statement: *port map (output register signal 1, output register signal 2, comparator output signal).* Additionally, if the module library

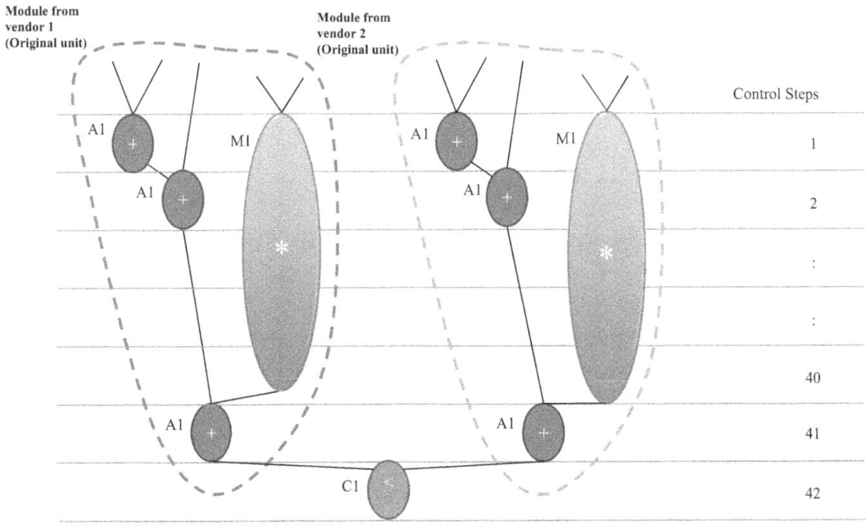

Figure 6.13 Schedule diagram of a DMR-based Trojan security system

comprises design objectives (such as area, power, delay) corresponding to at least two different IP vendors, the presence of Trojan security feature is present in the IP core design.

Note: The main difference between reliability and Trojan security properties is that reliability does not require modules from two distinct IP vendors. On the other hand, Trojan security requires module from two distinct vendors.

6.3.2 IP core validation

Once all the features are extracted, the feature-set of each DSP IP core generated corresponding to competing vendors as well as IP whose ownership is to be determined (i.e., IP_{ID}) are created. Subsequently, the feature-set of competing IP vendor is matched with the feature-set of IP_{ID} as per (6.2). If match percent (μ) of 100% is obtained, then the ownership of IP_{ID} is awarded to that particular IP vendor.

$$\mu = (\text{Number of features matching with features of } IP_{ID}) / (\text{Total number of features in feature set}) \times 100 \tag{6.2}$$

In a typical scenario, the feature-set used in CFE approach (Sengupta and Kachave, 2018; Inoue *et al.*, 2011) is enough for resolving the IP core ownership conflict. However, we would like to discuss various exceptional scenarios and procedure to follow to resolve the ownership conflict in those scenarios.

I. **None of the competing IP vendors has a 100% match**: In such a scenario, the ownership of IP_{ID} will not be awarded to any of the vendors competing for claiming the ownership of IP_{ID}.

II. **More than one competing IP vendors have a 100% match**: In such a scenario (called "deadlock in IP resolution"), the feature-set should be expanded by identifying additional properties that may distinguish between two or more competing IP vendors. This is performed by further investigating each and every IP core to identify should there be additional properties that could be considered to break this deadlock. Based on this further investigation, new features are added in the feature set. In other words, the feature-set is expanded to incorporate these new features. Further, feature extraction rules need to be devised corresponding to these newly added features. Next, feature extraction is performed now for these newly features from the IP cores who are in deadlock. Subsequently, feature matching is performed with the updated feature-set. Finally, if there is no exceptional scenario, the ownership is awarded to the IP vendor whose feature-set matches 100% with that of IP_{ID}.

6.3.3 Important characteristics of customized CFE

- Customized CFE approach (Sengupta and Kachave, 2018) is applicable to a scenario where each competing IP vendor has a unique HLS tool for claiming ownership of IP_{ID}. However, in scenarios when two or more IP vendors are claiming ownership based on same HLS tool obtained through third party, then the devised "feature-set" of customized CFE is not suffice for resolving ownership conflict.
- As already discussed in Section 6.2, customized CFE approach requires only executable version of HLS tool. Further, it does not require source code of these HLS tools. Thus, there is no concern for IP vendors about source code or breach of confidential information. Additionally, if some competing IP vendor refuses to provide executable version of HLS tool, the vendor is discarded from ownership-awarding process. This is because, typically, the rightful owner of an IP would not hesitate in providing executable version of his/her HLS tool, as ownership of the IP core is at stake.
- Customized CFE is applicable for IP vendors of DSP IP cores. However, IP vendors for FPGA would not fall within the scope of Sengupta and Kachave (2018).

6.4 Analysis on case studies

This section reports the analysis on case studies for the automated customized CFE approach (Sengupta and Kachave, 2018) and (Koushanfar *et al.*, 2005). The customized CFE approach (Sengupta and Kachave, 2018) was tested on express DSP benchmark suite from Express Benchmark Suite (http://www.ece.ucsb.edu/EXPRESS/benchmark/). The benchmarks primarily comprise DSP applications. The customized CFE approach was executed for data-intensive IP cores (such as ARF, BPF, and DCT), control-intensive IP core (such as FFT and FIR), as well as multimedia data-intensive IP core (such as JPEG IDCT).

6.4.1 Results of the customized CFE approach

For intensive testing of the generic applicability of Sengupta and Kachave (2018), diverse IP vendors with their different HLS tools (from academia as well as industry) were selected. Moreover, latest HLS tools of vendors (published in last 10 years) having versatile design objectives (such as area-execution time, power-execution time), different DSE methodologies (such as PSO, GA, and BFOA), and properties (such as Trojan security, fault tolerance, and watermarking) were also included during the analysis. The IP vendors with their HLS tools considered are listed below:

1. IP Vendor 1: Hybrid PSO-GA-based HLS tool (Harish Ram *et al.*, 2011)
2. IP Vendor 2: Fault tolerant-based HLS tool (Inoue *et al.*, 2011)
3. IP Vendor 3: Fault secure-based HLS tool (Sengupta and Sedaghat, 2015)
4. IP Vendor 4: Watermarking-based HLS tool (Koushanfar *et al.*, 2005)
5. IP Vendor 5: Trojan security-based HLS tool (Rajendran *et al.*, 2013)
6. IP Vendor 6: PSO-based HLS tool (Mishra and Sengupta, 2014)
7. IP Vendor 7: BFOA-based HLS tool (Sengupta and Bhadauria, 2015)

Tables 6.1 to 6.6 show the results of analysis obtained for various DSP and multimedia cores. As evident from these tables, customized CFE approach (Sengupta and Kachave, 2018) was able to resolve the ownership conflict for each of these cores. Further, even if several IP core features overlap between competing IP vendors, yet customized CFE was still able to resolve the ownership conflict. This shows that customized CFE approach is strongly capable of resolving ownership conflict with the current feature set. It is interesting to note that one feature is enough to distinguish between competing IP vendors. For example, out of the ten features in the feature-set, even if nine features of both IP cores (corresponding to the two competing IP vendors) overlap with IP under test, but only one feature remains distinct to each other, it may be enough identifying the actual IP owner. This is because the competing IP vendor with nine overlapping features will have 90% match with IP_{ID} while the actual IP vendor will have 100% match with IP_{ID}. For example, as shown in Table 6.1, even though IP_{V1}, IP_{V5}, and IP_{V7} have eight features common in them, the ownership conflict among these tools with respect to IP_{ID} was resolved based on the remaining two features and ownership was awarded to IP core vendor 5. The results obtained for FIR benchmark show an exceptional case as discussed in section III.B, where none of the competing vendor's "feature-set" matches 100% with the "feature-set" of IP_{ID}. In such a case, ownership is not awarded to any of the competing vendors.

Table 6.7 shows the relative comparison between CFE (Sengupta and Kachave, 2018) and hardware watermarking (Koushanfar *et al.*, 2005) for DSP IP core protection based on the design overhead (storage registers). As evident from the results, CFE (Sengupta and Kachave, 2018) incurs 0% design overhead as compared to hardware watermarking (Koushanfar *et al.*, 2005). This is because hardware watermarking requires embedding vendor signature into the IP core design, which may incur storage overhead. On the other hand, CFE is performed

Table 6.1 Feature-set of IP_{ID} and IP_V for ARF benchmark (Note: IP_{Vn} = IP core generated by competing HLS tool of IP vendor "n")

				Benchmark: ARF (28 nodes)				
IP features	**IP_{ID}**	**IP_{V1} (Hybrid PSO-GA HLS (Harish Ram *et al.*, 2011))**	**IP_{V2} (Fault secure HLS (Sengupta, 2015))**	**IP_{V3} (Watermark-HLS (Koushanfar *et al.*, 2005))**	**IP_{V4} (Trojan Secure-HLS (Rajendran *et al.*, 2013))**	**IP_{V5} (BFOA-HLS (Bhadauria, 2015))**	**IP_{V6} (Fault Tolerant-HLS (Inoue *et al.*, 2011))**	**IP_{V7} (PSO-HLS (Mishra and Sengupta, 2014))**
Schedule algorithm	LIST	LIST	LIST	LIST	LIST	LIST	LIST	LIST
Resource type	A,M	A, M	A, M, C	A, M, C, S	A, M, C, S	A, M	A, M, C, S	A, M
Chaining	Present	Present	Absent	Absent	Absent	Present	Absent	Present
Bus width support	32 bit	16 bit	16 bit	32 bit	16 bit	32 bit	16 bit	8 bit
Data pipelining	No	No	No	No	No	No	No	Yes
Multicycling	Present	Present	Present	Absent	Absent	Present	Absent	Present
Design objective	Area-execution time/Power-execution time	Area-power-latency	Area-execution time/Power-execution-time	Area-execution time/Power-execution time	Area-execution time/Power-execution time	Area-execution time/Power-execution time	Area-latency	Area-execution time/Power-execution time
Fault reliability	Absent	Absent	Present	Absent	Absent	Absent	Present	Absent
Loop support	DFG	DFG	DFG	DFG	DFG	DFG	DFG	DFG
Trojan security	Absent	Absent	Absent	Absent	Present	Absent	Absent	Absent
Match percent	NA	80	60	70	50	100	40	80

Table 6.2 Feature-set of IP_{ID} and IP_V for BPF benchmark

	Benchmark: BPF (29 nodes)							
IP features	**IP_{ID}**	**IP_{V1} (Hybrid PSO-GA HLS (Harish Ram *et al.*, 2011))**	**IP_{V2} (Fault secure HLS (Sengupta, 2015))**	**IP_{V3} (Watermark-HLS (Koushanfar *et al.*, 2005))**	**IP_{V4} (Trojan Secure-HLS (Rajendran *et al.*, 2013))**	**IP_{V5} (BFOA-HLS (Bhadauria, 2015))**	**IP_{V6} (Fault Tolerant-HLS (Inoue *et al.*, 2011))**	**IP_{V7} (PSO-HLS (Mishra and Sengupta, 2014))**
Schedule algorithm	LIST	LIST	LIST	LIST	LIST	LIST	LIST	LIST
Resource type	A,M	A, M	A, M, C	A, M, C, S	A, M, C, S	A, M	A, M, C, S	A, M
Chaining	Present	Present	Absent	Absent	Absent	Present	Absent	Present
Bus width support	16 bit	16 bit	16 bit	32 bit	16 bit	32 bit	16 bit	8 bit
Data pipelining	Absent	Absent	Absent	Absent	Absent	Absent	Absent	Present
Multi-cycling	Present	Present	Present	Absent	Absent	Present	Absent	Present
Design objective	Area-power-latency	Area-power-latency	Area-execution-time/Power-execution time	Area-execution time/Power-execution time	Area-execution time/Power-execution time	Area-execution time/Power-execution time	Area-latency	Area-execution time/Power-execution time
Fault reliability	Absent	Absent	Present	Absent	Absent	Absent	Present	Absent
Loop support	DFG	DFG	DFG	DFG	DFG	DFG	DFG	DFG
Trojan security	Absent	Absent	Absent	Absent	Present	Absent	Absent	Absent
Match percent	NA	100	60	50	50	80	50	70

Table 6.3 Feature-set of IP_{ID} and IP_V for DCT benchmark

	Benchmark: DCT (42 nodes)							
IP features	**IP_{ID}**	**IP_{V1} (Hybrid PSO-GA HLS (Harish Ram *et al.*, 2011))**	**IP_{V2} (Fault secure HLS (Sengupta, 2015))**	**IP_{V3} (Watermark-HLS (Koushanfar *et al.*, 2005))**	**IP_{V4} (Trojan Secure-HLS (Rajendran *et al.*, 2013))**	**IP_{V5} (BFOA-HLS (Bhadauria, 2015))**	**IP_{V6} (Fault Tolerant-HLS (Inoue *et al.*, 2011))**	**IP_{V7} (PSO-HLS (Mishra and Sengupta, 2014))**
Schedule algorithm	LIST	LIST	LIST	LIST	LIST	LIST	LIST	LIST
Resource type	A,M	A, M	A, M, C	A, M, C, S	A, M, C, S	A, M	A, M, C, S	A, M
Chaining	Present	Present	Absent	Absent	Absent	Present	Absent	Present
Bus width support	32 bit	16 bit	16 bit	32 bit	16 bit	32 bit	16 bit	8 bit
Data pipelining	Absent	Absent	Absent	Absent	Absent	Absent	Absent	Present
Multicycling	Present	Present	Present	Absent	Absent	Present	Absent	Present
Design objective	Area-execution time/Power-execution time	Area-Power-latency	Area-execution time/Power-execution time	Area-execution time/Power-execution time	Area-execution time/Power-execution time	Area-execution time/Power-execution time	Area-latency	Area-execution time/Power-execution time
Fault reliability	Absent	Absent	Present	Absent	Absent	Absent	Present	Absent
Loop support	DFG	DFG	DFG	DFG	DFG	DFG	DFG	DFG
Trojan security	Absent	Absent	Absent	Absent	Present	Absent	Absent	Absent
Match percent	NA	80	60	70	50	100	40	80

Table 6.4 Feature-set of IP_{ID} and IP_V for FFT benchmark

				Benchmark: FFT (56 nodes)				
IP features	**IP_{ID}**	**IP_{V1} (Hybrid PSO-GA HLS (Harish Ram *et al.*, 2011))**	**IP_{V2} (Fault secure HLS (Sengupta, 2015))**	**IP_{V3} (Watermark-HLS (Koushanfar *et al.*, 2005))**	**IP_{V4} (Trojan Secure-HLS (Rajendran *et al.*, 2013))**	**IP_{V5} (BFOA-HLS (Bhadauria, 2015))**	**IP_{V6} (Fault Tolerant-HLS (Inoue *et al.*, 2011))**	**IP_{V7} (PSO-HLS (Mishra and Sengupta, 2014))**
Schedule algorithm	LIST	LIST	LIST	LIST	LIST	LIST	LIST	LIST
Resource type	A, M, C	A, M	A, M, C	A, M, C, S	A, M, C, S	A, M	A, M, C, S	A, M, C
Chaining	Absent	Present	Absent	Absent	Absent	Present	Absent	Present
Bus width support	16 bit	16 bit	16 bit	32 bit	16 bit	32 bit	16 bit	8 bit
Data pipelining	Absent	Absent	Absent	Absent	Absent	Absent	Absent	Present
Multicycling	Present	Present	Present	Absent	Absent	Present	Absent	Present
Design objective	Area-execution time/Power-execution time	Area-power-latency	Area-execution time/Power-execution time	Area-execution time/Power-execution time	Area-execution time/Power-execution time	Area-execution time/Power-execution time	Area-latency	Area-execution time/Power-execution time
Fault reliability	Present	Absent	Present	Absent	Absent	Absent	Present	Absent
Loop support	CDFG,DFG	DFG	CDFG,DFG	CDFG,DFG	CDFG,DFG	DFG	DFG	CDFG,DFG
Trojan security	Absent	Absent	Absent	Absent	Present	Absent	Absent	Absent
Match percent	NA	50	100	60	60	50	60	60

Table 6.5 Feature-set of IP_{ID} and IP_V for FIR benchmark: FIR (23 nodes)

IP features	IP_{ID}	IP_{V1} (Hybrid PSO-GA HLS (Harish Ram *et al.*, 2011))	IP_{V2} (Fault secure HLS (Sengupta, 2015))	IP_{V3} (Watermark-HLS (Koushanfar *et al.*, 2005))	IP_{V4} (Trojan Secure-HLS (Rajendran *et al.*, 2013))	IP_{V5} (BFOA-HLS (Bhadauria, 2015))	IP_{V6} (Fault Tolerant-HLS (Inoue *et al.*, 2011))	IP_{V7} (PSO-HLS (Mishra and Sengupta, 2014))
Schedule algorithm	LIST	LIST	LIST	LIST	LIST	LIST	LIST	LIST
Resource type	A, M, C	A, M	A, M, C	A, M, C, S	A, M, C, S	A, M	A, M, C, S	A, M, C
Chaining	Present	Present	Absent	Absent	Absent	Present	Absent	Present
Bus width support	8 bit	16 bit	16 bit	32 bit	16 bit	32 bit	16 bit	8 bit
Data pipelining	Absent	Absent	Absent	Absent	Absent	Absent	Absent	Absent
Multicycling	Present	Present	Present	Absent	Absent	Present	Absent	Present
Design objective	Area-execution time/Power-execution time	Area-power-latency	Area-execution time/Power-execution time	Area-execution time/Power-execution time	Area-execution time/Power-execution time	Area-execution time/Power-execution time	Area-latency	Area-execution time/Power-execution time
Fault reliability	Absent	Absent	Present	Absent	Absent	Absent	Present	Absent
Loop support	CDFG,DFG	DFG	CDFG,DFG	CDFG,DFG	CDFG,DFG	DFG	DFG	CDFG,DFG
Trojan security	Present	Absent	Absent	Absent	Present	Absent	Absent	Absent
Match percent	NA	50	60	50	60	60	20	90

Table 6.6 Feature-set of IP_{ID} and IP_V for JPEG IDCT benchmark

Benchmark: JPEG IDCT (112 nodes)								
IP features	**IP_{ID}**	**IP_{V1} (Hybrid PSO-GA HLS (Harish Ram *et al.*, 2011))**	**IP_{V2} (Fault secure HLS (Sengupta, 2015))**	**IP_{V3} (Watermark-HLS (Koushanfar *et al.*, 2005))**	**IP_{V4} (Trojan Secure-HLS (Rajendran *et al.*, 2013))**	**IP_{V5} (BFOA-HLS (Bhadauria, 2015))**	**IP_{V6} (Fault Tolerant-HLS (Inoue *et al.*, 2011))**	**IP_{V7} (PSO-HLS (Mishra and Sengupta, 2014))**
Schedule algorithm	LIST	LIST	LIST	LIST	LIST	LIST	LIST	LIST
Resource type	A, M	A, M	A, M, C	A, M, C, S	A, M, C, S	A, M	A, M, C, S	A, M
Chaining	Present	Present	Absent	Absent	Absent	Present	Absent	Present
Bus width support	8 bit	16 bit	16 bit	32 bit	16 bit	32 bit	16 bit	8 bit
Data pipelining	Present	Absent	Absent	Absent	Absent	Absent	Absent	Present
Multicycling	Present	Present	Present	Absent	Absent	Present	Absent	Present
Design objective	Area-execution time/Power-execution time	Area-power-latency	Area-execution time/Power-execution time	Area-execution time/Power-execution time	Area-execution time/Power-execution time	Area-execution time/Power-execution time	Area-latency	Area-execution time/Power-execution time
Fault reliability	Absent	Absent	Present	Absent	Absent	Absent	Present	Absent
Loop support	DFG	DFG	DFG	DFG	DFG	DFG	DFG	DFG
Trojan security	Absent	Absent	Absent	Absent	Present	Absent	Absent	Absent
Match percent	NA	70	50	50	40	80	30	100

Table 6.7 Storage overhead comparison of customized CFE approach w.r.t. watermarking (Koushanfar et al., *2005) for IP protection during HLS*

Benchmark	Customized CFE for resolving ownership conflict for IP generated using HLS	Watermarking IP protection HLS approach (Koushanfar *et al.*, 2005)
	Storage registers	Storage registers
ARF	0	11
BPF	0	11
DCT	0	11
FFT	0	10
FIR	0	11
JPEG_IDCT	0	25

after IP core has been generated. Therefore, customized CFE does not incur design overheads. Further, due to the same reason, CFE does not impact design performance of an IP core; however, signature insertion during hardware watermarking may impact the performance of an IP core.

6.5 Conclusion

This chapter discussed an alternative ownership protection methodology called CFE from Sengupta and Kachave (2018), applicable for resolving ownership conflict of DSP IP cores. Further, the chapter described several features of the "feature-set" that are used as means of resolving ownership conflict with 0% degradation in hardware performance, and incur 0% storage overhead compared to the popular hardware watermarking-based ownership protection for DSP cores (Koushanfar *et al.*, 2005).

The future directions could target devising novel "expanded feature-set" that can resolve ownership conflict of DSP IP core in case of "deadlocks." Further, feature extraction rules for "scheduling algorithm" feature can be modified to include other scheduling algorithms such as force directed scheduling.

6.6 Exercises

1. What are the differences between CFE and watermark?
2. What are the differences between CFE and IP metering?
3. What are the different phases of CFE?
4. How does feature extraction work?
5. Explain feature extraction algorithm for chaining.
6. Explain feature extraction algorithm for multicycling.
7. Explain feature extraction algorithm for pipelining.
8. Explain feature extraction algorithm for Trojan security.
9. Explain feature extraction algorithm for transient fault security.
10. Explain IP validation process.

11. Explain feature extraction algorithm for scheduling.
12. Where the feature extraction is applied?
13. How does CFE for IP core protection differ from normal CFE?
14. What other feature of an IP core generated through HLS tool, be used for feature extraction?
15. What is the difference between generic feature and unique feature in CFE of IP protection?
16. Explain some of the HLS tools/approaches on which CFE can be applied at.
17. What is the trend of average extraction time of different features in feature set of CFE?
18. Explain some of the underlying assumptions of CFE approach during IP core protection.
19. What would happen if the feature set used is not capable of finding the actual IP owner?
20. Why CFE does not incur any hardware overhead?

References

Bhunia S., Abramovici M., Agrawal D., *et al.*, 'Protection against hardware Trojan attacks: Towards a comprehensive solution', *Proceedings of IEEE Design & Test*, 2013, vol. 99, pp. 1–1.

D'Angelo S., Metra C., and Sechi G., 'Transient and permanent fault diagnosis for FPGA-based TMR systems,' *Defect and Fault Tolerance in VLSI Systems, 1999. DFT '99. International Symposium on, Albuquerque*, NM, 1999, pp. 330–338.

Express Benchmark Suite, University of California San Diego, http://www.ece.ucsb.edu/EXPRESS/benchmark/, Last accessed on 2018.

Franke K. and Srihari S. N., 'Computational forensics: An overview'. In: Srihari S.N., Franke K. (eds) *Computational Forensics. IWCF 2008. Lecture Notes in Computer Science*, 2008, vol. 5158. Springer, Berlin, Heidelberg.

Gupta S., 'Loop shifting and compaction for the high-level synthesis of designs with complex control flow', *Technical Report CECS-TR*-03-14, 2003, UC Irvine.

Harish Ram D. S., Bhuvaneswari M. C., and Logesh S. M., 'A novel evolutionary technique for multi-objective power, area and delay optimization in high level synthesis of datapaths,' *Proceedings of the IEEE Computer Society Annual Symposium on VLSI (ISVLSI '11),* 2011, pp. 290–295.

Inoue T., Henmi H., Yoshikawa Y., and Ichihara H., 'High-level synthesis for multi-cycle transient fault tolerant datapaths.' *2011 IEEE 17th International On-Line Testing Symposium*, 2011, pp. 13–18.

International Chamber of Commerce, 'Estimating the global economic and social impacts of counterfeiting and piracy,' International Chamber of Commerce, Paris, France,Tech. Rep., 2011.

Kachave D. and Sengupta A., 'Protecting ownership of reusable IP core generated during high level synthesis', *Proceedings of IEEE International Symposium on Nanoelectronic and Information Systems (iNIS)*, Dec. 2016, pp. 80–82.

Kachave D. and Sengupta A., 'Forensic engineering for resolving ownership conflict of reusable IP cores generated during high level synthesis,' *IEEE VLSI Circuits & Systems Letter*, 2017, vol. 2 (2), pp. 15–18.

Koushanfar F. and Qu G., 'Hardware metering.' *Proceedings of the 38th Annual Design Automation Conference*, ACM 2001, pp. 490–493.

Koushanfar F, Hong I., and Potkonjak M., ''Behavioral synthesis techniques for intellectual property protection,'' *ACM Transactions on Design Automation of Electronic Systems*, 2005, vol. 10 (3), pp. 523–545.

Kurra N. K. S. and Panda P. R., 'The impact of loop unrolling on controller delay in high level synthesis', *Proceedings of IEEE Design, Automation and Test Europe (DATE)*, California, 2007, pp. 391–396.

Lao Y. and Parhi K. K., 'Protecting DSP circuits through obfuscation,' *2014 IEEE International Symposium on Circuits and Systems (ISCAS)*, Melbourne VIC, 2014, pp. 798–801.

Li L. and Zhou H., 'Structural transformation for best-possible obfuscation of sequential circuits,' *2013 IEEE International Symposium on Hardware-Oriented Security and Trust (HOST)*, Austin, TX, 2013, pp. 55–60.

McFarland M. C., 'Tutorial on high-level synthesis', *Proceedings of 25th ACM/IEEE Design Automation Conference*, California, 1988, pp. 330–336.

Mishra V. K. and Sengupta A., 'MO-PSE: Adaptive multi-objective particle swarm optimization based design space exploration in architectural synthesis for application specific processor design', *Elsevier Journal on Advances in Engineering Software*, 2014, vol. 67, pp. 111–124.

Nomura S., Sinclair M. D., Ho C.-H., Govindaraju V., de Kruijf M., and Sankaralingam K. 'Sampling + DMR: practical and low-overhead permanent fault detection,' *Proceedings of the 38th annual international symposium on Computer architecture* (ISCA'11). ACM, New York, NY, USA, 2011, pp. 201–212.

Paulin P. G. and Knight J. P., 'Scheduling and binding algorithms for high-level synthesis', *Proceedings of 26th ACM/IEEE Design Automation Conference*, California, 1989, pp. 1–6.

Rajendran J., Zhang H., Sinanoglu O., and Karri R., 'High-level synthesis for security and trust,' *IEEE 19th International On-Line Testing Symposium (IOLTS)*, 2013, Chania, pp. 232–233.

Sengupta A., 'Intellectual property cores: Protection designs for CE products,' *IEEE Consumer Electronics Magazine*, 2016, vol. 5 (1), pp. 83–88.

Sengupta A., 'Hardware security of CE devices [Hardware matters],' *IEEE Consumer Electronics Magazine*, 2017, vol. 6 (1), pp. 130–133.

Sengupta, A. and Bhadauria S., 'Bacterial foraging driven exploration of multi cycle fault tolerant datapath based on power-performance tradeoff in high level synthesis,' *Expert Systems with Applications*, 2015, vol. 42 (10), pp. 4719–4732.

Sengupta A. and Bhadauria S., 'Exploring low cost optimal watermark for reusable IP cores during high level synthesis,' *IEEE Access*, 2016, vol. 4, pp. 2198–2215.

Sengupta A. and Kachave D., 'Forensic engineering for resolving ownership problem of reusable IP core generated during high level synthesis,', *Elsevier Journal on Future Generation Computer Systems*, 2018. https://doi.org/10.1016/j.future.2017.08.001

Sengupta A. and Kundu S., 'Securing IoT hardware: Threat models and reliable, low-power design solutions,' *IEEE Transactions on Very Large Scale Integration (VLSI) Systems*, 2017, vol. 25 (12), pp. 3265–3267.

Sengupta A. and Roy D., 'Antipiracy-aware IP chipset design for CE devices: A robust watermarking approach [Hardware matters],' *IEEE Consumer Electronics Magazine*, 2017, vol. 6 (2), pp. 118–124.

Sengupta A. and Sedaghat R., 'System and methodology for development of a system architecture using optimization parameters,' 21 December 2010, U.S. patent US12974925.

Sengupta A. and Sedaghat R., 'Integrated scheduling, allocation and binding in High Level Synthesis using multi structure genetic algorithm based design space exploration,' *2011 12th International Symposium on Quality Electronic Design*, Santa Clara, CA, 2011, pp. 1–9.

Sengupta, A. and Sedaghat R., 'Swarm intelligence driven design space exploration of optimal k-cycle transient fault secured datapath during high level synthesis based on user power–delay budget,' *Elsevier Journal on Microelectronics Reliability*, 2015, vol. 55 (6), pp. 990–1004.

Sengupta A., Roy D., and Mohanty S. P., 'Triple-phase watermarking for reusable IP core protection during architecture synthesis,' *IEEE Transactions on Computer-Aided Design of Integrated Circuits and Systems*, 2018, vol. 37 (4), pp. 742–755.

Sengupta A., Bhadauria S., and Mohanty S. P., 'Embedding low cost optimal watermark during high level synthesis for reusable IP core protection,' *2016 IEEE International Symposium on Circuits and Systems (ISCAS)*, Montreal, QC, 2016, pp. 974–977.

Sengupta A., Bhadauria S., and Mohanty S. P., 'TL-HLS: Methodology for low cost hardware Trojan security aware scheduling with optimal loop unrolling factor during high level synthesis,' *IEEE Transactions on Computer-Aided Design of Integrated Circuits and Systems*, 2017, vol. 36 (4), pp. 655–668.

Sengupta A., Roy D., Mohanty S. P., and Corcoran P., 'DSP design protection in CE through algorithmic transformation based structural obfuscation,' *IEEE Transactions on Consumer Electronics*, 2017, vol. 63 (4), pp. 467–476.

Wong J. L., Kirovski D., and Potkonjak M., 'Computational forensic techniques for intellectual property protection,' *IEEE Transactions on Computer-Aided Design of Integrated Circuits and Systems*, 2004, vol. 23 (6), pp. 987–994.

Yasin M., Rajendran J., Sinanoglu O., and Karri R.. 'On improving the security of logic locking,' *IEEE Transactions on Computer-Aided Design of Integrated Circuits and Systems,* 2016, vol. 35 (9), pp. 1411–1424.

Yu S.-Y. and McCluskey E. J., 'Permanent fault repair for FPGAs with limited redundant area,' *Proceedings 2001 IEEE International Symposium on Defect and Fault Tolerance in VLSI Systems*, San Francisco, CA, 2001, pp. 125–133.

Zhang J., 'A practical logic obfuscation technique for hardware security,' *IEEE Transactions on Very Large Scale Integration (VLSI) Systems*, 2016, vol. 24 (3), pp. 1193–1197.

Zhang X. and Tehranipoor M., 'Case study: Detecting hardware Trojans in third-party digital IP cores'. *Proceedings of IEEE International Symposium on Hardware-Oriented Security and Trust*, California, 2011, pp. 67–70.

Chapter 7
Structural obfuscation of DSP cores used in CE devices

Previous chapters, watermarking, fingerprinting, and forensic engineering have been discussed for resolving various ownership-related problems. This chapter discusses structural obfuscation approaches to thwart IP piracy and reverse engineering (RE). This approach, when effective, can save billions of dollars of revenue losses in CE and semiconductor industry. Especially, a multi-high-level transformation-based structural obfuscation process has been presented for DSP IP cores as hardware-hardening technique. The chapter is structured as follows: Section 7.2 highlights the fundamentals of obfuscation with stress on structural obfuscation; Section 7.3 discusses different compiler transformation-driven structural obfuscation methodology. Section 7.4 discusses low-cost structurally obfuscated design exploring technique. Section 7.5 demonstrates a multistage structural obfuscation technique through a motivational example. Section 7.6 presents the results of a case study.

7.1 Introduction

Today's CE devices, ranging from modern smart phone, smart TV, home appliance, set-top box, tablet, digicam to recent smart speaker, are designed using SoC platforms (Thavalengal and Corcoran, 2016; Mohanty, 2015; Kim *et al.*, 2015). These SoCs comprise various system modules such as memory (SRAM, Flash), A-to-D converter, custom processor, or co-processor, DSP kernels, A/V codecs, and wireless modems. Among these modules, DSP kernels in the form of IP cores are primarily responsible for data/power-intensive computation at high speed, minimal silicon area, and low power in an SoC (Li *et al.*, 2015). Thus, DSP IP cores are the heart of every SoC-based CE devices. For example, ConnX D2 DSP IP core used in telecom infrastructure of voice-over internet protocol (VoIP) and wireless mobile device (ConnX, 2009). In a cellular telephone system, it is used in speech encoding/decoding process. Low-end DSP IP cores are included in digital camera, portable phones, consumer audio, headsets, while mid-end DSP cores are included in cellular phone, fax/voice receiver. Finally, high-end DSP IP cores are included in various CE systems including wireless base station and cable modems. Further, reusability of DSP IP core maximizes design productivity to cope with the increasing design complexity in electronic systems (Castillo *et al.*, 2007), thus maintaining a balance in global supply chains for SoC-based CE devices (Sengupta and Sedaghat, 2011, 2013).

These reusable DSP IP cores warrant a substantial investment in research, workforce, money, and time by an IP vendor/designer. Therefore, these mandate protection from emerging IP threats such as cloning and/or infringement, and RE (Roy and Sengupta, 2017; Sengupta and Bhadauria, 2016; Wong *et al.*, 2004; Koushanfar, 2012; Alkabani and Koushanfar, 2007; Sengupta and Kundu, 2017; Sengupta *et al.*, 2018). According to literature sources, the total fiscal revenue loss due to piracy and counterfeiting in 2013 for G20 nations was $96–130 billion which will increase to $199–270 billion in 2022. It is projected that 10% of the internationally sold electronics products are forged (Guajardo *et al.*, 2008; Lao and Parhi, 2015). An attacker can "silently" reuse a DSP IP core without legal permission from the IP owner to save design time, cost, and effort (Sengupta, 2016). He/she can easily identify the functionality of a DSP IP core from its architecture through RE. RE of an IP core is a process of identifying its design, structure, and functionality (Rajendran *et al.*, 2013). Using RE, one can identify the device technology (Intel Transistors, 2012), extract the gate-level netlist (Torrance and James, 2009), and infer the IP functionality (ExtremeTech, 2012). Though according to the Semiconductor Chip Protection Act of 1984 (SCPA), RE is a legal process for teaching, analysis, and evaluation purpose; however, an attacker can use RE process for IP piracy/attacks which is illegal. An attacker can also insert malicious Trojan logic into an IP core, if he/she can identify the functionality of a DSP IP core through RE. Therefore, it is very crucial to design a DSP IP core in a way that it provides substantial resiliency against RE (Yang *et al.*, 2013; Parhi, 2012; Lao and Parhi, 2015; Walz and Schrder, 2016; Sengupta, 2017). A thematic attack scenario of protected DSP IP core for different CE devices is shown in Figure 7.1 (Mohanty, 2003; Sengupta, 2016). One way to achieve resilience against RE is to obscure the functionality of a DSP IP core by transforming its architecture. This transformation should change the standard structure of a DSP IP core into an unobvious architecture while preserving correct functionality. Therefore, the functionality will not be easily identifiable from its structure, thereby increasing the RE complexity for an adversary. This process is known as structural obfuscation. More broadly, obfuscation is a process of obscuring or disguising the actual functionality of the design (product) in order to protect its intellectual property content. Though hardware obfuscation is motivated from software obfuscation, the former is more complex than the latter, despite both of them fundamentally having the same objective. Software obfuscation aims to conceal the implementation code of an algorithm using techniques such as dead code elimination, code unrolling, while hardware obfuscation aims to obscure the circuit (hardware) functionality/structure. This process is classified, as it is achievable through several complex techniques, depending on which form (design abstraction layer. Refer Figure 7.2) the hardware circuit is described at, e.g., RTL codes, netlist, layout, fabricated chip. Sometimes, the software obfuscation methods after customization can be applied for hardware obfuscation. However, a completely exclusive method may be needed for hardware obfuscation depending on the hardware design abstraction level. In summary, a hardware obfuscator (process or algorithm) is responsible for transforming a design

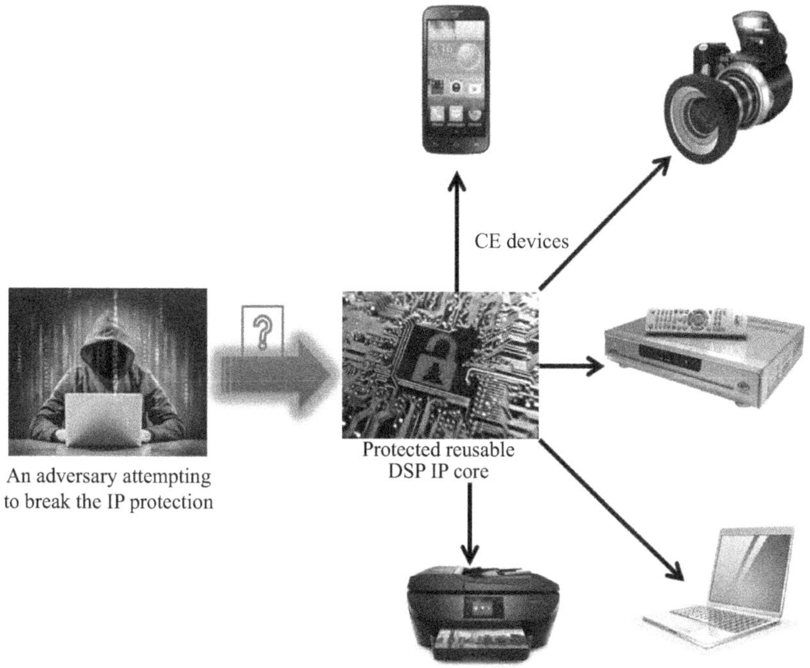

Figure 7.1 A thematic representation of secured DSP IP core for CE devices that is resilient from adversary

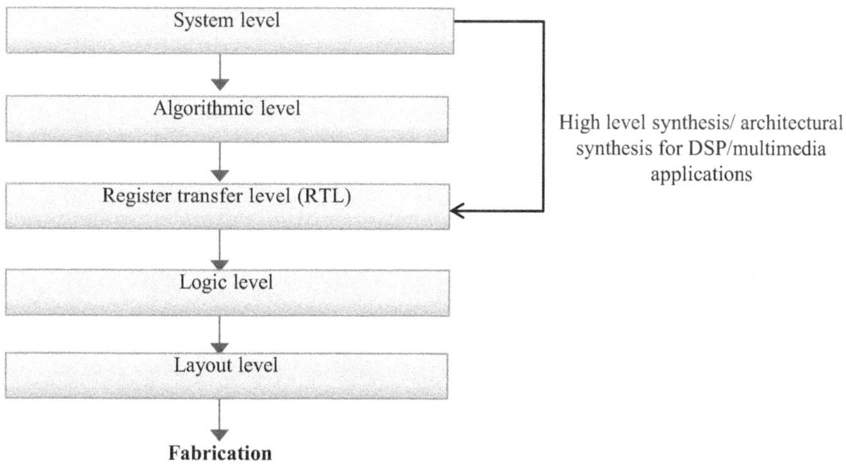

Figure 7.2 Abstraction levels of integrated circuit design

representation "D" into its obfuscated form "O(D)" that has the same functionality as "D," yet is unintelligible to an adversary targeting to recover "D" from "O(D)." More details on the threat model of obfuscation are explained in the following sections of this chapter.

7.1.1 Threat model

Hardware structural obfuscation transforms the original hardware design architecture of an IP core into a nonobvious structural architecture. Thus enhancing the RE complexity as it is more tough for an attacker to discover the actual functionality of an IP core from its transformed structure. Consequently, it provides defense against threats resulting due to **RE** such as **IP piracy, IP infringement (counterfeit),** and **Trojan insertion**.

7.1.2 Benefits of providing security at higher design abstraction level

The designing process of a DSP application in the form of an IP core is very complex. Therefore, the DSP IP core design flow is divided into multiple design abstraction levels such as system level, algorithmic level, register transfer level, logic level, and layout level (as shown in Figure 7.2). Every layer has its exclusive design automation or CAD method which enables the design of DSP IP cores with lesser design time. Architectural synthesis (also known as HLS) is one such automatic CAD process, which is responsible for generating equivalent register-transfer-level hardware description languages such as VHDL, Verilog of DSP applications. Examples of a few well-known commercial and academic HLS tools for DSP core/multimedia are shown in Figure 7.3 (Sengupta, 2016; Mohanty, 2003). As shown in Figure 7.3, each HLS tool is represented as tool name, followed by the owner name (highlighted in bold) and the launching year of the tool. However, none of these tools incorporate security through obfuscation.

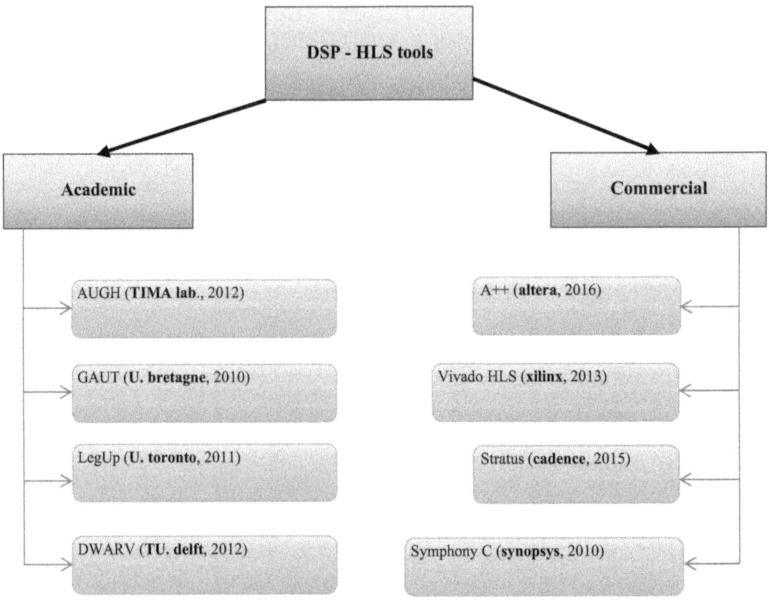

Figure 7.3 Classification and examples of HLS tools based on license

The resilience of a DSP IP core against RE can be provided in any abstraction level of the design flow. However, providing resilience against RE attack through structural obfuscation during architectural synthesis is beneficial as the structural change performed in higher design abstraction level obfuscates the datapath and control unit structures of the application. Thus, it also obfuscates the gate structure (netlist) of the application. Additionally, architectural synthesis for digital IPs is way matured to achieve hardware area minimization, power optimization, performance enhancement, and fault-tolerant IP design solution with lesser design complexity. Therefore, structural obfuscation during architectural synthesis is capable of generating a strongly obscured design which is functionally equivalent but appears remarkably different than the original structure. The only way to use this type of obfuscated design is as a "black-box" module (Chakraborty and Bhunia, 2009).

7.2 Obfuscation for IP core protection—a broad view

This section discusses the fundamentals of different obfuscation-based IP core protection mechanisms with emphasis on high-level transformation-based structural obfuscation mechanisms that are available in the literature.

Obfuscation-based IP protection can be achieved in three different ways: (1) by transforming the HDL code, known as "source code-based obfuscation," (2) by using key-based logic locking, known as "logic obfuscation," and (3) by hiding the functionalities of a design, known as "structural obfuscation."

7.2.1 Code obfuscation techniques

To thwart an attacker from intercepting the HDL code, it either alters (Brzozowski and Yarmolik, 2007) the source code into a compound version or encrypts (Barria *et al.*, 2016) it. In Brzozowski and Yarmolik (2007), HDL code obfuscation is achieved using diverse alteration approaches which are challenging to reverse engineer. In Barria *et al.* (2016), the source code is encrypted using "Dead Code Addition" approach on a Crypter.

7.2.2 Logic obfuscation techniques

Logic obfuscation (also known as logic encryption) locks the implementation of a design by implanting supplementary components into it. These components are controlled by key gates. Applying incorrect key will produce functionally different output than the correct output of the design. Logic obfuscation is of two types: (1) sequential and (2) combinational (Rajendran *et al.*, 2013). In sequential logic obfuscation, the additional components could exist in a key-controlled obfuscation mode. For example, in Chakraborty and Bhunia (2009), dummy states are added with the original states of the FSM of a design to obfuscate it. These dummy states are controlled by an external key. Applying incorrect key leads to obfuscation mode or invalid state whereas applying correct key leads to normal mode. In Desai *et al.* (2013), the design is locked based on internally generated key sequence (known as Code-Word). This technique hides the design using this Code-Word in the transition function. On applying the incorrect key, the FSM will deviate from the correct transition path.

In combinational logic obfuscation, the additional components could be XOR/XNOR gates (Roy *et al.*, 2008; Rajendran *et al.*, 2012; Rajendran *et al.*, 2015), AND/OR gates (Dupuis *et al.*, 2014), Lookup Table (LUT) (Baumgarten *et al.*, 2010), muxes (Wang *et al.*, 2016; Zhang, 2016), or combination of these elements (Lee and Touba, 2015). EPIC (Roy *et al.*, 2008) is one of the primary work where the framework of logic obfuscation is proposed. In this approach, XOR/XNOR gates are inserted as key-gates in the random location of the design. Random insertion of key-gates into a design does not always guarantee that wrong output is produced for all input patterns on application of incorrect key. Additionally, in this approach, the key can be sensitized by input pattern. Rajendran *et al.* (2015) inserted XOR/XNOR key-gates in the design in such a way that the Hamming distance (HD) between the incorrect output and correct output became maximum. In Lee and Touba (2015), XOR/XNOR key-gates are combined with mux key-gates to achieve a higher HD between correct and incorrect output. Key-controlled LUT-based logic obfuscation is proposed in Baumgarten *et al.* (2010). In this approach, the complete IP is divided into multiple fixed and configurable parts. The configurable part is then realized as a key-controlled LUT. As the logic type of the reconfigurable part can be changed after fabrication, an adversarial foundry cannot misuse the IP. Several attack strategies are available against these logic obfuscation techniques, such as key-sensitization (Rajendran *et al.*, 2015), Boolean satisfiability (SAT) and hill-climbing search-based attack (Subramanyan *et al.*, 2015). These logic obfuscation techniques have proven to be powerful for combinational/sequential circuits. However, in the context of DSP cores, these techniques incur more design overhead due to extra key-gates been implanted into the design, as well as lack of framework to select low-cost design architecture. Moreover, if the key is leaked somehow, the IP designer cannot prevent any attack until the IP is redesigned for a new key. More details on this technique are provided in Chapter 8.

7.2.3 Structural obfuscation techniques

On the other hand, structural obfuscation tries to hide the correct functionalities of a design through transformations, arbitrary assignment of logic components, inserting dummy wires, or irregular routing (Vijayakumar *et al.*, 2017). In other words, structural obfuscation changes the architecture of an IP core from its standard structure to an unobvious structure while preserving its functionality. Therefore, even if an adversary physically tampers or REs the IP core to obtain its design, it will be harder for him/her to discover its functionality by inspecting its structure. Structural obfuscation does not insert any additional component as key-gate into the design. Therefore, there is no such design overhead for structural obfuscation. In structural obfuscation technique, a design is secured until its functionality is hidden from an adversary. However, if the application is traced by the attacker, structural obfuscation cannot protect an IP core.

High-level transformations such as pipelining, interleaving, unfolding, folding, tree height transformation (THT), redundant operation elimination, and look-ahead transformations can be applied during architectural synthesis of a DSP IP core to achieve a trade-off between multiple design objectives such as area, power, and delay. However, these transformations can also be used as structural obfuscation techniques (Sengupta and Roy, 2017; Parhi, 2012). These high-level transformations modify the structure of a DSP IP core significantly while keeping the actual functionality unchanged. These alterations result in multiple ambiguous architectures of the same DSP IP core. Authors in Lao and Parhi (2015) have performed structural obfuscation for DSP circuits through a single high-level transformation technique, i.e., folding (Note: Though authors in Lao and Parhi (2015) have performed both structural and functional obfuscation, we are only focusing on the structural obfuscation portion in this chapter). The folding transformation generates various meaningful modes which increase the complexity for an attacker to distinguish the correct functionalities within different modes. Additionally, different folding factors cause different equivalent circuits which lead to ambiguity for an attacker. According to Lao and Parhi (2015), it can be observed that due to folding transformation, circuits with different functionality may have the same structure, and circuits with same functionality may have dissimilar structure. Since folding factor has a crucial impact on the design cost, selecting an appropriate folding factor may lead to an optimized DSP circuit design. However, no optimal folding factor evaluation is presented in the aforesaid approach. Thus, it may lead to higher design overhead. Additionally, single high-level transformation-based (Lao and Parhi, 2015) structural obfuscation for DSP applications is less secured compared to a multistage high-level transformation-based structural obfuscation (Sengupta *et al.*, 2017a).

Sengupta *et al.* (2017a) performed a multiple compiler-driven high-level transformations-based structural obfuscation for DSP core/kernels during architectural synthesis. Unlike Lao and Parhi (2015), this approach is capable of obfuscating loop-based DSP applications also. Further, due to multiple high-level transformations, the obfuscated DSP IP core impacts gate-level structure significantly different than its original structure, thus enhancing the RE complexity. Additionally, this approach performs optimization to select a low-cost architecture that meets the design objectives such as latency, area, and all-inclusive design cost. Therefore, a compiler-driven, low-cost, multistage transformation-based structural obfuscation with stronger robustness is presented in Sengupta *et al.* (2017a). A detailed explanation of this approach is discussed in Section 7.3.

7.3 Compiler transformation-driven structural obfuscation

As mentioned earlier, due to the insertion of additional logic elements, the design as well as lack of framework to select low cost design architecture, overhead of logic obfuscation is significantly larger than the structural obfuscation in the context of DSP IP cores. Additionally, logic obfuscation techniques are vulnerable to multiple attack strategies such as key-sensitization (Rajendran *et al.*, 2015), Boolean SAT, and hill-climbing search-based attack (Subramanyan *et al.*, 2015). Further, the structural obfuscation presented in Lao and Parhi (2015) is less secured against the RE attacks due to its single-stage transformation-based obfuscation technique. The multistage structural obfuscation approach presented in Sengupta *et al.* (2017a) hides the functionality of a DSP application by transforming its design into an unobvious structure through several successive compiler-driven transformation techniques. Though compiler-driven transformation is a known concept for design optimization, to the best of authors' knowledge, these transformation techniques have been never used before to perform structural obfuscation. These transformations increase the complexity for an adversary to discover its functionality. Five high-level transformation approaches, i.e., redundant operation elimination (ROE), logic transformation (LT), THT, loop unrolling (LU), and loop-invariant code motion (LICM) are used to obfuscate both loop and non-loop-based DSP IP cores (Sengupta *et al.*, 2017a). However, loop-based transformations (i.e., LU and LICM) are not applicable to non-loop-based DSP applications. The structurally obfuscated design is further explored using PSO-based design exploration (DSE) to obtain the low-cost design architecture (Sengupta *et al.*, 2017b). Lastly, a low-cost, structurally obscured DSP IP is created. The illustration of low-cost multistage-driven structural obfuscation process is shown in Figure 7.4. Detailed explanation of each high-level transformation process used in Sengupta *et al.* (2017a) is explained in Section 7.3.

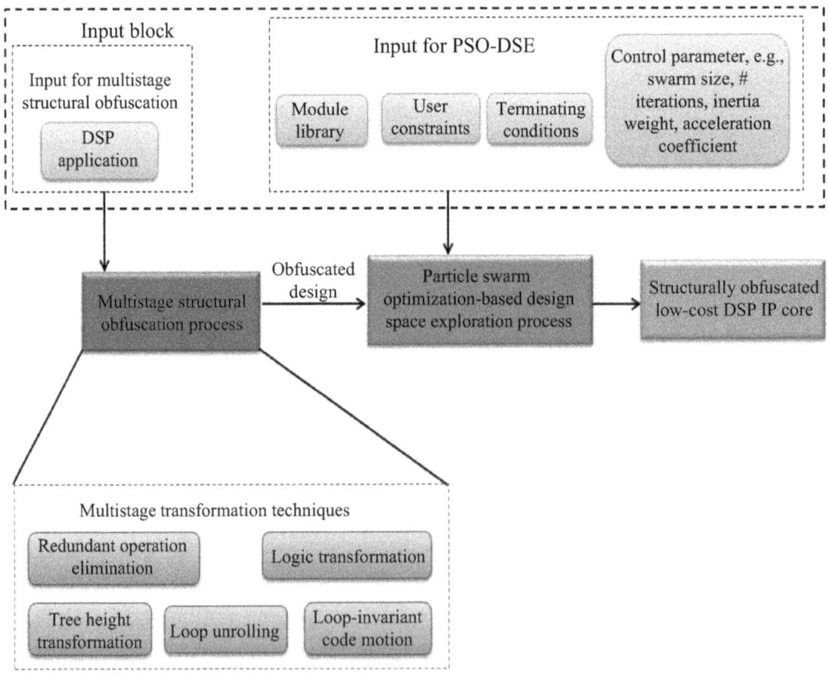

Figure 7.4 Overview of low-cost multistage-driven structural obfuscation process

7.3.1 Formulation and evaluation models

7.3.1.1 Formulation

To determine a low-cost, structurally obfuscated design for a given DSP core expressed as C-code/DFG or loop-based CDFG within user-specified area-delay constraints.

Minimize: Minimize structurally obfuscated design cost ($A_{total}{}^{obf}$, $L_{total}{}^{obf}$) for an optimal design solution (X_i), where $(X_i) = N(R_1)$, $N(R_2)$, … , $N(R_D)$, at the behavioral level; and $N(R_D)$ is the number of resource-type R_D.

Subject to: $A_{total}{}^{obf} \leq A_{cons}$ and $L_{total}{}^{obf} \leq L_{cons}$ and RE-resilient IP core through structural obfuscation, where $A_{total}{}^{obf}$ and $L_{total}{}^{obf}$ are the total area and latency of the structurally obfuscated design and A_{cons} and L_{cons} are the user-provided area and latency constraints, respectively.

7.3.1.2 Evaluation models

a. Design area evaluation model

Total design area ($A_{total}{}^{obf}$) consumed by a structurally obfuscated reusable DSP IP core is expressed by the following model (Sengupta *et al.*, 2017a):

$$A^{obf}_{total} = \sum_{i=1}^{D} A(R_i) * R_i + A(\text{buffer}) * N(\text{buffer}) + A(\text{mux}) * N(\text{mux}) \tag{7.1}$$

Total design area is evaluated by adding total area of hardware resource units (such as adder, subtractor, multiplier), interconnecting units (such as multiplexer and de-multiplexer), and storage units (such as register). The total design area is assessed with respect to module library of 15 nm NanGate-based technology (Open Cell Library, 2017).

b. Design delay evaluation model

Total design latency ($L_{total}{}^{obf}$) of structurally obfuscated reusable IP core is expressed by the following model (Sengupta *et al.*, 2017a):

$$L^{obf}_{total} = \left(L^{obf}_{body} * \left\lfloor \frac{I}{UF} \right\rfloor \right) + (I \bmod UF) * L^{obf}_{first} \tag{7.2}$$

In the above expression, $L_{body}{}^{obf}$ is the single iteration execution delay of the loop body of obfuscated CDFG; I is the total loop iteration (loop count); $L_{first}{}^{obf}$ is the delay to execute the first iteration of obfuscated CDFG. The total design latency is evaluated with respect to the module library of 15 nm NanGate-based technology (Open Cell Library, 2017) at 1 GHz clock frequency.

c. Design cost evaluation function

The solution design cost is estimated using the following function (Sengupta *et al.*, 2017a):

$$C_f(X_i) = w_1 \frac{A^{obf}_{total} - A_{cons}}{A^{obf}_{max}} + w_2 \frac{L^{obf}_{total} - L_{cons}}{L^{obf}_{max}} \tag{7.3}$$

where $C_f(X_i)$ is the design cost of the structurally obfuscated DSP IP core with resource configuration X_i, A_{max}^{obf} and L_{max}^{obf} are the maximum possible area and latency in the design space, respectively, w_1 and w_2 are user-specified area and latency weightage factor, respectively; to provide equal preference, both are preset as 0.5.

7.3.2 Multistage high-level transformation techniques

As mentioned earlier, the multistage structural obfuscation methodology (Sengupta *et al.*, 2017a) is achieved using ROE-, LT-, THT-, LU-, and LICM-based techniques. In this approach, a DSP core is taken as input as C-Code/CDFG or DFG. Each of these transformation techniques is then applied to the input application in order to transform it. Each transformation technique is explained and demonstrated through an example. For example, a nonsecured CDFG (present in Figure 7.5) is accepted as input for demonstration purpose. (Note: Application in Figure 7.5 is chosen, as it enables all the five multistage transformations to be demonstrated over a single example. However, any application could have been used.) The application is then scheduled based on user-defined resource configuration (example: three adders and four multipliers). The respective nonobfuscated scheduled CDFG and its corresponding circuit representation are depicted in Figures 7.6 and 7.7, respectively. The circuit presented in Figure 7.7 employs 6 numbers of 4:1 switches, 4 numbers of 2:1 switches, 10 numbers of input registers, 1 number of output

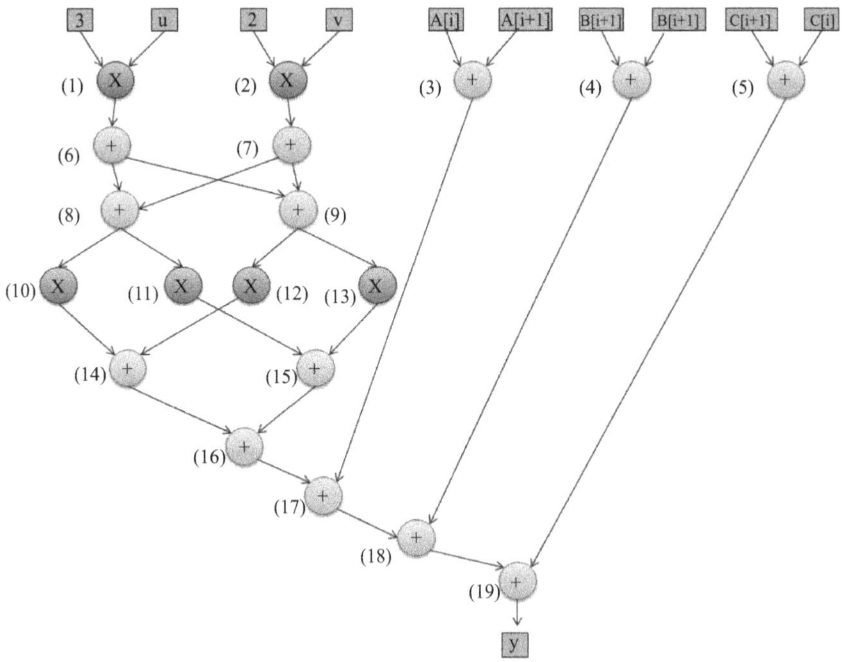

Figure 7.5 Original non-obfuscated CDFG

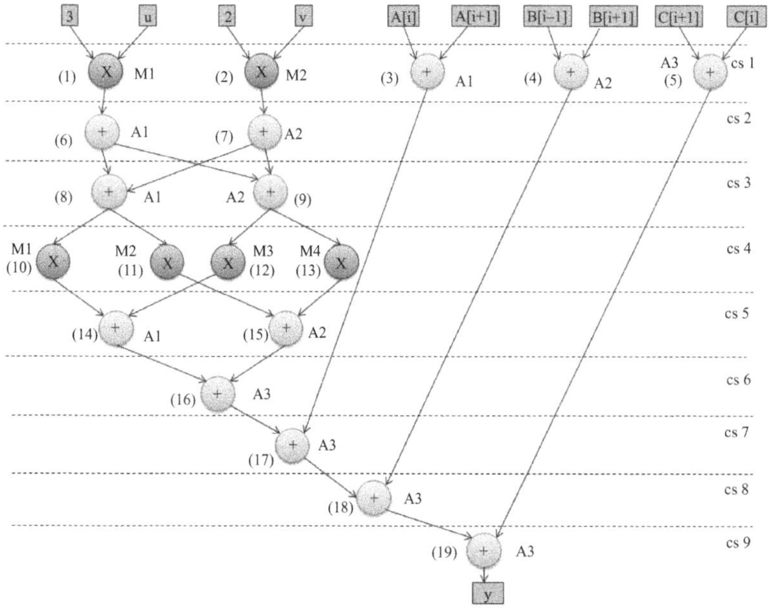

Figure 7.6 Scheduled CDFG of DSP IP design with three adders and four multipliers

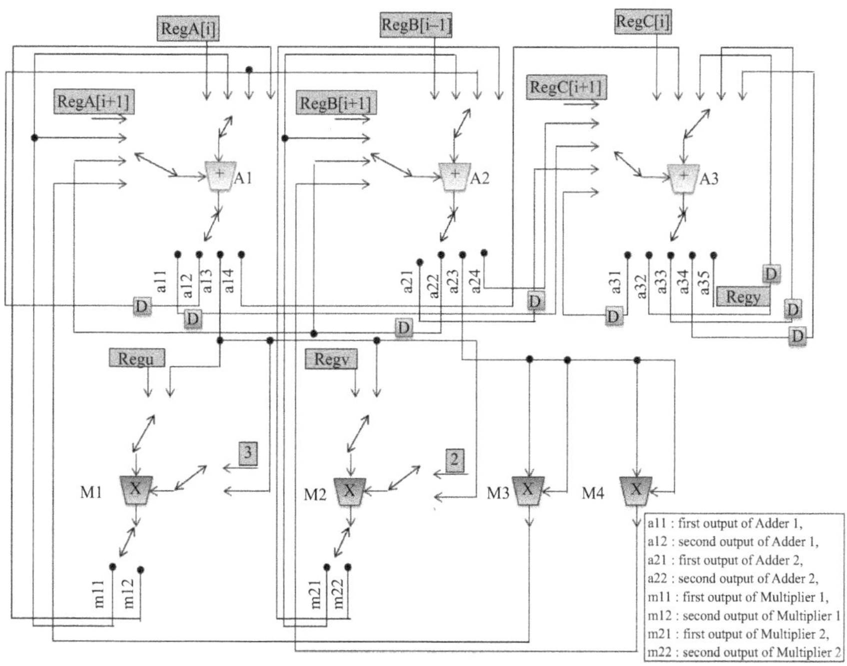

Figure 7.7 DSP circuit of a non-obfuscated design

register, and 8 numbers of delay elements. This nonobfuscated CDFG is used as an example to explain each transformation technique in the following subsections.

1. Redundant operation elimination, process: This compiler-driven transformation technique removes the redundant or duplicate operations from a set of operations in a DSP application and then performs necessary adjustment if required. In the context of DFG/CDFG, each operation is represented by a node. Therefore, in the input DFG/CDFG of a DSP core, a node can be recognized as a redundant node, if another node with the same pair of parents and operation type (i.e., adder, multiplier, comparator, etc.) is also available in that DSP application. According to Sengupta *et al.* (2017a), if a pair of such nodes is found, then the nodes with higher node number are removed from the application. However, to keep the correct functionality of the application, necessary adjustment is required in the graph. Lastly, ROE-based structurally obfuscated DSP circuit is generated as intermediate output which is fed to the next obfuscation stage.

   For instance, in the nonobfuscated design depicted in Figure 7.5, the redundant operations are nodes 9, 11, 12, 13, 15, which are removed using ROE method as depicted in Figure 7.8. In order to preserve the functionality of the circuit, the inputs of nodes 14 and 16 are adjusted accordingly, as shown in Figure 7.8.

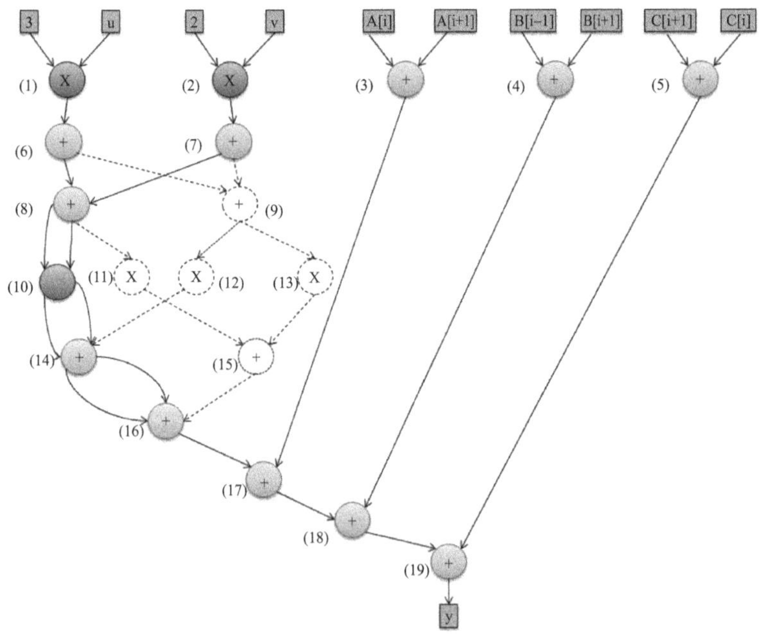

Figure 7.8 ROE-based obfuscated design

2. Logic transformation process: This compiler-driven transformation technique modifies the functionality of an application by generating a different logically equivalent function. In the context of DFG/CDFG, logic transformation alters the nodes of the input graph such that the structure of the graph appears significantly different compared to the original while satisfying the correct functionality of the application. Finally, LT-based structurally obfuscated DSP circuit is produced as output. This transformation is powerful to reflect change at the gate level after HLS.

   For example, Figure 7.9 represents the logic transformation-driven obfuscated form of the input graph (Figure 7.8); newly added/modified nodes are shown through node numbers 20 & 7 and the modified dependencies are marked with dotted line. Post HLS, the design is functionally equivalent but structurally obfuscated.
3. THT process: This transformation technique changes the critical path dependency of the input application accepted as DFG/CDFG, thus modifying the height of the graph. More specifically, THT divides the critical path computation into multiple sub-computations and then executes each sub-computation in parallel. Thus, THTs generate structurally dissimilar yet functionally equivalent graphs. Finally, THT-based structurally obfuscated DSP circuit is produced as output post-HLS design process.

   For example, in Figure 7.10, THT-based structurally obfuscated form of the input graph (shown in Figure 7.9) is shown. It reduces the height of the obfuscated graph from 10 to 8 compared to its input design. The computation of nodes 17 and 18 in obfuscated design is executed earlier compared to the

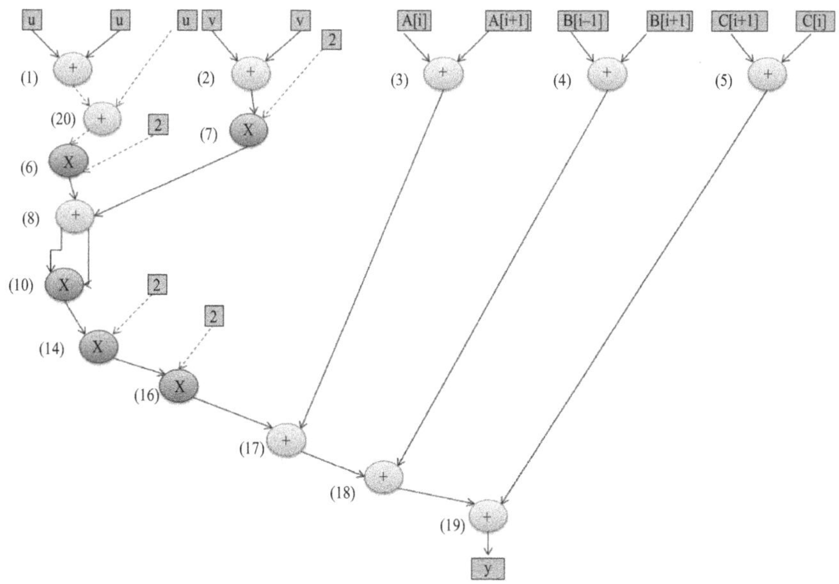

Figure 7.9 Logic transformation-based obfuscated design

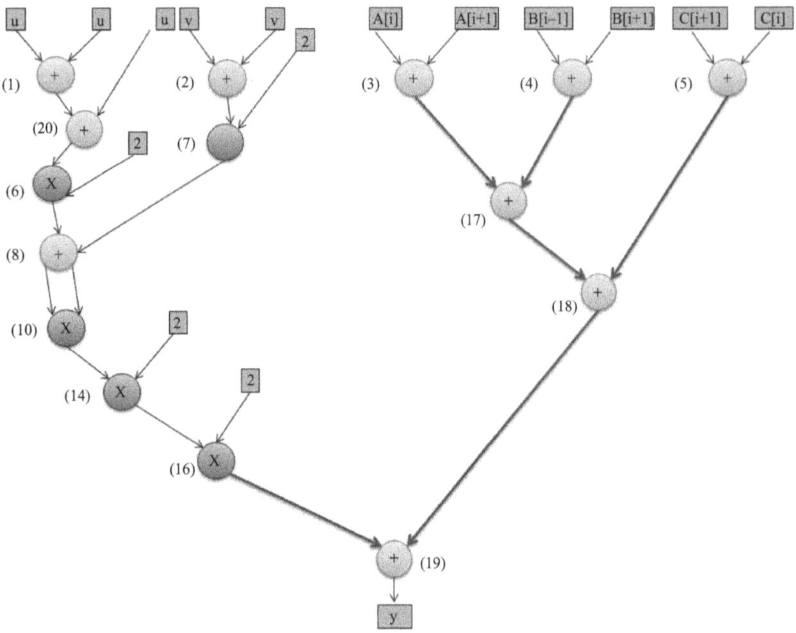

Figure 7.10 THT-based obfuscated design

original design. The dependencies of the obfuscated graphs are adjusted to maintain the correct functionality. The modified dependencies are marked with deep lines.

4. LU process: This loop transformation approach unfolds or unrolls a loop-based application considering the unrolling factor. LU can be performed by executing the same calculation present inside a loop multiple times. Lastly, LU-driven structurally obfuscated DSP circuit is generated as output, post HLS.

   For instance, Figure 7.11 unrolls the input CDFG based on the unrolling factor 2. In Sengupta *et al.* (2017a), a low-cost unrolling factor can be explored through PSO-driven DSE process (Sengupta *et al.*, 2017b). LU minimizes the execution time as well as, concurrently, structurally obfuscates the IP design.

5. Loop-invariant code motion process: This loop transformation process identifies those loop-independent operations which are presented inside the loop and moves them out the loop. In the context of CDFG, the nodes which are independent of loop are shifted out from the loop body. More specifically, LICM transformation moves out those nodes whose iterative execution inside the loop or single execution outside the loop does not make any difference in the final output of the application. Therefore LICM produces structurally different, yet functionally same designs. Lastly, LICM-based structurally obfuscated DSP circuit is produced as output after processing through HLS.

   For instance, Figure 7.12 depicts the LICM-based structurally obfuscated version of the input CDFG. The dotted box indicates the loop-independent

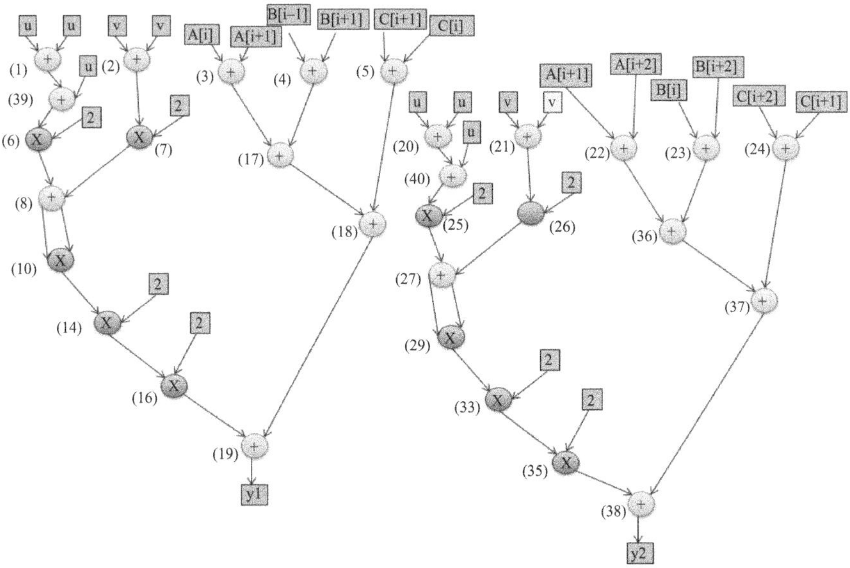

Figure 7.11 Obfuscated design-based on LU (unrolled twice)

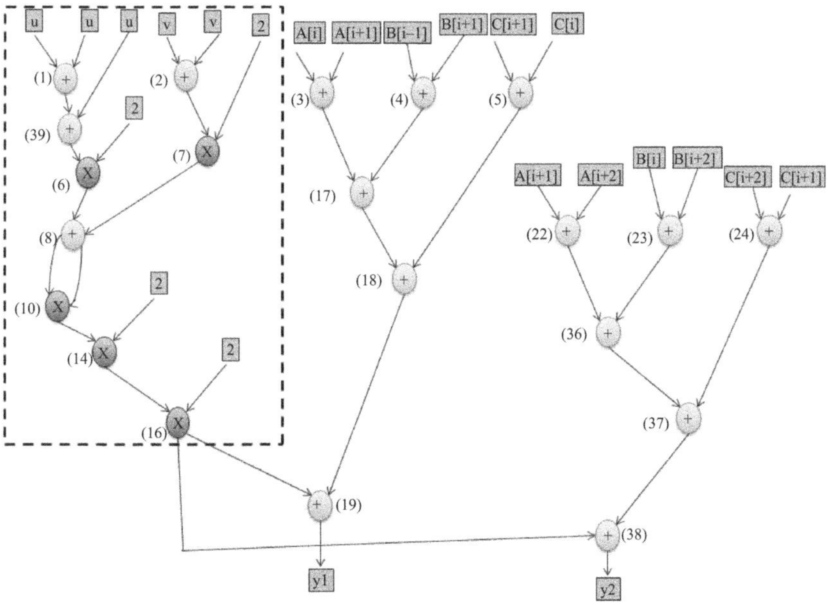

Figure 7.12 LICM-based obfuscated design (unrolled twice)

nodes. As per Figure 7.12, for unrolling factor 2, loop-independent operations are performed once, whereas loop-dependent operations are performed twice.

7.4 Low-cost structural obfuscation for DSP IP core

Sengupta *et al.* (2017a) explored a low-cost DSP design solution through PSO-based exploration (Sengupta *et al.*, 2017b). It accepts the structurally obfuscated design of the DSP core as C-code/loop-based CDFG, module library, and user-provided delay-area restrictions with terminating criteria and control parameters of PSO (such as social factor, cognitive factor, inertia weight) as inputs and produces structurally obfuscated, low-cost DSP circuit as output. The description of the PSO-DSE process is explained in the following subsections.

7.4.1 Overview on PSO

PSO is an optimization process driven through population-based searching that iteratively explores an optimal solution from a pool of available solutions, known as search space. Each possible resource configuration is encoded as a particle and the fitness of each particle is assessed through an objective function to find an optimal solution. Each particle's velocity guides that particle toward an optimal solution. The velocity of each particle is calculated using the current global best "gbest" and its own best location "lbest." After discovering a better global or local best, velocity and position of a particle are updated, thus moving toward the optimal solution.

In the context of the low-cost DSP design solution, each particle represents a resource configuration and an unrolling factor value as a design solution. The first three particles are initialized with the minimum (min), maximum (max), and mid-resources configuration along with its unrolling factor value. If the initial population size is more than three particles, then the rest of the particles are initialized randomly within the range of min and max resource configuration value. The fitness of each particle is calculated using the cost function shown in (7.3). PSO-DSE is also explained in detail in Chapter 3.

7.4.2 Movement of particle

As mentioned in Section 7.4.1, the movement of a particle is guided through its velocity. In other words, velocity of a particle is used to generate a new particle position with better fitness value. In the PSO-DSE process, each dimension (d) of a particle velocity (V_{d_i}) is updated based on the following expression (Sengupta *et al.*, 2017b):

$$V_{d_i}^{+} = \omega V_{d_i} + b_1 r_1 \left[X_{d_{lb_i}} - X_{d_i}\right] + b_2 r_2 \left[X_{d_{gb}} - X_{d_i}\right] \tag{7.4}$$

where V_{d_i} and $V^{+}{}_{d_i}$ are current and new velocities of ith particle in dth dimension, respectively; X_{d_i} indicates the resource value or unrolling factor of ith particle in dth dimension; $X^{+}{}_{d_{lb_i}}$ is the local best position of ith particle in dth dimension;

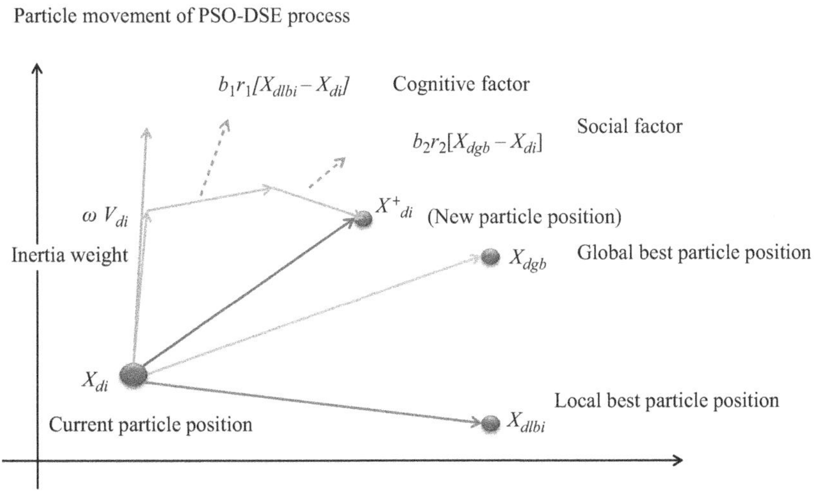

Figure 7.13 Particle movement in PSO-DSE process (Note: X_{di} indicates dth dimension of ith particle including datapath resource configuration, unrolling factor, or mixture of both)

and $X_{d_{gb}}$ is global best of dth dimension; ω is the inertia weight, b_1 and b_2 are social and cognitive factors, respectively; r_1 and r_2 are random numbers between 0 and 1. The new particle position is generated based on the following equation (Sengupta *et al.*, 2017b):

$$X_{d_i}{}^+ = X_{d_i} + V_{d_i}{}^+ \tag{7.5}$$

where $X^+{}_{d_i}$ is new resource value or unrolling factor of ith particle in the dth dimension. The movement of particle for PSO-DSE process is shown in Figure 7.13.

7.4.3 Terminating condition of PSO

This DSE process searches for a low-cost design solution for DSP circuit iteratively. However, this iterative searching process terminates based on the following two conditions. The first condition is: if the number of iteration reaches the maximum number of predefined iteration count (I_{max}), and second is: if no improvement in global best is found for a predefined (δ) number of iteration. According to Sengupta *et al.* (2017b), the values of I_{max} and δ can be set as 100 and 10, respectively.

7.5 A case study for multistage structural obfuscation

For the ease of demonstration of the multistage structural obfuscation process, we are taking the same nonobfuscated CDFG depicted in Figure 7.5 as an input. In the graph, lighter nodes denote adder and darker nodes denote multiplier. The integer value on the side of each node designates the respective node number.

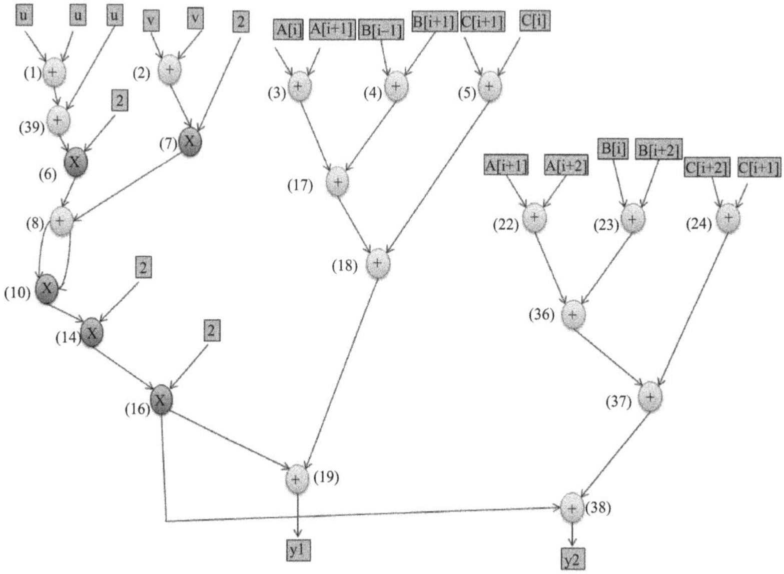

Figure 7.14 Final multistage-driven obfuscated design

As shown in Figure 7.5, three arrays, i.e., A, B, and C with two constant values, i.e., *u* and *v* are taken as primary inputs to the design. The arrays are loop-dependent variables and the constant values are loop-independent variables. Before applying any transformation technique, the total number of nodes in the graph is 19 and the height of the graph is 9. Let us assume this loop-based CDFG has loop size as 8. After applying multiple successive transformations on the input DSP graph, the final structurally obfuscated design is shown in Figure 7.14. Due to ROE, node numbers 9, 11, 12, 13, and 15 are removed from the original design; due to LT, node numbers 6, 7, 14, 16, and 39 are altered; due to THT, the height of the graph is decreased from 9 to 8; based on unrolling factor 2, the CDFG is unrolled two times; and finally due to LICM, the loop-invariant nodes are kept outside the loop. Therefore, the total number of nodes present in the original design is incremented from 19 to 21 in the final obfuscated DSP design (in the form of CDFG).

As mentioned earlier, this structurally obfuscated design is next fed as input with the module library and PSO control parameters to explore a low-cost design solution of DSP circuit through PSO. Figure 7.15 represents the final low-cost, structurally obscured scheduled IP design using three adders and one multiplier (obtained through PSO-DSE). The DSP circuit representation of the same is depicted in Figure 7.16. It uses eight 8:1 switches, 18 input registers, 2 output registers, and 13 delay elements. The changes (due to structural obfuscation) in the DSP circuit for this application at register-transfer/module level and gate level are reported in Table 7.1. (Note: Similarly, functionally equivalent obfuscated DSP IP core can be obtained for well-known applications such as FIR, IIR, HWT, JPEG

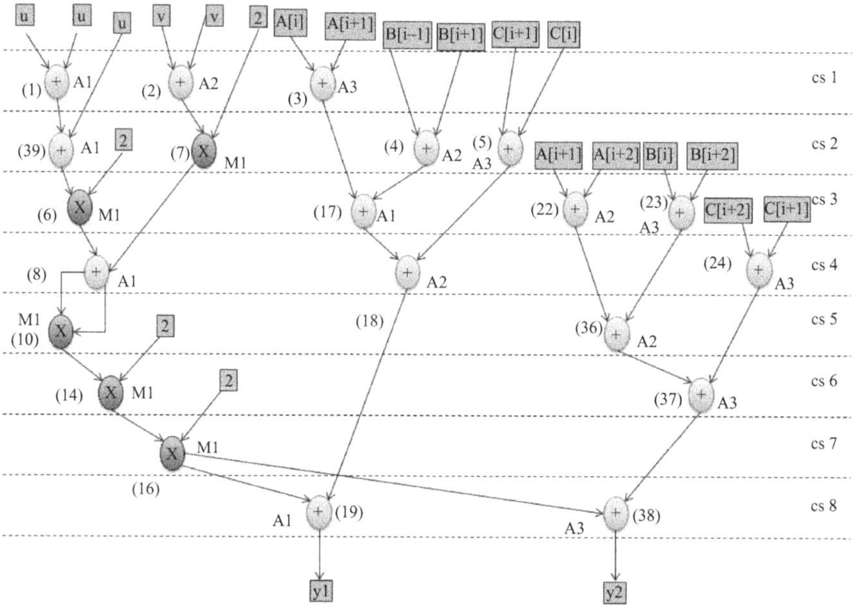

Figure 7.15 *Scheduled low-cost obfuscated DSP IP design with three adders and one multiplier (Note: Resource configuration is explored through PSO-DSE)*

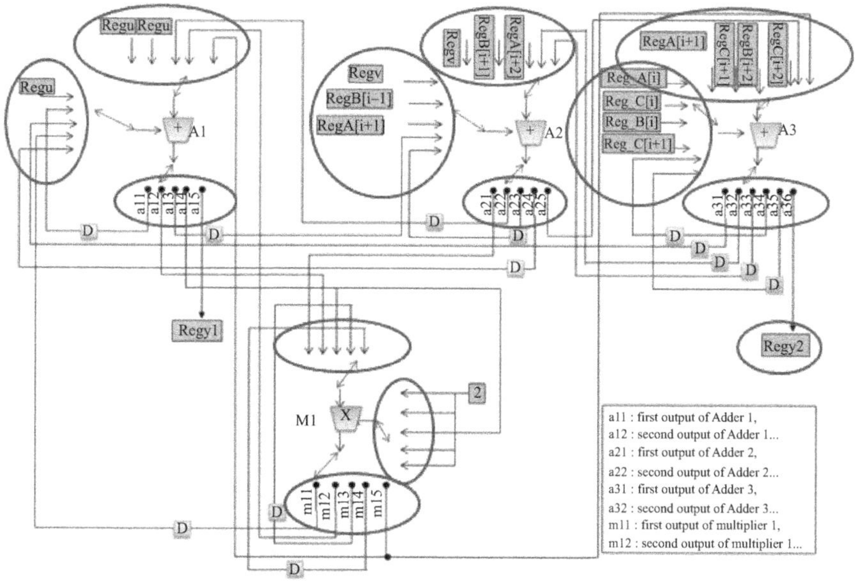

Figure 7.16 *Functionally equivalent DSP circuit (low-cost). (Note: Encircled portion indicates DSP circuits components transformed due to obfuscation, without change of functionality)*

Table 7.1 Module- and gate-level comparison of multistage obfuscated (Sengupta et al., 2017a) design with original design

Types of elements	Module (RT) level		Gate level		
	Original design	**Multistage obfuscated design**	**Original design**	**Multistage obfuscated design**	**Gates affected**
Resource elements	Three adders, four multipliers	Three adders, one multiplier	1,264	496	768
Switches elements	6 4:1, 4 2:1	8 8:1	1,408	3584	2,176
Storage elements	10 input, 1 output	18 input, 2 output	704	1280	576
Delay elements	8	13	512	832	320
Total	–	–	**3,888**	**6,192**	**3,840**

CODEC, DCT, and IDCT. A case study of obfuscated JPEG CODEC is described in Chapter 9).

7.6 Analysis of case studies

The multistage structural obfuscated mechanism section presented in Sengupta *et al.* (2017a) is separated into three sub-sections: (1) result of multistage structural obfuscation in terms of design latency, design area, embedding cost and security on case studies/test cases, (2) comparative study between single-stage structural obfuscation (Lao and Parhi, 2015), and (3) multistage structural obfuscation (Sengupta *et al.*, 2017a).

For the analysis, a 15nm NanGate technology library (Open Cell Library, 2017) is employed to assess both design area and design latency of a DSP circuit. While calculating the design cost using (7.3) both $\O_1$ and $\O_2$ are assigned 0.5 to deliver equivalent priority to both silicon area and design latency. However, it may be possible to tune the weighing factors to provide different preference on silicon area and design latency. Further, following optimal settings from Sengupta *et al.* (2017b) are employed for PSO engine: b_1 and b_2 (acceleration coefficients) $= 2$; ω (inertia weight) $=$ linearly diminishing between 0.9 and 0.1; r_1 and r_2 (random numbers) $= 1$; swarm size $p = 3/5/7$; $\delta = 10$ or $I_{max} = 100$ as terminating condition.

7.6.1 Result of multistage structural obfuscation

The results of low-cost multistage structural obfuscation approach in terms of design latency, design area, and embedding cost with different particle size are

Table 7.2 Results for low-cost multistage obfuscated (Sengupta et al., 2017a) DSP core for different particle size

Benchmark	Particle size	Design solution	Obfuscated design area (μm^2)	Obfuscated design latency (ps)	Obfuscated design cost
2D Autoregression lattice filter (ARF)	3	4A, 2M	241.83	1,900.80	−0.37
	5	4A, 2M	241.83	1,900.80	−0.37
	7	4A, 2M	241.83	1,900.80	−0.37
Differential equation	3	2A, 3M, 1C, 16UF	330.99	4,498.80	−0.33
	5	2A, 3M, 1C, 16UF	330.99	4,498.80	−0.33
	7	2A, 3M, 1C, 16UF	330.99	4,498.80	−0.33
Auto-correlation	3	4A, 8M,1C, 8UF	726.17	3,557.15	−0.46
	5	7A, 8M,1C, 8UF	800.68	2,940.45	−0.47
	7	7A, 8M,1C, 8UF	800.68	2,940.45	−0.47
FIR (6-tap)	3	2A, 2M,1C, 8UF	214.99	1,477.85	−0.36
	5	2A, 2M,1C, 8UF	214.99	1,477.85	−0.36
	7	2A, 2M,1C, 8UF	214.99	1,477.85	−0.36
DHMC	3	4A, 4M, 1C, 6UF	419.07	8,581.61	−0.63
	5	4A, 4M, 1C, 6UF	419.07	8,581.61	−0.63
	7	4A, 4M, 1C, 6UF	419.07	8,581.61	−0.63

shown in Table 7.2. It is shown from Table 7.2 that no reduction in the quality (cost) of design solution is found with the increment of particle size, for except "Auto Correlation" benchmark. In "Auto Correlation" benchmark, for particle size 5, the design cost is better than particle size 3, but it remains the same for both swarm size 5 and 7. Further, the low-cost design solution for all obfuscated DSP benchmarks generated through the PSO-DSE process also satisfies the user-provided design constraints. Design cost is a normalized metric that denotes the total area and total latency of a DSP circuit design. Design cost also known as implementation cost of a DSP design is a normalized function of area and latency.

The strength of the multistage obfuscation can be calculated using a metric called "Power of Obfuscation" (Sengupta and Roy, 2017). The *power of obfuscation* is calculated based on the structural divergence between the original DSP circuit design and its transformed version. Power of Obfuscation (P^{obf}) indicates a normalized value between 0 and 1, and is expressed using the following expression:

$$P^{obf} = \frac{\sum_{i=1}^{5} n_i}{n_T} \tag{7.6}$$

where n_i is the unique number of modified nodes due to ith stage transformation method; n_T is the total number of nodes of the original (before applying any transformation method) DSP circuit. A higher value of P^{obf} indicates stronger security of an obfuscated DSP design. A modified node is defined based on the following four cases (Sengupta *et al.*, 2017a):

- If parents of a node in the structurally transformed CDFG are changed than its original
- If child of a node in the structurally transformed CDFG is changed than its original
- If the operation type of a node in the structurally transformed CDFG is changed than its original
- If a node existing in the original CDFG does not present in the obfuscated design

7.6.2 Comparative study and discussion

Table 7.3 reports a comparative study of an original (nonsecured) design with its structurally obfuscated version with respect to obfuscated design area, latency, and cost. For instance, the design latency, area, and cost of original autoregressive filter (ARF) benchmark are 2,573.5 ps, 241.80 μm^2, and −0.26, respectively, whereas, for structurally obfuscated ARF benchmark, the design area, design latency, and design cost are 241.80 μm^2, 1900.80 ps, and −0.37, respectively. It can be seen from Table 7.3 that an obfuscated design may achieve lesser execution time compared to the original design. This is because due to multistage transformations the obfuscated design may reduce the total number of operations available in the

Table 7.3 Comparison of multistage obfuscated (Sengupta et al., 2017a) design with original design

Benchmark	**Original design (non-obfuscated)**			**Multistage obfuscated design (Sengupta *et al.*, 2017a)**		
	Area (μm^2)	**Latency (ps)**	**Cost**	**Area (μm^2)**	**Latency (ps)**	**Cost**
ARF	241.8	2,573.5	−0.26	241.8	1,900.8	−0.37
DIF-EQN	333.0	7,246.5	−0.18	333.0	4,498.8	−0.33
AUTO-CO	736.4	3,399.2	−0.48	736.4	2,940.5	−0.46
FIR	215.0	1,661.3	−0.31	215.0	1,477.8	−0.35
DHMC	419.1	8,872.1	−0.27	419.1	8,581.6	−0.63

Table 7.4 Comparison of multistage obfuscated (Sengupta et al., 2017a) design with single-stage obfuscated (Lao and Parhi, 2015) design w.r.t. area, latency, and cost

Benchmark	**Single-stage obfuscated design (Lao and Parhi, 2015)**			**Multistage obfuscated design (Sengupta *et al.*, 2017a)**		
	Area (μm^2)	**Latency (ps)**	**Cost**	**Area (μm^2)**	**Latency (ps)**	**Cost**
ARF	788.8	1,511.5	−0.08	241.8	1,900.8	−0.37
DIF-EQN	3,590.1	2,326.4	−0.15	333.0	4,498.8	−0.33
AUTO-CO	3,196.1	1,144.9	−0.32	736.4	2,940.5	−0.46
FIR	399.5	1,315.7	−0.25	215.0	1,477.8	−0.35
DHMC	4,818.9	2,891.1	−0.28	419.1	8,581.6	−0.63

graph. Therefore, the structurally transformed graph may optimize the latency after scheduling.

Table 7.4 reports the comparative study between the multistage-based obfuscation technique (Sengupta *et al.*, 2017a) with the single-stage-based obfuscation technique (Lao and Parhi, 2015) with respect to design area, latency, and obfuscation design cost. For instance, for ARF benchmark (Mohanty *et al.*, 2008; Sengupta and Mohanty, 2016) the design area, design latency, and design cost of single-stage obfuscation approach are 788.80 μm^2, 1,511.50 ps, and −0.08, respectively, while the low-cost multistage obfuscation approach explores low-cost design solution through PSO-DSE for the same benchmark (design area, latency, and cost are reported earlier). Therefore, the low-cost multistage obfuscation approach achieves 47% reduction of design cost on average for standard benchmarks (DSP Benchmark Suite, 2017; Benchmark1, 2017; Benchmark2, 2017), compared to single-stage obfuscation approach (Lao and Parhi, 2015). As mentioned earlier, the reduction of design cost is achieved for multistage obfuscation technique because it performs PSO-DSE-based optimization on area-delay trade-off to explore the low-cost design solution. The graphical representation of design cost comparison between single-stage and multistage obfuscation is shown in Figure 7.17.

Table 7.5 reports the comparative study between the multistage-based obfuscation technique (Sengupta *et al.*, 2017a) with the single-stage-based obfuscation technique (Lao and Parhi, 2015) in terms of security through the *power of obfuscation* metric. It can be observed from the table that multistage-based obfuscation process delivers tougher IP protection. This is because for all the DSP applications, multistage-based obfuscation approach has higher P^{obf} value. For instance, for single-stage-based obfuscated ARF, the P^{obf} is 0.43, whereas for multistage-based obfuscated ARF, it is 0.55. In summary, the multistage obfuscation approach (Sengupta *et al.*, 2017a) achieves 29% better robustness on average compared to single-stage obfuscation approach (Lao and Parhi, 2015). The graphical representation of P^{obf} comparison between single-stage and multistage obfuscation is shown in Figure 7.18.

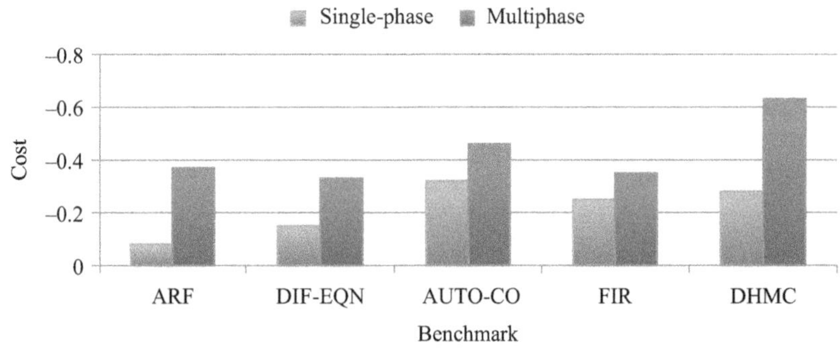

Figure 7.17 Comparison of design cost between single-stage and multistage obfuscation for DSP Cores

Table 7.5 Comparison of multistage obfuscated (Sengupta et al., 2017a) design with single-stage obfuscated (Lao and Parhi, 2015) design in terms of P^{obf}

Benchmark	P^{obf} for multi-stage obfuscation	P^{obf} for single-stage obfuscation	P^{obf} improvement %
ARF	0.55	0.43	21.82
DIF-EQN	0.72	0.50	30.56
AUTO-CO	0.75	0.50	33.33
FIR	0.75	0.50	33.33
DHMC	0.69	0.50	27.54

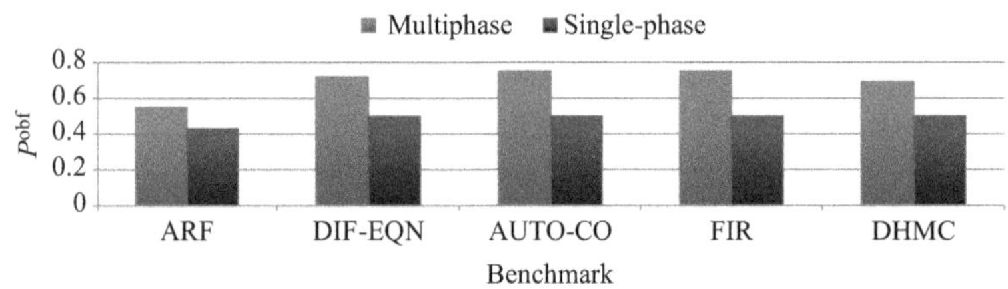

Figure 7.18 Comparison of P^{obf} between single-stage and multistage obfuscation

7.7 Conclusion

This chapter presents structural obfuscation mechanisms to protect DSP IP core. The mechanisms perform structural obfuscation through multiple compiler and loop-based transformation techniques. Additionally, some obfuscation approaches offer lower design cost and stronger robustness than the others do. The robustness of the multistage obfuscation mechanism discussed in this chapter can be improved by adding more transformation techniques on the DSP IP core. Further, the combination of structural obfuscation approach and functional obfuscation can also enhance the security level for DSP applications. This would enable to hide an application and its implementation from an adversary.

7.8 Exercises

1. What is structural obfuscation?
2. What are the advantages/disadvantages of structural obfuscation?
3. What are the compiler-driven techniques for structural obfuscation?
4. Which stage of HLS is structural obfuscation of DSP core performed?
5. What is folding-based structural obfuscation?
6. How does structural obfuscation attain the reduction of design latency?
7. How does structural obfuscation complicate RE attack?
8. Where is the impact of structural obfuscation seen in a design?
9. What is power of obfuscation and how to enhance it?
10. What are the other possible ways of structural obfuscation for DSP core?
11. Implement a structurally obfuscated CODEC IP core.
12. How to optimize latency of an obfuscated design besides high-level transformations?
13. Explain the threat model in case of structural obfuscation.
14. How is structural obfuscation different from its functional counterpart?
15. What is the role of PSO-DSE in obfuscation?
16. How is a particle encoded in structural obfuscation process?
17. Explain the different transformations used for obfuscation.
18. Explain the advantages and drawbacks of each transformation.
19. Which transformation is capable of providing maximum obfuscation of DSP cores? Why?
20. How is design overhead due to obfuscation evaluated for DSP cores?

References

Alkabani Y. M. and Koushanfar F., 'Active hardware metering for intellectual property protection and security,' *Proceedings of USENIX Security Symposium,* 2007, pp. 20:1–20:16.

Barria C., Cordero D., Cubillos C., and Osses R., 'Obfuscation procedure based in dead code insertion into crypter,' *in International Conference on Computers Communications and Control (ICCCC),* May 2016, pp. 23–29.

Baumgarten A., Tyagi A., and Zambreno J., 'Preventing IC piracy using reconfigurable logic barriers,' *IEEE Design Test of Computers*, 2010, vol. 27 (1), pp. 66–75.

Benchmark1, 2017, 'Express Benchmark Suite,' [Online]. Available: https://www.ece.ucsb.edu/EXPRESS/benchmark/, University of California Santa Barbara, last accessed on December 2017.

Benchmark2, 2017, 'Vipul Kumar Mishra, Anirban Sengupta, "Swarm Inspired Exploration of Architecture and Unrolling Factors for Nested Loop Based

Application in Architectural Synthesis”,’ *IET Electronics Letters*, 2015, vol. 51 (2), pp. 157–159.

Brzozowski M. and Yarmolik V. N., ‘Obfuscation as intellectual rights protection in VHDL language,’ *Computer Information Systems and Industrial Management Applications (CISIM),* June 2007, pp. 337–340.

Castillo E., Meyer-Baese U., Garcia A., Parrilla L., and Lloris A., ‘IPP@HDL: Efficient intellectual property protection scheme for IP cores,’ *IEEE Transactions on Very Large Scale Integration Sys.*, 2007, vol. 15 (5), pp. 578–591.

Chakraborty R. S. and Bhunia S., ‘HARPOON: An Obfuscation-Based SoC Design Methodology for Hardware Protection,’ *IEEE Transactions on CAD of Integrated Circuits and Systems*, 2009, vol. 28 (10), pp. 1493–1502.

Desai A.R., Hsiao M.S., Wang C., Nazhandali L., and Hall S., ‘Interlocking obfuscation for anti-tamper hardware,’ *Proceedings of Cyber Security and Information Intelligence Research Workshop (CSIIRW),* 2013, pp. 8:1–8:4.

Dupuis S., Ba P. S., Di Natale G., Flottes M. L., and Rouzeyre B., ‘A novel hardware logic encryption technique for thwarting illegal overproduction and Hardware Trojans,’ *IEEE International Online Testing Symposium (IOLTS)*, 2014, pp. 49–54.

ExtremeTech, iphone 5 A6 SoC reverse engineered, reveals rare hand-made custom CPU, and tri-core GPU (2012).

Guajardo J., Kumar S. S., Schrijen G. J., and Tuyls P., ‘Brand and IP protection with physical unclonable functions,’ *Proceedings of IEEE International Symposium on Circuits and Systems*, May 2008, pp. 3186–3189.

Kim J., Jung E. S., Lee Y. T., and Ryu W., ‘Home appliance control framework based on smart TV set-top box,’ *IEEE Transactions on Consumer Electronics*, 2015, vol. 61 (3), pp. 279–285.

Koushanfar F., *‘Hardware Metering: A Survey’*, New York, NY: Springer New York, 2012, pp. 103–122.

Lao Y. and Parhi K. K., ‘Obfuscating DSP circuits via high-level transformations,’ *IEEE Transactions on Very Large Scale Integration Systems*, 2015, vol. 23 (5), pp. 819–830.

Lee Y. W. and Touba N. A., ‘Improving logic obfuscation via logic cone analysis,’ *Latin-American Test Symposium (LATS)*, 2015, pp. 1–6.

Li M., Zhang P., Zhu C., *et al.*, ‘High efficiency VLSI implementation of an edge--directed video up-scaler using high level synthesis,’ *Proceedings of IEEE International Conference on Consumer Electronics (ICCE)*, 2015, pp. 92–95.

Mohanty S. P., ‘Energy and transient power minimization during behavioral synthesis,’ Doctoral Dissertation, Department of Computer Science and Engineering, University of South Florida, USA, 2003.

Mohanty S. P., ‘*Nanoelectronic Mixed-Signal System Design*,’ McGraw-Hill Education, New York, 2015, no. 0071825711.

Mohanty S. P., Ranganathan N., Kougianos E., and Patra P., ‘*Low-Power High-Level Synthesis for Nanoscale CMOS Circuits*,’ Springer, New York, 2008, ISBN-10: 0387764739, ISBN-13: 978-0387764733.

NanGate 15 nm open cell library. [Online]. Available: http: //www.nangate.com/?pageid=2328, last accessed on June 2017.

Parhi K. K., 'Verifying equivalence of digital signal processing circuits,' in *Conference Record of the Forty Sixth Asilomar Conference on Signals, Systems and Computers (ASILOMAR)*, Nov. 2012, pp. 99–103.

Rajendran J., Pino Y., Sinanoglu O., and Karri R., 'Security analysis of logic obfuscation,' *in DAC*, 2012, Jun. 2012, pp. 83–89.

Rajendran J., Sam M., Sinanoglu O., and Karri R., 'Security analysis of integrated circuit camouflaging,' *in Proceedings of ACM SIGSAC Conference on Computer & Communications Security*, 2013, pp. 709–720.

Rajendran J., Zhang H., Zhang C., *et al.*, 'Fault analysis-based logic encryption,' *IEEE Transactions on Computers*, 2015, vol. 64 (2), pp. 410–424.

Roy D. and Sengupta A., 'Low overhead symmetrical protection of reusable IP core using robust fingerprinting and watermarking during high level synthesis,' *Future Generation Computer Systems*, 2017, vol. 71, pp. 89–101.

Roy J. A., Koushanfar F., and Markov I. L., 'EPIC: Ending piracy of integrated circuits,' *in DATE*, March 2008, pp. 1069–1074.

Sengupta A., 'Intellectual property cores: Protection designs for CE products,' *IEEE Consumer Electronics Magazine*, 2016, vol. 5 (1), pp. 83– 88.

Sengupta A., 'Hardware security of CE devices,' *IEEE Consumer Electronics Magazine*, 2017, vol. 6 (1), pp. 130–133.

Sengupta A. and Bhadauria S., 'Exploring low cost optimal watermark for reusable IP cores during high level synthesis,*'* *IEEE Access*, 2016, vol. 4, pp. 2198–2215.

Sengupta A. and Kundu S., 'Securing IoT hardware: Threat models and reliable, low-power design solutions' *IEEE Transactions on VLSI Systems*, 2017, vol. 25 (12), pp. 3265–3267.

Sengupta A. and Mohanty S. P., 'High-level synthesis of digital integrated circuits in the nanoscale mobile electronics era,' in Nano-CMOS and Post-CMOS Electronics: Circuits and Design, Edited by S. P. Mohanty and A. Srivastava, The Institute of Engineering and Technology (IET), London, 2016, ISBN-10: 184919999X, ISBN-13: 978-1849199995.

Sengupta A. and Roy D., 'Protecting an intellectual property core during architectural synthesis using high-level transformation based obfuscation,' *IET Electronics Letters*, 2017, vol. 53 (13), pp. 849–851.

Sengupta A. and Roy D., 'Antipiracy-aware IP chipset design for CE devices: A robust watermarking approach', *IEEE Consumer Electronics Magazine*, 2017, vol. 6 (2), pp. 118–124.

Sengupta A. and Sedaghat R., 'Integrated scheduling, allocation and binding in high level synthesis using multi structure genetic algorithm based design space exploration,' *International Symposium on Quality Electronic Design*, 2011, pp. 1–9.

Sengupta A. and Sedaghat R., 'System and methodology for development of a system architecture using optimization parameters', Google Patents, 2013, https://www.google.com/patents/US8397204.

Sengupta A., Roy D., and Mohanty S. P., 'Triple-phase watermarking for reusable IP core protection during architecture synthesis,' *IEEE Transactions on CAD of Integrated Circuits and Systems*, 2018, vol. 37 (4), pp. 742–755.

Sengupta A., Roy D., Mohanty S. P., and Corcoran P., 'DSP design protection in CE through algorithmic transformation based structural obfuscation,' *IEEE Transactions on Consumer Electronics*, 2017a, vol. 63 (4), pp. 467–476.

Sengupta A., Bhadauria S., and Mohanty S. P., 'TL-HLS: Methodology for low cost hardware Trojan security aware scheduling with optimal loop unrolling factor during high level synthesis,' *IEEE Transactions on CAD of Integrated Circuits and Systems*, 2017b, vol. 36 (4), pp. 655–668.

Subramanyan P., Ray S., and Malik S., 'Evaluating the security of logic encryption algorithms,' *IEEE International Symposium on Hardware Oriented Security and Trust (HOST)*, 2015, pp. 137–143.

Intel's 22-nm tri-gate transistors exposed (2012). [Online]. Available: http://www.chipworks.com/blog/technologyblog/2012/04/23/intels-22-nm-tri-gate-transistors-exposed

Thavalengal S. and Corcoran P., 'User authentication on smartphones: Focusing on iris biometrics,' *IEEE Consumer Electronics Magazine*, 2016, vol. 5 (2), pp. 87–93.

DSP Benchmark Suite. [Online]. Available: http://www.ece.ucsb.edu/EXPRESS/benchmark/, last accessed on June 2017.

Tensilica's ConnX D2 DSP Engine Wins EDN Top 100 Electronic Products Award for 2009. [Online]. Available: https://ip.cadence.com/news/310/330/Tensilica-s-ConnX-D2-DSP-Engine-Wins-EDN-Top-100-Electronic-Products-Award-for-2009.

Torrance R. and James D., 'The state-of-the-art in IC reverse engineering', *Proceedings of Workshop on Cryptographic Hardware and Embedded Systems (CHES)* (Springer, Berlin, Heidelberg, 2009), pp. 363–381.

Vijayakumar A., Patil V. C., Holcomb D. E., Paar C., and Kundu S., 'Physical design obfuscation of hardware: A comprehensive investigation of device and logic-level techniques,' *IEEE Transactions on Information Forensics and Security*, 2017, vol. 12 (1), pp. 64–77.

Walz S. and Schrder Y., 'A privacy-preserving system architecture for applications raising the energy efficiency,' *Proceedings of IEEE International Conference on Consumer Electronics (ICCE),* 2016, pp. 62–66.

Wang X., Jia X., Zhou Q., *et al.*, 'Secure and low-overhead circuit obfuscation technique with multiplexers,' *in GLSVLSI*, May 2016, pp. 133–136.

Wong J. L., Kirovski D., and Potkonjak M., 'Computational forensic techniques for intellectual property protection,' *IEEE Transactions on CAD of Integrated Circuits and Systems*, 2004, vol. 23 (6), pp. 987–994.

Yang H., Basutkar N., Xue P., Kim K., and Park Y. H., 'Software defined DVT-T2 demodulator using scalable DSP processors,' *IEEE Transactions on Consumer Electronics*, 2013, vol. 59 (2), pp. 428–434.

Zhang J., 'A practical logic obfuscation technique for hardware security,' *IEEE Transactions on Very Large Scale Integration Systems*, 2016, vol. 24 (3), pp. 1193–1197.

Chapter 8

Functional obfuscation of DSP cores used in CE devices

The previous chapter detailed many approaches for structural obfuscation but we now move forward to another class of obfuscation called "functional obfuscation." The use of either of the techniques is the choice of design engineers. This chapter presents several methods to thwart IP piracy and RE attacks through functional obfuscation. More specifically, we will discuss IP functional locking blocks (ILBs)-based logic obfuscation for DSP cores used in CE devices as hardware hardening technique (Sengupta *et al.*, 2018). Moreover, a PSO-based DSE is performed to generate a low-overhead, functionally obfuscated (FO) design solution for DSP cores (Sengupta and Sedaghat, 2011, 2013). The rest of the chapter is organized as follows: Section 8.2 discusses different attack scenarios and threat model. Section 8.3 explains selected functional obfuscation techniques that are available in the literature. Section 8.4 discusses the design process of functional obfuscation for DSP IP cores used in CE devices. Section 8.5 presents the security analysis of functional obfuscation methodology for DSP IP core design. Section 8.6 discusses PSO-based optimization for FO design. Section 8.7 presents analysis on case studies/test cases.

8.1 Introduction

In the current IC supply chain model, multiple parties are involved to handle the increasing design complexity and cost. To avoid establishing and maintaining charge of a foundry with advanced fabrication facility, fabless semiconductor companies export their designed ICs (such as DSP cores) to another company having fabrication facility. Additionally, sometimes system integrator while designing a SoC, imports DSP IP cores from 3PIP vendors to meet the time-to-market requirement (Castillo *et al.*, 2007). Thus, in this globalized business model for DSP cores, different countries and companies have different IP regulation policies and models. Therefore, in this scenario, chip designing process is susceptible to many serious IP security threats such as piracy, overbuilding, counterfeiting, RE, and insertion of hardware Trojan (Sengupta and Kundu, 2017; Sengupta *et al.*, 2018b; Wong *et al.*, 2004; Alkabani and Koushanfar, 2007). This necessitates improvement in the chip-designing process to thwart these threats.

IP piracy is a process of illegal usage or selling of IP cores. An untrusted foundry, present in the design flow of an IP core, can steal a design, clone it illegally, and then resell it to an other entity. Thus, it can bypass a substantial amount of research and development process, workforce, money, and time invested by an original IP designer. Moreover, he/she can claim the ownership of that IP core (Koushanfar, 2012; Roy and Sengupta, 2017; Sengupta and Roy, 2017; Sengupta *et al.*, 2018b). Similarly, an untrusted foundry may insert malicious logic in the form of hardware Trojan into the design to introduce malfunctioning of DSP IP cores (Sengupta *et al.*, 2017b). Insertion of hardware Trojan can serve many benefits to the adversary based on its type. It can leak crucial information of a DSP IP core used in the defense system, damage the reputation of a well-established company, or degrade the performance of an IP core, etc. According to The Economic Impacts of Counterfeiting and Piracy (2017), the total fiscal revenue loss due to piracy and counterfeiting in 2013 for G20 nations was $96–130 billion which may rise to $199–270 billion in 2022. It is identified that 10% of the globally sold electronics products are counterfeited (Guajardo *et al.*, 2008; Lao and Parhi, 2015).

RE of a DSP/IP core is a process of gaining the complete understanding of its functionality, design, and structure (Vijayakumar *et al.*, 2017; Torrance and James, 2009; Sengupta *et al.*, 2017a). Therefore, people can use RE process for both honest and fraudulent intentions. People can also use RE as a tool for education, research and development, and analysis and evaluation purpose of an existing product. In many countries, RE with honest intentions is considered as a legal process until it violates any design copyrights and patents issues (Biggerstaff, 1989). However, RE can also be used for dishonest intentions such as overbuilding, piracy, or counterfeiting a DSP/IP core design or inserting a hardware Trojan, which may lead to huge profit for an adversary. Therefore, RE for dishonest intentions incubates or invites these aforementioned potential threats. The possibility of RE attack during various stages of IC design, based on Rajendran *et al.* (2015), is shown in Figure 8.1.

One popular hardware hardening technique to counteract RE attack for DSP IP core is functional/logic obfuscation. Functional obfuscation (also known as logic encryption) inserts additional logic components known as key-components into the design to hide its functionality and implementation. These additional logic components accept key bits as input and produce the output based on these key values. Therefore, applying correct key combination will produce correct result while applying wrong keys will lead to exhibit an incorrect functionality of the design. Thus functional obfuscation thwarts RE process for fraudulent intentions.

8.2 Attack scenarios and threat model

This section discusses the possible attack scenarios and threat model of functional obfuscation.

8.2.1 Possible attack scenarios

To generate a locked netlist, key components or key-gates can be inserted anywhere in the circuit. However, incorrect placement of key components may invite

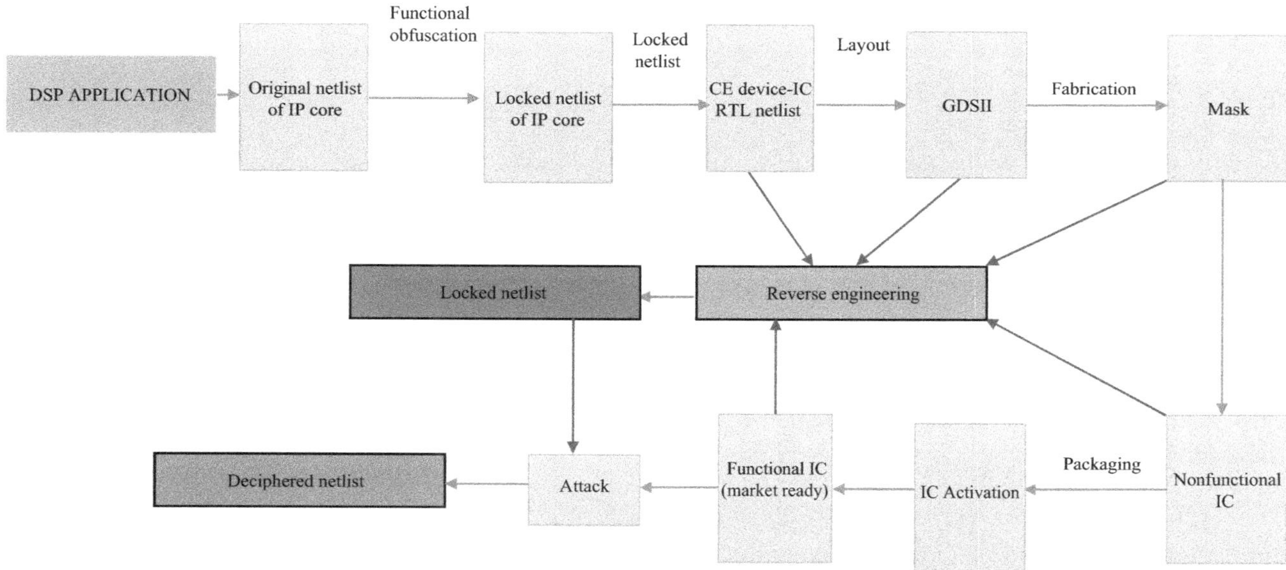

Figure 8.1 Possibility of RE attack during various stages of IC design

several attacks. In this subsection, we discuss about the attacks related to placements of key components as well as other attacks.

1. Isolated key-gates-based key-sensitization attack: A key-gate in an locked netlist is considered as isolated if there is no path between that key-gate and any other key-gates (Yasin *et al.*, 2016). An attacker can easily sensitize the key value of an isolated key-gate. For example in Figure 8.2(a), there is no path between key-gates K1 and K2. Therefore, K1 and K2 are isolated key-gates. An attacker can sensitize the value of K1 as 0 to the output 1 (O1) by applying "100XXX" i/p pattern.
2. Run of key-gates-based key-sensitization attack: A set of key-gates in an locked netlist is considered as run of key-gates if they are connected in a back-to-back fashion (Yasin *et al.*, 2016). Few run of key-gate structures are vulnerable for locked netlist as it reduces the effort of an attacker to identify the correct key by increasing the possible correct key combinations. The reason is that some run of key-gate structure may be replaceable with a single key-gate, thus, diminishing security strength. For example, in Figure 8.2(b), K1 and K2 are connected back to back, thus forming a run of key-gates. In this case both "01" and "10" are correct key values. Moreover, K1 and K2 can be replaced by a single key-gate (XOR) with "1" as the correct key value.

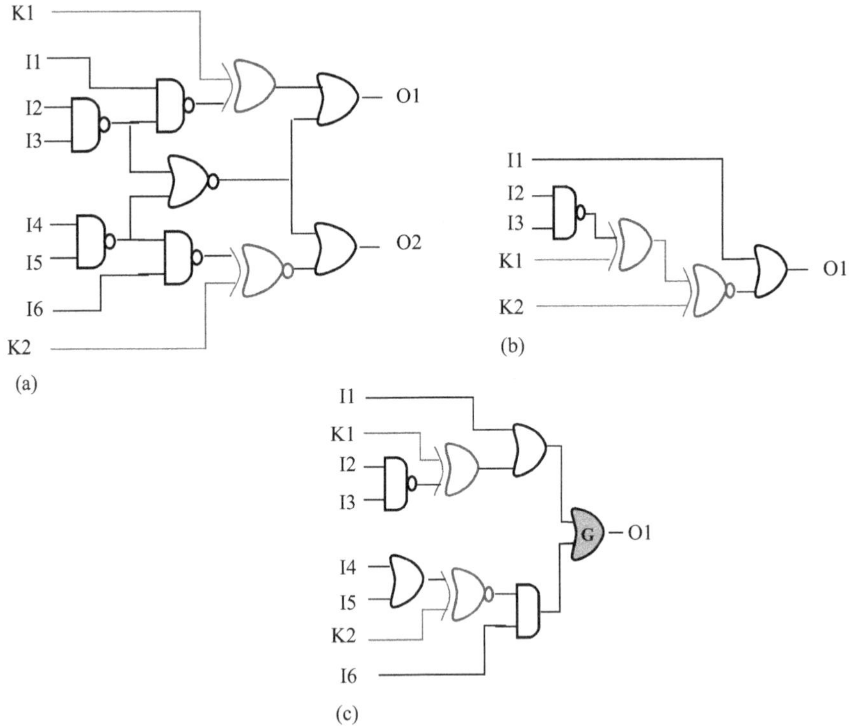

Figure 8.2 (a) Isolated key gates K1 and K2; (b) run of key gates K1 and K2; and (c) concurrently mutable key gates K1 and K2

3. Mutable key-gates-based key-sensitization attack: If two or more key-gates converge but have no common path between them, sensitization may still occur (Yasin *et al.*, 2016). Therefore, if key-gates K1 and K2 converge to another gate so that a specific key-bit can be obtained by muting the other, then both key-gates are called concurrently mutable key-gates. In this type of scenario, an attacker can easily trace the key value of K1 by muting K2. For example, in Figure 8.2(c), K1 and K2 converge at the gate G. The value of K1 can be sensitized to O1 by muting the effect of K2. Applying $I_6 = 0$ in the input pattern will mute K2. Similarly, the value of K2 can be sensitized to O1 by muting the effect of K1. Applying $I_1 = 1$ in the input pattern will mute K1.
4. IP piracy and Trojan attack: If somehow the correct key combination of a locked netlist is identified by an attacker, he/she can easily discover the correct functionality of that netlist. This may enhance the possibility of IP piracy/counterfeit and Trojan insertion attacks. An attacker can either resell/redistribute the IP core illegally to make a profit or can insert a Trojan logic at safe places to introduce malfunctioning of an IP core.
5. SAT attack: This attack is based on the following model:
   - An attacker may be present in an untrusted foundry whose objective is to identify the correct key value of the locked netlist.
   - The attacker has access to the GDSII (graphic data system) layout file of the locked netlist. He/she will try to perform RE to extract the locked netlist from the GDSII file. This is because fabrication is performed at the foundry containing layout details provided by the designer.
   - Attacker also has an activated functional IC bought from the open market, which can be used to identify the correct key value by providing a set of input patterns and inspecting their corresponding correct output as black box.

   The core of the SAT attack algorithm works iteratively by finding distinguishing input patterns (DIPs) that identify a subset of wrong key combinations that produce incorrect output, until the correct input output pair is obtained (Subramanyan *et al.*, 2015; Xie and Srivastava, 2018). First, the SAT attack algorithm formulates a SAT formula and then the SAT solver generates an assignment to input variable (which is the distinguishing input pattern). Once the distinguishing input pattern is formed, it is then fed to the activated functional IC and correct output is observed. This distinguishing input/output pair can be used to eliminate the wrong key combinations, till no new one is found.
6. Removal attack: In this attack, an adversary tries to remove all the key component from the locked netlist to generate an unlocked circuit. Removal attack identifies all the key components from a locked netlist and removes them. An attacker can then resell/redistribute the IP core illegally to make a profit or can insert a Trojan logic at safe places to introduce malfunctioning of the unlocked netlist.

8.2.2 Threat model

The functional obfuscation approach aims to protect an IC/IP core (such as a DSP core or combinational/sequential circuit) by generating a locked netlist of it, in order to enhance the complexity of RE attack. In this threat model, it is assumed that an adversary may be present in an untrusted foundry.

8.3 Selected functional obfuscation approaches

This section discusses few selected functional obfuscation-based IC/IP core protection mechanisms with emphasis on the functional obfuscation of DSP cores.

As mentioned earlier, functional obfuscation hides the implementation and functionality of a design by inserting key-components into it. The original circuit integrated with key-components is known as locked design. These key-components could be designed using FSM (Chakraborty and Bhunia, 2009), XOR/XNOR gates (Roy *et al.*, 2008; Rajendran *et al.*, 2015), AND/OR gates (Dupuis *et al.*, 2014), lookup table (LUT) (Baumgarten *et al.*, 2010), muxes (Wang *et al.*, 2016; Zhang, 2016), ILBs (Sengupta *et al.*, 2018), or a combination of these elements (Lee and Touba, 2015). All these key-components are inserted into the design in such a way that the correct functionality of the circuit will only depend on the correct key value.

Authors in Sengupta *et al.* (2018) propose ILBs to generate FO DSP cores. Among other functional obfuscation techniques available in the literature, this approach targets to generate a locked design for DSP cores. This approach inserts 8-bit key value-based ILB on each bit of output data. It can generate several different structures of ILBs (of similar area/delay) based on the key value. The ILBs are designed using various combinations of AND, NAND, NOT, XOR, and XNOR gates. Each ILB structure produces correct output for a unique key value. Moreover, these ILBs are capable to provide multi pair-wise security feature against sensitization attack and defense against SAT attack through a light-weight custom AES block integrated with the key bits. Thus, this approach ensures strong counter against the RE attack. Additionally, a PSO-driven DSE process is performed to explore optimal design architecture for obfuscated DSP cores. Therefore, it is capable to generate a low-cost, low-power, and secured FO netlist for DSP IP core. Further detailed explanation of this approach is discussed in the next section. Other popular functional obfuscation techniques (except Lao and Parhi (2015) which partially uses functional locking) available in the literature mostly target to generate a locked netlist for sequential or combination circuits but not DSP cores. Authors in Lao and Parhi (2015) perform both structural and functional obfuscations. Structural obfuscation is performed through folding-based transformation technique and functional obfuscation is achieved by using the correct initialization key. Incorrect initialization key value fails to enable the reconfigurator, resulting in activating either a meaningful but nonfunctional mode or a non-meaningful mode.

Authors in Chakraborty and Bhunia (2009) insert dummy states as key-components within the original states of the FSM of a design. FSM enters into an invalid state (obfuscation mode) with incorrect key value and will reach a valid state (normal mode) with correct key value. Desai *et al.* (2013) generated a locked design based on Code-Word, which is similar to an internally generated key sequence based on a specific state transition function. For incorrect key value, the

FSM will deviate from the correct transition path and will resume normal transition with the correct key value. Both these approaches generate a locked design for sequential circuits.

In the combinational circuits, Roy *et al.* (2008) proposed the framework for generating locked design using XOR/XNOR gates for the first time. However, in this approach, key-components are inserted at random locations of the design which becomes vulnerable to key-sensitization and removal attacks. Additionally, random insertion of key-component may not always guarantee to produce incorrect output for the locked design on applying incorrect key value. Yasin *et al.* (2016) proposed selection algorithm for the key-component location which will maximize the Hamming distance between incorrect and correct output. Additionally, this approach provides pairwise security through prohibiting isolated key-gates, runs of key-gates, and mutable convergent key-gates to prevent key-sensitization attack. For example, in Figure 8.3(a), both the key-gates K1 and K2 are pairwise secure. Therefore, sensitizing K1 in the output bit is not possible without controlling the K2, and vice versa. This mandates to apply brute force on the key bits.

One-way random functions (ORF) circuit based countermeasure against SAT based attack is shown in Figure 8.3(b) (Yasin *et al.*, 2016). In this approach, a public AES has been used as ORF to defend SAT attack instead of PUF because PUF has following drawbacks: (1) it cannot add a significant number of clauses as it has a small circuit; and (2) it is susceptible to modeling attacks (Pappu *et al.*, 2002). Removal attack on AES is handled by synthesizing the public AES along with the locked netlist, such that it changes the complete structure of the public AES. Therefore, it becomes difficult for an attacker to identify the AES block separately from the locked design.

Lee and Touba (2015) proposed key-components as a combination of XOR/ XNOR gates and multiplexers to enhance the HD between incorrect and correct output. Baumgarten *et al.* (2010) used LUT as key-component to achieve the locked design. In this functional obfuscation approach, the complete design is subdivided into multiple fixed and reconfigurable parts, and then the LUTs are only

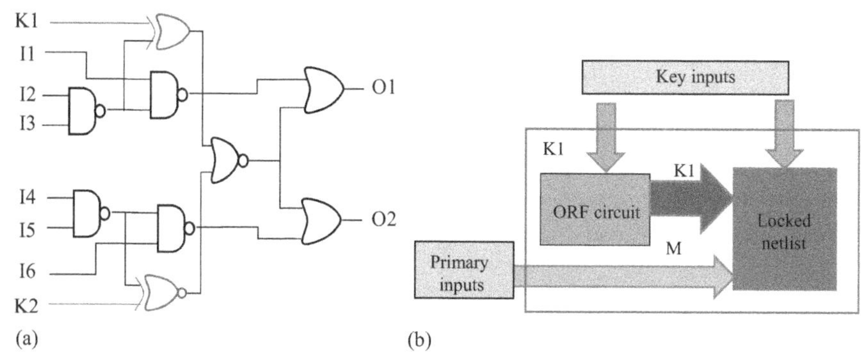

Figure 8.3 (a) Sensitization attack on K1 and K2 is not possible as K1 and K2 are pairwise secured and (b) ORF circuit based countermeasure against SAT based attack

used to implement reconfigurable parts. After fabrication, once the design is returned from the foundry, these key-controlled LUTs are changed to a different logic. Thus, misuse of a design by an untrusted foundry can be prevented.

Functional obfuscation is vulnerable to many attacks, such as key-component removal attack, key-sensitization attack (Yasin *et al.*, 2016), Boolean SAT attack and hill-climbing search-based attack (Subramanyan *et al.*, 2015), an approximate deobfuscation algorithm-based SAT attack (Kaveh *et al.*, 2017), etc. Moreover, inserting additional key-components into a design incur power and area overhead resulting in a higher design cost.

8.4 Design of functionally obfuscated DSP core

As mentioned earlier, Sengupta *et al.*'s approach (2018) is the only one which protects a DSP IP core completely through functional obfuscation. To achieve the FO design, this approach inserts different ILBs as key-component in the netlist design of a DSP core. Additionally, to explore optimal design architecture, PSO-driven DSE is integrated with the functional obfuscation approach. In this section, we discuss formulation and low-cost functional obfuscation approach for DSP core (Sengupta *et al.*, 2018).

8.4.1 Formulation

To generate a low-cost, low-power FO netlist for a DSP application (given in the form of a C-code or CDFG), which accepts module library and PSO control parameters as inputs.

8.4.2 Low-cost obfuscation method for DSP core

The design process of low-cost functional obfuscation for DSP core consists of two major phases, i.e., (1) designing functionally locked DSP core and (2) exploring low-cost design configuration through the PSO-DSE process. Both these phases perform iteratively until a globally optimal design configuration is explored. Figure 8.4 shows the two major phases, their corresponding steps, and the primary inputs to perform low-cost functional obfuscation. As shown in Figure 8.4, the primary inputs are a DSP application in the form of a C-code/CDFG, ILB design (based on each 8-bit key value), module library, and PSO control parameters (such as acceleration coefficient, inertia weight, swarm size, terminating criteria). In the process of PSO-DSE in HLS (detailed PSO-DSE framework is explained in Section 8.6), each iteration explores a design configuration and ILB structure, as a particle in the swarm. Based on this design configuration, a gate-level datapath structure (post-processing through standard HLS framework) of the input CDFG application is generated, which is then locked using ILB (where structures of some ILBs are decided on encrypted output from AES block and remaining without encryption. More details are provided in VC). This generates an FO design of the input DSP core. Then the total power and delay of the locked design are calculated to evaluate the design cost through a fitness function. Subsequently, particle positions are updated to explore a new fitter particle, and the

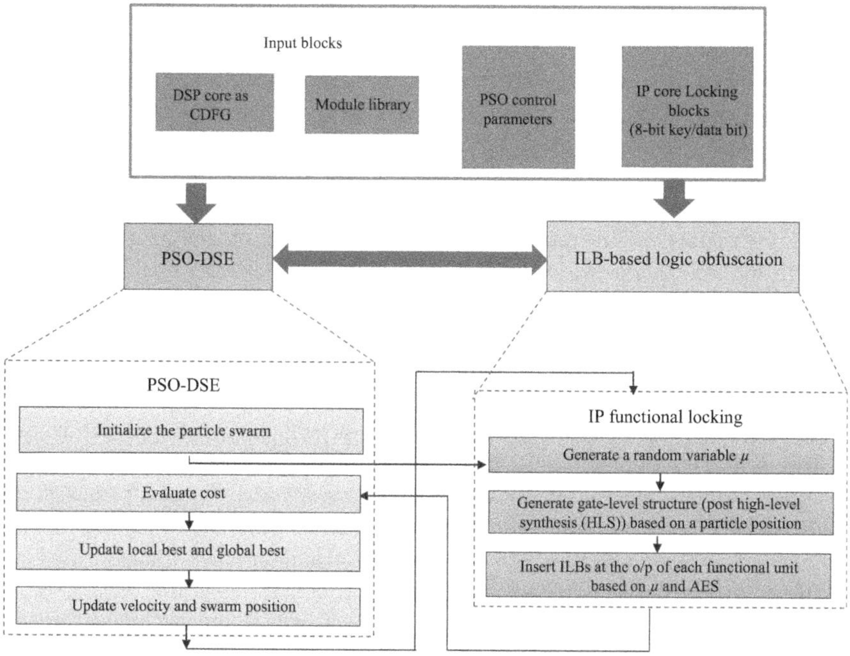

Figure 8.4 Framework for functional obfuscation of DSP cores

complete process is repeated till the termination condition is met. Thus, the final particle position after meeting the terminating condition is used to generate a low-cost obfuscated design of a DSP core. Therefore, the final design architecture not only optimizes the total power, delay, and implementation cost but also minimizes the total gate count of the complete locked netlist, which includes total number of functional units, registers/latches, muxes/demuxes, and ILBs.

Lack of an optimization framework, for generating a locked netlist of DSP IP core, results in huge design overhead. This is because DSP cores are generally large in size and can be designed using multiple alternative hardware design solutions. Any random selection of design solution will lead to higher design overhead. In this approach (Sengupta *et al.*, 2018), particles are encoded with both hardware resource configuration and ILB block insertion procedure to obtain the optimal locked design. Thus, the optimal obfuscated design solution in terms of the number of gate count, power consumption, and execution delay of the locked netlist is achieved through a PSO-DSE process. In the next three subsections, we discuss the overviews of ILBs, its security features, and insertion technique, respectively.

1. *Overview: IP core-locking blocks:* This subsection explains a brief overview of IP core-locking block. ILBs are key components inserted into a nonobfuscated design to generate a FO design. Each ILB consists of 8-bit key value inserted into each bit of output data. ILBs are designed using the different combination of AND, NAND, NOT, XOR, and XNOR gates. However, structures of ILB

depend on the key values. Innumerable different structures of ILBs with the same area are possible. Each of these ILBs generates correct output if a correct key value is provided. In Sengupta *et al.* (2018), a subset of key-bits are fed as input to an AES block and based on the encrypted output the ILB structures are configured. However, rest of the key-bits are fed directly to the ILBs. This is because it can help prevent SAT attack for DSP core (which is not possible now but as a proactive countermeasure for future). The ILBs are designed using various combination of AND, NAND, NOT, XOR, and XNOR gates and with multiple features (explained in the next subsection) to handle different attacks.

2. *Features: IP core-locking blocks:* This subsection discusses the major security features of ILB design. As mentioned earlier, the ILBs are inserted in each output data bit of the gate structure of DSP IP core to generate a locked netlist. These ILB structures are designed mainly to avoid key-sensitization and removal attacks. Key-sensitization attack is a process of identifying the key value by applying different input patterns and checking their corresponding outputs (Yasin *et al.*, 2016). Therefore, multiple features are considered while designing the ILBs to avoid any possible scenario of key-sensitization and removal attacks.

   Multi pair-wise security: All key-bits in a locked netlist are considered as multi pair-wise secure if an attacker cannot sensitize anyone of them to the output data bit without controlling or knowing the remaining all other key-bit values (Yasin *et al.*, 2016). In a locked netlist, key-bits are inaccessible to an attacker, therefore, he/she cannot feed input to it. Thus, in multi pair-wise secure key-bits, identifying one key value of a key-bit depends on the key value of all remaining key-bits, therefore an attacker cannot crack any key value of the locked design. Different sample ILB structures of Sengupta *et al.* (2018) are shown in Figure 8.5. According to Figure 8.5, all the ILBs have 8 key-bits and all are multi pair-wise secure. In other words, to sensitize any key-bit

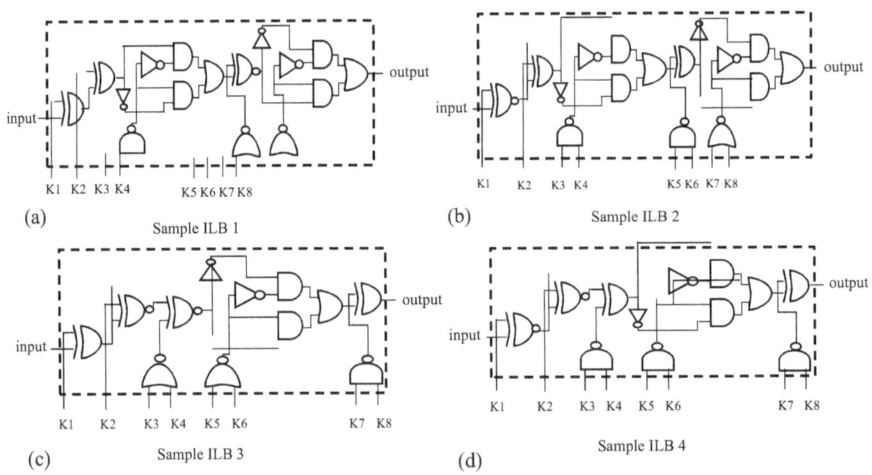

Figure 8.5 Different sample ILB structures designed based on key values (coming either from AES block or directly)

of an ILB to the output, an attacker has to know or control the correct value of other seven key-bits. Therefore, to identify the correct value of other seven key-bits, an attacker has to apply brute-force attack. For example, to sensitize key-bit K1 of the ILB shown in Figure 8.5(a) to the output bit, an attacker has to control the other seven key-bits' value, i.e., "0110000" which is challenging. Thus, multi pair-wise security feature enhances key-sensitization attack for the ILBs.

Prohibiting key-gate isolation: To avoid key-sensitization attack, a locked netlist design should be prohibited from isolated key-gate. Different ILBs structures (shown in Figure 8.5) designed in Sengupta *et al.* (2018) consist of eight intertwined key-gates dependent on each other. This ensures no isolated key-gate is present in the ILBs. Thus, prohibiting isolated key-gate from the locked netlist enhances key-sensitization attack for the ILBs.

Protection against run of key-gates: In Sengupta *et al.* (2018), the key-gates of ILBs are interlaced among each other, therefore, replacing all key-gates (eight input key-gates) with a single one is highly difficult.

Nonmutable convergent key-gates: If two key-gates converge at some other gate in such a way that none of the key-bits can be muted, we can call these key-gates as nonmutable convergent key-gates (Yasin *et al.*, 2016). As in Sengupta *et al.* (2018), all the ILBs consist of eight nonmutable key-bits, therefore, to sensitize one key-bit to the output remaining seven key-bits need to be muted, which is a complicated process. Thus, nonmutable key-gate placement in the locked netlist enhances key-sensitization attack for the ILBs.

3. *Insertion: IP core-locking blocks*: In the process of inserting ILBs, the first step is to generate a random integer number "μ" (through PSO-DSE process) such that:

$$1 \leq \mu \leq T_{\text{ILB}} \tag{8.1}$$

where T_{ILB} indicates the total number of ILB structures/templates available to insert in the design (e.g., Sengupta *et al.* (2018) used $T_{\text{ILB}} = 4$). Post HLS, after obtaining gate-level datapath structures, ILBs are inserted in each output data bit of functional unit based on the "μ" value to generate a locked netlist. In this process, same ILB is inserted "μ" times into the design and then after "μ" repetition next ILB from T_{ILB} is selected. This process will continue until all the output data bits of functional units are connected to ILBs. For instance, a gate-level structure of FIR benchmark (Mohanty, 2015), designed using 1 adder and 1 multiplier, is functionally obfuscated through "μ" = 2. The complete gate structure of FO (through ILBs) design is shown in Figure 8.6. As "μ" = 2 is used while inserting ILBs, ILB1 (shown in Figure 8.5(a)) is inserted for the first two output data bit of adder, and then ILB2 (shown in Figure 8.5(b)) is inserted for the next two output data bit, and so on until all the output data bit of adder and multiplier are locked through ILBs. Insertion of ILBs in the design can be performed randomly also to lock the design, but according to Sengupta *et al.* (2018), this may incur huge power and cost overhead with compromised security. On the contrary, in this approach different "μ" values are explored through power-delay trade-off, thus optimizing the overheads. Additionally, no output data bit remains unobfuscated (unlocked) in this approach, thus ensuring strong security.

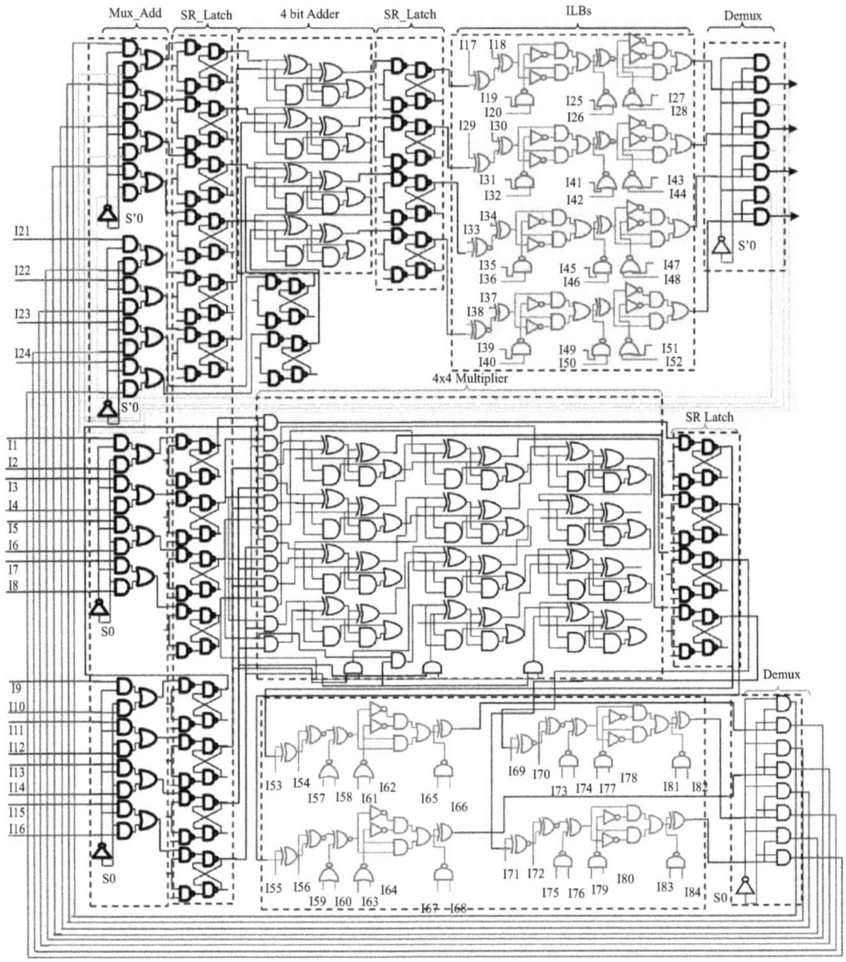

Figure 8.6 Obfuscated gate structure of 4-bit FIR designed using (1A, 1M, $\mu = 2$), locked with 64-bit keys (I53–I84, I25–I52, I17–I20)

8.5 Security of functionally obfuscated DSP core design

In this section, we discuss different aspects of security for FO DSP core design. In the first subsection, we discuss about total possible key-space for a locked netlist through ILBs. In the second subsection, we discuss about the security analysis of designed ILBs from an attacker's perspective. In the last subsection, we discuss about different possible attack scenarios and how the designed ILBs thwart them.

8.5.1 Keyspace

The number of possible key-space (K_s) for a locked netlist through ILBs can be represented using the following equation:

$$K_s = 2^{(b*m*f)} \tag{8.2}$$

where "K_s" provides a measure of "Strength of Obfuscation," "b" is the number of key-bits per ILB, "m" is the output bit-width of each functional unit, and "f" is the number of functional units used to design the datapath. For example, as shown in Figure 8.6, a 4-bit FIR is designed using one adder and one multiplier, which is locked using an 8-bit key-based ILBs. Thus, for this example, $b = 8$, $m = 4$, and $f = 2$. Therefore, using (8.2), the possible key-space of this locked netlist is determined as $2^{(8*4*2)} = 1.8 * 10^{19}$.

8.5.2 Security analysis

In the security analysis of the locked netlist, it is assumed that an attacker has access to following three things:

(i) A FO netlist
(ii) Activated functional circuit (brought from the market)
(iii) Advance RE facility

In this scenario, through RE, an attacker can determine the total number of input key bits, he/she then searches in the key-space to identify the combination of key values. For instance, as shown in Figure 8.6, an attacker can determine that the design contains 64 numbers of input key-bits. He/she can perform brute force attack to identify a correct combination of 64 key values, which require testing of total 2^{64} different combinations. During brute force attack, if we assume that 1 billion (10^9) keys can be applied per second to the locked netlist, it will take approximately 585 years to identify the correct key value.

8.5.3 Countermeasures against attacks

With the same assumptions that an attacker has access to three aforementioned things, in this section we will discuss countermeasures against key-sensitization attacks (Yasin *et al.*, 2016), IP piracy attacks (Sengupta, 2016; Castillo *et al.*, 2007), Trojan insertion (Sengupta, 2017; Sengupta *et al.*, 2017b), SAT attacks (Subramanyan *et al.*, 2015; Xie and Srivastava, 2018), and their countermeasures.

(1) Key-sensitization attack: In this attack, the attacker tries to determine/sensitize the correct key value by applying multiple input patterns. Key-sensitization attack will be easy for an attacker if key-gates are isolated, not pairwise secured, mutable, and connected in a back-to-back fashion. An attacker can apply a particular input pattern on the functional IC brought from the market and observe its corresponding, then apply the same input in the locked netlist with different key values to determine the correct one. Therefore, to prevent the key-sensitization attack placement of key-component is very important. Key-gates should be placed in a locked netlist in such a fashion that there is a dependency between key-gates. This ensures "prohibition of key isolation" and "multi pair-wise security" features (discussed in Section 8.2).

Multiple key-gates should not be connected in such a fashion that it can be replaceable using a single key-gate, which increases the number of valid keys in the key-space. This ensures "prohibition of run of key-gates." Additionally, sensitization attack can also be performed if key-gates can be muted (discussed in Section 8.2.1). The ILBs are designed in such a way (with intertwining among gate structures for eight key inputs) that key-gates are multi pair-wise secure, no isolated key-gates are available in an ILB, no key-gates can be muted and enhances attack complexity to determine and replace run of key-gates using a single key. Thus, Sengupta *et al.*'s (2018) approach provides resilience against key-sensitization attacks.

(2) IP piracy and Trojan attack: The ILB-based functional obfuscation methodology hinders an attacker's effort (as key-gates structures are heavily intertwined among themselves for 8-bit key inputs) to detect the correct key through RE. Only a valid combination of key-value can actuate the locked netlist.

(3) SAT attack: SAT solver is not scalable for multiplier circuit (Subramanyan *et al.*, 2015; Finke, 2015; Chakraborty *et al.*, 2017). Even for a small-size 3-bit multiplier, the equivalent CNF has 20 clauses (Finke, 2015). Generally, a DSP core comprises several large bit-size multipliers as an integral part. Therefore, DSP IP core designs are SAT-attack-resistant. However, authors in Sengupta *et al.* (2018) provide a proactive countermeasure (assuming efficient SAT attacks may be formulated for DSP cores in future), solution motivated from Yasin *et al.* (2016), where the subset of the key-bits are encrypted through a custom lightweight. *AES block (not public design)* with fixed secret key. Based on the encrypted key-bits, subset of ILBs are designed and inserted in the obfuscated netlist. This prevents an attacker from determining the correct key value if he/she does not have the key of AES block (shown in Figure 8.7). Sengupta *et al.* (2018) designed their own *custom lightweight AES architecture* comprising 429 logic elements. This AES block utilizes <1% of final design logic elements while implementation using Altera Quartus (2018) 13.0 version in Cyclone II FPGA (2018) after integration with the obfuscated netlist of the DSP core. Top-level design of custom AES block-integrated obfuscated 4-bit FIR and its corresponding gate structure implemented in Altera Quartus are shown in Figures 8.8 and 8.9, respectively. A partially extracted portion of AES-integrated obfuscated netlist file is shown in Figure 8.10, which demonstrates that individual logic components are indistinguishable.

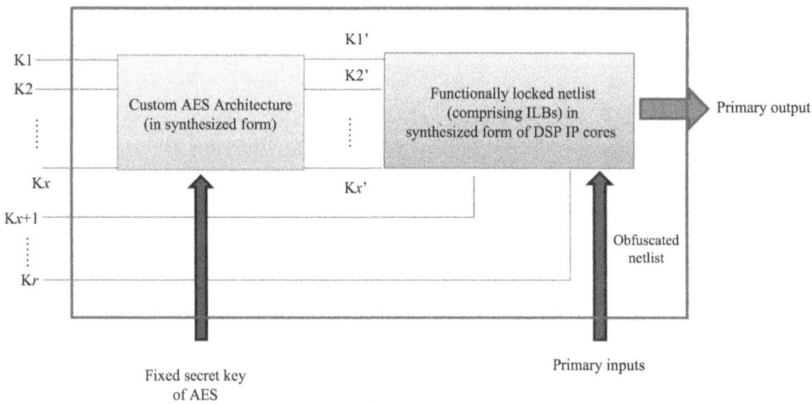

Figure 8.7 Countermeasure against SAT and removal attacks on obfuscated DSP IP core design

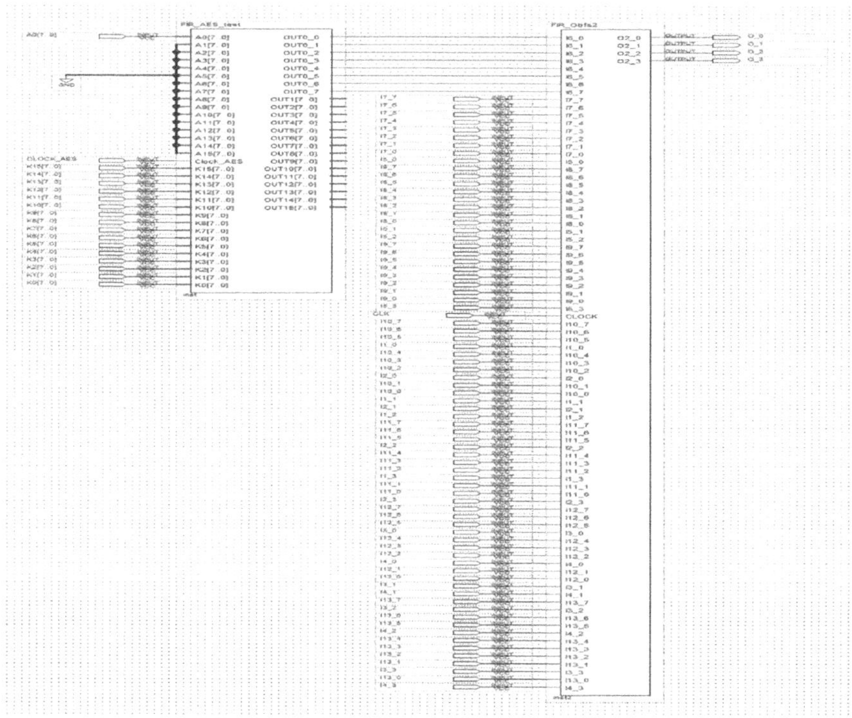

Figure 8.8 *Top-level implemented design of custom AES block integrated obfuscated 4-bit FIR (Note: K0–K15 indicate 128 bit key of custom AES block, A0–A15 indicate primary input of AES block and O_0-O_3 indicate final output of FIR; in this case only 1 ILB, i.e., 8-bit key A0[7..0] is encrypted through AES, therefore A1–A15 is grounded, similarly only 8 bit output of AES, i.e., OUT0_0-OUT-7 is connected as input to FIR block, i.e., I6_0-I6_7, rest of the key inputs of FIR block is feed directly, not via. AES block)*

The simulation result of the FO FIR design is shown in Figure 8.11. Some examples of lightweight AES architectures available in the literature are Hamalainen *et al.* (2006) and Morioka and Satoh (2002). Post-synthesis result for FIR DSP core after integrating custom AES block is shown in Table 8.1. It is computationally infeasible to determine the inputs of AES from its output when the key is unknown. To prevent SAT attack, following steps are performed:

(i) Designer synthesizes an AES block with a fixed secret key.
(ii) Subsets of keys of functional obfuscation are encrypted through the AES block.
(iii) The outputs of the AES block (encrypted keys) are connected to few ILBs in the locked netlist depending on the number of keys are encrypted through AES.

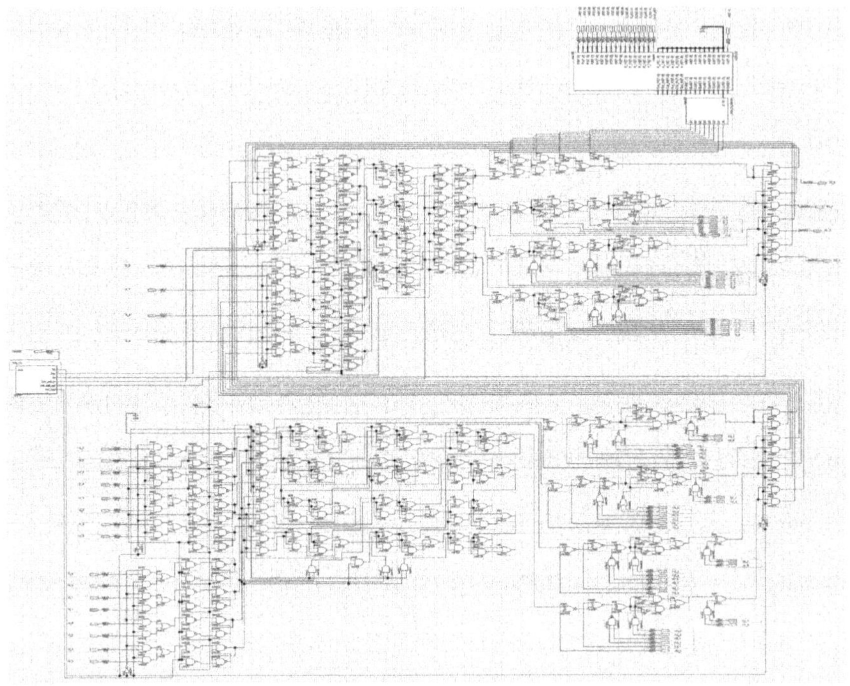

Figure 8.9 Implemented gate structure of obfuscated 4-bit FIR where subset of key bits are encrypted through custom AES block

(iv) Based on the fixed secret key of the AES block and the input to AES, a designer can identify the output of AES and configures the ILBs accordingly by re-organizing the internal gate structure of ILB. (Note: The internal reorganization of the gate within ILB does not add any extra gate but only changes its configuration and connectivity.)

Therefore, the total area of ILB and design cost remain almost similar. For example, based on an encrypted output of key bits (first 8-bits, viz., 11001101) through AES block, a reconfigured ILB can be designed as shown in Figure 8.12. As a custom AES block is designed in Sengupta *et al.* (2018), its structure is unknown to an outsider. Therefore, distinguishing and removing AES block from the post-synthesized functionally locked gate design is not feasible. However, it is important to note that more the number of key inputs of functional obfuscation fed into the AES block, more is the number of available exhaustive options for reconfigurable ILBs. For example, if the number of key inputs of functional obfuscation fed through AES block is 8 bit, then the exhaustive reconfigurations possible is 2^8; while if it is 16 bits, the exhaustive reconfigurations possible is $2^{16} = 65{,}536$ (which is large). Further, the attack time grows exponentially which enhances the difficulty for an attacker in determining the correct key value. However, instead of using custom lightweight AES block, alternative lightweight solutions such as anti-SAT block (Xie and

```
// Location: LCCOMB_X57_Y51_N10
cycloneiv_lcell_comb \inst132|6|Y[0] (
// Equation(s):
// \inst132|6|Y [0] = (GLOBAL(\inst132|26|Str_RegA~clkctrl_outclk ) & ((\inst132|5|O[0]~11_combout ))) # (!GLOBAL(\inst132|26|Str_RegA~clkctrl_outclk ) & (\inst132|6|Y [0]))

	.dataa(\inst132|6|Y [0]),
	.datab(gnd),
	.datac(\inst132|5|O[0]~11_combout ),
	.datad(\inst132|26|Str_RegA~clkctrl_outclk ),
	.cin(gnd),
	.combout(\inst132|6|Y [0]),
	.cout());
// synopsys translate_off
defparam \inst132|6|Y[0] .lut_mask = 16'hF0AA;
defparam \inst132|6|Y[0] .sum_lutc_input = "datac";
// synopsys translate_on

// Location: LCCOMB_X58_Y52_N26
cycloneiv_lcell_comb \inst132|26|Sel_mux~3 (
// Equation(s):
// \inst132|26|Sel_mux~3_combout  = (\inst132|26|Sel_mux [0]) # ((!\inst132|26|count [6] & (\inst132|26|count [5] $ (!\inst132|26|Equal0~2_combout ))))

	.dataa(\inst132|26|count [5]),
	.datab(\inst132|26|count [6]),
	.datac(\inst132|26|Sel_mux [0]),
	.datad(\inst132|26|Equal0~2_combout ),
	.cin(gnd),
	.combout(\inst132|26|Sel_mux~3_combout ),
	.cout());
// synopsys translate_off
defparam \inst132|26|Sel_mux~3 .lut_mask = 16'hF2F1;
defparam \inst132|26|Sel_mux~3 .sum_lutc_input = "datac";
// synopsys translate_on
```

Figure 8.10 Randomly extracted portions of obfuscated netlist file of FIR integrated with custom AES block

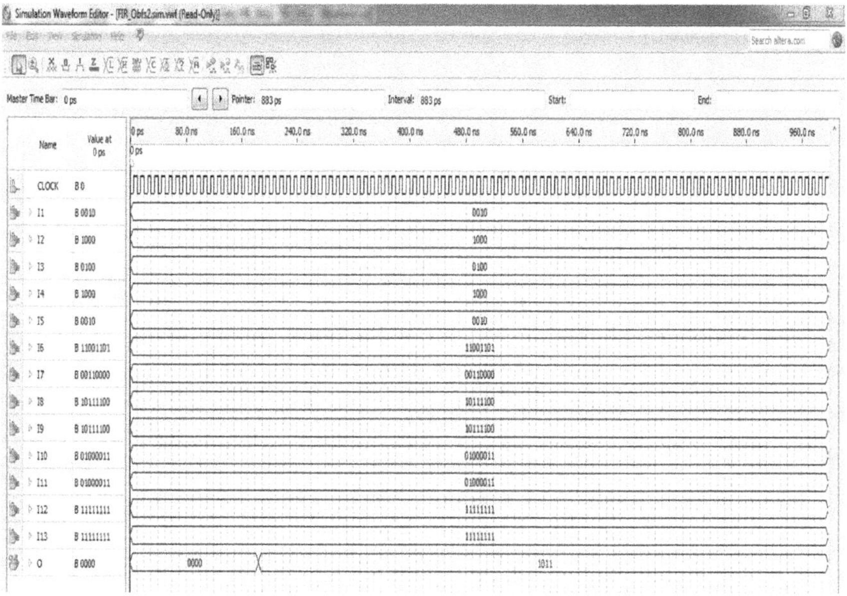

Figure 8.11 Simulation result of FO FIR design

Table 8.1 Resource utilization summary of post synthesis obfuscated netlist with lightweight AES using Altera Quartus of Intel Cyclone-II: EP2C70F896C6 FPGA

Resource usage	Obfuscated FIR with AES
Logic elements	548/68,416
Dedicated logic registers	32/68,416

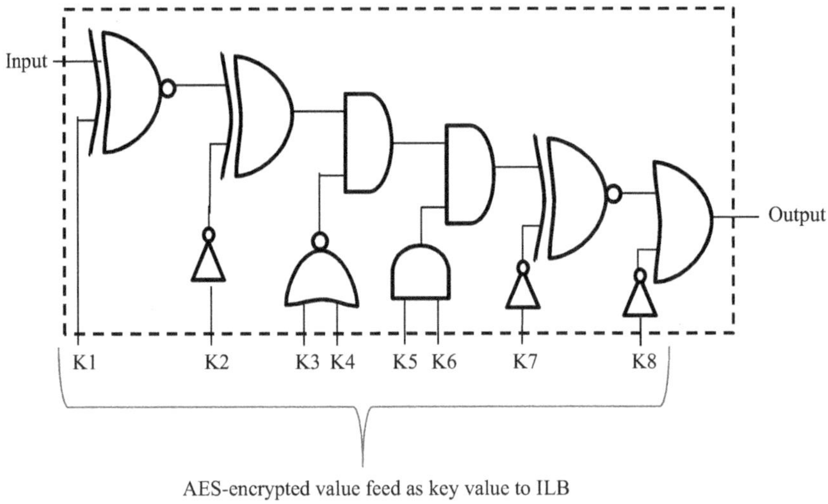

Figure 8.12 Example of a reconfigured ILB based on a sample encrypted output (11001101....) from AES circuit. Shows only the first encrypted 8 keys acting as actual keys inputs to a ILB in design netlist

Srivastava, 2018) may also be used to prevent SAT attack. However, as mentioned in Yasin *et al.* (2017), detection and removal of the anti-SAT block could be possible through "signal probability skew (SPS)" attack.

(4) ILB and AES removal attack: The removal of ILBs and AES from the locked netlist is difficult because:

(i) The ILBs along with AES block inserted into the design for locking is resynthesized. This synthesis makes each component of the AES circuit along with the ILB-based locked netlist indistinguishable, which is not feasible to remove by an attacker. This is because of two reasons: (1) first, as mentioned earlier, the subset of ILB structures inserted in the netlist is reconfigured by the designer based on the AES-encrypted output corresponding to the secret key. Therefore, no fixed template of ILBs are available, the configuration of ILB depends on AES output which in turn depends on the secret key and input of the AES, resulting into several possible structures. Moreover, the number of possibilities also increase with the number of key-inputs fed to ILBs through

AES. Therefore, knowing the structures of all the possible ILBs for an attacker is not practical in this scenario. (2) Second, since Sengupta *et al.* (2018) have designed their own custom lightweight AES architecture which is not publicly available, component of AES structure after post-synthesis become indistinguishable from the locked netlist of DSP core.

(ii) The sequencing of ILBs to generate a locked netlist of DSP core depends on the "μ" variable explored through PSO-DSE process. Therefore, identifying ILB configuration cannot be determined in advance.

8.6 Optimization engine for functional obfuscation of DSP cores

The ILB-based functional obfuscation approach for DSP cores (Sengupta *et al.*, 2018) uses the PSO-DSE framework to achieve low-cost obfuscated netlist for DSP IP cores. The PSO-DSE process used in this approach comprises three sub-processes, viz., encoding and initialization of particle, evaluating the fitness of a particle, and generating fitter particle position.

8.6.1 Particle encoding

To optimize the obfuscated design cost, each particle is encoded through the array of available resource types and "μ" value. For example, *i*th encoded particle (P_i) in the particle population can be represented as $P_i = \{n(R_1), n(R_2), \ldots, n(R_d), \mu\}$, where $n(R_d)$ indicates the number of resources of type R_d; and "d" is the total number of resource types available; "μ" is defined in (8.1). Based on the aforementioned encoding template, the first three particles (P_1, P_2, and P_3) of the PSO-DSE process are initialized with the minimum, maximum, and mid-resource values, respectively, which can be represented as follows:

$$\begin{aligned}
P_1 &= \{\min(R_1),\ \min(R_2), \ldots,\ \min(R_D), \mu\} \\
P_2 &= \{\max(R_1),\ \max(R_2), \ldots,\ \max(R_D), \mu\} \\
P_3 &= \{(\min(R_1) + \max(R_1))/2, (\min(R_2) + \max(R_2))/2, \ldots, \\
&\qquad (\min(R_D) + \max(R_D))/2, \mu\}
\end{aligned}$$

If the initial number of population is more than three, rest of the particles are initialized as follows:

$$\begin{aligned}
P_j = [&\{(\min(R_1) + \max(R_1))/2 \pm a\}, \{(\min(R_2) + \max(R_2))/2 \pm a\}, \ldots, \\
&\{(\min(R_D) + \max(R_D))/2 \pm a\}, \mu]
\end{aligned}$$

In the above expression, "a" indicates a random integer number within the range of maximum and minimum resource values of resource type R_d.

8.6.2 Particle fitness

After performing particle encoding and particle initialization, the next process is to calculate the fitness of each particle to determine the fittest one. To calculate the fitness of each particle, a FO gate structure is generated based on that particle position. Then the total area/power consumption and delay of that obfuscated design are determined, based on which, the fitness of that particle is calculated using the following equation:

$$C_f(P_i) = \O_1 \frac{P^{FO}}{P^{FO}_{\max}} + \O_2 \frac{D^{FO}}{D^{FO}_{\max}} \tag{8.3}$$

where $C_f(P_i)$ indicates normalized weighted cost/fitness of a particle P_i, $\O_1$, and $\O_2$ denote user-provided weightage factor for area/power and delay components, respectively. The total power consumption and execution delay of a FO design are indicated through P^{FO} and D^{FO}, respectively. $P_{\max}{}^{FO}$ and $D_{\max}{}^{FO}$ indicate the maximum possible power (estimated as power consumption of maximum number of hardware resources available in the module library) and $D_{\max}{}^{FO}$ maximum possible delay (estimated as delay utilizing minimum number of hardware resources) of FO design, respectively.

8.6.3 Updating particle

After calculating the fitness of each particle of the swarm, the fittest particle (having the lowest cost value) till the current iteration known as "local best" is updated. Further, the global best is updated among all the local best particle positions. Based on the local and global best, each dimension (d) of a particle velocity (V_{d_i}) can be updated using following equation:

$$V_{d_i}^{+} = \omega V_{d_i} + b_1 r_1 \left[R_{d_{lb_i}} - R_{d_i}\right] + b_2 r_2 \left[R_{d_{gb}} - R_{d_i}\right] \tag{8.4}$$

where $V_{d_i}^{+}$ and V_{d_i} are the new and current velocity of ith particle in the dth dimension, respectively; R_{d_i} is resource value of ith particle in dth dimension; $R_{d_{lb_i}}$ is the local best position of ith particle in dth dimension; and $R_{d_{gb}}$ is the global best of dth dimension; ω is the inertia weight, b_1 and b_2 are social and cognitive factors, respectively; r_1 and r_2 are random numbers between 0 and 1. The new particle position is generated based on the following expression:

$$R_{d_i}^{+} = R_{d_i} + V_{d_i}^{+} \tag{8.5}$$

where $R_{d_i}^{+}$ is the updated resource value of ith particle in dth dimension.

All the aforementioned three sub-processes of PSO-DSE execute iteratively to search for a low-cost obfuscated solution until any one of the following terminating criteria is met:

(i) The global best is not updated for 10 consecutive iterations.
(ii) The number of iteration reached the user-defined maximum number of iteration limits.

After reaching the termination criteria, the current global best design solution is the final low-cost solution for functional obfuscated design among all the possible design solutions.

8.7 Analysis of case studies

As discussed earlier, excluding (Sengupta *et al.*, 2018) robust functional obfuscation (RF-Ob), other approaches do not perform functional obfuscation for DSP IP cores. Thus, Sengupta *et al.* (2018) had re-implemented the functional obfuscation technique of Yasin *et al.* (2016)—robust functional locking (RFL)—and tested for DSP IP cores followed by performing comparative analysis in terms of security and design overhead. However, in Yasin *et al.* (2016), the target applications are the sequential and combinational circuits, available in ISCAS-85 benchmark circuits (Hansen *et al.*, 1999). On the other hand, RF-Ob (Sengupta *et al.*, 2018) target DSP cores, available in EXPRESS benchmark suites (DSP Benchmark Suite, 2017). A 15nm NanGate-based technology scales (Open Cell Library, 2017) was used to evaluate both power consumption and execution delay of a FO DSP IP core. The design cost is calculated using (8.3).

The analysis of case studies/test cases of DSP cores discussed in this chapter is divided into three sub-sections: (1) security analysis, (2) design overhead analysis, and (3) comparative analysis between Sengupta *et al.* (2018) and Yasin *et al.* (2016).

8.7.1 Security analysis

The term "strength of obfuscation (SoO)" is used as a metric of security to determine how difficult it is to reverse engineer an FO DSP IP design for an attacker. The SoO is calculated based on (8.2). The time required to determine the correct key combination is determined with the assumption of 1 billion key combinations were applied per second. The SoOs of both Sengupta *et al.* (2018) and Yasin *et al.* (2016) for DSP cores are reported in Table 8.2. For example, in FIR benchmark, total 2.13×10^{96} different input combinations of key value are possible according to Sengupta *et al.* (2018) approach to decipher the FO netlist through the brute-force attack. Similarly, in Yasin *et al.* (2016) approach, total 2.23×10^{43} different combinations of key values are possible. Therefore, Sengupta *et al.* (2018) approach enhances SoO by a factor of 9.57×10^{52} compared to Yasin *et al.*'s (2016). The comparison of Sengupta *et al.* (2018) and Yasin *et al.* (2016) in terms of SoO is shown in Figure 8.13.

Table 8.2 also reports the time required to determine the correct key combination from a locked netlist of DSP IP core. For example, for the same benchmark (FIR benchmark), in Sengupta *et al.* (2018) approach, total 6.75×10^{79} years are required to identify the correct key combination through the brute-force attack. Similarly, in Yasin *et al.*'s (2016) approach, total 7.07×10^{26} years are required to identify the correct key combination. Therefore, Sengupta *et al.* (2018) approach enhances attack time by a factor of 9.57×10^{52} compared to Yasin *et al.*'s (2016) approach. The comparison of Sengupta *et al.* (2018) and Yasin *et al.* (2016) in terms of attack time is shown in Figure 8.14.

Further, AES block conjunction with ILBs is capable of increasing the SAT attack time exponentially for DSP IP cores. As mentioned in Sengupta *et al.* (2018), if the number of key inputs fed onto AES block increases linearly, the SAT attack time increases exponentially.

Table 8.2 Comparison of SoO and attack time between RF-Ob (Sengupta et al., 2018) and RFL (Yasin et al., 2016) for DSP cores

DSP core benchmarks [19]		RF-Ob (Sengupta *et al.*, 2018)			RFL (Yasin *et al.*, 2016)			SoO enhancement (by factor of)	Increase in Attack time (by factor of)
Name	Size	No. of key-bits encoded	SoO	Attack time (in years)	No. of key-bits encoded	SoO	Attack time (in years)		
IIR	9,919	192	6.28 e+57	1.99 e+41	96	7.92 e+28	2.51 e+12	7.92 e+28	7.92 e+28
FIR	16,047	320	2.13 e+96	6.75 e+79	144	2.23 e+43	7.07 e+26	9.57 e+52	9.57 e+52
ARF	14,833	256	1.15 e+77	3.65 e+60	112	5.19 e+33	1.64 e+17	2.23 e+43	2.23 e+43
JPEG IDCT	42,710	1344	3.83 e+404	1.21 e+388	432	1.10 e+130	3.49 e+113	3.46 e+274	3.46 e+274
Mesa interpolate	48,853	832	2.86 e+250	9.07 e+233	464	4.76 e+139	1.51 e+123	6.01 e+110	6.01 e+110

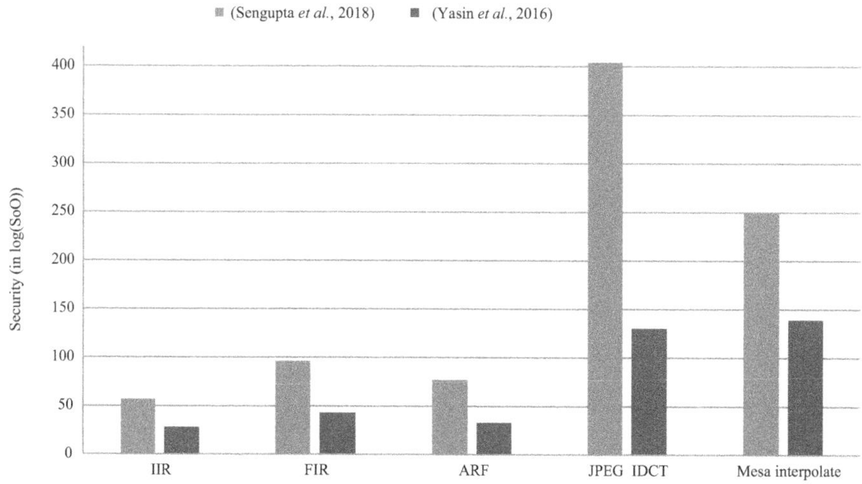

Figure 8.13 Comparison of strength of obfuscation for DSP cores

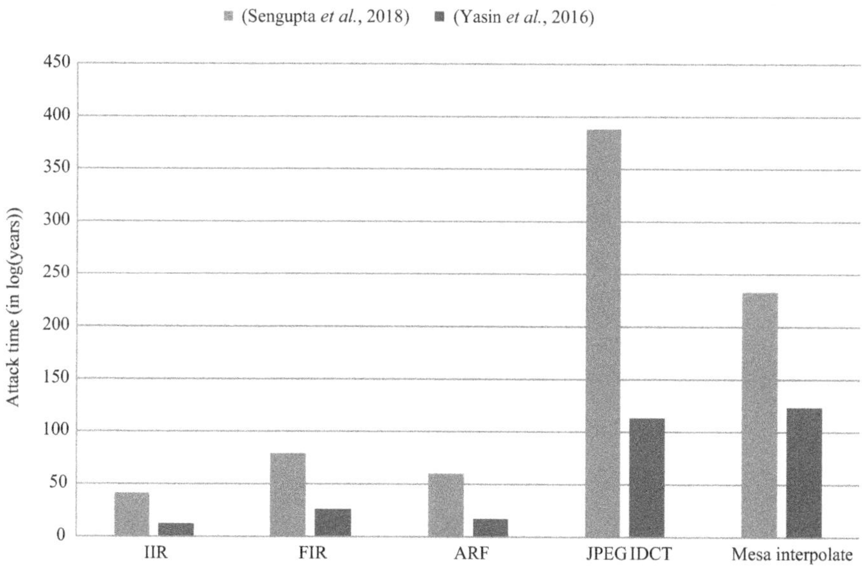

Figure 8.14 Comparison of attack time for DSP cores

8.7.2 Overhead analysis

In this sub-section, we analyze the design overhead of RF-Ob (Sengupta *et al.*, 2018) and RFL (Yasin *et al.*, 2016) in terms of leakage power and cost due to the insertion of key components, compared to its equivalent nonobfuscated (baseline) design. The nonobfuscated design, also known as baseline design, indicates an unsecured, unlocked DSP IP core design where no key components are inserted. Table 8.3 reports the power consumption comparison of Sengupta *et al.* (2018) and

Table 8.3 Comparison of power of RF-Ob (Sengupta et al., 2018) and RFL (Yasin et al., 2016) w.r.t. baseline for DSP cores

Benchmark	**Baseline design**		**Design of RF-Ob (Sengupta *et al.*, 2018)**			**Design of RFL (Yasin *et al.*, 2016)**		
	Configuration	**Power (in μW)**	**Configuration**	**Power (in μW)**	**Power overhead (in %)**	**Configuration**	**Power (in μW)**	**Power overhead (in %)**
IIR	1A, 3M	14.870	1A, 2M, $\mu=4$	20.146	35.48%	2A, 4M	24.850	67.11%
FIR	4A, 4M	33.871	3A, 2M, $\mu=4$	41.864	23.59%	4A, 5M	45.274	33.66%
ARF	2A, 4M	34.835	2A, 2M, $\mu=3$	39.234	12.62%	3A, 4M	43.967	26.21%
JPEG IDCT	5A, 5M	115.400	11A, 10M, $\mu=2$	172.523	49.50%	12A, 15M	178.843	54.97%
Mesa interpolate	5A, 3M	98.242	8A, 5M, $\mu=4$	132.924	35.30%	13A, 16M	155.673	58.45%

Yasin *et al.* (2016) with baseline design and their corresponding overhead percentage. The power overhead occurs for both the functional obfuscation techniques due to insertion of key components. However, the overhead is comparatively lesser for Sengupta *et al.* (2018) than that of Yasin *et al.* (2016) because low-cost optimal DSP core design solution is explored for Sengupta *et al.* (2018) through PSO-DSE framework (Sengupta *et al.*, 2017b). More explicitly, Yasin *et al.* (2016) incur higher power overhead due to the lack of a DSP architecture optimization framework. In Sengupta *et al.* (2018), not only power overhead is optimized through PSO-DSE process but also the total gate count after inserting ILBs, total execution delay, and the overhead design cost of the locked netlist are additionally optimized. Lack of an architecture optimization framework while generating an obfuscated DSP IP core incurs huge design overhead cost. The reason is because mostly DSP cores are huge in size and can be designed using various alternative design architectures. Selecting low-cost design architectural solution from a pool of alternatives necessitates integration of an optimization framework. In Sengupta *et al.* (2018), the low-cost design solution that optimizes the overall gate count of locked netlist (including the ILB blocks) is explored through PSO-DSE (discussed in detail earlier in Section 8.6).

The average power overheads of Sengupta *et al.* (2018) and Yasin *et al.* (2016) for DSP core benchmarks are 31.30% and 48.08%, respectively, compared to a baseline design. Table 8.4 reports the design cost comparison of Sengupta *et al.* (2018) and Yasin *et al.* (2016) with baseline design and their corresponding overhead percentage. The average design cost overheads of Sengupta *et al.* (2018) and Yasin *et al.* (2016) for DSP benchmarks are 11.82% and 23.28%, respectively, compared to a baseline design. For both the cases, the reason for achieving lesser overhead is the same, as discussed before.

8.7.3 Comparative analysis

In this subsection, the comparative analysis between Sengupta *et al.* (2018) and Yasin *et al.* (2016) is shown in terms of leakage power consumption and obfuscated design cost. Additionally, exploration time of PSO and design delay before and after inserting the ILBs are also discussed. The graphical representation of the leakage power consumption of Sengupta *et al.* (2018) and Yasin *et al.* (2016) is shown in Figure 8.15. It can be seen that for all the tested DSP benchmarks (Sengupta *et al.*, 2018) achieves lesser power consumption than related approaches. An average reduction of 11.07% in static power consumption is achieved in Sengupta *et al.* (2018) compared to Yasin *et al.* (2016). The graphical representation of the obfuscated design cost of Sengupta *et al.* (2018) and Yasin *et al.* (2016) is shown in Figure 8.16. It can be observed that for all the DSP benchmarks (Sengupta *et al.*, 2018) achieves lesser design cost than related approaches. An average reduction of 8.87% in obfuscated design cost is achieved in Sengupta *et al.* (2018) compared to Yasin *et al.* (2016). As both Sengupta *et al.* (2018) and Yasin *et al.* (2016) include AES block to handle (potential) SAT attack, therefore the total power and design cost reduction in Sengupta *et al.* (2018) obtained due to PSO-DSE remains intact.

Table 8.4 Comparison of design cost of RF-Ob (Sengupta et al., 2018) and RFL (Yasin et al., 2016) w.r.t. baseline for DSP cores

Benchmark	**Baseline design**		**Design of RF-Ob (Sengupta *et al.*, 2018)**			**Design of RFL (Yasin *et al.*, 2016)**		
	Configuration	**Cost**	**Configuration**	**Cost**	**Cost overhead (in %)**	**Configuration**	**Cost of (in μW)**	**Cost overhead (in %)**
IIR	1A, 3M	0.6118	1A, 2M, $\mu=4$	0.6810	11.31%	2A, 4M	0.7427	21.39%
FIR	4A, 4M	0.5134	3A, 2M, $\mu=4$	0.5638	9.81%	4A, 5M	0.5853	14.00%
ARF	2A, 4M	0.4467	2A, 2M, $\mu=3$	0.5259	17.73%	3A, 4M	0.5281	18.22%
JPEG IDCT	5A, 5M	0.3141	11A, 10M, $\mu=2$	0.3629	15.53%	12A, 15M	0.4455	41.83%
Mesa interpolate	5A, 3M	0.2954	8A, 5M, $\mu=4$	0.3093	4.70%	13A, 16M	0.3573	20.95%

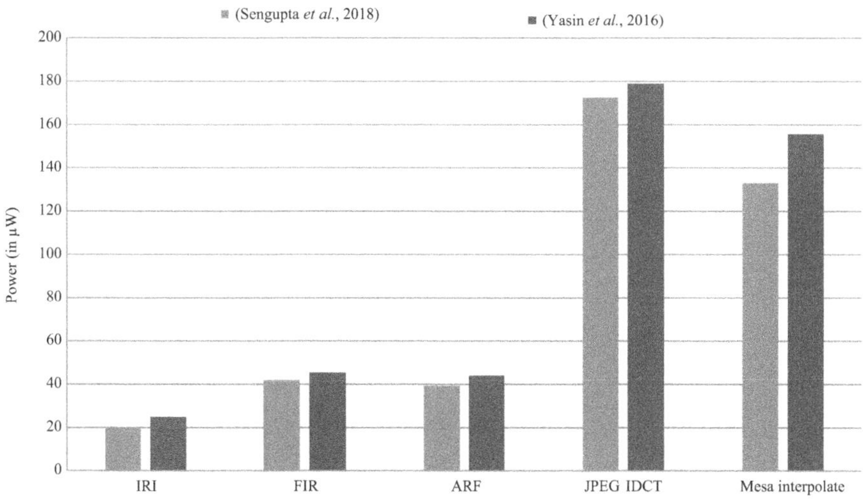

Figure 8.15 Comparison of power consumption

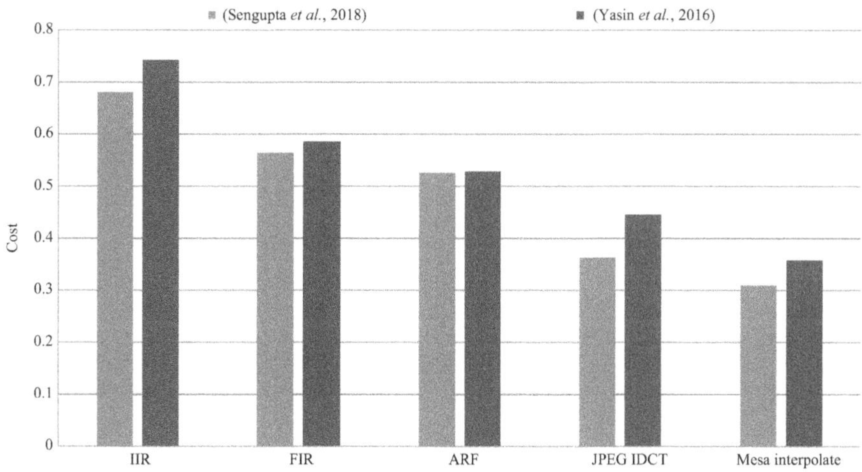

Figure 8.16 Comparison of design cost

Figure 8.17 presents the overall design delay overhead due to insertion of ILBs in Sengupta *et al.* (2018). It can be observed from Figure 8.17, the delay incurred due to the insertion of ILB is marginal. Moreover, the total delay (including the delay of ILBs) of the obfuscated design is optimized through the PSO-DSE framework. Figure 8.18 presents the exploration time (in seconds) of the PSO approach for generating low-cost obfuscated netlist for DSP cores. It can also be noted that for very large size DSP IP cores such as JPEG IDCT and MESA (around 50K plus gates), the exploration time is not too large. Therefore, the optimization framework used in Sengupta *et al.* (2018) does not suffer from scalability issue.

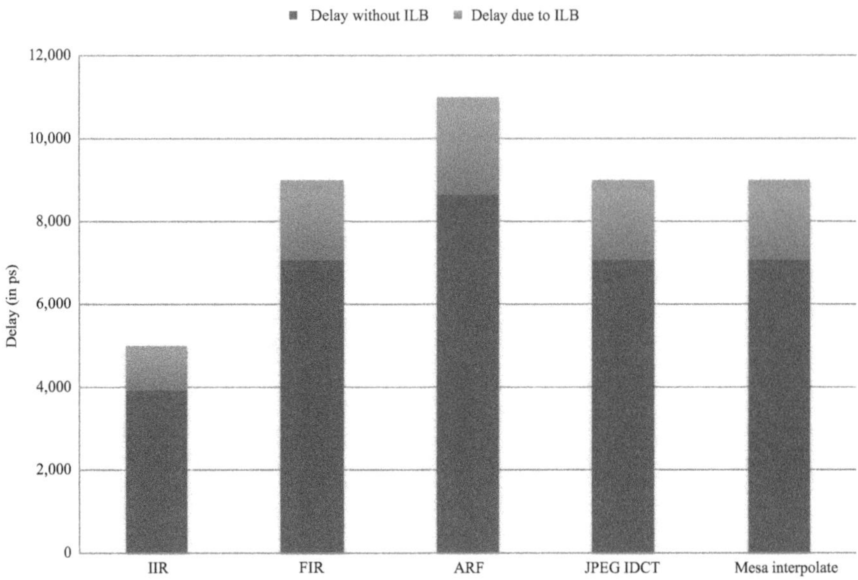

Figure 8.17 Delay due to the insertion of ILBs

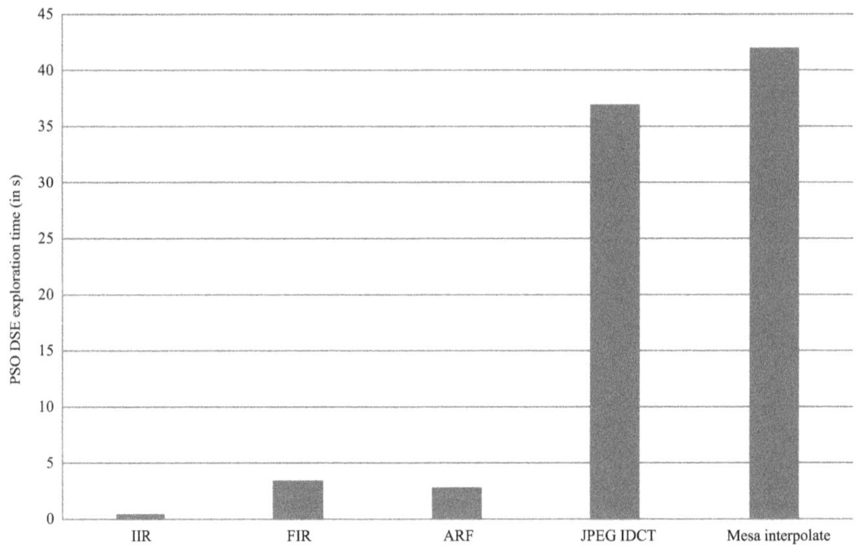

Figure 8.18 Exploration time of PSO DSE

8.8 Conclusion

This chapter presents a low-cost, functional obfuscation mechanism for DSP IP core. This mechanism inserts reconfigured ILBs to lock a netlist of DSP IP core. Moreover, it integrates AES block to prevent SAT attack as a proactive countermeasure.

Moreover, it was observed that Sengupta *et al.* (2018) yielded power reduction, design cost reduction, and SoO enhancement over other similar approaches. The research on functional obfuscation is still an open area and more research needs to be conducted for the protection of other hardware circuits such as combinational/sequential circuit benchmarks against SAT attacks. There are new attacks such as approximate SAT and signal probability skew that are being launched to nullify the effect of anti-SAT blocks used in conjunction with combinational/sequential circuit benchmarks. There is significant scope of work in the future in this direction.

8.9 Exercises

1. What is functional obfuscation?
2. What is the difference between structural obfuscation and functional obfuscation in the context of DSP cores?
3. What are the disadvantages of functional obfuscation?
4. What is threat model of functional obfuscation?
5. How/where is locking performed?
6. What are the properties of ILBs?
7. Why functional obfuscation is vulnerable to SAT attack for combinational/sequential circuits?
8. Why functional obfuscation is not vulnerable to SAT attack for DSP core?
9. What is CNF?
10. How to find CNF for basic gates?
11. How to determine DIP?
12. How does AES protect against removal attacks of ILBs?
13. How is AES customized during the insertion with obfuscation design?
14. How is AES removal attack prevented?
15. Design an obfuscated FFT ($n = 8$).
16. What are the possible alternatives of AES for functional obfuscation?
17. What is key sensitization attack?
18. Design an ILB with AND, NAND, NOT, XOR, and XNOR gates.
19. What is the difference between pairwise security and multi pair-wise security?
20. What is the role of random variable "μ" in functional obfuscation?
21. What is run of key gates? How is it a drawback in the context of obfuscation?
22. Design a lightweight AES that uses less than 1% FPGA resources.
23. How do you determine the design cost of a FO DSP design?

References

Alkabani Y. M. and Koushanfar F., 'Active hardware metering for intellectual property protection and security,' *Proceedings of USENIX Security Symposium,* 2007, pp. 20:1–20:16.

Baumgarten A., Tyagi A., and Zambreno J., 'Preventing IC piracy using reconfigurable logic barriers,' *IEEE Design Test of Computers*, 2010, vol. 27 (1), pp. 66–75.

Biggerstaff T. J., 'Design recovery for maintenance and reuse,' *Computer*, 1989, vol. 22 (7), pp. 36–49.

Castillo E., Meyer-Baese U., Garcia A., Parrilla L., and Lloris A., 'IPP@HDL: Efficient intellectual property protection scheme for IP cores,' *IEEE Transactions on Very Large Scale Integration Systems*, 2007, vol. 15 (5), pp. 578–591.

Chakraborty R. S. and Bhunia S., 'HARPOON: An obfuscation-based SoC design methodology for hardware protection,' *IEEE Transactions on CAD of Integrated Circuits and Systems*, 2009, vol. 28 (10), pp. 1493–1502.

Chakraborty S., Gupta A., and Jain R. 'Matching multiplications in bit-vector formulas,' Springer International Publishing, Cham, 2017, pp.131–150.

Cyclone II FPGA, Available: https://www.altera.com/products/fpga/cyclone-series/cyclone-ii/overview.html, last accessed on 2018.

DSP Benchmark Suite, Available: http://www.ece.ucsb.edu/EXPRESS/benchmark/, 2017. last accessed on June 2017.

Desai A.R., Hsiao M.S., Wang C., Nazhandali L., and Hall S., 'Interlocking obfuscation for anti-tamper hardware,' in *Proceedings of Cyber Security and Information Intelligence Research Workshop (CSIIRW)*, 2013, pp. 8:1–8:4.

Dupuis S., Ba P. S., Di Natale G., Flottes M. L. and Rouzeyre B., 'A novel hardware logic encryption technique for thwarting illegal overproduction and hardware Trojans,' *IEEE InternationalOn-Line Testing Symposium (IOLTS)*, 2014, pp. 49–54.

Finke M., 'Equisatisfiable SAT encodings of arithmetical operations', 2015, Available: http://www.martin-finke.de/documents/Masterarbeit_bitblast_Finke.pdf.

Guajardo J., Kumar S. S., Schrijen G. J., and Tuyls P., 'Brand and IP protection with physical unclonable functions,' *Proceedings of IEEE International Symposium on Circuits and Systems*, 2008, pp. 3186–3189.

Hamalainen P., Alho T., Hannikainen M., and Hamalainen T.D., 'Design and implementation of low-area and low-power AES encryption hardware core', *IEEE 9th EUROMICRO Conference on Digital System Design: Architectures, Methods and Tools*, 2006, Washington, pp. 577–58.

Hansen M. C., Yalcin H. and Hayes J. P., 'Unveiling the ISCAS-85 benchmarks: A case study in reverse engineering,' *IEEE Design & Test of Computers*, 1999, vol. 16 (3), pp. 72–80.

Koushanfar F., '*Hardware Metering: A Survey*', New York, NY: Springer New York, 2012, pp. 103–122.

Lao Y. and Parhi K. K., 'Obfuscating DSP circuits via high-level transformations,' *IEEE Transactions on Very Large Scale Integration Systems*, 2015, vol. 23 (5), pp. 819–830.

Lee Y. W. and Touba N. A., 'Improving logic obfuscation via logic cone analysis,' *Latin-American Test Symposium (LATS)*, 2015, pp. 1–6.

Mohanty S. P., *Nanoelectronic Mixed-Signal System Design*. McGraw-Hill Education, New York, 2015, no. 0071825711.

Morioka S. and Satoh A., 'An optimized S-Box circuit architecture for low power AES design', *4th International Workshop on Cryptographic Hardware and Embedded Systems (CHES)*, 2002, pp. 172–186.

NanGate 15 nm open cell library, Available: http://www.nangate.com/?pageid=2328, 2017, last accessed on June 2017.

Pappu R., Recht B., Taylor J., and Gershenfeld N., 'Physical one-way functions', *Science*, 2002, vol. 297 (5589), pp. 2026–2030.

Quartus Altera, Available: https://dl.altera.com/13.0sp1, last accessed on 2018.

Rajendran J., Zhang H., Zhang C., *et al.*, 'Fault analysis-based logic encryption,' *IEEE Transactions on Computers*, 2015, vol. 64 (2), pp. 410–424.

Roy D. and Sengupta A., 'Low overhead symmetrical protection of reusable IP core using robust fingerprinting and watermarking during high level synthesis,' *Future Generation Computer Systems*, 2017, vol. 71, pp. 89–101.

Roy J. A., Koushanfar F., and Markov I. L., 'EPIC: Ending piracy of integrated circuits,' *in DATE*, 2008, pp. 1069–1074.

Sengupta A., 'Intellectual property cores: Protection designs for CE products,' *IEEE Consumer Electronics Magazine*, 2016, vol. 5 (1), pp. 83– 88.

Sengupta A., 'Hardware security of CE devices: Threat models and defence against IP Trojans and IP piracy,' *IEEE Consumer Electronics Magazine*, 2017, vol. 6 (1), pp. 130–133.

Sengupta A. and Kundu S., 'Securing IoT hardware: Threat models and reliable, low-power design solutions,' *IEEE Transactions on VLSI Systems*, 2017, vol. 25 (12), pp. 3265–3267.

Sengupta A. and Roy D., 'Antipiracy-aware IP chipset design for CE devices: A robust watermarking approach', *IEEE Consumer Electronics Magazine*, 2017, vol. 6 (2), pp. 118–124.

Sengupta A. and Sedaghat R., 'Integrated scheduling, allocation and binding in high level synthesis using multi structure genetic algorithm based design space exploration,' *International Symposium on Quality Electronic Design*, 2011, pp. 1–9.

Sengupta A. and Sedaghat R., 'System and methodology for development of a system architecture using optimization parameters', *US Patent by United Sates Patent and Trademark Office (USPTO)*, 2013, Patent no. US 8,397,204.

Sengupta A., Roy D., Mohanty S. P., and Corcoran P., 'DSP design protection in CE through algorithmic transformation based structural obfuscation,' *IEEE Transactions on Consumer Electronics*, 2017a, vol. 63 (4), pp. 467–476.

Sengupta A., Bhadauria S., and Mohanty S. P., 'TL-HLS: Methodology for low cost hardware Trojan security aware scheduling with optimal loop unrolling factor during high level synthesis,' *IEEE Transactions on CAD of Integrated Circuits and Systems*, 2017b, vol. 36 (4), pp. 655–668.

Sengupta A., Kachave D., and Roy D., 'Low cost functional obfuscation of reusable IP cores used in CE hardware through robust locking,' *IEEE Transactions on Computer-Aided Design of Integrated Circuits and Systems*, 2018, doi: 10.1109/TCAD.2018.2818720.

Sengupta A., Roy D., and Mohanty S. P., 'Triple-phase watermarking for reusable IP core protection during architecture synthesis,' *IEEE Transactions on CAD of Integrated Circuits and Systems*, 2018b, vol. 37 (4), pp. 742–755.

Shamsi K., Li M., Meade T., Zhao Z., Pan D. Z., and Jin Y., 'Appsat: Approximately deobfuscating integrated circuits' *IEEE International Symposium on Hardware Oriented Security and Trust (HOST)*, 2017, pp. 95–100.

Subramanyan P., Ray S., and Malik S., 'Evaluating the security of logic encryption algorithms,' *IEEE International Symposium on Hardware Oriented Security and Trust (HOST)*, 2015, pp. 137–143.

The Economic Impacts of Counterfeiting and Piracy, last modified: 02.03.2017. 2017. Available: http://www.inta.org/Communications/Documents/Forms/AllItems.aspx.

Torrance R. and James D., 'The state-of-the-art in IC reverse engineering', *Proceedings of Workshop on Cryptographic Hardware and Embedded Systems (CHES)* (Springer, Berlin, Heidelberg), 2009, pp. 363–381.

Vijayakumar A., Patil V. C., Holcomb D. E., Paar C., and Kundu S., 'Physical design obfuscation of hardware: A comprehensive investigation of device and logic-level techniques,' *IEEE Transactions on Information Forensics and Security*, 2017, vol. 12 (1), pp. 64–77.

Wang X., Jia X., Zhou Q., *et al.*, 'Secure and low-overhead circuit obfuscation technique with multiplexers,' *GLSVLSI*, 2016, pp. 133–136.

Wong J. L., Kirovski D., and Potkonjak M., 'Computational forensic techniques for intellectual property protection,' *IEEE Transactions on CAD of Integrated Circuits and Systems*, 2004, vol. 23 (6), pp. 987–994.

Xie Y. and Srivastava A., 'Anti-SAT: Mitigating SAT attack on logic locking,' *IEEE Transactions on Computer-Aided Design of Integrated Circuits and Systems*, 2018, doi: 10.1109/TCAD.2018.2801220.

Yasin M., Rajendran J. J., Sinanoglu O., and Karri R., 'On improving the security of logic locking,' *IEEE Transactions on Computer-Aided Design of Integrated Circuits and Systems*, 2016, vol. 35 (9), pp. 1411–1424.

Yasin M., Mazumdar B., Sinanoglu O., and Rajendran J., 'Removal attacks on logic locking and camouflaging techniques,' *IEEE Transactions on Emerging Topics in Computing*, 2017, doi: 10.1109/TETC.2017.2740364.

Zhang J., 'A practical logic obfuscation technique for hardware security,' *IEEE Transactions on Very Large Scale Integration Systems*, 2016, vol. 24 (3), pp. 1193–1197.

Chapter 9

Obfuscation of JPEG CODEC IP core for CE devices

Joint Picture Export Group (JPEG) is the most commonly used image compression standard in the world. One cannot comprehend a CE system that does not process a JPEG. Without JPEG there is no smartphone photography, no social media. So, authors feel strongly that this important multimedia core "JPEG" needs to give well-deserving credit in terms of securing it when security/protection of "DSP core" has been a major focus of all the discussions in this book so far. The chapter is organized as follows: Section 9.2 provides an overview of DCT-based JPEG compression and decompression process. Section 9.3 explains the design process of generating structurally obfuscated JPEG CODEC IP core; Section 9.4 provides a detailed insight on the implementation process of obfuscated JPEG codec IP core in a CAD synthesis tool. Section 9.5 provides implementation and analysis of JPEG CODEC IP core as well as compressed images through the devised JPEG CODEC IP core.

9.1 Introduction

Most of the modern CE device comprises a dedicated lens or camera to capture and/or display digital images such as smartphone, tablets, scanner, laptop, smartwatch (Thavalengal and Corcoran, 2016; Kim *et al.*, 2015; Corcoran and Andrae, 2014; Tang *et al.*, 2016). Due to the enhancement of camera lenses, recording components, and displaying components, current digital imaging systems are capable of capturing and displaying high-resolution images (Corcoran *et al.*, 2001; Andorko *et al.*, 2011). As high-resolution images contain too much detailed information, they are large in size. Storing or transmitting these large size images is a critical issue for storage space and transmission bandwidth, respectively. Reducing the size of an image while storing and/or transferring is one of the popular and commercially successful techniques to address this crisis. JPEG standard formed in 1992 proposed DCT-based image compression. Image compressions are of two types: (1) lossy and (2) lossless. In lossy image compression, less important information of an image is discarded permanently (Mohanty *et al.*, 2003; Mohanty, 1999). Camera (Corcoran *et al.*, 2001; Andorko *et al.*, 2011) in a smartphone, tablets, laptop, smartwatch, etc. uses lossy image compression (Hnesh and Demirel, 2016). In lossless image compression, no loss of information occurs. Camera

in medical imaging (Bilgin and Marcellin, 1998), satellite imaging, forensic imaging, etc. uses lossless image compression (Li *et al.*, 2017; Scarmana and McDougall, 2015).

DCT-based JPEG image compression is lossy by nature (Obukhov and Kharlamov, 2008). DCT segregates an image into multiple subparts or blocks based on the visual quality of the image and then converts each block to the frequency domain from spatial domain. It discards small high-frequency components; therefore, DCT-based JPEG image compression method is lossy.

Image compression and decompression (CODEC) is a major data-intensive/power-intensive computational process in the SoC of almost all smart CE devices. Therefore, in these CE products, JPEG-based CODEC is a vital function implemented through a dedicated IP core to perform the data-intensive/power-intensive computational process for image compression and decompression (Roy and Sengupta, 2018). The use of JPEG CODEC in the form of an IP core not only balances the trade-off between hardware area and processing speed but also enhances the design productivity. For these aforementioned advantages, JPEG CODEC IP core may face several threats such as IP piracy (Sengupta *et al.*, 2018a), IP infringement (Sengupta and Roy, 2017a; Roy and Sengupta, 2017; Wong *et al.*, 2004; Alkabani and Koushanfar, 2007), malicious Trojan insertion (Sengupta *et al.*, 2017a), RE (Sengupta and Roy, 2017b; Lao and Parhi, 2015; Vijayakumar *et al.*, 2017; Sengupta *et al.*, 2017b), which therefore demands its protection (Sengupta, 2017). The motivation for designing a protected JPEG CODEC IP core is explained in the rest of the section.

Motivation: Standard JPEG CODEC design process is not capable (also does not aim) to obscure the functionality from an adversary by transforming the architecture. This does not deter RE process for an adversary. Successful RE attack enables an adversary (in the foundry) to identify the functionality of the design to counterfeit the netlist of the design and make several copies of it without the knowledge of the IP vendor/owner. Further, if an adversary is able to identify the functionality of the design from its functional structure, he/she can make a malicious modification (i.e., insert Trojan) into the design and create malfunctioning. Both these threats can be avoided if RE process is made harder for an adversary. This is possible by structurally obfuscating the JPEG CODEC design such that the functionality of it becomes unobvious to an adversary, i.e., architecture designed to be hidden to an adversary. This strongly hinders the RE process, thus thwarting the threats. Therefore, a JPEG CODEC is considered secure, if its architecture is not obvious, i.e., its functionality is not easily discoverable by inspecting its structure (Sengupta *et al.*, 2018b). A thematic attack scenario of obfuscated JPEG CODEC IP core is shown in Figure 9.1.

JPEG CODEC is selected as a case study in this chapter because it is a very vital core widely used in several CE applications. Keeping in view the recent surge of IP counterfeits and hardware Trojan, it is essential to secure JPEG CODEC IP cores against RE threats as failing to do so may lead to serious functional failures (due to Trojan) and/or significant loss to CE industry (due to IP counterfeit). There has not been much discussion to addresses this concern; therefore, this chapter provides a detailed discussion on this crucial aspect.

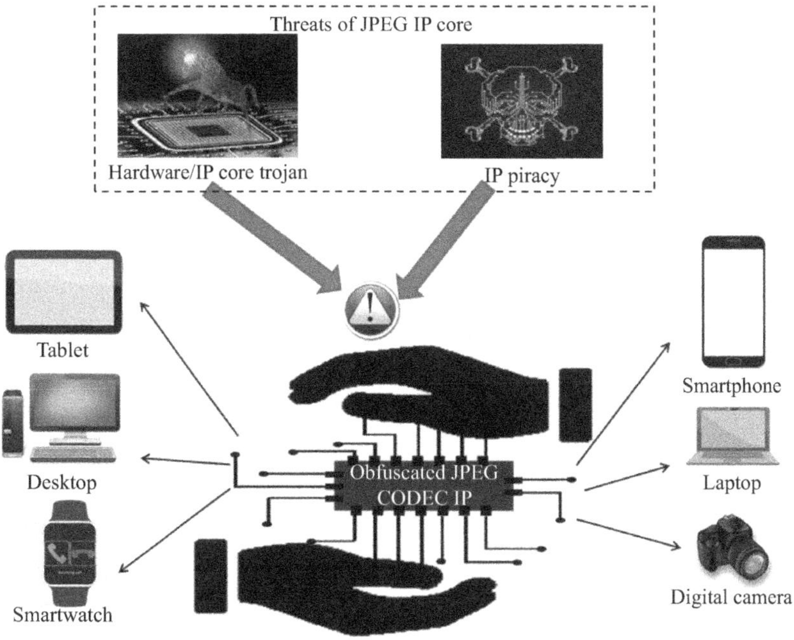

Figure 9.1. A pictorial representation of obfuscated JPEG CODEC IP core use in different camera-based CE devices

9.2 Overview of JPEG compression and decompression

Typically, JPEG is a lossy image compression technique based on DCT. DCT is a Fourier-like transformation technique proposed by Ahmed *et al.* (1974). However, unlike Fourier transformation which represents a signal as a mixture of sine and cosine expansion, DCT represents a signal in only cosine expansion. Based on the image's visual quality, DCT segregates it into multiple parts or spectral sub-bands. It converts an input image from the spatial domain to the frequency domain. The DCT is known for its high "energy compaction" property, meaning that the transformed signal can be easily analyzed using few low-frequency components. In simpler terms, DCT discards small high-frequency components, i.e., sharp transitions in intensity for grayscale images. Therefore, DCT-based image compression method is lossy, where less important information of an image is discarded. The process of discarding less important information is known as quantization. Image compression ratio with image quality can be controlled through quantization factor. The output bitstream is generated from the quantized coefficients through different encoding mechanisms and is then stored into a storage device. Reconstruction of the compressed image starts with decoding the bitstream data. De-quantization and inverse DCT (IDCT) is applied back to back on the decoded data to generate the decompressed image. The complete end-to-end flow of DCT-based JPEG compression and decompression is shown in Figure 9.2 (Mohanty *et al.*, 2003;

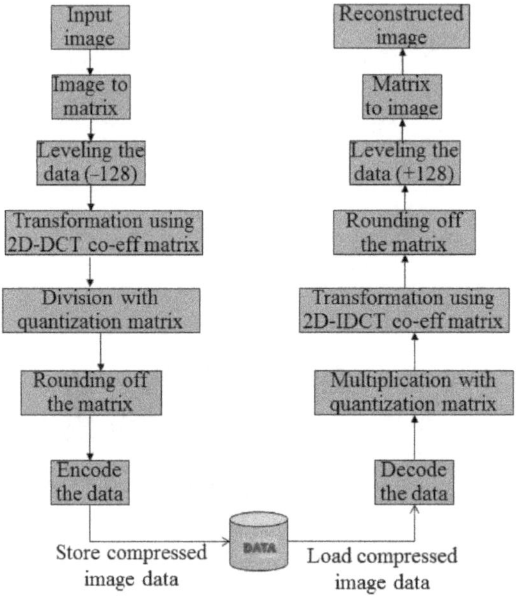

Figure 9.2. End-to-end flow of DCT-based JPEG compression and decompression process

Mohanty, 1999). Each step of JPEG image compression and decompression process is discussed in the rest of the section.

9.2.1 DCT-based JPEG image compression process

9.2.1.1 Processing of the input image

In an 8-bit grayscale image, the pixel intensity ranges from [0–255], where 0 indicates pure black and 255 indicates pure white. Therefore, each grayscale image is represented using 256 shades of gray. To reduce the dynamic range (required in DCT processing), the midpoint of this range, i.e., 128 is subtracted from each pixel value of the image. Therefore, the modified data range, i.e., [−128, 127] is centered on zero. The generic 2D-DCT coefficient matrix (T) for a sample of $N \times N$ can be expressed as follows (DCT matrix, 2018):

$$T(i,j) = \begin{cases} \dfrac{1}{\sqrt{N}}, \ \mathit{if}\ i = 0 \\ \sqrt{\dfrac{2}{N}} \cos\left[\dfrac{(2j+1)i\pi}{2N}\right], \ \text{if } i > 0 \end{cases} \tag{9.1}$$

where "N" is the size of the DCT matrix, "i" and "j" is the *i*th row, *j*th column component of the 2D-DCT matrix. In standard JPEG process, "N" is equal to 8 and *i* and *j* vary from 0 to 7. The 2D-DCT coefficient matrix for an 8 × 8 block is shown in Figure 9.3 (also explained in Chapter 3). According to Figure 9.3, all

$$
\mathbf{T} = \begin{bmatrix}
C_4 & C_4 & C_4 & C_4 & C_4 & C_4 & C_4 & C_4 \\
C_1 & C_3 & C_5 & C_7 & -C_7 & -C_5 & -C_3 & -C_1 \\
C_2 & C_6 & -C_6 & -C_2 & -C_2 & -C_6 & -C_6 & -C_2 \\
C_3 & -C_7 & -C_1 & -C_5 & C_5 & C_1 & C_7 & -C_3 \\
C_4 & -C_4 & -C_4 & C_4 & C_4 & -C_4 & -C_4 & C_4 \\
C_5 & -C_1 & -C_7 & C_3 & -C_3 & -C_7 & C_1 & -C_5 \\
C_6 & -C_2 & C_2 & -C_6 & -C_6 & C_2 & -C_2 & C_6 \\
C_7 & -C_5 & C_3 & -C_1 & C_1 & -C_3 & C_5 & -C_7
\end{bmatrix}
$$

Figure 9.3. 2D-DCT coefficient matrix

$$
\mathbf{M} = \begin{bmatrix}
m_{11} & m_{12} & m_{13} & m_{14} & m_{15} & m_{16} & m_{17} & m_{18} \\
m_{21} & m_{22} & m_{23} & m_{24} & m_{25} & m_{26} & m_{27} & m_{28} \\
m_{31} & m_{32} & m_{33} & m_{34} & m_{35} & m_{36} & m_{37} & m_{38} \\
m_{41} & m_{42} & m_{43} & m_{44} & m_{45} & m_{46} & m_{47} & m_{48} \\
m_{51} & m_{52} & m_{53} & m_{54} & m_{55} & m_{56} & m_{57} & m_{58} \\
m_{61} & m_{62} & m_{63} & m_{64} & m_{65} & m_{66} & m_{67} & m_{68} \\
m_{71} & m_{72} & m_{73} & m_{74} & m_{75} & m_{76} & m_{77} & m_{78} \\
m_{81} & m_{82} & m_{83} & m_{84} & m_{85} & m_{86} & m_{87} & m_{88}
\end{bmatrix}
$$

Figure 9.4. Generic 8×8 block of input image

elements of the first row ($i = 0$) are equal to $1/\sqrt{8}$ according to (9.1). As "T" is an orthogonal matrix, the inverse of "T" (i.e., T^{-1}) is equal to the transpose of "T" (i.e., T^{trans}).

9.2.1.2 Perform DCT on input image

To perform DCT on the input image, the image should be subdivided into multiple nonoverlapping 8 × 8 blocks. Each block can be processed independently, therefore block-wise parallelism is possible for transformation, which is the key feature of 8 × 8 DCT. A generic 8 × 8 block "M" of input image matrix is shown in Figure 9.4, where m_{ij} express pixel intensity of ith row, jth column component of the "M" block. The DCT on the "M" block can be performed through following matrix multiplications:

$$X = D * T^{\text{trans}} \tag{9.2}$$

In the above expression, D is calculated as follows:

$$D = T * M \tag{9.3}$$

The "D" matrix ($d_{11}, d_{12}, \ldots, d_{88}$) transforms the rows of the "M" block. Further, "X" matrix ($X_{11}, X_{12}, \ldots, X_{88}$) transforms the columns of the "D" matrix. Thus, transforms both rows and columns of the "M" block. Therefore, "X" is the DCT block of "M" block.

9.2.1.3 Perform quantization

The transformed block "X" is now ready to compress through quantization. As human eye is not good in distinguishing the high-frequency brightness deviation in an image, a large amount of information reduction is possible from the high-frequency components of an image. This is achieved by dividing each element of transformed block "X" by the corresponding element of the quantization matrix (Q) and then rounding to the nearest integer. As a result, most of the high-frequency components are rounded to zero, which indicates discarding less important high-frequency components. Additionally, many of the remaining components become small positive or negative numbers, which take fewer bits to represent. The quantization matrix is designed in such a way that it provides more resolution to higher perceivable frequency components over lower perceivable frequency components. The designing process of standard quantization matrix ($\mathbf{Q_{50}}$) involves subjective experiment of human visual system. In the range of 1 to 100, standard quantization matrix provides quality level as 50. It thus provides a balance between image compression ratio and image quality. The standard quantization matrix is shown in Figure 9.5. However, image quality and compression ratio can vary by changing the quantization matrix. To obtain a quantization matrix with quality level greater than $\mathbf{Q_{50}}$, each element of $\mathbf{Q_{50}}$ should be multiplied with (100−quality level)/50 and then rounding to the nearest integer. Similarly, to obtain a quantization matrix with quality level lower than $\mathbf{Q_{50}}$, each element of $\mathbf{Q_{50}}$ should be multiplied with 50/quality level and then rounding to the nearest integer.

$$\mathbf{Q_{50}} = \begin{bmatrix} 16 & 11 & 10 & 16 & 24 & 40 & 51 & 61 \\ 12 & 12 & 14 & 19 & 26 & 58 & 60 & 55 \\ 14 & 13 & 16 & 24 & 40 & 57 & 69 & 56 \\ 14 & 17 & 22 & 29 & 51 & 87 & 80 & 62 \\ 18 & 22 & 37 & 56 & 68 & 109 & 103 & 77 \\ 24 & 35 & 55 & 64 & 81 & 104 & 113 & 92 \\ 49 & 64 & 78 & 87 & 103 & 121 & 120 & 101 \\ 72 & 92 & 95 & 98 & 112 & 100 & 103 & 99 \end{bmatrix}$$

Figure 9.5. Standard quantization matrix ($\mathbf{Q_{50}}$)

9.2.1.4 Data encoding

The 2D quantized 8 × 8 block (X′) is now ready to be compressed through encoding. The block is converted to a 1D array through "zigzag" scanning. Thereafter, employing *run-length encoding (RLE),* enables grouping similar frequencies together. The encoded data is then converted into binary to generate a bitstream. Finally, the bitstream data of the compressed image is stored into a storage device.

9.2.2 DCT-based JPEG image decompression process

9.2.2.1 Data decoding

Image reconstruction process starts with data decoding. The stored data is first fetched from the storage device and decoded through RLE. The quantized 2D 8 × 8 block (X′) is reconstructed through "inverse zigzag" scanning from the 1D array.

9.2.2.2 Perform inverse quantization

The reconstructed X′ is now ready to be decompressed through inverse quantization. Decompression is achieved by multiplying each element of reconstructed block by the corresponding element of quantization matrix, to obtain block (X″).

9.2.2.3 Perform IDCT on decompressed image

Similar to applying DCT on the input block in the JPEG compression process, IDCT is applied on the decompressed block (X″). The IDCT on the X″ block can be performed through following matrix multiplications:

$$O = E * T \tag{9.4}$$

In the above expression, E is calculated as follows:

$$E = T^{\text{trans}} * X'' \tag{9.5}$$

The "E" matrix (e_{11}, e_{12}, ..., e_{88}) transforms the rows of the X″ block. Further, "O" matrix (o_{11}, o_{12}, ..., o_{88}) transforms the columns of the "E" matrix. Thus the above transformations, transform both the rows and columns of the X″ block. Therefore, "O" is the IDCT block of X″ block. After applying IDCT on the decompressed block, each element is rounded to the nearest integer.

9.2.2.4 Reconstruct the decompressed image

To reconstruct the decompressed image from the inversely transformed decompressed block, 128 is added to each of the element of the decompressed block to keep the range of output image pixel between [0, 255] instead of [−128, 127].

9.3 Design process of structurally obfuscated JPEG IP core

9.3.1 Threat model, problem formulation, and optimization framework

9.3.1.1 Threat model

To protect a JPEG CODEC IP core from RE and Trojan insertion through structural obfuscation. In order to maximize protection ability, impede the correct functionality of the application by generating a functionally equivalent but structurally non-obvious architecture of JPEG CODEC IP Core. (Note: Motivation of using structural obfuscation to guard against RE was thoroughly explained in Chapter 7).

9.3.1.2 Problem formulation

In order to protect a JPEG CODEC IP core from RE threat, design a structurally obfuscated, low-cost hardware. The hardware design cost is evaluated by calculating the total hardware area and total design latency. The hardware design cost adopted from Sengupta *et al.* (2018b) is formulated as:

$$C_f(X_i) = w_1 \frac{L_{total}^{\text{JPEG}}}{L_{\max}^{\text{JPEG}}} + w_2 \frac{A_{\text{total}}^{\text{JPEG}}}{A_{\max}^{\text{JPEG}}} \tag{9.6}$$

where $Cf(X_i)$ is the hardware design cost for design solution Xi; $L_{\text{total}}{}^{\text{JPEG}}$ and $A_{\text{total}}{}^{\text{JPEG}}$ represent the total design latency and total hardware area of the obfuscated JPEG IP core, respectively; $L_{\max}{}^{\text{JPEG}}$ and $A_{\max}{}^{\text{JPEG}}$ indicate the maximum possible design latency and hardware area, respectively, using the specified module library; weightage factor of two design parameters (latency-area) is controlled via w_1 and w_2.

9.3.1.3 Optimization framework

To obtain a low-cost hardware for JPEG IP core, PSO-driven DSE framework is used. This particle-based heuristic approach iteratively searches for a new global optima based on local minima and current global optima. Detailed explanation of the PSO-DSE process and formulation models is provided in Chapter 3 and in Sengupta *et al.* (2017a). Benefits of the PSO-DSE process are as follows:

(i) Capable of obtaining global optima while bypassing local minima
(ii) Capable of incorporating stochasticity during exploration process
(iii) Capable of preserving a proper balance between exploitation and exploration throughout the iterations

9.3.2 Constructing non-obfuscated DFG for JPEG compression

The generic steps to design a structurally obfuscated JPEG CODEC IP core are shown in Figure 9.6. In this section, we discuss how to construct a standard (non-obfuscated) DFG from the matrix expression of JPEG compression. Equations (9.2) and (9.3) represent the function of applying 2D-DCT on the input image block M. (Note: Detailed explanation of generating DCT matrix and its corresponding graph is given in Chapter 3.) This function is expanded and modeled as a mathematical expression to design an equivalent DFG. Further, this DFG is fed to an HLS engine to obtain a JPEG compression IP core. Based on (9.2), the corresponding mathematical expression for the first pixel (X_{11}) of the transformed block is modeled as:

$$\begin{aligned} X_{11} = (&c_4 * d_{11} + c_4 * d_{12} + c_4 * d_{13} + c_4 * d_{14} \\ &+ c_4 * d_{15} + c_4 * d_{16} + c_4 * d_{17} + c_4 * d_{18}) \end{aligned} \tag{9.7}$$

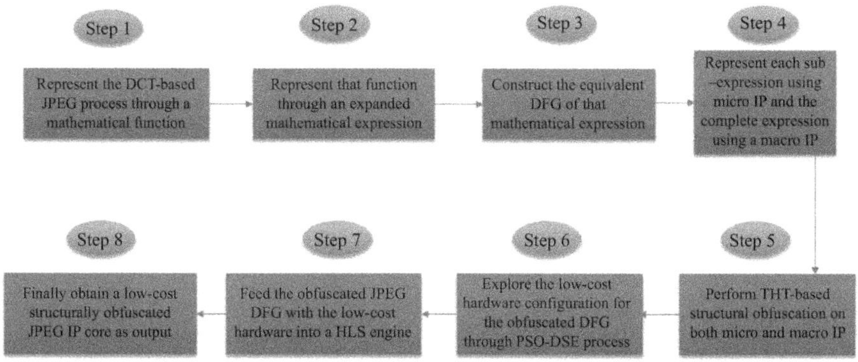

Figure 9.6. Steps to design a low-cost obfuscated JPEG IP core

whereas based on (9.3), $d_{11}, d_{12}, \ldots, d_{18}$ are modeled as:

$$d_{11} = (c_4 * m_{11} + c_4 * m_{21} + c_4 * m_{31} + c_4 * m_{41} + c_4 * m_{51} + c_4 * m_{61} + c_4 * m_{71} + c_4 * m_{81}) \quad (9.8)$$

$$d_{12} = (c_4 * m_{12} + c_4 * m_{22} + c_4 * m_{32} + c_4 * m_{42} + c_4 * m_{52} + c_4 * m_{62} + c_4 * m_{72} + c_4 * m_{82}) \quad (9.9)$$

Similarly,

$$d_{18} = (c_4 * m_{18} + c_4 * m_{28} + c_4 * m_{38} + c_4 * m_{48} + c_4 * m_{58} + c_4 * m_{68} + c_4 * m_{78} + c_4 * m_{88}) \quad (9.10)$$

Similarly, mathematical expression for the rest of the pixels of the transformed block can also be modeled, where the pixel values (m_{11}–m_{88}) and DCT coefficient will change depending on the pixel coordinate of the transformed block. However, the overall pattern of operations of the expression will be same. Based on this pattern, a DFG is generated where different pixel values and DCT coefficients are fed as the input. As in (9.7), we have modeled the mathematical expression for the first pixel (X_{11}) of the transformed block, and the corresponding DFG is shown in Figure 9.7. In the process of computing X_{11} through the complete DFG, termed as macro IP, each d_{ij} (represented through (9.8–9.10)) is computed through a sub-DFG termed as micro-IP. For example, micro-IP8, represented through blue node executes d_{18}. The zoom in representation of micro-IP8 is also shown in Figure 9.7. Totally 16 nodes (node no. 113–128) are part of micro-IP, where darker nodes indicate multiplication operation and lighter nodes indicate addition operation. Similarly, micro-IP1-7 executes d_{11}–d_{17}. The macro IP contains total 136 nodes; among these, the output of node number 135 produces the transformed pixel intensity of X_{11}. As mentioned earlier, quantization is applied on the each pixel of the transformed block, where each element of the transformed block is divided with the corresponding element of the quantization matrix. In this DFG, the output of node number 135 is multiplied with the inverse value of corresponding number (i.e., instead of division, multiplication is performed with the inverse of the

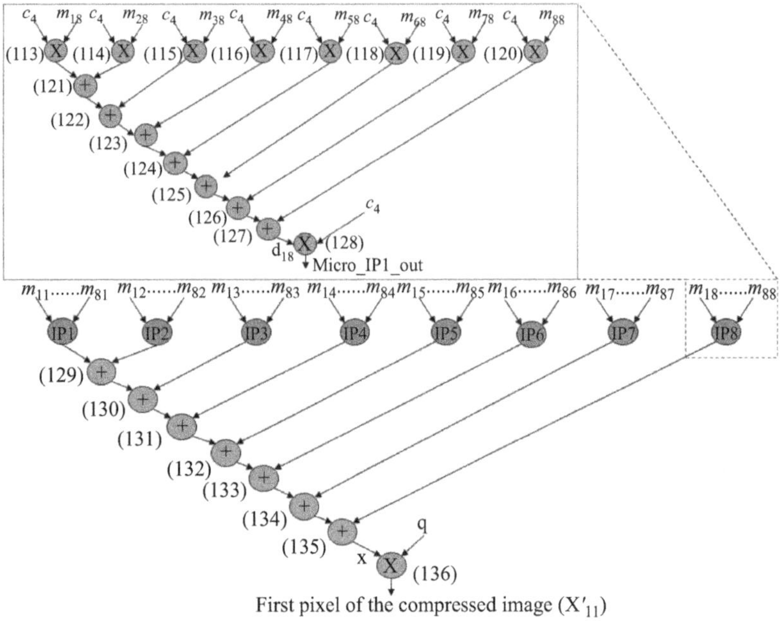

Figure 9.7. Non-obfuscated DFG of JPEG compression for calculating first compressed image pixel (X′ 11)

number) of the quantization matrix (**Q**). This q value is fed as another input of the DFG based on the quality level of the compressed image. For higher quality level, the q value will be less and vice versa while calculating the same pixel. Subsequently, the output of node number 136 produces the compressed pixel intensities (X_{11}') as the final output. This standard structure of JPEG DFG is common for most of the applications; therefore, identifying functionality from its structure is easy for an attacker. This invites threats such as RE, insertion of Trojan logic, counterfeits as fraudulent means of designing/maligning JPEG IP Core. This can be shielded to a certain extent by hiding functionality of the design. In the next subsections, we first discuss how to design a structurally obfuscated DFG for JPEG compression, followed up with its obfuscated hardware design.

9.3.3 Generating structurally obfuscated JPEG compression IP core

Structural obfuscation hinders the RE process by transforming the standard architecture of JPEG hardware to an unobvious design. Therefore, identifying the original functionality of the structurally obfuscated design is harder for an attacker. THT-based structural obfuscation is applied to the standard structure of JPEG DFG (Sengupta *et al.*, 2017b). It is a compiler-driven optimization technique which parallelizes the dependent critical path operations by dividing it into temporary sub-computations. However, this technique has not been used to generate a

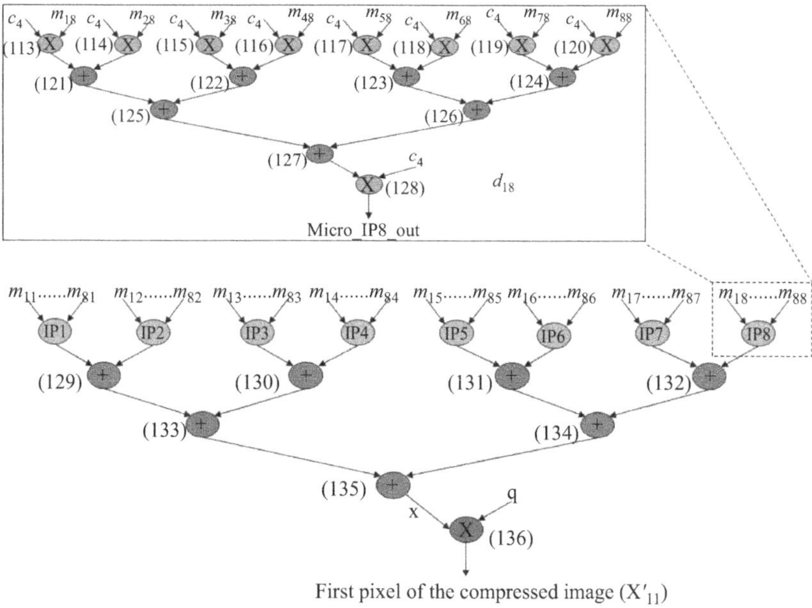

Figure 9.8. Obfuscated DFG of JPEG compression for calculating first compressed image pixel (X′11)

structurally obfuscated DFG for JPEG. The standard structure of JPEG DFG shown in Figure 9.7 is structurally obfuscated through THT. The transformed structure of JPEG DFG is shown in Figure 9.8. The zoom-in representation of obfuscated micro-IP8 is also shown in the figure. Due to this transformation, each micro-IP, as well as macro-IP, is modified. For example, in micro-IP8, 12 nodes among 16 nodes are affected. They are node no. 115, 116, 117, 118, 119, 120, 122, 123, 124, 125, 126, and 127. Therefore, in the eight micro-IPs, 96 nodes are affected. Additionally in the macro IP, six nodes are affected which include 130, 131, 132, 133, 134, and 135. A node is considered affected if its parents or child is changed or the operation type of that node is changed compared to the original one. These changes produce a different structure of datapath, interconnectivity logic, multiplexers/demultiplexers, and controller logic, once processed through high-level synthesis.

The design process of obfuscated, low-cost JPEG compression IP core is shown in Figure 9.9. As shown in this figure, the primary input is the non-obfuscated (standard) DFG of JPEG compression (designed in the previous section). Then it is structurally obfuscated through THT (explained initially in this section). A low-cost hardware configuration for obfuscated JPEG DFG is explored through PSO-DSE which is used to design a JPEG compression IP core in HLS framework (Sengupta and Sedaghat, 2011; Coussy *et al.*, 2009). In order to perform this process, PSO-DSE accepts additional inputs such as resource constraints, module library, and control parameters. For a detailed explanation of the PSO-DSE process, see Chapter 3 and Sengupta *et al.* (2017a). Finally, a dedicated IP core for JPEG

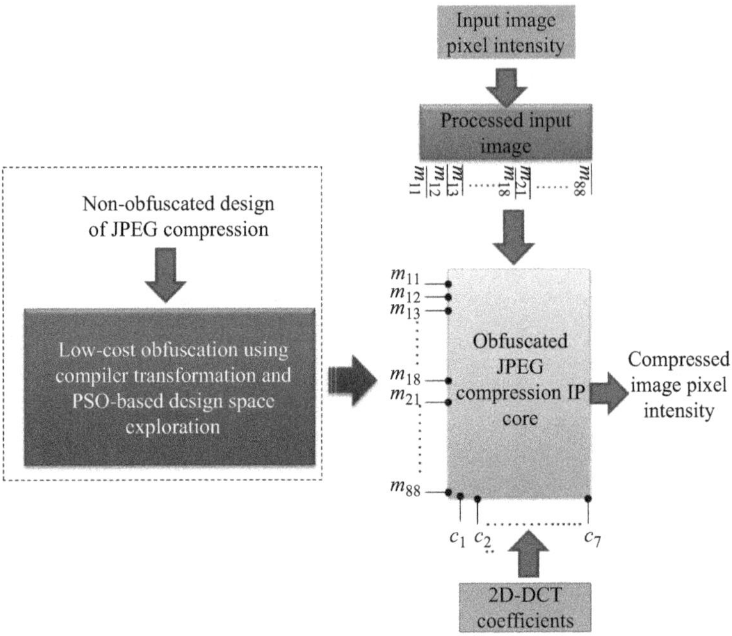

Figure 9.9. Obfuscated low-cost IP core for JPEG compression

compression is developed which is structurally obfuscated and low cost by nature. This IP core accepts leveled-off pixel intensity, 2D-DCT coefficient, and quantization matrix as inputs and produces transformed compressed pixel intensity as output.

9.3.4 Generating structurally obfuscated JPEG decompression IP core

In this section, we discuss how to construct an obfuscated DFG from the matrix expression of JPEG decompression. The compressed block X′ is decompressed through inverse quantization. In this process, each element of X′ is multiplied with the corresponding element of the quantization matrix of the same quality level, to obtain decompressed block X″. Then 2D-IDCT is performed on the decompressed block X″. Equations (9.4) and (9.5) represent the function of applying 2D-IDCT on the decompressed block X″. Similar to JPEG compression process, this function is expanded and modeled as a mathematical expression to design an equivalent DFG. Further, this DFG is fed to a HLS engine to obtain a JPEG decompression IP core. Based on (9.4), the corresponding mathematical expression for the first pixel (o_{11}) of the inversely transformed block is modeled as follows:

$$o_{11} = (c_4 * e_{11} + c_1 * e_{12} + c_2 * e_{13} + c_3 * e_{14} + c_4 * e_{15} + c_5 * e_{16} + c_6 * e_{17} + c_7 * e_{18}) \tag{9.11}$$

In the above expression, $e_{11}, e_{12}, \ldots, e_{18}$ can be expressed as :

$$e_{11} = (c_4 * X''_{11} + c_1 * X''_{21} + c_2 * X''_{31} + c_3 * X''_{41} + c_4 * X''_{51} + c_5 * X''_{61} + c_6 * X''_{71} + c_7 * X''_{81}) \tag{9.12}$$

$$e_{12} = (c_4 * X''_{12} + c_1 * X''_{22} + c_2 * X''_{32} + c_3 * X''_{42} + c_4 * X''_{52} + c_5 * X''_{62} + c_6 * X''_{72} + c_7 * X''_{82}) \tag{9.13}$$

Similarly,

$$e_{18} = (c_4 * X''_{18} + c_1 * X''_{28} + c_2 * X''_{38} + c_3 * X''_{48} + c_4 * X''_{58} + c_5 * X''_{68} + c_6 * X''_{78} + c_7 * X''_{88}) \tag{9.14}$$

Similarly, the mathematical expression for rest of the pixels of the inversely transformed block can be modeled, where the pixel values of X''_{11}–X''_{88} and IDCT coefficients will change depending on the pixel coordinate of the inversely transformed block. Since the overall pattern of operations of the expression is same as DCT-based JPEG compression process, we can conclude that the DFG of JPEG DCT and IDCT are almost same except for few things. For example, in case of compression we assumed that each element of the compressed block is multiplied with the corresponding element (q) of the quantization matrix, to perform quantization (through node number 136). However, quantization through node number 136 is not required in DFG of JPEG IDCT. Instead, the last node (node 136) can be used to perform de-levelization by adding 128 to the pixel intensity of the decompressed block. Thus, the total number of nodes in the DFG for both JPEG compression and decompression remains the same. Therefore, THT-based structurally obfuscated DFG can be similarly generated for the JPEG decompression also.

The design process of obfuscated, low-cost JPEG decompression IP core is almost similar to JPEG compression IP core. The obfuscated JPEG decompression IP core is shown in Figure 9.10. As shown, the decompression IP core accepts processed (i.e., after multiplying each element of compressed block with the corresponding element of the quantization matrix) compressed pixel intensity and 2D-DCT coefficient as inputs and produces inversely transformed decompressed pixel intensity as output.

9.4 Implementation of JPEG CODEC IP core

9.4.1 Designing obfuscated JPEG compression IP core

The complete design flow of JPEG compression process using devised low-cost, obfuscated JPEG compression IP core is shown in Figure 9.11. In this complete design flow, partial steps are performed through hardware and the remaining steps are performed through software. More specifically, 2D-DCT transformation and quantization steps are performed through devised JPEG compression IP core and leveling of input image pixel intensity and data encoding is performed through software. Each 8×8 block leveled-off pixel intensity, stored in a hardware queue,

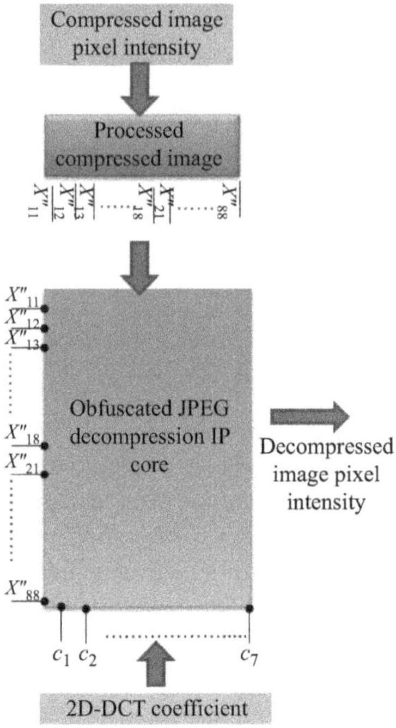

Figure 9.10. Obfuscated low-cost IP core for JPEG decompression

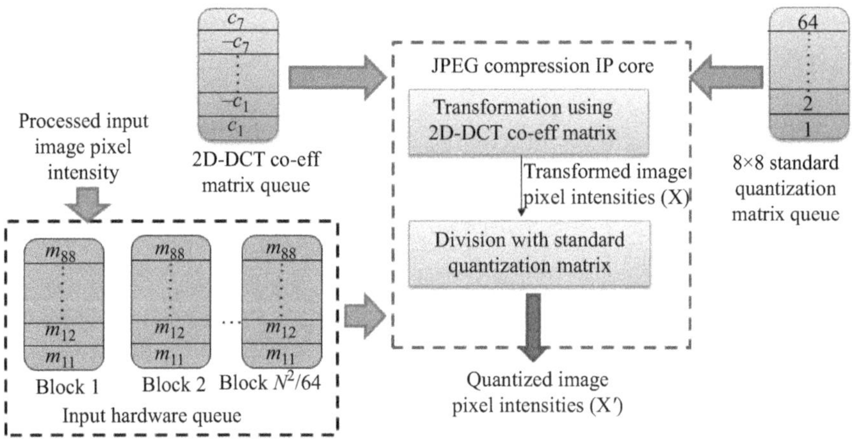

Figure 9.11. Obfuscated low-cost IP core for JPEG decompression

is provided as input to the devised JPEG compression IP core, along with quantization matrix and 2D-DCT coefficient matrix. In this section, we emphasize the implementation of hardware portion of JPEG compression IP core design process. The JPEG compression IP core is designed using HLS framework (Sengupta and

Sedaghat, 2011; Coussy *et al.*, 2009), where the obfuscated DFG of JPEG compression (shown in Figure 9.8) is first scheduled based on a resource configuration. As mentioned earlier, the resource configuration which generates a low-cost design is explored through PSO-DSE process. Here, the DFG is scheduled based on the list-scheduling algorithm with three adders (Adder_A1, Adder_A2, and Adder_A3) and three multipliers (Multiplier_M1, Multiplier_M2, and Multiplier_M3). This indicates, based on the dependency, each CS of the scheduled DFG can have maximum three additions and three multiplication operations out of 136 operations available in the DFG. In each CS, every scheduled operation is assigned to its corresponding hardware and operations assigned to a same hardware are bound (shared) together. Table 9.1 represents all the 136 operations, their CS, and corresponding assigned hardware. Total 30 CS are required to schedule all the operations. Both adders and multipliers are designed in IEEE 754 half precision floating-point format. Therefore, all the operators have bit-width 16 bits. The implemented JPEG compression IP core consists of two units: data path unit and controller unit. The

Table 9.1 Scheduling of operation and their allocated hardware for obfuscated JPEG IP core

C.S.	Opn. allocated to M1	Opn. allocated to M2	Opn. allocated to M3	Opn. allocated to A1	Opn. allocated to A2	Opn. allocated to A3
1	1	2	3	–	–	–
2	4	5	6	9	–	–
3	7	8	17	10	11	–
4	18	19	20	12	13	–
5	21	22	23	25	26	14
6	24	33	34	27	29	15
7	35	36	37	28	41	–
8	38	39	40	42	30	–
9	49	50	51	43	44	45
10	52	53	54	57	46	31
11	55	56	65	58	59	47
12	66	67	68	60	61	–
13	69	70	71	73	74	62
14	72	81	82	75	77	63
15	83	84	85	76	89	–
16	86	87	88	90	78	–
17	97	98	99	91	92	93
18	100	101	102	105	94	79
19	103	104	113	106	107	95
20	114	115	116	108	109	–
21	117	118	119	121	122	110
22	120	16	32	123	125	111
23	48	64	80	124	129	–
24	96	112	–	126	130	–
25	–	–	–	127	131	133

data path unit represents the required number of resources and their interconnections while the controller unit synchronizes those resources by changing their signals (e.g., enables, selectors, strobes). In other words, controller unit is responsible for controlling the data flow through the data path by activating and deactivating different components of data path in different clock cycle, thus synchronizing each component to generate an error-free output. The generic block diagram representation of data path unit of low-cost obfuscated JPEG IP is shown in Figure 9.12. As shown in this figure, the data-path consist of three adders and multipliers, ten multiplexer of type 32:1, two multiplexer of type 16:1, five demultiplexer of type 1:32, one demultiplexer of type 1:16, forty registers of 16-bits and 18 latches. The primary inputs are represented through A1,... A8,... Q1,... Q8 and the dependent inputs are represented through i_1, i_2, ..., i_{135}, while i_{136} represents the output. The implemented schematic representation of the data path unit and the complete IP core are shown in Figures 9.13 and 9.14, respectively, and the complete controller unit signal description is available in the "Appendix." Both data path and controller is implemented using Quartus CAD tool in Intel Cyclone II (Cyclone II FPGA, 2018) FPGA.

9.4.2 Designing obfuscated JPEG decompression IP core

The complete design flow of JPEG decompression process using devised low-cost obfuscated JPEG decompression IP core is shown in Figure 9.15. Similar to the JPEG compression design flow, few steps are performed through software and rest is performed through hardware. More specifically, data decoding and inverse quantization are performed through software and 2D-IDCT transformation and leveling of output image pixel intensity are performed through devised JPEG decompression IP. Finally, each 8×8 block with leveled-off decompressed pixel intensity is stored in an output hardware queue.

Similar to JPEG compression, during DFG scheduling, same resource configuration is used. The same number of components, i.e., three adders and multipliers of 16-bits, ten multiplexers of type 32:1, two multiplexers of type 16:1, five demultiplexers of type 1:32, one demultiplexer of type 1:16, Forty registers of 16-bits and 18 latches are used in the data path unit. Similar to the JPEG compression IP core, both data-path and controller of JPEG decompression IP core are also implemented using Quartus CAD tool in Intel Cyclone II (Cyclone II FPGA, 2018) FPGA.

9.4.3 End-to-end JPEG CODEC through designed hardware/IP core

In this section, we discuss the end-to-end JPEG CODEC process through the devised JPEG IP core. This discussion has three segments. In the first segment, we discuss about each step of image compression starting from an input image to generating a quantized image through an illustrative example. In the second segment, we present about data encoding and decoding process through RLE algorithm using an illustrative example. In the last segment, we discuss about each step of image decompression starting from a de-quantized image to generate the post-compression reconstructed image pixel through another illustrative example.

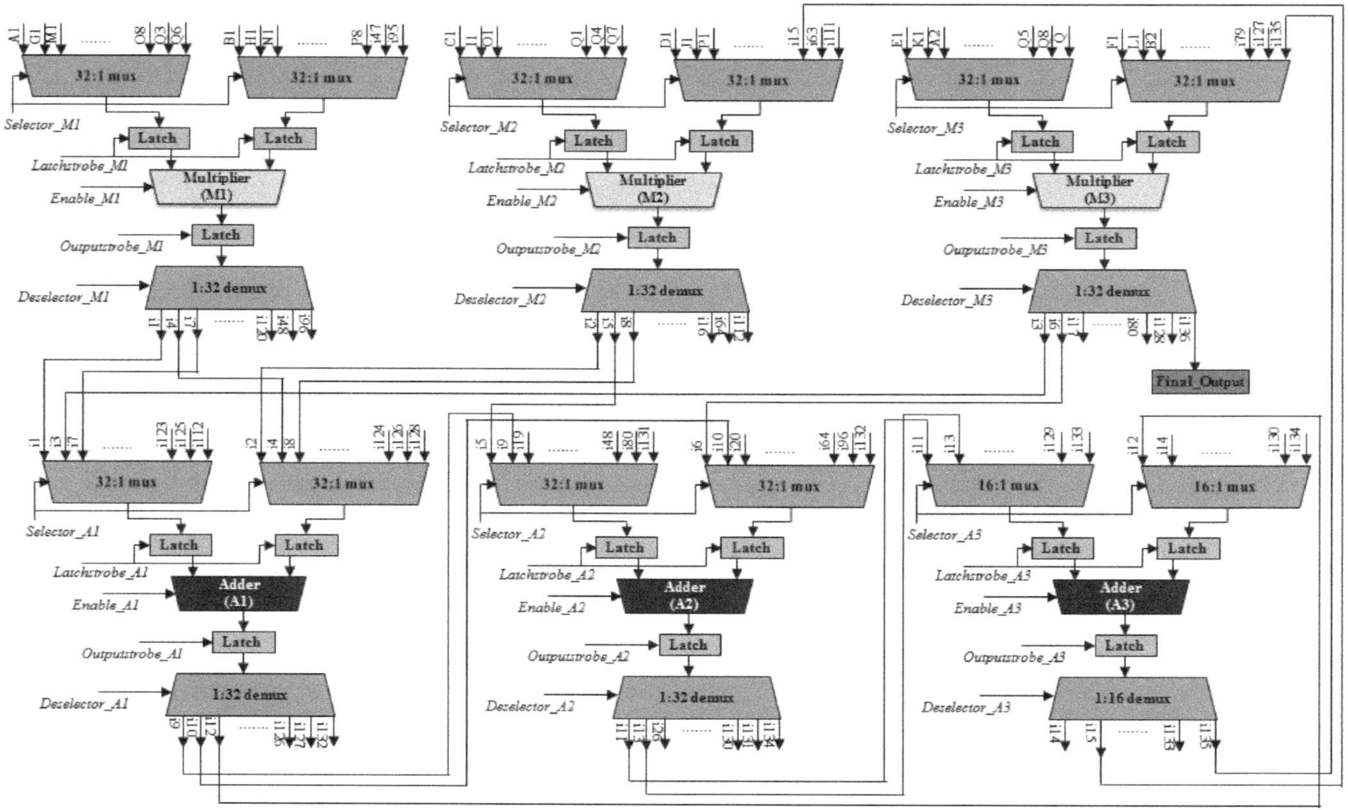

Figure 9.12. Block diagram representation of data-path unit of devised low-cost obfuscated JPEG CODEC IP core

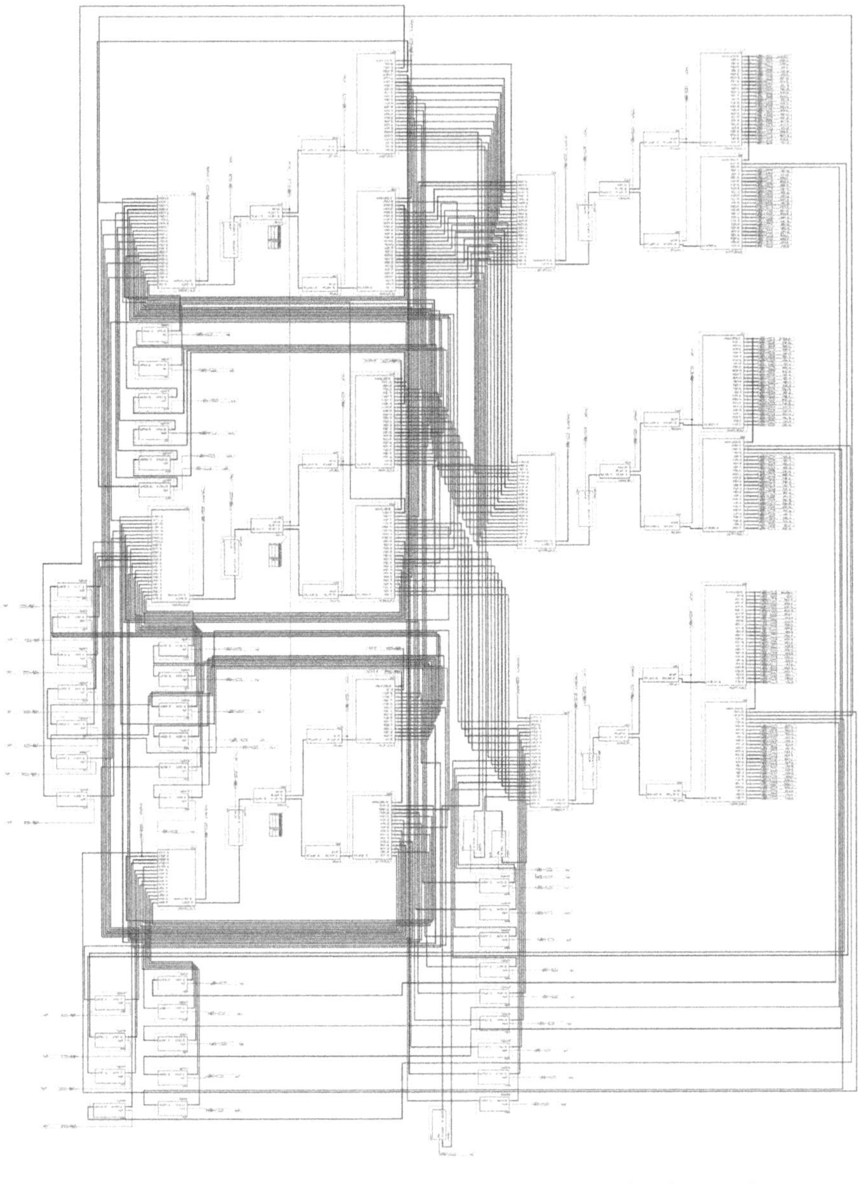

Figure 9.13. Schematic representation of data-path unit of obfuscated JPEG CODEC IP core

A grayscale image of 512 × 512 taken from "dataset of standard 512 × 512 grayscale test images" (Standard dataset, 2018) is selected as an input image. The input image is shown in Figure 9.16(a). To perform DCT on the input image, multiple leveled-off, nonoverlapping 8 × 8 blocks of the image pixel are formed from the input image. Out of multiple nonoverlapping 8 × 8 blocks, a randomly selected

Figure 9.14. Schematic representation of complete obfuscated JPEG CODEC IP core

block for the input image is shown in Figure 9.16(b). This block (containing 64-pixel values) is provided as one of the primary inputs to the devised obfuscated JPEG compression IP core (shown in Figure 9.16(c)). The designing process of obfuscated JPEG compression IP core is discussed in Section 9.4.1. The designed IP core

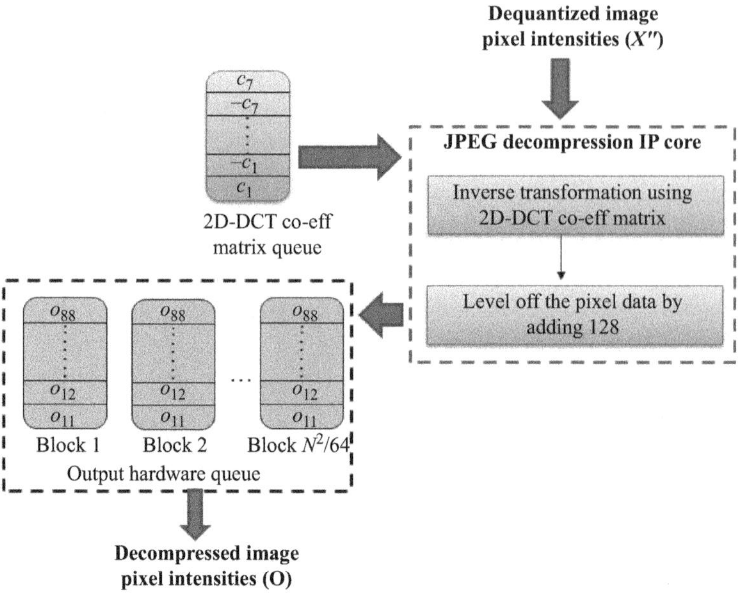

Figure 9.15. Obfuscated low-cost IP core for JPEG decompression

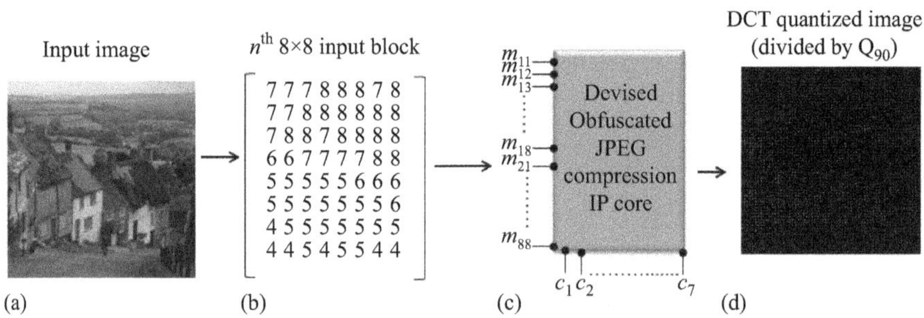

Figure 9.16. Motivational example of JPEG compression through compression IP core

performs DCT and quantization on the input block and produces transformed and quantized image as output. In this example, we have performed compression based on quality level 90, i.e., using quantization matrix $\mathbf{Q_{90}}$. This process is repeated until all the 8×8 blocks of the input image is transformed and quantized. The generated transformed and quantized image is shown in Figure 9.16(d).

Data encoding, which is a form of lossless data compression, is performed on each quantized block, before storing the pixel data into a storage device. It involves arranging the data of the quantized block in "zigzag" fashion then employing RLE algorithm on it. Figure 9.17(a) shows the quantized block corresponding to input image block (shown in Figure 9.16(b)). Zigzag scanning (the format of zigzag

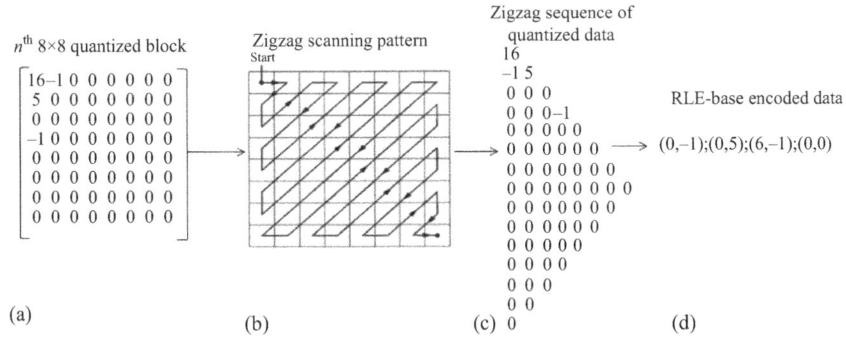

Figure 9.17. Demonstrative example of data encoding through RLE algorithm

scanning is shown in Figure 9.17(b)) is performed on the quantized block and the output is shown in Figure 9.17(c). The quantized data sequenced in zigzag fashion is encoded through RLE algorithm. RLE replaces **each nonzero AC coefficient** value with a pair (RUNLENGTH, AMPLITUDE), where "RUNLENGTH" is the **number of zeroes** that came before that nonzero AC coefficient and "AMPLITUDE" is the value of that particular nonzero coefficient. The encoded version of the quantized data using RLE is shown in Figure 9.17(d), where (0, 0) always indicates the end of the block. The remaining elements (from the 64 available slots) of the 1D array in RLE encoded data are not stored, as they are all zeroes. (Note: The first value of the matrix, i.e., 16 is the DC coefficient; it is not encoded through RLE.) Each of these encoded values is converted to its equivalent binary form to generate the corresponding bitstream of the block and then stored in a storage device. To regenerate the quantized block from the bitstream data, decoding needs to be performed. To achieve the decoding of data, all the aforementioned steps have to be performed in reverse order. Thus while obtaining (decoding) 1D array from RLE encoded data, the vacant slots from the 64 elements (of 8×8 block) are filled up with zeroes. The inverse zigzagging is performed to obtain 2D array of 8×8 block of quantized block.

To reconstruct the image from its transformed and quantized form, each block is de-quantized using the same quantization matrix used for compression. As in this example, we have performed compression using quantization matrix $\mathbf{Q_{90}}$, the de-quantization (inverse quantization) is performed based on quantization matrix $\mathbf{Q_{90}}$ only. The generated de-quantized image is shown in Figure 9.18(a). The 8×8 de-quantized block (containing 64-pixel values) is provided as one of the primary input to the devised obfuscated JPEG decompression IP core (shown in Figure 9.18(b)). The designing process of obfuscated JPEG decompression IP core is discussed in Section 9.4.2. The designed IP core performs IDCT and leveling-off of the de-quantized block and produces 8×8 decompressed image pixel intensity values as output. The corresponding decompressed image pixel intensity of the same 8×8 block (shown in Figure 9.16(b)) is shown in Figure 9.18(c). Again, this process is repeated until all the 8×8 de-quantized blocks are inversely transformed and leveled off. Finally, the post-compression reconstructed image is generated by combining all

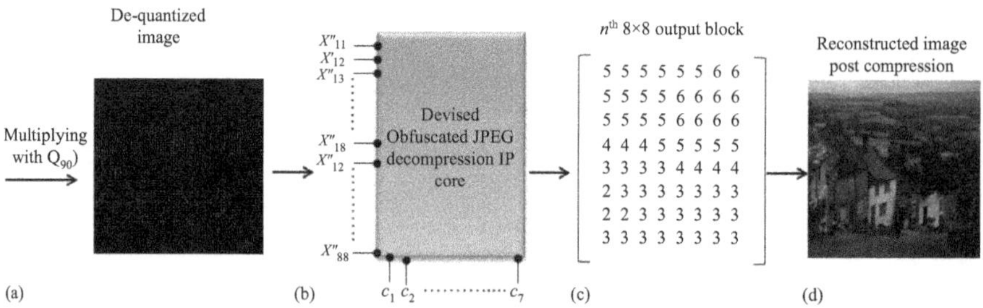

Figure 9.18. Informative example of JPEG decompression through devised JPEG decompression IP core

the 8 × 8 inversely transformed and leveled-off blocks. The reconstructed image of the corresponding input image (shown in Figure 9.16(a)) is shown in Figure 9.18(d).

9.5 Analysis on case studies

To verify the devised IP cores for JPEG compression and decompression, images (bit depth equal to 4 bits) provided as input are taken from two different image datasets. They are "dataset of standard 512 × 512 grayscale test images" (Standard dataset, 2018) and "NASA images" (NASA images, 2018). Both the devised IP cores are designed, implemented, and simulated in Intel Quartus (Intel Quartus, 2018) v7.2 software with respect to Cyclone II family (Cyclone II FPGA, 2018), device no. EP2C35F672C6 at 50-MHz clock frequency. However, it can be emulated in any FPGA device. Table 9.2 reports the device utilization summary of devised JPEG compression IP core for the aforementioned target device. The devised JPEG CODEC IP core which is structurally obfuscated and low-cost in nature is compared with another JPEG CODEC IP core which is structurally obfuscated but not optimized in nature indicating the design configuration of the obfuscated IP core is not explored through any optimization framework/engine. Therefore, deficiency of any low-cost DSE framework results in higher design cost overhead. Table 9.3 reports the comparative analysis between these two aforementioned IP cores in terms of design area, latency, and cost. A 15 nm technology scale based on NanGate library (Open Cell Library, 2018) is used to evaluate the design area and the design latency of both IP cores, where area is in μm^2 and latency is calculated in picoseconds. Based on the weighted normalized design area and design latency of the obfuscated JPEG IP core, design cost is calculated using (9.6).

Therefore, design cost of the obfuscated JPEG IP core is unitless. According to the results shown in Table 9.3, the devised JPEG CODEC IP core design achieves more than 5% lower design cost than the non-optimized obfuscated JPEG CODEC IP core design. The reason for achieving lower design cost is due to the involvement of PSO-DSE engine, which explores the low-cost design configuration. Therefore, it is referred as "low-cost" JPEG CODEC IP core.

Table 9.2 Device utilization summary of devised JPEG CODEC IP cores w.r.t. Intel Cyclone II FPGA

Device utilization summary	**Total used**	**Total available**	**Used %**
Logic elements	12,148	33,216	37
Four input function	5,018	11,072	45
Three input function	6,608	11,072	60
≤Two input function	494	11,072	4
Total register	1,826	34,593	5
Total pins	322	475	68

Table 9.3 Comparison of low-cost obfuscated IP core with nonoptimized obfuscated JPEG CODEC IP core in terms of area, latency, and cost

Design metrics	**Non-optimized obfuscated JPEG IP core**	**Devised low-cost obfuscated IP core**
Resource configuration	4A, 4M	3A, 3M
Design area (μm^2)	397.94	298.45
Design latency (ps)	5,860.40	7,603.22
Design cost	0.3884	0.3671

Table 9.4 Comparison of low-cost obfuscated IP core with non-obfuscated JPEG CODEC IP core in terms of area, latency, cost, SoO

Design metrics	**Non-obfuscated JPEG IP core**	**Devised low-cost obfuscated IP core**	**Structure changed due to obfuscation**	**Improved SoO of devised design (%)**
Resource configuration	4A, 8M, 12 (8:1) Mux, 12 (16:1) Mux, 6 (1:8) Demux, 6 (1:16) Demux, 25 Reg	3A, 3M, 10 (32:1) Mux, 2 (16:1) Mux, 5(1:32) Demux, 1(1:16) Demux, 13 Reg	**10064 Gates**	–
Design area (μm^2)	483.66	298.45	–	–
Design latency (ps)	6,533.08	7,603.22	–	–
Design cost	0.4582	0.3671	–	–
SoO	0	0.7574	–	**76%**

The devised JPEG CODEC IP core which is structurally obfuscated and low-cost in nature is also compared with another JPEG CODEC IP core which is not structurally obfuscated indicating that no structural transformation technique is performed in the DFG of JPEG CODEC to hide its structural identity. Therefore, deficiency of any structural transformation technique results in nonsecured design which is vulnerable to RE attack. Table 9.4 reports the comparative analysis between these two aforementioned IP cores in terms of design area, latency, cost, and SoO. The SoO of JPEG CODEC IP core is measured through a metric, which is partially adopted from Sengupta *et al.* (2018b), expressed as follows:

$$\text{SoO} = \frac{a_{\text{JPEG}}}{a_{\text{JPEG}}^{\text{Total}}} \tag{9.15}$$

where a_{JPEG} indicates that the number of nodes in the DFG of the JPEG CODEC is altered due to structural transformation; ${a_{\text{JPEG}}}^{\text{Total}}$ indicates the total available nodes in the DFG of the JPEG CODEC before applying THT technique. The robustness of obfuscated JPEG CODEC IP core is measured through SoO metric, a higher SoO value indicates stronger robustness in the range of [0, 1]. In other words, SoO metric indicates how much the structure of an obfuscated IP core differs from its original structure. The more a design differs from its original structure the harder it will be for an attacker to identify the application through RE. Therefore, insertion of any malicious logic in the IP design or performing IP piracy is extremely difficult without knowing the actual functionality of the application. According to the results shown in Table 9.4, the devised obfuscated JPEG CODEC IP core design achieves more than 76% SoO than a non-obfuscated JPEG CODEC IP core design. The structural obfuscation is obtained using THT which is responsible for obfuscating the original design of JPEG CODEC by transforming the height of the data flow graph representing the JPEG CODEC, thus modifying the nodes in the DFG of the JPEG CODEC. Modification of a node in a DFG indicates whether, (1) the source/parents of a node is changed; (2) the child of a node is changed; or (3) operation type of a node is changed, compared to the original. The effect of modification of nodes is also reflected in the gate structure/netlist of the design. According to the results shown in Table 9.4, more than 10,000 gates are affected due to the modification of nodes in the DFG of the JPEG CODEC. This is achieved due to the transformation in gate-level architecture of functional resources, multiplexers, demultiplexers, and registers. This enormous transformation in gate-level architecture of the complete JPEG CODEC IP core makes it an unobvious structure. Therefore, identifying the correct functionality of this

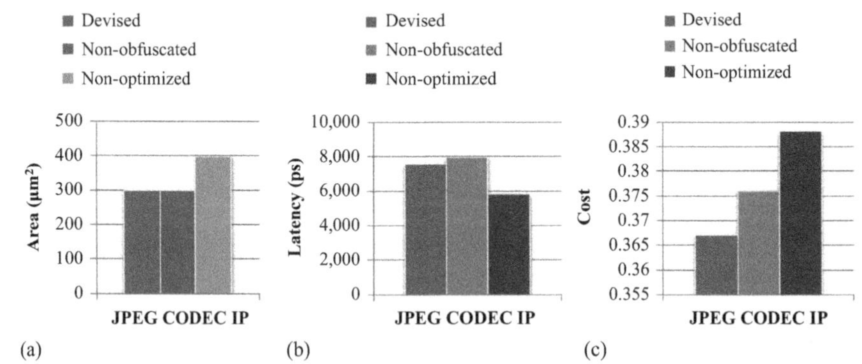

Figure 9.19. Comparison of devised low-cost obfuscated IP core with non-obfuscated and nonoptimized JPEG CODEC IP core in terms of (a) area; (b) latency; and (c) cost

unobvious structure is extremely complex and tedious for an attacker. Figure 9.19(a)–(c) shows the comparison between devised, non-obfuscated, and nonoptimized JPEG CODEC IP core in terms of area, latency, and cost, respectively.

To compress standard input images using devised JPEG CODEC IP core, totally six images are chosen from "dataset of standard 512 × 512 grayscale test images" (Standard dataset, 2018) and "NASA images" (NASA images, 2018) datasets. The compression is performed for two different quality levels, i.e., quality level 90 ($\mathbf{Q_{90}}$) and 70 ($\mathbf{Q_{70}}$). Additionally, the mean square error (MSE) and the peak signal-to-noise ratio (PSNR) of the corresponding compressed images obtained through devised JPEG CODEC IP core design are also calculated. As PSNR is commonly used to calculate the quality of the reconstructed image of lossy compression, we have used PSNR to evaluate image degradation of compressed images obtained through devised JPEG CODEC IP core. PSNR is defined for two images A and B of size $M \times N$ as (Kougianos *et al.*, 2016; Obukhov and Kharlamov, 2008):

$$\mathrm{PSNR}(A, B) = 20\log_{10}\frac{MAX_A}{\sqrt{MSE(A, B)}} \tag{9.16}$$

where A is the original image, B is the reconstructed image, MAX_A is the maximum pixel value in image A, and *MSE* (A, B) is a mean square error between image A and B. *MSE*(A, B) can be defined as (Kougianos *et al.*, 2016; Obukhov and Kharlamov, 2008):

$$MSE(A, B) = \frac{1}{M * N}\sum_{i=0}^{M-1}\sum_{j=0}^{N-1}\|A(i,j) - B(i,j)\|^2 \tag{9.17}$$

Tables 9.5 and Table 9.6 report the storage size, reduction percentage, MSE, and PSNR of the compressed image for quality level 90 and 70, respectively. It can be observed that a higher compression efficiency is achieved for $\mathbf{Q_{70}}$ than $\mathbf{Q_{90}}$ while compromising the image quality (indicated by PSNR).

Table 9.5 Compressed size, reduction percentage, MSE, and PSNR of compressed image for $\boldsymbol{Q_{90}}$ *(Images 1 to 6 have been extracted from Standard dataset (2018) and NASA images (2018)*

Images	Original size (bits)	Compressed size (bits)	Compression efficiency (%)	MSE	PSNR
Image 1	1,048,576	236,568	77.44	3.84	17.68
Image 2	1,048,576	200,000	80.93	3.50	18.08
Image 3	1,048,576	170,640	83.73	2.00	20.51
Image 4	1,048,576	194,144	81.49	2.63	19.32
Image 5	1,048,576	210,760	79.90	3.25	18.40
Image 6	1,048,576	216,136	79.39	2.20	20.09

Table 9.6 Compressed size, reduction percentage, MSE, and PSNR of compressed image for Q_{70} (Images 1 to 6 have been extracted from Standard dataset (2018) and NASA images (2018))

Images	Original size (bits)	Compressed size (bits)	Compression efficiency (%)	MSE	PSNR
Image 1	1,048,576	106,768	89.82	4.85	16.66
Image 2	1,048,576	99,120	90.55	4.23	17.26
Image 3	1,048,576	90,584	91.36	2.91	18.88
Image 4	1,048,576	91,952	91.23	3.50	18.08
Image 5	1,048,576	102,112	90.26	3.83	17.69
Image 6	1,048,576	105,168	89.97	3.11	18.60

9.6 Conclusion

This chapter discussed a case study of structural obfuscation using JPEG CODEC IP core which is available in almost every CE devices. The design mechanism and implementation process of low-cost obfuscated on-chip IP core design for 2D-DCT-based JPEG image compression and decompression are explained here. This structural obfuscation hides an application and its implementation from an adversary by transforming its standard architecture to a non-obvious one. In this process, THT is used to structurally obfuscate the design and low-cost hardware configuration is explored through PSO-DSE engine. The devised IP core is synthesized and simulated through Intel Quartus Cyclone II FPGA. The designed IP core can be used as a dedicated IP core in any camera-based CE devices to perform image CODEC. Further, the robustness of the structurally obfuscated design is measured through SoO metric, which indicates that more than 76% SoO is achieved compared to a standard non-obfuscated JPEG CODEC design. Additionally, as structural obfuscation does not incur any design overhead, therefore, it is an extremely useful method of protection for Trojan insertion, IP piracy/counterfeit. The SoO of the devised JPEG CODEC IP core discussed in this chapter can be enhanced by performing multiple transformation techniques on the DFG of the JPEG CODEC IP core as a part of future work. Additionally, combination of functional obfuscation technique with structural obfuscation can also enhance the robustness for DSP applications. This would enable to hide an application and its implementation from an adversary.

9.7 Exercises

1. What is image pixel intensity levelization?
2. What is the function of DCT in JPEG compression?
3. How is DCT performed of JPEG?
4. When is quantization performed and how?

5. What is standard quantization matrix?
6. Where does the loss of data happen in lossy JPEG compression?
7. What is the utility of rounding function in JPEG?
8. How to construct quantization matrix Q90, Q70, and Q10 from the standard quantization matrix?
9. Explain the decompression steps.
10. Explain the role of IDCT in JPEG decompression.
11. Explain the role of delevelization.
12. How is obfuscated JPEG codec generated?
13. Why is JPEG CODEC IP obfuscated?
14. How obfuscation protects against RE attacks?
15. What is the logic inside micro-IP of JPEG CODEC?
16. What are the generic steps of obfuscation of JPEG CODEC IP core?
17. What is the benefit of PSO in obfuscation of JPEG?
18. What are MSE and PSNR? How are they evaluated?
19. Explain the role of DCT matrix in compression.
20. Derive DCT coefficient matrix and calculate its values.
21. Derive the mathematical function of hardware design of JPEG CODEC.
22. Explain the end-to-end design flow of JPEG.

References

Ahmed N., Natarajan T., and Rao K. R., 'Discrete cosine transform,'*IEEE Transactions on Computers*,1974, vol. C-23 (1), pp. 90–93.

Alkabani Y. M. and Koushanfar F., 'Active hardware metering for intellectual property protection and security,' *Proceedings of USENIX Security Symposium*, Berkeley, CA, 2007, pp. 20:1–20:16.

Andorko I., Corcoran P., and Bigioi P., 'A dual image processing pipeline camera with CE applications,' in *Proceedings of IEEE International Conference Consumer Electronics (ICCE)*, 2011, Las Vegas, NV, USA, pp. 737–738.

Bilgin A. and Marcellin M.W., 'Applications of reversible integer wavelet transforms to lossless compression of medical image volumes,' in *Proceedings of IEEE International Symposium on Information Theory*, 1998, Cambridge, MA, USA, pp. 411.

Corcoran P. M. and Andrae A., 'On thin-clients and the cloud: Can smartphones and tablets really reduce electricity consumption?' *Proceedings of IEEE International Conference on Consumer Electronics*, 2014, pp. 81–84.

Corcoran P. M., Bigioi P., and Steinberg E., 'Wireless transfer of images from a digital camera to the Internet via a standard GSM mobile phone,' *IEEE Transactions on Consumer Electronics*, 2001, vol. 47 (3), pp. 542–547.

Coussy P., Gajski D. D., Meredith M., and Takach A., 'An introduction to high-level synthesis,'*IEEE Design & Test of Computers*, 2009, vol. 26 (4), pp. 8–17.

Cyclone II FPGA, Available: https://www.altera.com/products/fpga/cyclone-series/cyclone-ii/overview.html, last accessed on 2018.

'Dataset of standard 512 × 512 grayscale test images', Available: http://decsai.ugr.es/cvg/CG/base.htm, last accessed on 2018.

'DCT matrix', Available: http://www.engr.colostate.edu/ECE513/SP09/lectures/lectures11_12.pdf, last accessed on 2018.

Hnesh A. M. G. and Demirel H., 'DWT-DCT-SVD based hybrid lossy image compression technique,' *Proceedings of International Image Processing*, Appl. Syst. (IPAS), Hammamet, Tunisia, 2016, Nov. 2016, pp. 1–5.

Intel Quartus, Available: https://dl.altera.com/13.0sp1, last accessed on 2018.

Kim J., Jung E. S., Lee Y. T., and Ryu W., 'Home appliance control framework based on smart TV set-top box,' *IEEE Transactions on Consumer Electronics*, 2015, vol. 61 (3), pp. 279–285.

Kougianos E., Mohanty S. P., Coelho G., Albalawi U., and Sundaravadivel P., "Design of a high-performance system for secure image communication in the internet of things (Invited Paper)", *IEEE Access Journal*, 2016, vol. 4, pp. 1222–1242.

Lao Y. and Parhi K. K., 'Obfuscating DSP circuits via high-level transformations,'*IEEE Transactions on Very Large Scale Integration Systems*, 2015, vol. 23 (5), pp. 819–830.

Li S., Yin H., Fang X., and Lu H., 'Lossless image compression algorithm and hardware architecture for bandwidth reduction of external memory,' *IET Image Processing*, 2017, vol. 11 (6), pp. 379–388.

Mohanty S. P., 'Watermarking of digital images,' Masters Thesis, Department of Electrical Engineering, Indian Institute of Science, Bangalore, India, 1999.

Mohanty S. P., Ranganathan N., and Namballa R. K., 'VLSI implementation of invisible digital watermarking algorithms towards the development of a secure JPEG encoder,' *Proceedings of the IEEE Workshop on Signal Processing System (SIPS)*, 2003, pp. 183–188.

'NanGate 15 nm open cell library'. Available: http://www.nangate.com/?pageid=2328, last accessed on 2018.

'Nasa image and video library', Available: https://images.nasa.gov/n#/, last accessed on 2018.

Obukhov A. and Kharlamov A., 'Discrete cosine transform for 8 × 8 blocks with CUDA,' 2008, Nvidia whitepaper document.

Roy D. and Sengupta A., 'Low overhead symmetrical protection of reusable IP core using robust fingerprinting and watermarking during high level synthesis,' *Future Generation Computer Systems*, 2017, vol. 71, pp. 89–101.

Roy D. and Sengupta A., 'Obfuscated JPEG image decompression IP core for protecting against reverse engineering,' *IEEE Consumer Electronics Magazine*, 2018, vol. 7 (3), pp. 104–109.

Scarmana G. and McDougall K., 'Exploring the application of some common raster scanning paths on lossless compression of elevation images,' *Proceedings of IEEE International Geoscience and Remote Sensing Symposium (IGARSS)*, 2015, Milan, Italy, pp. 4514–4517.

Sengupta A., 'Hardware security of CE devices,' *IEEE Consumer Electronics Magazine*, 2017, vol. 6 (1), pp. 130–133.

Sengupta A. and Roy D., 'Antipiracy-aware IP chipset design for CE devices: A robust watermarking approach', *IEEE Consumer Electronics Magazine*, 2017a, vol. 6 (2), pp. 118–124.

Sengupta A. and Roy D., 'Protecting an intellectual property core during architectural synthesis using high-level transformation based obfuscation,' *IET Electronics Letters*, 2017b, vol. 53 (13), pp. 849–851.

Sengupta A. and Sedaghat R., 'Integrated scheduling, allocation and binding in high level synthesis using multi structure genetic algorithm based design space exploration,' *International Symposium on Quality Electronic Design*, 2011, pp. 1–9.

Sengupta A., Bhadauria S., and Mohanty S. P., 'TL-HLS: Methodology for low cost hardware Trojan security aware scheduling with optimal loop unrolling factor during high level synthesis,' *IEEE Transactions on CAD of Integrated Circuits and Systems*, 2017a, vol. 36 (4), pp. 655–668.

Sengupta A., Roy D., Mohanty S. P., and Corcoran P., 'DSP design protection in CE through algorithmic transformation based structural obfuscation,' *IEEE Transactions on Consumer Electronics*, 2017b, vol. 63 (4), pp. 467–476.

Sengupta A., Roy D., and Mohanty S. P., 'Triple-phase watermarking for reusable IP core protection during architecture synthesis,' *IEEE Transactions on CAD of Integrated Circuits and Systems*, 2018a, vol. 37 (4), pp. 742–755.

Sengupta A., Roy D., Mohanty S. P., and Corcoran P., 'Low-cost obfuscated JPEG CODEC IP core for secure CE hardware,' *IEEE Transactions on Consumer Electronics*, 2018b, vol. 64 (3), pp. 365–374.

Tang Q., Groba M., Juarez E., Sanz C., and Pescador F., 'Real-time power-consumption control system for multimedia mobile devices,' *IEEE Transactions on Consumer Electronics*, 2016, vol. 62 (4), pp. 362– 370.

Thavalengal S. and Corcoran P., 'User authentication on smartphones: Focusing on iris biometrics,' *IEEE Consumer Electronics Magazine*, 2016, vol. 5 (2), pp. 87–93.

Vijayakumar A., Patil V. C., Holcomb D. E., Paar C., and Kundu S., 'Physical design obfuscation of hardware: A comprehensive investigation of device and logic-level techniques,' *IEEE Transactions on Information Forensics and Security*, 2017, vol. 12 (1), pp. 64–77.

Wong J. L., Kirovski D., and Potkonjak M., 'Computational forensic techniques for intellectual property protection,' *IEEE Transactions on CAD of Integrated Circuits and Systems*, 2004, vol. 23 (6), pp. 987–994.

Chapter 10

Advanced encryption standard (AES) and its hardware watermarking for ownership protection

In this era of consumer electronics, cyber security is one of the key challenges. Any security, privacy, or protection methods that are deployed rely on cryptography. Advance encryption standard (AES) is one of the heavily used cryptography algorithms for its advantages. This chapter is dedicated to the process of AES and its hardware design in the form of IP core. Several hardware security techniques rely on AES IP core as an important block. Additionally, since this is such an important core, its self-protection against forgery/piracy is also crucial. Thus, this chapter will also discuss about AES IP core protection using watermarking.

10.1 Introduction

Cryptography is very widely used in everyday life, ranging from established applications such as wireless local area network, procuring items with a credit or debit card, installing a software update, smart cards, banking application, voice of internet protocol (VOIP) to emergent domains, such as electronic health system, internet of things security, smart city security, and hardware security (Paar and Pelzl, 2009; Bhunia and Tehranipoor, 2018; Sengupta, 2013, 2015, 2017; Sengupta *et al.*, 2017; Sengupta and Bhadauria, 2016; Sengupta and Sedaghat, 2011; Mishra *et al.*, 2017; Sengupta and Sedaghat, 2013; Sengupta and Roy, 2017a, 2017b). Cryptography comprises cryptographic algorithms that are symmetrical by nature. There is tremendous usage of symmetric ciphers, especially for the encryption of data and integrity check of messages. Furthermore, cryptographic protocol deals with the application of cryptographic algorithms. An example of cryptographic protocol is transport layer security (TLS) scheme, widely employed in browser advanced encryption standard (AES) is a very popular symmetric cipher used today. The AES, also known by its original name Rijndael, is a specification for the encryption of electronic data recognized by the U.S. National Institute of Standards and Technology (NIST). In the context of AES (Rijndael family), three members were designated by NIST, each with a block size of 128 bits, but dissimilar key lengths: 128, 192, and 256 bits. It is a mandatory standard in several industry/commercial systems and US government applications such as Wi-Fi encryption standard IEEE 802.11i, TLS, and VOIP (Paar and Pelzl, 2009).

In this chapter, we focus on the following aspects: (a) AES algorithm—description of each step and its flow diagram, (b) corresponding hardware mapping of AES block cipher—overview of how each step can be mapped to a specific hardware sub-block, (c) key scheduler—description of each step, (d) corresponding hardware mapping of key scheduler used in AES, (e) watermarking encoding of AES hardware—protecting an AES hardware against threats such as false claim of ownership and piracy, (f) design of watermarked AES hardware—description of the watermark embedding process, and (g) impact of several different watermark signature types on AES hardware and its implementation on FPGA—to highlight the overhead due to watermark.

10.2 AES algorithm

10.2.1 Overview of AES

AES algorithm has three different variations of key sizes, i.e., 128 bits or 192 bits or 256 bits. The input data size is 128 bits and the output encrypted data is also 128 bits. The input data is called the plain text and the output encrypted data is called cypher text. Figure 10.1 shows the AES input/output parameters. The core structure of AES comprises the following components: (i) byte substitution layer, (ii) diffusion layer, (iii) key addition layer, and (iv) key schedule (Kosaraju *et al.*, 2006; Mohanty *et al.*, 2007). Three key lengths are supported in AES. The number of internal rounds of the cipher depends on the key length. For example, in case of 128 bit key length, the number of internal rounds (n_r) is 10, while for 192 bit key length, the number of internal rounds is 12, and finally for 256 bit key length, the value of n_r is 14, respectively.

10.2.2 AES algorithm—description and custom hardware design

The first major phase of AES algorithm is 'byte substitution layer (S-box)' which is responsible for non-linear transformation using lookup tables for each element of the state. The main significance of this step is to add '*confusion*' to the data elements. In other words, the transformation achieved in individual state bits is rapidly propagated into the datapath. The second major phase is called '*Diffusion*' which consists of two sub-layers, viz. '*ShiftRows*' that permutes the data at the byte level

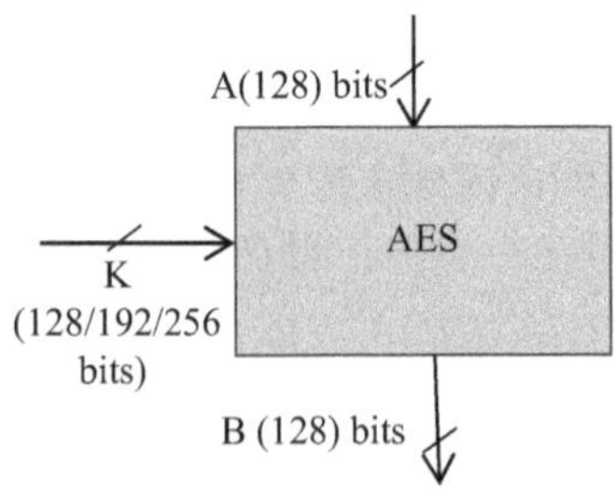

Figure 10.1 AES block cipher

and '*MixColumn*' is a matrix based operation that is responsible for mixing of blocks of four bytes of data. The significance of '*Diffusion*' is to introduce diffusion property to the data element. The third major phase is '*key addition*' layer which is responsible for XOR-ing the 128 bit key (or its derived key) to the state of each element obtained after performing '*MixColumn*' operation. Finally, the '*key schedule'* calculates round keys or subkeys ($k_0, k_1, \ldots, k_{nr}$) from the original AES key. Figure 10.2 shows the major steps of the AES algorithm in terms of design flow highlighting each block responsible for performing specific operation. As an example, we use key length of 128 bits for demonstration (Paar and Pelzl, 2009).

10.2.2.1 Byte substitution (S-box) layer

S-box is the non-linear component of AES which is accountable for bit manipulation. It consists of 16 S-boxes all of which are identical to each other. All 256 possible input elements have a unique mapping to one output element. S-box is shown in Figure 10.3 and is adopted from Paar and Pelzl (2009). For example, if input byte to the S-box is $(B2)_{hex}$, then based on the mapping available in S-box, the output byte is $(37)_{hex}$. In other words, $S((B2)_{hex}) \rightarrow (37)_{hex} \rightarrow S(10110010) \rightarrow 00110111$. This clearly shows how bit manipulation is achieved using byte substitution layer of AES. As shown in Figure 10.2, bytes $X_0, X_1, X_2, \ldots, X_{15}$ are substituted with $Y_0, Y_1, Y_3, \ldots, Y_{15}$ using S-box. Every four bytes of data X_0, X_1,

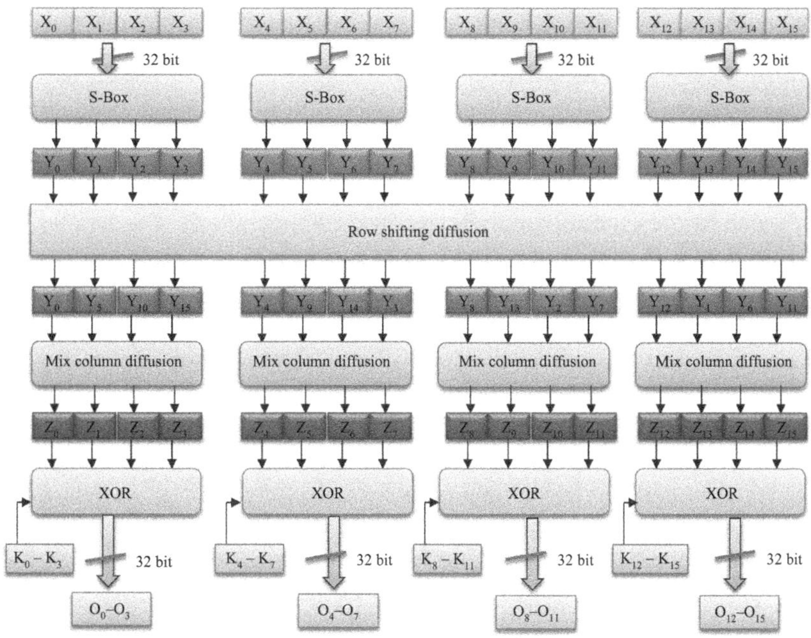

Figure 10.2 Round 'i' (1–10) of AES algorithm for 128 bit key and 128 bit I/O

								Y								
X	0	1	2	3	4	5	6	7	8	9	A	B	C	D	E	F
0	63	7C	77	7B	F2	6B	6F	C5	30	01	67	2B	FE	D7	AB	76
1	CA	82	C9	7D	FA	59	47	F0	AD	D4	A2	AF	9C	A4	72	C0
2	B7	FD	93	26	36	3F	F7	CC	34	A5	E5	F1	71	D8	31	15
3	04	C7	23	C3	18	96	05	9A	07	12	80	E2	EB	27	B2	75
4	09	83	2C	1A	1B	9E	5A	A0	52	3B	D6	B3	29	E3	2F	84
5	53	D1	00	ED	20	FC	B1	5B	6A	CB	BE	39	4A	4C	58	CF
6	D0	EF	AA	FB	43	4D	33	85	45	F9	02	7F	50	3C	9F	A8
7	51	A3	40	8F	92	9D	38	F5	BC	B6	DA	21	10	FF	F3	D2
8	CD	0C	13	EC	5F	97	44	17	C4	A7	7E	3D	64	5D	19	73
9	60	81	4F	DC	22	2A	90	88	46	EE	B8	14	DE	5E	0B	DB
A	E0	32	3A	0A	49	06	24	5C	C2	D3	AC	62	91	95	E4	79
B	E7	C8	37	6D	8D	D5	4E	A9	6C	56	F4	EA	65	7A	AE	08
C	BA	78	25	2E	1C	A6	B4	C6	E8	DD	74	1F	4B	BD	8B	8A
D	70	3E	B5	66	48	03	F6	0E	61	35	57	B9	86	C1	1D	9E
E	E1	F8	98	11	69	D9	8E	94	9B	1E	87	E9	CE	55	28	DF
F	8C	A1	89	0D	BF	E6	42	68	41	99	2D	0F	B0	54	BB	16

Figure 10.3 S-box (Paar and Pelzl, 2009)

X_2, X_3 gets transformed into Y_0, Y_1, Y_2, Y_3. Similarly, for next four bytes X_4, X_5, X_6, X_7 gets transformed into Y_4, Y_5, Y_6, Y_7. Further similarly, next four bytes X_8, X_9, X_{10}, X_{11} gets transformed into Y_8, Y_9, Y_{10}, Y_{11}. Finally, last four bytes X_{12}, X_{13}, X_{14}, X_{15} gets transformed into Y_{12}, Y_{13}, Y_{14}, Y_{15}.

10.2.2.2 Row-shifting diffusion layer

The significance of the *ShiftRows* transformation is to upsurge the diffusion process of AES. This transformation works as follows:

1. First row is retained unaffected.
2. Second row is cyclically moved three bytes to the right.
3. Third row is cyclically moved two bytes to the right.
4. Fourth row is cyclically moved one byte to the right.

The output of *byte substitution* layer is fed as an input to the *ShiftRows* diffusion layer as shown below. Let us assume the output of *byte substitution* layer is Y_0, Y_1, Y_2, $Y_3, \ldots, Y_{15}$. It is arranged in matrix form as shown below in Table 10.1. The transformation applied on each row is indicated through notes on the top of the arrow. The outputs for every four input byte are indicated in each column. For example, corresponding to the input four bytes Y_0, Y_1, Y_2, Y_3, the transformed output four bytes are Y_0, Y_5, Y_{10}, Y_{15} as shown in Table 10.2. Similarly, the outputs are obtained for the remaining four bytes in each column.

10.2.2.3 MixColumn diffusion layer

This step is a major diffusion process in AES since each every input byte impacts four output bytes. It is responsible for mixing each column of the state matrix.

Table 10.1 State matrix before ShiftRow transformation

Y_0	Y_4	Y_8	Y_{12}
Y_1	Y_5	Y_9	Y_{13}
Y_2	Y_6	Y_{10}	Y_{14}
Y_3	Y_7	Y_{11}	Y_{15}

No change →
Left shift one pos. →
Left shift two pos. →
Left shift three pos. →

Table 10.2 State matrix after ShiftRow transformation

Y_0	Y_4	Y_8	Y_{12}
Y_5	Y_9	Y_{13}	Y_1
Y_{10}	Y_{14}	Y_2	Y_6
Y_{15}	Y_3	Y_7	Y_{11}

The input to the *MixColumn* diffusion layer is the output of *ShiftRows* diffusion layer. In order to perform *MixColumn* diffusion each 4-byte column is assumed as a vector and multiplied by a fixed 4×4 matrix. The matrix comprises of some constant values. The main important thing to note is that the matrix addition and multiplication is performed in Galois Field (GF) arithmetic (not integer arithmetic). The first four output bytes of *ShiftRows* diffusion acting as input to *MixColumn* diffusion are Y_0, Y_5, Y_{10}, Y_{15} which is assumed as a vector and multiplied by a fixed 4×4 matrix as shown below. Similarly, the next four bytes of *ShiftRows* diffusion acting as input to *MixColumn* diffusion are Y_4, Y_9, Y_{14}, Y_3. Furthermore, the next four bytes of *ShiftRows* diffusion acting as input to *MixColumn* diffusion are Y_8, Y_{13}, Y_2, Y_7. Finally, the last four bytes of *ShiftRows* diffusion acting as input to *MixColumn* diffusion are Y_{12}, Y_1, Y_6, Y_{11}. The respective outputs are Z_0, Z_1, Z_2, Z_3; Z_4, Z_5, Z_6, Z_7; Z_8, Z_9, Z_{10}, Z_{11}; Z_{12}, Z_{13}, Z_{14}, Z_{15}.

$$\begin{pmatrix} Z_0 \\ Z_1 \\ Z_2 \\ Z_3 \end{pmatrix} = \begin{pmatrix} 02 & 03 & 01 & 01 \\ 01 & 02 & 03 & 01 \\ 01 & 01 & 02 & 03 \\ 03 & 01 & 01 & 02 \end{pmatrix} \begin{pmatrix} Y_0 \\ Y_5 \\ Y_{10} \\ Y_{15} \end{pmatrix}$$

$$\begin{pmatrix} Z_4 \\ Z_5 \\ Z_6 \\ Z_7 \end{pmatrix} = \begin{pmatrix} 02 & 03 & 01 & 01 \\ 01 & 02 & 03 & 01 \\ 01 & 01 & 02 & 03 \\ 03 & 01 & 01 & 02 \end{pmatrix} \begin{pmatrix} Y_4 \\ Y_9 \\ Y_{14} \\ Y_3 \end{pmatrix}$$

$$\begin{pmatrix} Z_8 \\ Z_9 \\ Z_{10} \\ Z_{11} \end{pmatrix} = \begin{pmatrix} 02 & 03 & 01 & 01 \\ 01 & 02 & 03 & 01 \\ 01 & 01 & 02 & 03 \\ 03 & 01 & 01 & 02 \end{pmatrix} \begin{pmatrix} Y_8 \\ Y_{13} \\ Y_2 \\ Y_7 \end{pmatrix}$$

$$\begin{pmatrix} Z_{12} \\ Z_{13} \\ Z_{14} \\ Z_{15} \end{pmatrix} = \begin{pmatrix} 02 & 03 & 01 & 01 \\ 01 & 02 & 03 & 01 \\ 01 & 01 & 02 & 03 \\ 03 & 01 & 01 & 02 \end{pmatrix} \begin{pmatrix} Y_{12} \\ Y_1 \\ Y_6 \\ Y_{11} \end{pmatrix}$$

After performing matrix multiplication, the following equations are obtained on which GF arithmetic is performed.

$$Z_0 = (02 \cdot Y_0) + (03 \cdot Y_5) + (01 \cdot Y_{10}) + (01 \cdot Y_{15}), \quad (10.1)$$

$$Z_1 = (01 \cdot Y_0) + (02 \cdot Y_5) + (03 \cdot Y_{10}) + (01 \cdot Y_{15}), \quad (10.2)$$

$$Z_2 = (01 \cdot Y_0) + (01 \cdot Y_5) + (02 \cdot Y_{10}) + (03 \cdot Y_{15}), \quad (10.3)$$

$$Z_3 = (03 \cdot Y_0) + (01 \cdot Y_5) + (01 \cdot Y_{10}) + (02 \cdot Y_{15}). \quad (10.4)$$

Similarly, the rest of the equations can be derived from matrix multiplications, followed by GF arithmetic computation. Note: in GF arithmetic, the additions are simple bitwise XORs of the respective bytes. Multiplication by 01 is multiplication by the identity and does not include any specific operation. Multiplication by 02 can be realized as a multiplication by x, which is a left shift by one bit, and a modular reduction with $P(x) = (x^8 + x^4 + x^3 + x + 1)$ for 8 bit data or $(x^4 + x + 1)$ for 4 bit data. Note: In AES, 4-bit product is not possible after multiplication, thus $P(x) = (x^8 + x^4 + x^3 + x + 1) \rightarrow (100011011)$ is used. Similarly, multiplication by 03, which represents the polynomial $(x + 1)$, can be realized by a left shift by one bit and addition of the original value, subsequently by performing modular reduction with $P(x)$.

For instance, let us consider eqn. (10.1) above as: $Z_0 = (0_2 \cdot Y_0) + (0_3 \cdot Y_5) + (0_1 \cdot Y_{10}) + (0_1 \cdot Y_{15})$.

Furthermore, let us consider the state of input to *MixColumn* diffusion as: $(Y_0, Y_5, Y_{10}, Y_{15}) = (25, 25, 25, 25)$.

Now converting $(25)_{hex}$ in hexadecimal format to binary is: 00100101. This is also equivalent to polynomial: $x^5 + x^2 + 1$. Now inserting the binary of $(25)_{hex}$ = 00100101 in eqn. (10.1) we get:

$$\begin{aligned} Z_0 &= (02.25) + (03.25) + (01.25) + (01.25) \\ Z_0 &= (00000010.00100101) + (00000011.00100101) + (00000001.00100101) \\ &\quad + (00000001.00100101) \\ &= (01001010) + (01101111) + (00100101) + (00100101) \end{aligned}$$

(Note: Multiplying by 01 is same value; multiplying by 02 is left shift by 1 bit; multiplying by 03 is left shift by 1 bit and addition of the original value followed by a modular reduction with P(x) if required. The addition is achieved by XOR the bytes.)

$$\begin{aligned} &= (01001010)\text{XOR}(01101111)\text{XOR}(00100101)\text{XOR}(00100101) \\ &= 00100101. \end{aligned}$$

The above value is equivalent to $x^5 + x^2 + 1$. Since the order of power is less than 8, thus modular reduction is not required. Similarly, the rest of the values $(Z_1, Z_2, \ldots, Z_{15})$ can be computed.

10.2.2.4 Key addition layer

The inputs to the *key addition* layer are the output of *MixColumn* diffusion layer (current 16-byte state matrix) and a *subkey* which also comprises of 16 bytes. Bitwise XOR is performed on the two inputs to yield the output O_0–O_{15} for the next round (see Figure 10.2). This output is used as an input for byte substitution layer in next round. The example below illustrates the process in Tables 10.3 and 10.4. As seen below, keys (K_0–K_3) are bitwise XOR-ed with state matrix element (Z_0–Z_3) to obtain outputs (O_0–O_3) which is fed as an input for the next round. Similarly for the remaining elements of the state matrix, the output is computed (see Figure 10.2).

10.2.2.5 Key scheduler for 128-bit AES

The main role of key scheduler in AES is to accept the original 128 bit key and derive the subkeys for the next rounds of AES. The number of subkeys needed is obtained by adding one to the number of rounds. Hence, for key length of 128 bits, the number of rounds is $n_r = 10$, and there are 11 subkeys, each of 128 bits. The AES with a 192-bit key requires 13 subkeys of length 128 bits, and AES with a 256-bit key has 15 subkeys. The AES subkeys are calculated recursively, i.e., in order to derive subkey k_i, subkey k_{i-1} must be known. We will discuss the process of key scheduler in more details a little later. But now since all major phases of AES algorithm has been discussed, thus let us have a look at the hardware mapping process of AES in Figure 10.4. This is the generic hardware mapping process that is used to develop a custom hardware-based AES block. We will look into the hardware block of diffusion layer 2 (*MixColumn*) from the context of optimization now and digital watermarking later. The *MixColumn* hardware block can be designed as shown in Figure 10.5. However, in order to achieve delay and resource optimization, the hardware structure in Figure 10.5 can be re-designed/re-configured as shown in Figure 10.6.

The AES key schedule is word-oriented, where 1 word = 32 bits. Subkeys are stored in a key expansion array M that comprises of words. Figure 10.7 shows the *key scheduler* process for round '*i*'. Similarly, how the keys for round 1 have been generated, and the keys for next rounds using key scheduler are obtained. The g-function in the key scheduler process is responsible for non-linearity in the key scheduling. It accepts a four-byte input and produces a four-byte output. The substeps inside g-function are following: left rotation (e.g., V_0, V_1, V_2, V_3 changes to V_1, V_2, V_3, V_0), byte substitution using S-box, and adding round coefficient (RC). The RC is an element of the GF (2^8), i.e., an 8-bit value. It is only added to the

Table 10.3 State matrix before key addition operation

Z_0	Z_4	Z_8	Z_{12}
Z_1	Z_5	Z_9	Z_{13}
Z_2	Z_6	Z_{10}	Z_{14}
Z_3	Z_7	Z_{11}	Z_{15}

Bit-wise XOR with Keys K_0–K_{15} →

Table 10.4 State matrix after key addition operation

O_0	O_4	O_8	O_{12}
O_1	O_5	O_9	O_{13}
O_2	O_6	O_{10}	O_{14}
O_3	O_7	O_{11}	O_{15}

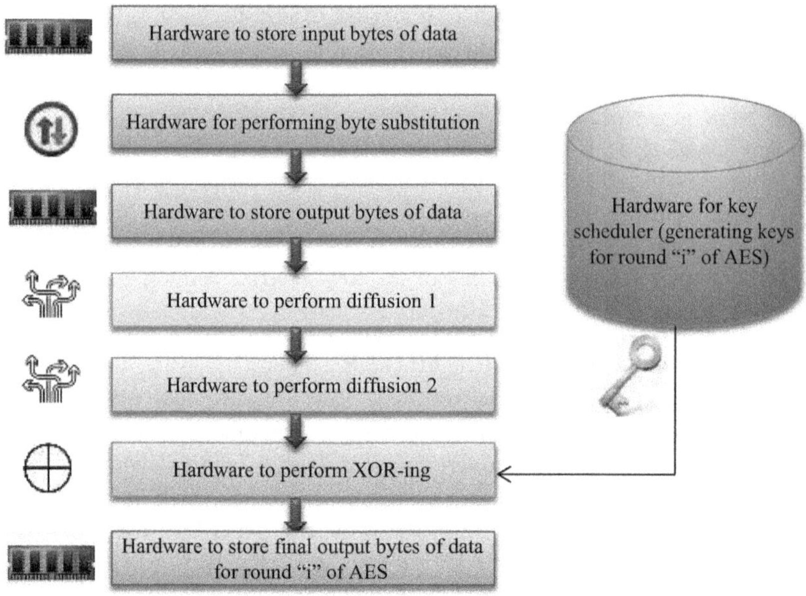

Figure 10.4 Hardware mapping of 128 bit AES algorithm

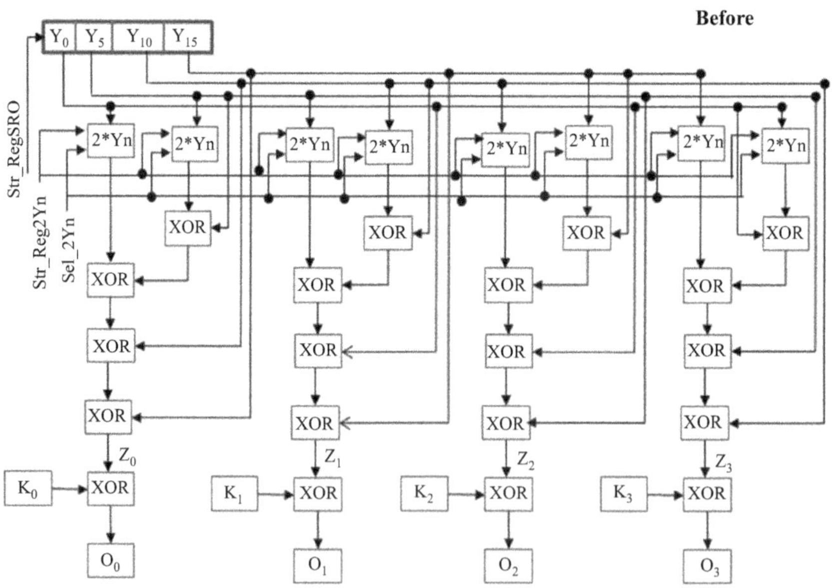

Figure 10.5 MixColumn (diffusion layer 2) hardware before optimization

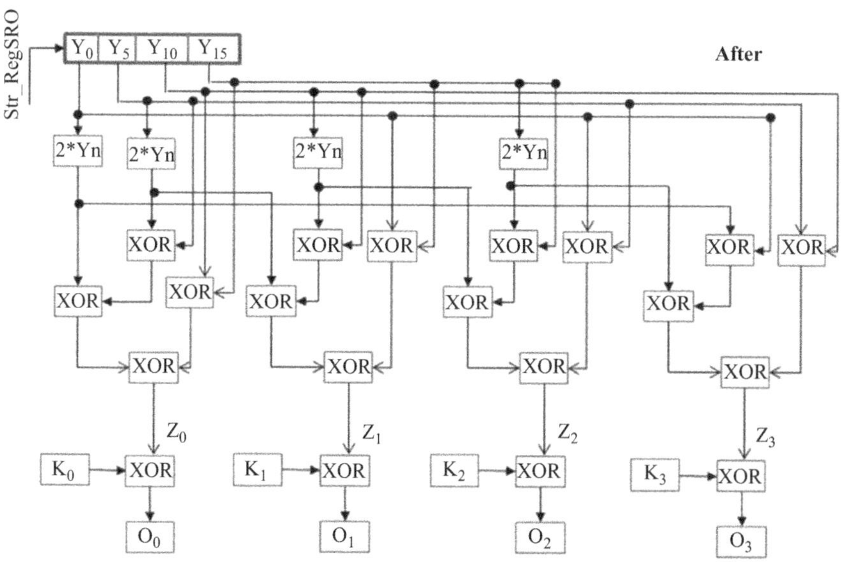

Figure 10.6 MixColumn (diffusion layer 2) hardware after optimization

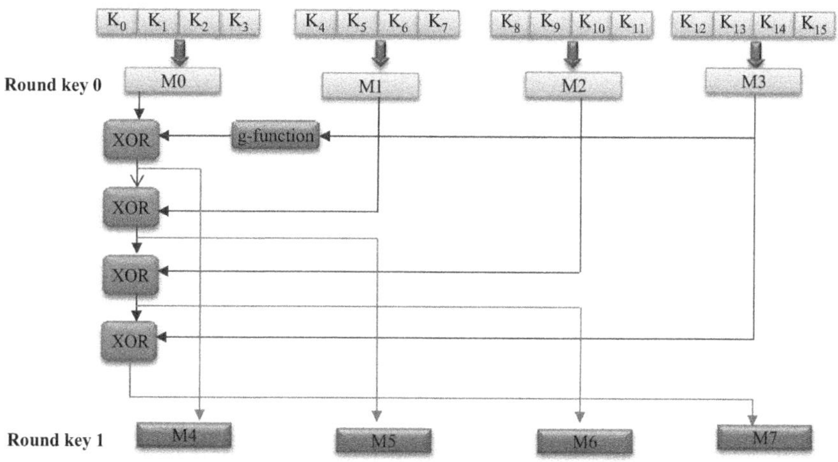

Figure 10.7 AES key schedule for 128-bit key size for round 'i'

leftmost byte in the function g. The RC value varies round to round and is added as per the following rule based on eqn. (10.5) in Paar and Pelzl (2009):

$$\begin{aligned}&\mathrm{RC}[1] = x^0 = (00000001)_2, \mathrm{RC}[2] = x^1 = (00000010)_2,\\&\mathrm{RC}[3] = x^2 = (00000100)_2, \ldots, \mathrm{RC}[10] = x^9 = (00110110)_2.\\&\mathrm{RC}[i] = x^{(x-1)} \bmod \mathrm{GF}(2^8)\end{aligned} \tag{10.5}$$

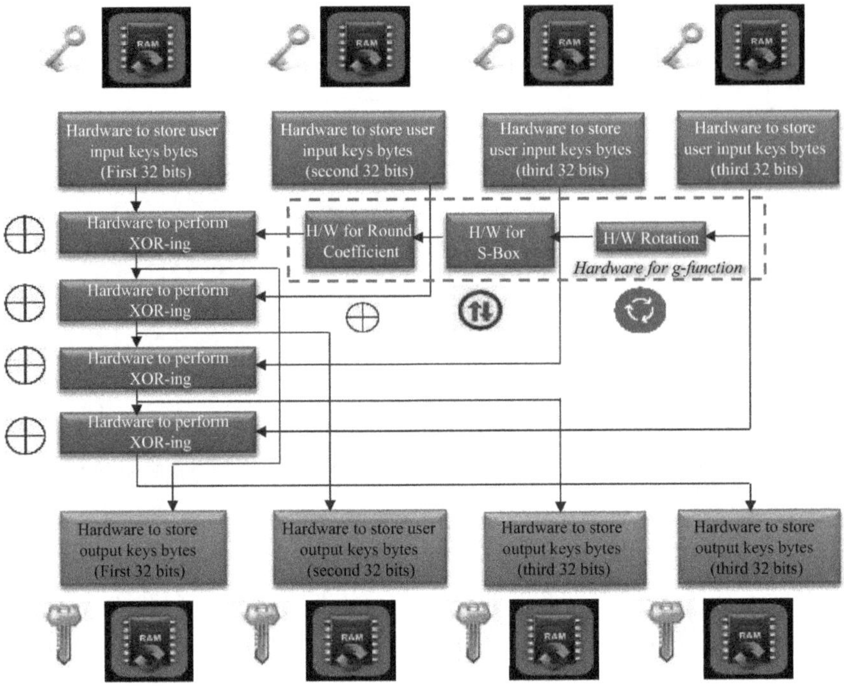

Figure 10.8 Hardware mapping of AES key schedule

But we know $GF(2^8) = x^8 + x^4 + x^3 + x + 1 = 100011011$.
Thus, for example when we compute $RC[9] = x^8$ using eqn. (10.5):

$$= 100000000 \bmod 100011011,$$
$$= (000011011)_{2},$$
$$= (1\,B)_{hex}.$$

Furthermore, $RC[10] = x^9 = 2 \times x^8$ which is equivalent to left shifting the value RC[9], i.e., $(000011011)_2 \rightarrow (00110110)_2$.

The respective hardware mapping of key scheduler process is shown in Figure 10.8, which indicates the hardware structures necessary for rotation, S-box implementation, RC, as well as the others units.

10.3 AES digital watermarking

We have learnt about the motivation of embedding watermark in an IP core or hardware. Similarly, in order to provide protection (provide deterrent) against false claim of ownership, piracy, forgery, digital watermark needs to be implanted in AES hardware design. In the chapter, we show how to implant watermark in a certain portion of AES hardware structure. The selected portion of AES structure where watermarking is embedded in '*MixColumn diffusion layer 2* hardware block' (shown in Figures 10.4 and 10.5, respectively).

This portion of AES hardware is selected for watermarking because of following desirable properties:

1. Option of embedding watermark without much design overhead.
2. Ease of implementation.
3. Seamless signature detection.
4. No chance of hardware functionality being affected.
5. Maximal opportunity of adding several signature digits corresponding to a watermark.

10.3.1 AES watermark encoding

In this chapter, we present a new encoding scheme for embedding signature digits 'α' and 'β' into the AES diffusion layer to design structure. The formal definition of digits 'α' and 'β' are as follows:

α digit = Perform double XOR blocks sharing in AES MixColumn structure and re-design;

β digit = Perform triple XOR blocks sharing in AES MixColumn structure and re-design.

For example, Figure 10.9a represents the portion of *MixColumn* structure responsible for computing Z0 (refer Figure 10.5). Embedding of 'α' indicates double sharing of two consecutive XOR blocks (indicated in Figure 10.9a), while embedding of 'β' indicates triple sharing of three consecutive XOR blocks (also indicated in Figure 10.9a). Figure 10.9b shows the resultant structure after embedding a 'α' digit in the original AES MixColumn structure. The resultant structure now contains two extra 2:1 multiplexers and one 1:2 demultiplexer (and storage registers) with a minimization of one XOR block. On the other hand, if a 'β' digit is implanted in the original AES MixColumn structure, then the resultant

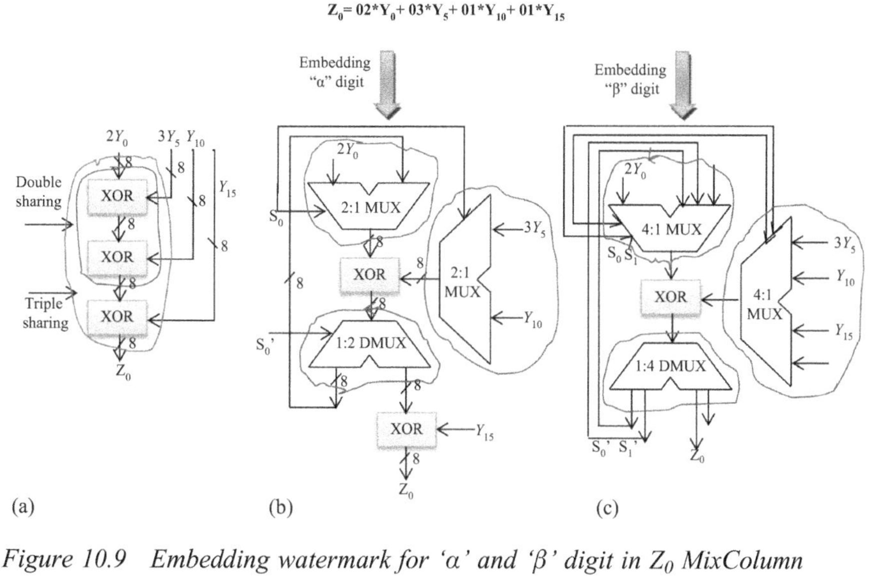

Figure 10.9 Embedding watermark for 'α' and 'β' digit in Z_0 MixColumn structure of AES

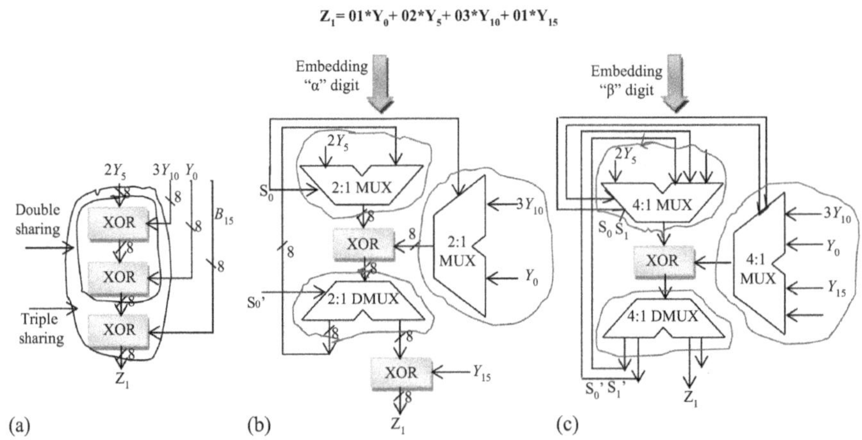

Figure 10.10 Embedding watermark for 'α' and 'β' digit in Z_1 MixColumn structure of AES

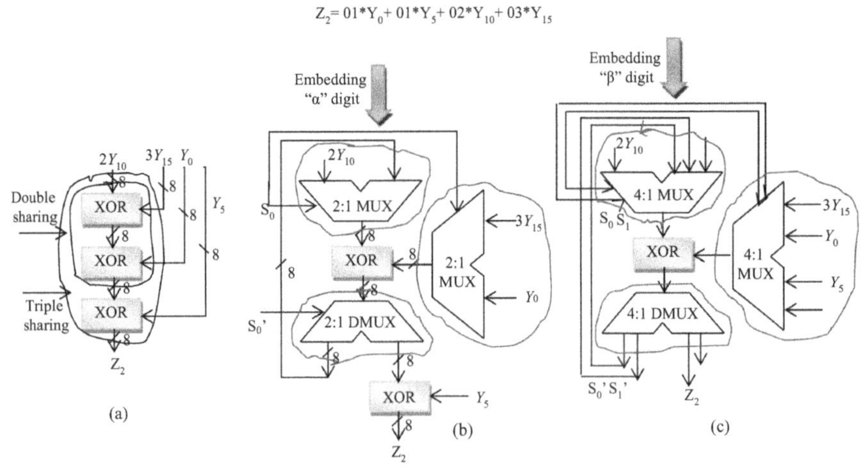

Figure 10.11 Embedding watermark for 'α' and 'β' digit in Z_2 MixColumn structure of AES

structure is shown in Figure 10.9c. The resultant structure now contains two extra 4:1 multiplexers and one 1:4 demultiplexer (and storage registers) with a minimization of two XOR block. Thus, evident different area-delay tradeoff is possible due to choice of different watermark signature digits. Moreover, the design overhead would also vary depending on the choice of the signature digits. We further present the portion of original *MixColumn* structure (with no watermark) responsible for computing Z_1, Z_2, and Z_3, along with its corresponding watermarked design in Figures 10.10, 10.11, and 10.12, respectively.

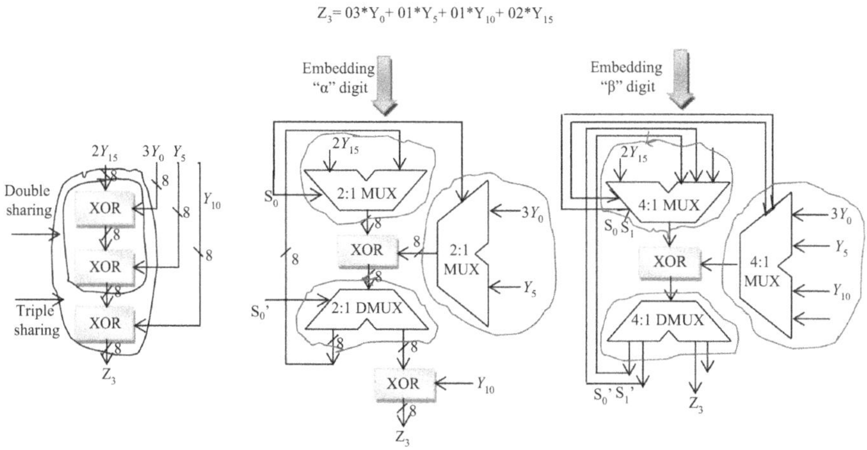

Figure 10.12 Embedding watermark for 'α' and 'β' digit in Z_3 MixColumn structure of AE

10.3.2 Process of embedding watermark in AES

1. Sort the output variables of *MixColumn* hardware structure in ascending order, i.e., Z_0, Z_1, $Z_2, \ldots, Z_{15}$ and store in list L[*k*].
2. Create a desired 16 digit watermark signature using random combinations of α and β digits.
3. For each occurrence of α or β digits, apply design changes (reconfiguration) to the *MixColumn* hardware structure corresponding to sorted output variables. Example, first α or β digit in a signature will affect the hardware structure corresponding to first sorted variable Z_0 (as it first in the sorted list L[*k*]), followed by next α or β digit affects the hardware structure corresponding to next sorted variable Z_1 and so on.
4. Insert registers after each multiplexer and demultiplexer in the modified *Mix-Column* hardware structure.

10.3.3 Signature detection

Signature detection is performed through the following two phases:

1. Inspection of the datapath of AES *MixColumn* structure to collect relevant hardware information.
2. Signature verification is performed to authenticate the existence of vendor's signature in the AES IP core design. Firstly, the encoded signature is decoded through the cognizance of the encoding rules. Subsequently, the presence of extra constraints as watermark is verified in the *MixColumn* hardware structure.

10.4 Case study of a watermarked AES hardware

The AES design was implemented in VHDL in Altera Quartus v.13 and run on AMD A8-4500M with 4 GB DDR3. The target hardware platform selected was Cyclone II—Device EP2C70F896C6.

For the sake of analysis, two different watermark signature sizes were chosen, viz. 8-digit and 16-digit for embedding into the AES hardware design. Three types of experiments were conducted: (a) variation of different 8-digit signature combinations and its impact on synthesis results of AES hardware, (b) variation of different 16-digit signature combinations and its impact on synthesis results of AES hardware, and (c) overhead obtained due to 8-digit signature and 16-digit signature compared to non-watermarked AES. Following parameters were observed and analyzed: total logic elements, total combinational functions, dedicated logic registers, total registers, and total pins. Results of implementation of AES for 8-digit and 16-digit watermark signature as well as non-watermark AES hardware is reported in Tables 10.5 and 10.6, respectively. A portion of watermarked AES hardware corresponding to signature $\alpha\alpha\beta\beta\alpha\alpha\beta\beta$ is shown in Figure 10.13. Figure 10.14 shows the comparison of non-watermarked AES128 design with three different combinations of 8-digit watermarked design in terms of logic elements. Figures 10.14 and 10.15 shows comparison of non-watermarked AES128 hardware with different combinations of 8-digit and 16-digit watermarked AES.

10.5 Conclusion

This chapter presented the following novel and interesting aspects: (a) AES algorithm—description of each step and its flow diagram, (b) corresponding hardware mapping of AES block cipher—overview of how each step can be mapped to a

Table 10.5 Results of implementation of AES (with and without watermark) for 8-digit signature

Parameters	AES128 (no watermark)	AES128 with watermark	AES128 with watermark	AES128 with watermark
8-Digit signature	–	$\alpha\alpha\beta\beta\alpha\alpha\beta\beta$	$\alpha\alpha\alpha\alpha\alpha\alpha\alpha\alpha$	$\beta\beta\beta\beta\beta\beta\beta\beta$
Family	Cyclone II	Cyclone II	Cyclone II	Cyclone II
Device	EP2C70F896C6	EP2C70F896C6	EP2C70F896C6	EP2C70F896C6
Total logic elements	1,117/68,416 (2%)	1,862/68,416 (3%)	1,812/68,416 (3%)	1,885/68,416 (3%)
Total combinational functions	1,112/68,416 (2%)	1,855/68,416 (3%)	1,806/68,416 (3%)	1,880/68,416 (3%)
Dedicated logic registers	20/68,416 (<1%)	32/68,416 (<1%)	26/68,416 (<1%)	27/68,416 (<1%)
Total registers	20	32	26	27
Total pins	385/622 (62%)	385/622 (62%)	385/622 (62%)	385/622 (62%)
Simulation time	360 ns	540 ns	500 ns	540 ns

Table 10.6 Implementation data of AES (with and without watermark) for 16-digit signature

Parameters	AES128 (no watermark)	AES128 with watermark	AES128 with alpha watermark	AES128 with beta watermark
16-Digit signature	–	$\alpha\alpha\beta\beta\alpha\alpha\beta\beta\alpha\alpha\beta\beta\alpha\alpha\beta\beta$	$\alpha\alpha\alpha\alpha\alpha\alpha\alpha\alpha\alpha\alpha\alpha\alpha\alpha\alpha\alpha\alpha$	$\beta\beta\beta\beta\beta\beta\beta\beta\beta\beta\beta\beta\beta\beta\beta\beta$
Family	Cyclone II	Cyclone II	Cyclone II	Cyclone II
Device	EP2C70F896C6	EP2C70F896C6	EP2C70F896C6	EP2C70F896C6
Total logic elements	1,117/68,416 (2%)	2,568/68,416 (4%)	2,483/68,416 (4%)	2,622/68,416 (4%)
Total combinational functions	1,112/68,416 (2%)	2,559/68,416 (4%)	2,478/68,416 (4%)	2,616/68,416 (4%)
Dedicated logic registers	20/68,416 (<1%)	32/68,416 (<1%)	26/68,416 (<1%)	27/68,416 (<1%)
Total registers	20	32	26	27
Total pins	385/622 (62%)	385/622 (62%)	385/622 (62%)	385/622 (62%)
Simulation time	360 ns	560 ns	500 ns	560 ns

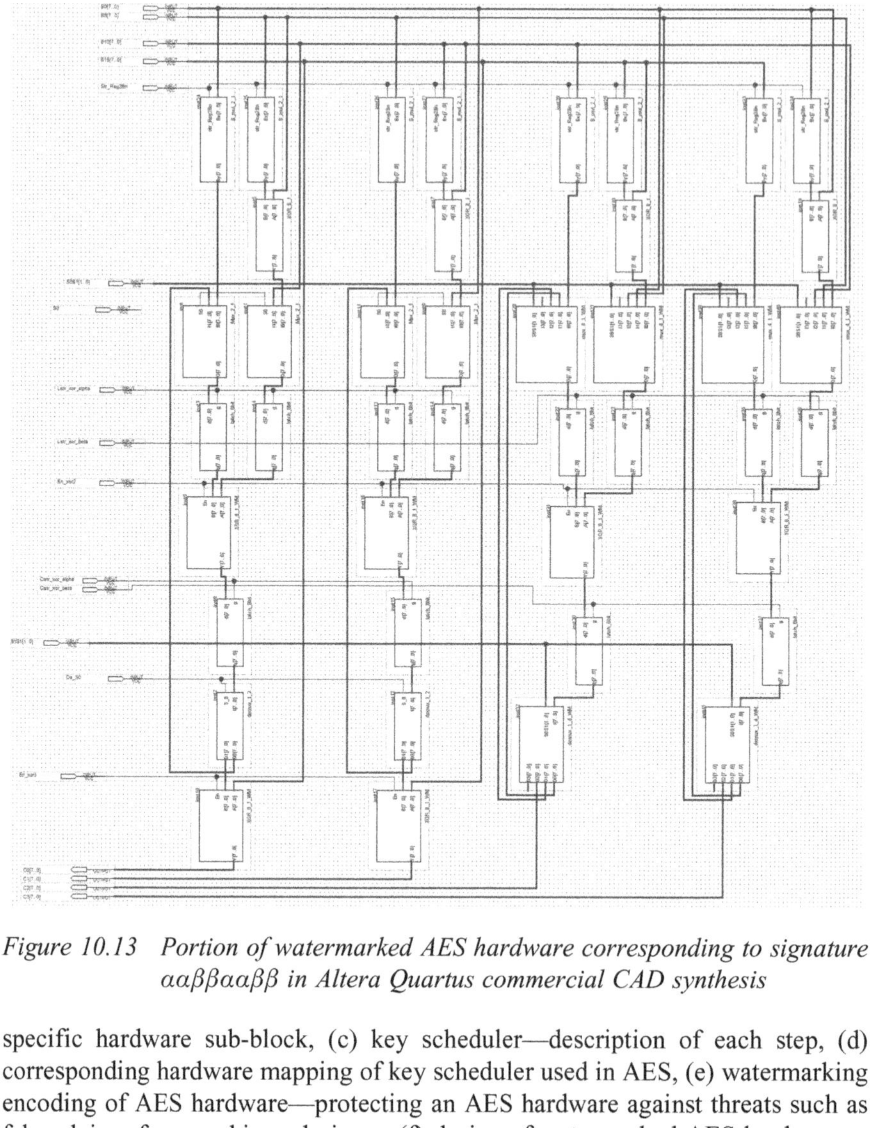

Figure 10.13 Portion of watermarked AES hardware corresponding to signature $\alpha\alpha\beta\beta\alpha\alpha\beta\beta$ in Altera Quartus commercial CAD synthesis

specific hardware sub-block, (c) key scheduler—description of each step, (d) corresponding hardware mapping of key scheduler used in AES, (e) watermarking encoding of AES hardware—protecting an AES hardware against threats such as false claim of ownership and piracy, (f) design of watermarked AES hardware—description of the watermark embedding process, and (g) impact of several different watermark signature types on AES hardware and its implementation on FPGA—to highlight the overhead due to watermark. There is no chapter that deals with the design flow of AES algorithm, its hardware design perspective as well as its digital watermarking process for different signature digits. Finally, it provides a lucid comparative study of watermarked AES hardware with a non-watermarked AES hardware. The content of this chapter is useful in understanding the details of AES described in conjunction with functional obfuscation of digital signal processing cores presented in Chapter 08.

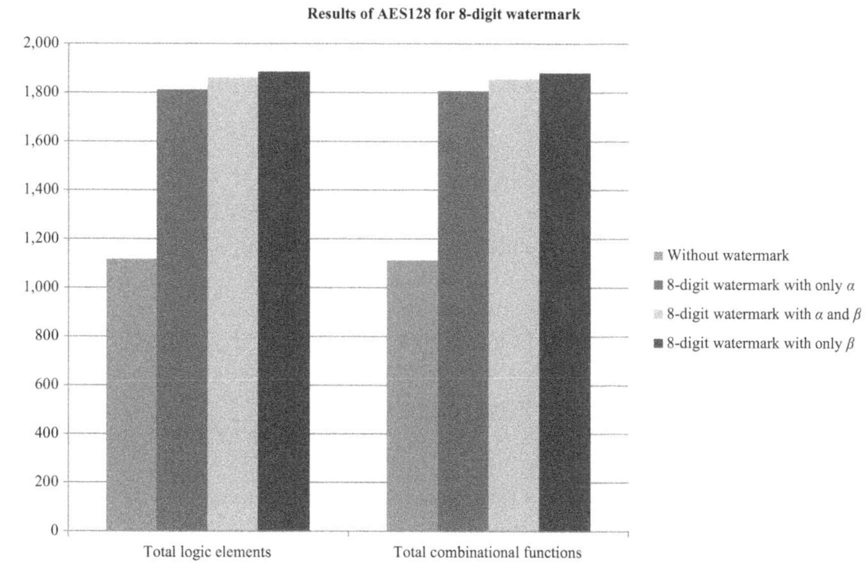

Figure 10.14 Comparison of non-watermarked AES128 design with three different combinations of 8-digit watermarked design

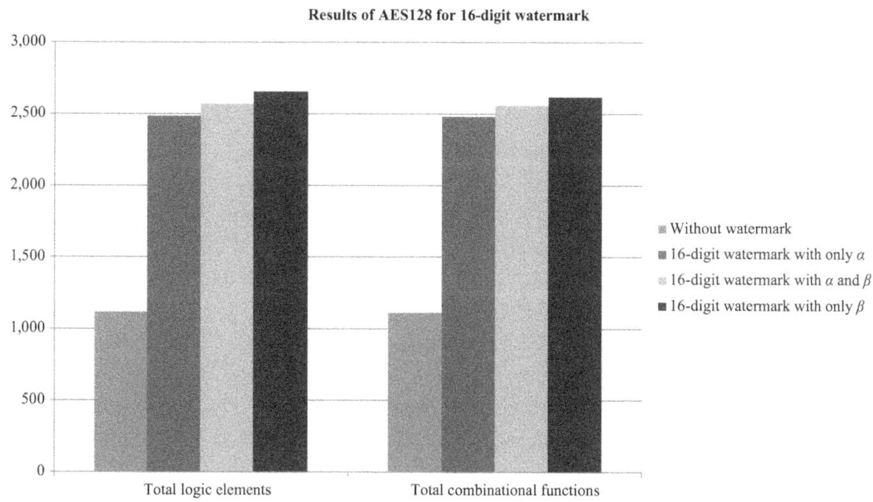

Figure 10.15 Comparison of non-watermarked AES128 design with three different combinations of 16-digit watermarked design

10.6 Exercises

1. What are the phases of AES?
2. How does S-box work?
3. Differentiate between diffusion layers 1 and 2?
4. How does key scheduling function?
5. What is the different number of rounds for different number of key length?
6. What is the function of modular reduction in AES and how does it perform?
7. How is column transformation performed?
8. How do you implement column transformation using H/W?
9. How modular reduction is implemented using H/W?
10. How is watermark embedded in AES?
11. Which stage of AES process is watermark inserted?
12. What is the impact of signature digits on watermarked AES design overhead?
13. How row-based diffusion is different than column based diffusion?
14. Explain state matrix in case of row-based diffusion?
15. Derive equations for MixColumn diffusion using GF arithmetic?
16. Explain state matrix in case of key addition layer?
17. How is hardware mapping of AEs performed? Explain the flow diagram?
18. Explain the MixColumn (diffusion layer 2) hardware perspective: before and after optimization?
19. How is RC calculated? Explain with example?
20. Explain hardware mapping of AES key schedule?

References

Bhunia S. and Tehranipoor M. M., 'Hardware security: A hands-on learning approach,' Morgan Kaufmann imprint, USA, 2018.

Kosaraju N. M., Varanasi M., and Mohanty S. P., 'A high-performance VLSI architecture for advanced encryption standard (AES) algorithm,' *Proceedings of the 19th International Conference on VLSI Design (VLSID)*, 2006, pp. 481–484.

Mishra P., Bhunia S., and Tehranipoor M., 'Hardware IP security and trust,' 2017, Springer, New York, USA.

Mohanty S. P., Sheth R., Pinto A., and Chandy M., 'CryptMark: A novel secure invisible watermarking technique for color images,' *Proceedings of the 11th IEEE International Symposium on Consumer Electronics (ISCE)*, 2007, pp. 1–6.

Paar C. and Pelzl J., 'Understanding cryptography – A textbook for students and practitioners,' Springer-Verlag, Berlin Heidelberg, eBook ISBN 978-3-642-04101-3, Number of Pages: XVIII, 372, Nov. 2009.

Sengupta A. (co-inventor: Reza Sedaghat), 'System and methodology for development of a system architecture using optimization parameters,' US Patent by United Sates Patent and Trademark Office (USPTO), March 12, 2013, Patent no. US 8,397,204.

Sengupta A., 'Protection of IP-core designs for CE products,' *IEEE Consumer Electronics Magazine*, 2015, vol. 5(1), pp. 83–89.

Sengupta A., 'Hardware security of CE devices: Threat models and defense against IP Trojans and IP piracy,' *IEEE Consumer Electronics Magazine*, 2017, vol. 6 (1), pp. 130–133.

Sengupta A. and Bhadauria S., 'Exploring low cost optimal watermark for reusable IP cores during high level synthesis,' *IEEE Access Journal, Invited paper*, 2016, vol. 4 (99), pp. 2198–2215.

Sengupta A. and Roy D., 'Antipiracy-aware IP chipset design for CE devices: A robust watermarking approach', *IEEE Consumer Electronics Magazine*, 2017a, vol. 6 (2), pp. 118–124.

Sengupta A. and Roy D., 'Protecting an intellectual property core during architectural synthesis using high-level transformation based obfuscation,' *IET Electronics Letters*, 2017b, vol. 53 (13), pp. 849–851.

Sengupta A. and Sedaghat R., 'Integrated scheduling, allocation and binding in high level synthesis using multi structure genetic algorithm based design space exploration system,' *Proceedings of 12th IEEE/ACM International Symposium on Quality Electronic Design (ISQED)*, Silicon Valley, California, USA, 2011, pp. 486–494.

Sengupta A., Bhadauria S., and Mohanty S. P., 'TL-HLS: Methodology for low cost hardware Trojan security aware scheduling with optimal loop unrolling factor during high level synthesis,' *IEEE Transactions on Computer Aided Design of Integrated Circuits & Systems (TCAD)*, 2017, vol. 36 (4), pp. 655–668.

Chapter 11

Hardware approaches for media and information protection and authentication

Technology scaling has allowed us to design high-performance devices with a low-power consumption. The advent of IoT has increased the versatility of data collection and there are many different ways of collecting and transferring data over the Internet. The data that is being collected is also in different forms, that is, text, images, videos, and audios. When they are shared, for a legit use, attackers can break the security and use them for illegible purposes or claim ownership to sell them commercially. This has been the trend lately where many counterfeit products are appearing in the market. This section discusses the digital watermarking, different schemes of digital watermarking, and how a media object can be secured using a watermark. They are also not completely resistant to attacks and measures need to be taken to secure the content that is being watermarked. The chapter also presents different issues with the watermark implementations, attacks and countermeasures to those attacks on watermarking. Section 11.1 presents a broad overview of the IP protection. Section 11.2 discusses the generic overview and components of any watermark system. Section 11.3 summarizes various types of watermarks. Section 11.4 discusses various applications of watermarking. Section 11.5 presents desired characteristics of watermarks. Sections 11.6 discusses the technical challenges of the watermarking. Section 11.7 discusses hardware-based watermarking systems available in the current literature. Section 11.8 discusses about watermarking in smart vehicles. Section 11.9 discusses about medical signal authentication. Section 11.10 highlights side-channel information leakage and its countermeasures. Section 11.11 outlines various forms of attacks on watermarks and watermarking systems. Section 11.12 presents the difficulties involved in making use of them in practice.

11.1 IP Protection—a broad overview

With the advent of new technology, new devices are being introduced into the world and high-performance applications are being designed. With better systems coming out every day, there is also a problem that they are used for various malicious applications like counterfeiting, which is a major problem in hardware and in media also. Many people nowadays are producing content and sharing it with the world

(Krishnamurthy, 2018; Mohanty, 2015). But the content that they own is being illegally used by many commercially, without paying royalties to the owner and in some cases, the content is being misused also. This raises many problems pertaining to the IPR and their protection. Figure 11.1 shows a broad overview on the area of counterfeiting. An image can be produced by the owner and watermarked but an attacker can attack, tamper, or remove the watermark from the image and can claim ownership. This creates many issues on the copyrights and ownership rights.

11.1.1 Digital rights management

Figure 11.2 shows different media and threat levels which are vulnerable to attacks. A cinema will have a high privacy, whereas a broadcast content will have low security. This has become a major issue lately where the content of the owner is being reproduced without paying royalties to him/her.

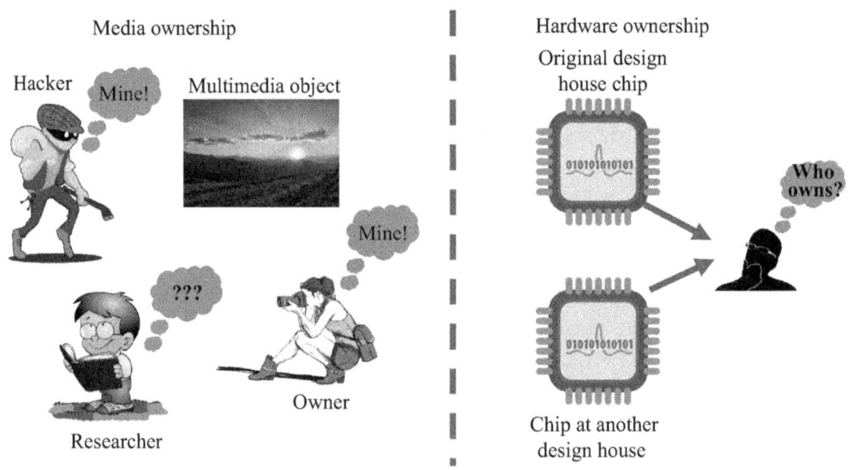

Figure 11.1 Broad overview of IP protection (Mohanty, 2009; Mohanty et al., 2004, 2005b, 2018)

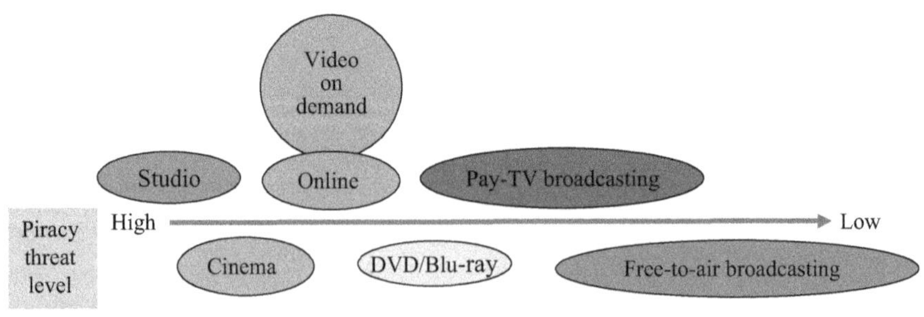

Figure 11.2 Multimedia video overview (Mohanty, 2009; Mohanty et al., 2004, 2005b, 2018)

Figure 11.3 Multimedia music overview (Mohanty, 2009; Mohanty et al., 2004, 2005b, 2018)

In such cases, the owner will have copyrights violated and misses out on the royalties that are entitled legally. When a content media like a digital video disc (DVD) or Blu-ray is produced with a video content, it is considered to be secured with the watermarks embedded in it. But there are also many attacks available that can remove or tamper the watermark on the content. Figure 11.3 shows different types of multimedia that are available on the Internet. Many of the audio services provide audio to the user at a monthly cost. But tampering with the audio that is broadcast over the Internet is not difficult with the technological advancements that are present at a current date. Digital rights management (DRM) is a major issue that many different areas of research are concentrating on. Watermarking the audio files, video files, images, and also text is a solution.

11.1.2 Copyright protection of multimedia—a brief history

There are several advantages to digital media over the analog world. Higher quality, ease of processing, and editing are some of them. The most common forms of digital media are compact disc (CD), DVD, TV broadcast, and computers. But there are also many disadvantages to the digital media. Digital media counterfeiting has been a very big problem lately. It can easily be duplicated, which has raised the need for more secure and effective DRM tools (Devillier, 2017; Kougianos *et al.*, 2009; Lah and Lewis, 2016; Ma, 2017; Ma *et al.*, 2016). Watermarking is one of the ways of securing the digital media such as audios, photographs, and videos. There are different kinds of watermarking like visible or invisible and can be inserted into different domains like spatial or frequency. There are various software implementations of digital watermarking but very few hardware implementations (Liu *et al.*, 2017; Sarreshtedari *et al.*, 2015; Zhao *et al.*, 2015). When designing a hardware implementation of watermarking, various issues are to be considered such as low-power consumption, high-performance, ease of integration, and reliability.

Emil Hembrooke invented electronic watermarking in 1954 (Cox and Miller, 2002). Since then there has been extensive research carried on by various experts from different domains, such as cryptography, computer science, communications, and signal processing. Digital watermarking was proposed to provide robust protection along with encryption and scrambling. Many of the following questions were raised by the community (Cox and Miller, 2002; Zhao *et al.*, 1999; Yeung, 1998):

1. What is digital watermarking and what are its properties?
2. What is the process of implementing digital watermarking?
3. What are its disadvantages?

Digital watermarking is the process of embedding information into digital media and generating a watermarked media (Liu *et al.*, 2017; Sarreshtedari *et al.*, 2015; Sequeira and Kundur, 2001; Zhao *et al.*, 2015) which evolved from steganography techniques. A difference between steganography and cryptography is shown in Figure 11.4. In the case of cryptography, a key is used to encrypt the message and convert into a cipher text and when it is decrypted, the original message will be exposed and will not be protected. In the case of steganography, the message is hidden in the plain text, and the Data is embedded with the original message (Mathivanan and Ganesh, 2017; Vijayan and Sreedivya, 2017). The way and intent of using the method differs between steganography and digital watermarking. A watermark can be perceived as an attribute of the carrier (cover). The information that will be watermarked into the media can be anything, including, and not limited to, tracking, license, authorship, copyright, and so on.

Digital watermarking has been around for a very long time and many applications are being developed that use the watermarking for security (Cox and Miller, 2002;

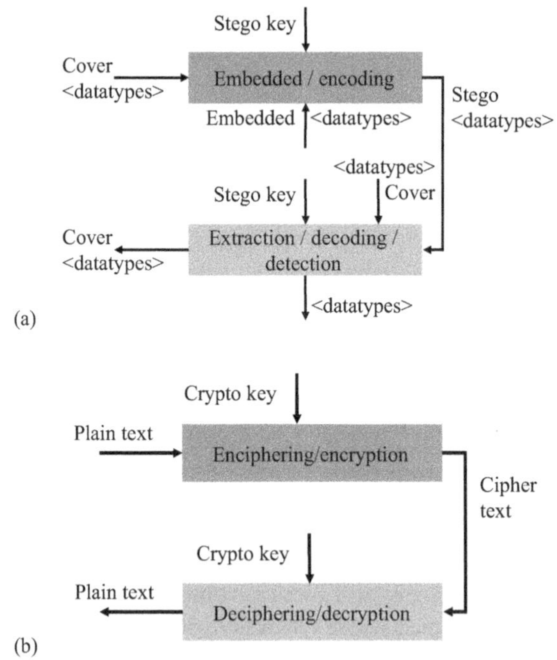

Figure 11.4 (a) Steganography versus (b) cryptography (Kahn, 1996; Pfitzmann, 1996)

Kamaruddin *et al.*, 2018; Kirovski and Malvar, 2001; Loan *et al.*, 2018; Memon and Wong, 1998; Mintzer *et al.*, 1997; Barnett, 1999), such as fingerprinting, ownership evidence, and authentication. No single watermarking technique can be universally used for all the applications, which makes the research and development of new techniques in watermarking more difficult and challenging. The watermarking technique depends on various factors including the type of application and deployment (Kamaruddin *et al.*, 2018). There are different attacks that can be launched on a media and its watermark, and the security resistance against such attacks and performance determines the success of watermarking scheme (Vellaisamy and Ramesh, 2014; Wang *et al.*, 2017b; Xiang *et al.*, 2014). The two major conflicting factors of watermarking are the performance of the embedded watermark and fulfilling the desired characteristics. Various benchmarks are available for watermarking techniques (Kaur *et al.*, 2012; Voloshynovskiy *et al.*, 2001).

There are three different parts of a watermark (Memon and Wong, 1998; Mohanty, 1999):

- The encoder
- The watermark
- The decoder

Algorithm that is used for marking will embed the watermark into the media object and the verification algorithm will authenticate if the object is watermarked. Spatial or frequency domain can be used to embed a watermark into the media, which can be visible or invisible. A visible watermark will be a translucent mark that the original media is overlaid on and is visible only to the consumer on careful observation. An invisible watermark cannot be seen even with careful observation and needs to be retrieved from the pixels where it is embedded into. To increase the robustness, multiple watermarks can be embedded into a single media object (Hua *et al.*, 2001; Mohanty *et al.*, 1999), which can be used for content authentication or copyright protection.

Before a watermarking scheme is used in a consumer market, different technical, legal, and social issues need to be addressed. Watermarking schemes have been already in use for a very long time. IBM visible watermarking techniques have been used in Vatican library project (Mintzer *et al.*, 1996). In order to develop a robust watermarking scheme or an algorithm, various properties need to be met. For example, application-specific watermarking algorithms need to be developed, which will require a formal body to be involved (ASCAP, 2018; Cox and Miller, 2002; Mintzer *et al.*, 1998; Broadcast Music Inc., 2018; Motion Picture Licensing Corporation, 2018). Copy Protection Technical Working Group (CPTWG) is a group that focuses on sharing knowledge and information on copyright protection (CPTWG, 2018). There has been extensive research in this area and many algorithms were proposed for watermarking. Hardware integration of watermarking has also been explored for better robustness and performance. Figure 11.5 shows a hardware watermarking chip integrated into a camera (Mohanty, 2009; Mohanty *et al.*, 2004, 2005b, 2018). The watermarks in the camera will be embedded in the spatial or DCT domain. The watermarking can be performed for either images or

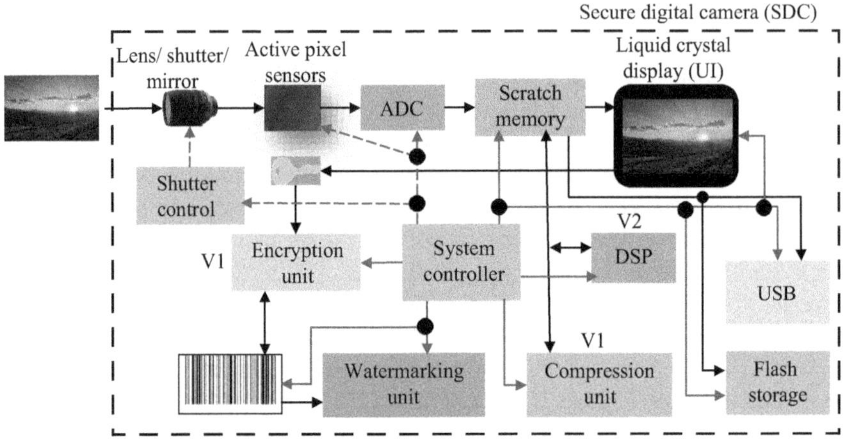

Figure 11.5 System architecture for a digital camera with DRM hardware components (Mohanty, 2009; Mohanty et al., 2004, 2005b, 2018)

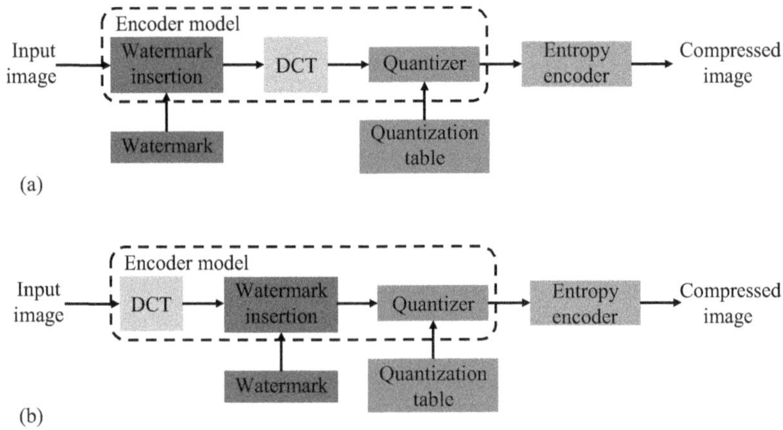

Figure 11.6 JPEG encoder with watermarking modules: (a) spatial domain and (b) DCT domain (Mohanty et al., 2003)

videos. An example of encoder in spatial and DCT domain modules for JPEG compression is presented in Figure 11.6.

11.1.3 Hardware versus media protection

Many questions arise when it comes to the copyright protection. There is a major issue of counterfeiting which should be protected against. Many different agencies are trying to protect the copyrights of the owners. But there are always attackers who illegally claim ownership by attacking the watermark protection of the original image or tampering with it. There is media protection and hardware protection when it comes to the Copyright Safeguarding and DRMs. Compared to media, hardware watermarking schemes are better. This is because they can be optimized for a better performance with a low-power consumption. The chip area of the watermarking module can also be optimized so that the overall cost of production is lower. Media protection in some cases is protecting the media on which the copy of original content is reproduced on. For example, DVD will have an information on the discs such that they will not be able to play the content outside the region of distribution. A disc produced in an area and region locked cannot be played outside that region in another drive. The contents of the disc will not be visible for reading. The hardware watermarking can be used for copyright protection and ownership authentication of the owner.

11.2 General framework for copyright protection

Primary use of watermark is to embed it into a digital media using different algorithms and extract it at a later time for authentication of ownership and so on. The digital media can be an image, a video, or an audio. For example, a watermark on media can be considered as a "seal" that is on an image to identify the ownership, the quality of media, and in some cases, the identity of purchaser (Wang *et al.*, 2018). Figure 11.7 shows the big picture of encoding and decoding techniques which consists of various parts such as encoder, comparator, and decoder (Memon and Wong, 1998; Mohanty, 1999). In case of a visible watermarking technique, the watermark used will be an image. But in case of an invisible watermarking technique, the watermark can be an image, a random, or a pseudo random number. The owner may have a unique watermark of his/her own and can also embed different watermarks in different media objects. When needed, the watermark can be extracted for authentication.

11.2.1 The encoder

Figure 11.7(a) shows the process of embedding a watermark into an original image. The block "E" in the figure represents an encoder. The original image (I) is fed to the encoder along with the watermark. The encoder is designed based on the watermark insertion algorithm that the owner wishes to use to secure the original image. Depending on the algorithm used, the original image is watermarked with the one that is given as an input to the encoder.

For example, let the original image input is denoted by I and watermark input is denoted by $W = \{w_1, w_2, \ldots\}$. The output watermarked image is denoted by $\hat{I}$.

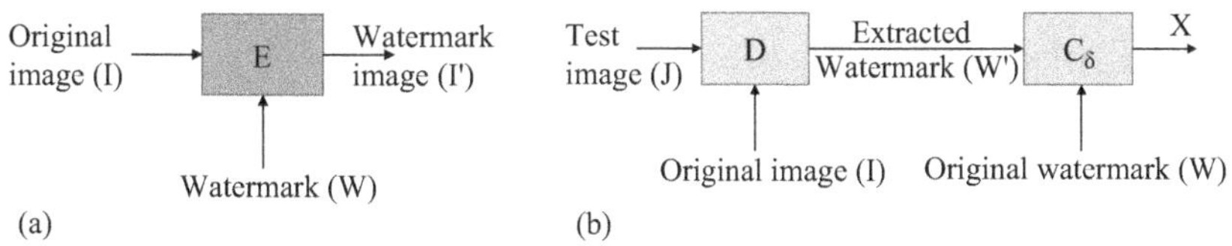

Figure 11.7 General framework for watermarking: (a) the encoder and (b) the decoder (Craver et al., 1998; Memon and Wong, 1998; Mohanty, 1999)

The encoder function is denoted by E and the inputs to the encoder for the original and watermark images are denoted as I and W, respectively. The output is given as:

$$E(I, W) = \hat{I} \tag{11.1}$$

The input watermark image W depends on various factors like size of I, its features, and identity.

11.2.2 The decoder

When the watermarked image is needed to be authenticated, it should sent through a decoder. Figure 11.6 shows the big picture of the decoding process. The decoding process will depend on various factors, including the process of inserting a watermark into the original image and nature of watermarking algorithm. The watermark extraction process is where the entire watermark is extracted from the image. But in some cases, only a specific watermark signal can be detected in the media, and this process is called a watermark detection. Watermark extraction can prove the authenticity of ownership on the media, but only detection can verify the ownership. The input to the decoder D is an image, that is watermarked, nonwatermarked, or even corrupted, denoted as J, which needs to be challenged for ownership. The decoder will recover the watermark W'. In this process, an additional image I can also be included, which is often the original and nonwatermarked version of J.

$$D(J, I) = W' \tag{11.2}$$

11.2.3 The comparator

After the watermark is extracted, it needs to be compared to the original watermarked image. The extracted watermark image from the decoder is sent to the comparator. The comparator is also given the original watermark as the input as shown in Figure 11.7(b). The comparator compares the extracted watermark W' and the original watermark W. The output of the comparator will be in binary, "1" if both the W' and W match and "0" if they do not.

$$C_\delta(W', W) = \begin{cases} 1, & c \leq \delta \\ 0, & \text{otherwise} \end{cases} \tag{11.3}$$

where $c = C_\delta(W', W)$ is the correlation between the two signatures and δ is some threshold. Thus, the watermarking scheme can be treated as a triplet (E, D, C_δ).

11.3 Types of digital watermarks

As mentioned in the previous sections, the research of digital watermarking has evolved and newer innovations are coming out every day. Figure 11.8 shows different types of watermarking techniques and their classification. This section describes different types of digital watermarking techniques classified based on different aspects of the media, watermark, and so on.

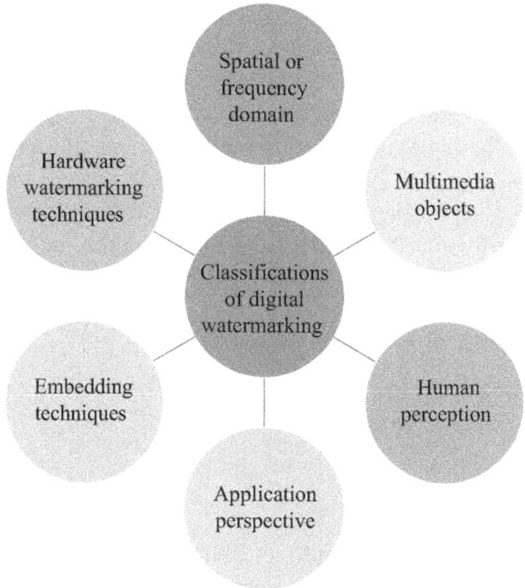

Figure 11.8 Types of watermarking techniques

11.3.1 Spatial versus frequency domain watermarking

The digital watermark can be embedded in frequency or spatial domains (Chen *et al.*, 2015; Mohanty *et al.*, 2018). There are also some alternatives implemented other than the frequency and spatial domain watermarking, like DCT, FT, and wavelet transform (WT). In all of the watermarking techniques, the spatial techniques are much simpler to implement and are less power hungry. For the frequency domain watermarking techniques also, there are several advantages (Cox *et al.*, 1996). Nonsalient components in the image or an audio file will be eliminated by the lossy compression. So if the watermark has to be preserved after a lossy compression, it should be embedded in a region of the data that is significant and will not be damaged in the lossy compression. Such processing is performed in the frequency domain. High-frequency components are subjected to more loss. So embedding the watermark in a low-frequency region is preferred while watermarking the media. In the case of spatial domain watermarking, cropping is a major issue that might affect the watermark. It is highly difficult to analyze the effects of distortion in spatial and frequency domains.

11.3.2 Based on multimedia objects

There are four different types of watermarking techniques based on the type of media:

- Image watermarking
- Video watermarking
- Audio watermarking
- Text watermarking

Image watermarking is the most focused area of research for the past several decades. Developing a common type of watermarking technique for all the types of media is highly difficult. Secure Spread Spectrum algorithm was considered to be better for image, audio, and video media (Cox *et al.*, 1996). But watermarking half-toned images has to be done in a different manner which was a contradiction to the previous claim (Memon and Wong, 1998). In such cases, the binary image is embedded with another binary watermark using visual cryptography. Text watermarking is much different compared to the image, audio, or video watermarking techniques (Kamaruddin *et al.*, 2018).

11.3.3 Based on human perception

Digital watermarking techniques can be classified into two types based on the human perception (Chen *et al.*, 2015; Mohanty *et al.*, 2018):

- Visible watermarking
- Invisible watermarking

In the case of a visible watermark, an image is overlaid on the original image that can be seen but is very translucent. Such watermark can only be seen when the media is carefully observed. An invisible watermark is not visible to the eye. In such cases, the pixel values are altered and the watermark is embedded but the original image itself is not altered. But when the appropriate techniques and algorithms are run, the invisible watermark can be retrieved and authenticated. But this watermark should be embedded into the image so that it is not damaged or removed when the image itself is modified. There are also different watermarking techniques that are available for implementation (Chen *et al.*, 2015; Hua *et al.*, 2001; Mohanty *et al.*, 1999, 2018). A dual watermark will implement both visible and invisible watermarking techniques. Invisible watermark techniques need the original image in order to detect the watermark when needed. But in the case of a visible watermarking technique, the original image is not needed for authentication. Invisible watermark can also be attacked by creating "counterfeit original" which is called invertible. A watermarking scheme (E, D, C_δ) is called quasi-invertible if, for any watermarked image $\hat{I}$, there exists a function E^{-1} such that

$$E^{-1}(\hat{I}) = (I', W') \tag{11.4}$$

$$C_\delta(D(\hat{I}), W') = 1 \tag{11.5}$$

where E^{-1} is a computationally feasible function, W' belongs to the set of allowable watermarks, and the images I and I' are perceptually similar.

11.3.4 From applications point of view

The digital watermarking techniques can be classified into two types based on the application perspective (Voyatzis and Pitas, 1999):

- Source
- Destination

In case of the source watermarking technique, the watermark which will identify the owner of the media will be embedded into all copies of the media. Authentication is the main aim of the source watermarking where the image received can be challenged to be real or tampered with. The technique where the watermark will identify the user but not the owner is called the destination watermarking technique. Here, each distributed copy embeds a unique watermark identifying a specific user. This type of watermark can be used to identify the media at the buyer side and will help reduce the counterfeiting.

11.3.5 Based on embedding techniques

Digital watermarking techniques can be classified as additive, multiplicative, or quantization based on the embedding techniques (Voloshynovskiy *et al.*, 2001). In the case of an additive technique, there are very weak dependencies between the watermark and the image. But in the case of a multiplicative scheme, the original data samples are multiplied by an independent factor. In the quantization watermarking schemes, the watermark and the original media will have strong dependencies but are statistically independent.

11.3.6 Hardware-based watermarking systems

There are many advantages to a hardware watermarking techniques. The module can be designed on an FPGA. Designing the module on an FPGA will give more room for optimization for performance and power consumption trade-off and also will be highly efficient (Gafsi *et al.*, 2016; Mohanty *et al.*, 2017). There are also cell-based ASIC designs that are available for the invisible and visible watermarking techniques. The choice between FPGA and cell-based IC is a trade-off between cost, power consumption, and performance (Mohanty *et al.*, 2004, 2005a, 2005b).

11.4 Applications of digital watermarks

This section discusses various applications of watermarking schemes, which include copyright protection, broadcast monitoring, misappropriation detection, ownership assertion, usage control, fingerprinting, information hiding, integrity verification, and so on (Chan *et al.*, 2015; Cheng *et al.*, 2014; Mohanty *et al.*, 2017; Zhang *et al.*, 2015).

11.4.1 Copyright protection

There are cases where the media objects need to be uploaded to the Internet or publicly available. There are also the cases of research. In such cases, the media might be subject to illegal distribution where no royalties will be paid to the owner and the media is distributed or used commercially. There are also applications where the owner might have to distribute the preview of the media for public display before the content is released. In those cases, the owner might desire to have the ownership mark where it can save the media to be used or distributed illegally.

A visible watermark can be used in such cases where a translucent image is spread across the significant portion of the original image or media.

11.4.2 Ownership assertion

The owner might make the media available over the Internet or in a public domain and desire to publicly indicate the ownership of the content. Then the observer will be interested in patronizing the institute that owns the media or object. For example, the books in a library will have a small print of the printers. In case of ownership assertion, an invisible watermark can also be used. A watermark or a key will be generated by the owner and can be embedded into the original media. In such scenarios, an invisible-robust watermark which is a function of original image can be used and more suited.

11.4.3 Authentication and integrity verification

For the authenticating and integrity verification purposes, an invisible-fragile watermark is used. When the media or object is used legally, for example, medical or news purposes, original image has to be verified that if it has been altered or tampered. If the image is captured and saved for a later use, it has to be verified that the image is not altered. For example, if an image is captured with a camera for news purposes, when it is being presented in an article, the image should be original and should not be altered. Similarly, the images of human fingerprints have been stored for various purposes, which has to be secure and authentic. In a library, invisible-fragile watermarking can be implemented to detect any alteration of images.

11.4.4 Fingerprinting

When an image is distributed into the public domain, the owner might want to increase the security of the media by implementing a unique watermark of his/her own into the media, which is called "fingerprint." After embedded into the media, a fingerprint can be retrieved and the origination of the image can be tracked. In this case, an invisible-robust watermark invulnerable to forgery, removal, and invalidation is most suitable.

11.4.5 Usage control

In cases of physical media such as DVD or audio CD, proprietary hardware is to be used for copying and viewing. In such a scenario, invisible-robust watermark is used to determine if the media can be played or copied. A watermark can also be used to determine the number of times the media is copied. There is also a chance that the watermark is changed by the hardware to increase the robustness and to prevent the illegal use of the object.

11.4.6 Broadcast monitoring

Broadcast monitoring is another application where watermarking can protect the content that is being distributed. Owner, broadcaster, and the user are the parties that are involved in this scenario. The owner and the protect rights can be identified using the invisible-robust watermarking in this case and the receiver end needs privacy monitoring stations. The monitoring station is responsible for watermark identification and not the owner in this case. Real-time detection also needs to be done without the original media content created being present in the scenario. A small visible watermark showing the ownership or TV broadcasting station logo can also be used.

11.4.7 Content labeling

Most of the new digital cameras have GPS modules integrated in the camera that is used to stamp the photographs with the location. Time and date can also be stamped into the photograph automatically. During the management and retrieval of the data, this additional information is much helpful. A universal product code (UPC) can also be inserted by the watermarking scheme in the camera to identify the source of the media. Invisible or visible watermarking scheme can be used for this purpose. For successful operation of this scheme, the DSP of the camera has to be integrated with the watermarking blocks.

11.4.8 Misappropriation detection

The misappropriate images and the identity of the person that has performed the act can be identified using the invisible-robust watermarks. The buyer of the shared content media, after buying the legit copy of the media, the mis-appropriator, may share the media for free, which can deprive the owner of legitimate licensing revenue. In this case, the watermark of the seller is retrieved and the availability of the seller's objects are checked. The mis-appropriator might edit the original multimedia objects and distribute them without the payment of royalties. In this case, the owner needs to insert an invisible-robust watermark showing to whom has that object been sold.

11.4.9 Anti-counterfeiting

Controlling the counterfeiting of media is another application of the watermarking techniques. Currency counterfeiting is a major issue that has to be addressed with high priority. With the availability of better performing scanners and printers, and open-source software, currency counterfeiting is becoming an issue every day. Inserting a digital watermark can save in such scenarios. The watermarking algorithms and schemes can be integrated into the printing hardware such that the currency can be saved from counterfeiting.

11.4.10 UAV safety

UAVs have been around for a long time and are used for various applications including military. In some cases, UAVs are used for surveillance where it will have video recording capabilities. The video recorded by the UAV can be used for various purposes, but UAV itself is subject to different kinds of attacks including denial of service, spoofing, and so on. If the attacker gets access to the UAV, the data that is being recorded by the UAV can be tampered with, and the video recordings can be rendered useless. In such scenarios, watermarking the video can to some extent mitigate the threats and protect the integrity of the video recordings.

11.4.11 Medical signals authentication

Medical field is a vast area and privacy protection is of the highest priority as it contains records of many patients. Many cases require confidential media such as videos, photos, or audio recordings, and also documents need to be stored where it may be subject to security threats. In such cases, invisible or visible watermark can be embedded into the media and can safeguard the media or documentation and mitigate the threats.

11.5 Desired characteristics of watermarks

This section discusses different characteristics of watermarks from a researcher point of view (Ahmaderaghi *et al*., 2018; Bender *et al*., 1996; Bi *et al*., 2013; Busch *et al*., 1999; Kougianos *et al*., 2009; Li *et al*., 2016; Makbol *et al*., 2016; Mohanty *et al*., 2004; Swanson *et al*., 1998). The general characteristics of digital watermarking—for example, cost, power consumption, tamper-resistance, perceptibility, orthogonality, robustness, unambiguity, universality, and so on—are discussed in this section. Trade-off between three different parameters is inevitable in the case of digital watermarking, data/information capacity, imperceptibility, and robustness. A single watermarking scheme cannot satisfy all three of these trade-off parameters simultaneously.

11.5.1 Perceptibility

The viewer should not be able to see the invisible watermark and the embedded watermark should not degrade the quality of the image. The watermark should be imperceptible in ideal scenarios. But in case of the imperceptible signal, if it goes through a lossy signal, the watermark signal will be removed or lost. A just noticeable distortion (JND) is usually observed by comparing two signals: watermarked and original (Cox and Miller, 1997; Delaigle *et al*., 1998, 2002; Kankanhalli *et al*., 1998). In case of visible watermark also, the watermark image should not degrade the quality of the original image.

11.5.2 Robustness

A watermark image that is embedded into the original image should be robust to different transformations, which include lossy compressions, digital and analog conversions, and signal distortions. In case of images and video watermarking, the watermark should be resistive to geometric distortions such as cropping, scaling, and translation. The robustness of watermark can be increased by embedding in a significant portion of a signal. But, to make the watermark imperceptible, it should be embedded in perceptually insignificant regions of an image. There are two trade-offs that conflict in designing the algorithm. Ownership establishment need not be the only goal of a robust watermark scheme in an image or a video unless they are implicated legally on the schemes (Craver *et al.*, 1998).

11.5.3 Tamper resistance

Signal distortions are not the only type of transformations that a watermarked image is subjected to but also attacks that intend to remove the watermark image from the original image (Voloshynovskiy *et al.*, 2001; Wang *et al.*, 2018). Hence, it is highly important that the invisible and visible watermarking schemes are resistant to such attacks. Detecting any attacks that are performed on the watermark should also be the quality of an invisible watermarking scheme. Using a private watermark, asymmetric encoder–decoder are some of the ways of protecting the image (Bender *et al.*, 1996; Cox and Miller, 1997).

11.5.4 Bit rate

Watermark bit rate is the amount of information of watermark image that can be encoded into an original image, which is an important aspect in case of public watermarking schemes (Cox and Miller, 1997). Host signal modification and the quantity can vary depending on the application and its type. Some application will require more information to be embedded and in other cases, it requires large amount of data insertions. A watermark algorithm can embed huge data amounts or can be highly robust to different kinds of attacks but they are trade-offs (Bender *et al.*, 1996; Swanson *et al.*, 1998).

11.5.5 Modifiability, multiplicity, cascadability, and orthogonality

In some cases, after a digital watermark is embedded into a media object, it might have to be changed (Cox and Miller, 1997), for example, to control the flow of usage so that the media is used at a limited rate. To achieve this, the original disc, once a copy of it is made, is rewritten with a new watermark on it so that copying it further becomes difficult. If any alteration to the original disc or copying has to be performed, the watermark can be removed first before doing so. Cascadability is the process of adding multiple watermarks to the media at different stages of production or reuse. Pseudo random numbers are used to create spread spectrum watermarking and such watermarks should be orthogonal to be detected individually.

11.5.6 Scalability

It is known that the speeds of the computer double every 18 months (Ross, 2003). Therefore a watermarking algorithm that currently requires high computation power might not seem such computationally demanding in the near future with a better performing computers. For example, the decoders that are initially designed are computationally not demanding but are not as efficient as the new and better ones as the new decoders will have more computational powers to perform the computations.

11.5.7 Unambiguity and universality

The owner of the media object can embed his/her own watermark, visible or invisible into the object for additional security. But the final goal of a watermark is to identify the owner of the media object. In detection process, the watermarking scheme should not have ambiguity and also should not lead to any deadlock situations (Swanson *et al.*, 1998). In order to achieve that, the original media object should meet some conditions (Craver *et al.*, 1996). There is a chance of watermarking an image that is already watermarked and claiming the ownership of the object. Also it can be desired that the same watermarking scheme be used for different purposes including images, videos, and audio media objects. This can be a lot helpful when all the media objects are implemented in a single application.

11.5.8 Pixel alteration and human intervention

In some cases, the value of the artwork that is being watermarked will be very high. In such cases, the watermark should not meddle with the quality of the object. Introducing the watermark or embedding it should not degrade the quality of the original image or artwork. And when a watermark is embedded into the object, the human intervention should be reduced to as minimal as possible. The whole process of watermarking technique should be automated. This can be a lot helpful in reducing the cost of production and the time taken to get the final watermarked product.

11.5.9 Reliability

A watermark, once it is embedded into the original media object, should be reliable in detecting and retrieving when necessary. This will pose some constraints on the capacity of the respective watermark used. There are three figures of merit (FoM) for measuring the reliability of the watermark. They are:

- False positive error rate
- False negative error rate
- Bit error rate

False positive error rate is the probability of determining the watermark when an nonwatermarked image is given as an input to the detection system and the ideal value should be 0. False negative error rate is the rate at which the detector misses the watermark when watermarked image is presented to it. Bit error rate is the number of correct bits that are detected in the watermark image.

11.5.10 Blindness

In the case of a blind watermark embedding scheme, the watermark pattern that is chosen to be embedded on the original media object is independent of the host. And while detecting such watermarks, the information regarding the original object will not be revealed or the knowledge of it will not be available. In some cases, the original media object will not be available when detecting the watermark for authentication or copyright purposes. In such cases, blind watermarking and detecting is highly useful and reliable. But there is a drawback or limitations to the amount of watermark data that can be embedded onto an object and retrieval can also be difficult (Swanson *et al.*, 1998).

11.5.11 Security

Securing the insertion phase of watermarking is of highest priority. To achieve more robustness and security with the watermarks, a secret key can be used such that the watermark that is embedded is dependent on the secret key. In such scenarios, the watermark will not be dependent on the algorithm but the secret key that is being used in the original image (Busch *et al.*, 1999; Swanson *et al.*, 1998). This can be very helpful where even if the adversary gets to know the algorithm that is used for embedding and detection of watermarks, the integrity of the watermark will not be affected.

11.5.12 Real-time operation

There are some applications where the watermarking is used for real-time applications. For example, there are many different TV broadcasting networks that do live TV shows or live sports events. In such cases, the watermarking of the media that is being shared should also be done real time. Such cases also demand the watermarking of audios, videos, and images to be watermarked in real time and broadcast to the public. The signal latency should also be as minimal as possible (Busch *et al.*, 1999; Khan *et al.*, 2010; Marcinak and Mobasseri, 2005; Wang *et al.*, 2008). The hardware that is used for watermarking should be designed with pipelining and parallel architectures for a better performance. The algorithm that is used for watermarking should also be designed to use parallel processing.

11.5.13 Cost and complexity

Implementing a watermark onto an image has a lot of limitations, one of which is the complexity and cost. When designing a watermarking hardware, the cost of the entire module will be calculated in terms of chip area and memory that is required by the algorithm. In case of the blind watermarking, having the original image on hand is not necessary and hence the memory requirements are also comparatively low. With the reduction in cost of watermarking, the overall production cost will also be affected, which leads to a low-cost production.

11.5.14 Energy consumption

With newer technologies being developed, power consumption bar is coming lower every day. This leads to the development of low-power consumption devices and applications. With the introduction of IoT and its applications, different devices require low-power consumption. Digital watermarking schemes also need to consume low-power consumption (Mohanty and Ranganathan, 2004, 2005). The technique may be software or hardware watermarking scheme but the power consumption of the overall process should be kept as low as possible. Different devices have unique specifications of hardware like power sources and running on battery. So, power optimization is highly necessary in terms of watermarking.

11.5.15 Integrability

Digital watermarking has come a very long way since its introduction. But so did the media community and devices that consume or distribute the content with watermarks with it. This has become a large system and watermarking algorithms should have integration capabilities into such huge systems. This is actually a high priority in case of hardware watermarking techniques where the compatibility of hardware module to be embedded in different applications should be increased. While implementing the watermarking chip, the target device, such as a DVD drive, digital camera, printer, and scanner, should be taken into consideration.

11.5.16 Characteristics specific to a watermark

As discussed in Section 11.3, there are different types of watermarking techniques. With the discussion that we had until now in the current section, the characteristics of different types of watermarks can be formulated.

11.5.16.1 Visible watermark

Visible watermarking technique is widely used in many applications and there are characteristics that need to be met for this type of watermark (Mintzer *et al.*, 1997; Mohanty *et al.*, 2000a, 2000b). The color and the monochrome images, when used for visible watermark, should be more obvious when used. When the visible watermark is embedded into an image, a significant portion of the image should be selected for the insertion so that the watermark is not deleted by cropping. The watermark should be transparent and the original image details should be visible. Watermark should be difficult to remove and the removal process should be more intensive compared to purchasing the original image from the owner.

11.5.16.2 Invisible-robust watermark

In the case of an invisible watermark, it should not be visible when embedded into the original image (Mintzer *et al.*, 1997). When an invisible watermark is embedded into the media, the quality of the original media should not be affected and degraded. An invisible watermark should not be affected to any signal distortions and is also resistant to different kinds of attacks targeted to remove the watermark illegally. Identification of the owner should be unambiguous by retrieving the watermark. The decoder of the watermark scheme should be scalable and also be integrated into different generations of computers and systems. Pixel modification should be kept to a minimum when inserting the watermark into a high-quality artwork. The whole process of insertion of watermark should be autonomous with a little human intervention.

11.5.16.3 Invisible-fragile watermark

Similar to the previously discussed invisible watermark, the invisible-fragile watermark should not be visible or noticeable and also should not degrade the quality of the original image. When the pixel values of the image are modified, the watermark should not be removed from the watermarked image and should be secure. This means that it is impossible to recover the changes, or regenerate the watermark after image alternations, even when the watermarking procedure and/or the watermark itself is known. In case of a high-quality image as an input to the insertion algorithm, the pixels that are modified must be kept to a minimum. In some applications, reversing the tampering of watermark might be necessary, which is called error correction.

11.5.16.4 Video watermark

In case of a video watermark, the presence of the watermark in the video should not affect the viewing quality of the video or playback. Compression of the video should not affect the video and watermark and vice versa. Detection of watermark should be reliable and with less errors and false detection, it should be kept to a minimum, ideally zero. Attacks on the video watermarking may be intentional or unintentional like frame dropping and variable rate encoding. Detection algorithm hardware should be designed for low-cost implementations.

11.5.16.5 Watermarking for broadcasting applications

Watermarking applications also include broadcasting applications (Busch *et al.*, 1999). The watermark used in that applications should be resistant to the attacks and the codec encoder should not be overworked. When the watermark is embedded into the image, it should not be affected by the low bit rate of MPEG-2, VHS tape recordings, and TV transmissions. For broadcasting applications, the capacity of 64 bits in 10-second video is used. Multiple watermarking schemes should be able to handle that with false positive detection probability less than 10^{-12} and watermark should be more secure.

11.5.16.6 Desired characteristics of hardware-based watermarking systems

Hardware watermarking chips are more secure and also can be optimized for better performance with low-power consumption (Mathai *et al.*, 2003a; Mohanty *et al.*, 2005a, 2005b). In such cases, the chip area occupied by the chip should be low to reduce the cost of the entire component. The power consumption of the hardware watermarking chip should also be less so that the modules can be integrated in battery operated applications. The integration of the chip into other devices should not be an issue. The operational latency of the chip should be low so that it can be used for real-time applications. Blind watermarking schemes are preferred in such applications.

11.6 Technical challenges for watermarking

There are many technical issues when it comes to the implementation of watermarking. Security, robustness, and imperceptibility are the trade-offs of the digital watermarking and this is what makes the research in this area interesting. For an increased imperceptibility, the watermark must be added to a high-frequency component in the media object. But to attain high robustness, the low-frequency components should be used for watermarking. This becomes more difficult with added characteristics such that the watermark data should be resistant to different kinds of attacks on them. This section discusses different kinds of challenges faced during the watermarking algorithm design (Delaigle *et al.*, 2002; Frattolillo, 2015; Kougianos *et al.*, 2009; Mohanty, 2015; Mohanty *et al.*, 1999; Peng *et al.*, 2010; Singh *et al.*, 2015, 2017; Vo *et al.*, 2017).

11.6.1 Properties of visual signals

The images and the videos are the visual signals. Behavior of such signals has to be studied and understood to develop more robust and secure watermarking schemes and hide them well in the signals. The visual signals are considered as the intensity versus space delays of image information, amplitude plots, and the intensity versus space and time displays. When these waves are studied more, the information regarding the properties of the signals will be revealed. Some of the properties are discussed in the following text.

11.6.1.1 Nonstationarity

The nonstationary property is highly common to almost all of the signals. There is a huge amount of segments of flat or slowly changing intensity in the image and video signals. Besides these there will also be edges and textured regions in the image and video signals. The edges are the important regions that need to be protected to maintain the quality of the image or video and the textured regions should be used for storing additional information.

11.6.1.2 Periodicity

There are frame-to-frame and line-to-line periodicities in the video and image signals. These are not precisely periodic in nature. There will be a little redundancy among the frames and the lines. In the case of the watermarking schemes, where additional content should be stored, this redundancy is taken as an advantage.

11.6.1.3 Power spectral density

The visual signals mostly will have the low-pass frequency spectrum in the long-time average sense or a global sense. In the short-term frequency analysis of these signals in a local sense, when a certain region of the image or frame is selected, they tend to possess all-pass or high-pass frequency spectra. This can be highly useful in the watermark embedding for adding additional information.

11.6.1.4 Properties of color signals

UV or IQ (chrominance) components of any kind of signal will have low energy with a lower bandwidth when compared with a luminance component Y. So when performing the data compression, a coarser quantization is used by subsampling them. This can be considered when developing algorithm for watermarking color and video signals.

11.6.2 Properties of the human visual system

Any watermarking scheme can be called a success when the human visual system (HVS) is also considered and made a best use of it. Various properties of HVS should also be considered when developing watermarking schemes for the images and video (Delaigle *et al.*, 2002; Kankanhalli *et al.*, 1998; Swanson *et al.*, 1998).

11.6.2.1 Brightness sensitivity

Low-intensity signals are easily perceived by the human eye when different background intensities are present. The surrounding intensity is increased, dark area intensity is reduced when the light areas intensity is increased. When the background is the mean of noise square, it tends to be more visible to the human eye. This is called the Weber's law. At low intensities, eye has more sensitivity; in case of highest intensity, it has reduced sensitivity.

11.6.2.2 Frequency sensitivity

When psychovisual studies were performed on the HVS it was discovered that it has a general bandpass characteristics with peak sensitivity of 3 or 4 cycles per degree and the reduced sensitivity at higher and lower spatial frequencies. Watermark intensity in the spatial frequency should be inversely proportional to the sensitivity of the corresponding frequencies.

11.6.2.3 Texture sensitivity

Background texture will control the visibility of distortion. When the background is a strong texture, the visibility will be less distorted. When the image block is highly textured, the energy is more evenly distributed among the DCT coefficients. When the image has a flat featured portion, energy is more concentrated in low-frequency components.

11.6.2.4 JND and minimally noticeable distortion

The visual signals are studied and the properties are then analyzed for a short time or spatiotemporally local analysis was performed on the input signals. When the properties are studied, it was observed that they can be used for deriving the perceptual distortion threshold. This is the function of time, space, or frequency. When the distortion by the watermarking algorithm is below the threshold value, signal quality will degrade. A critical distortion profile is expressed by the threshold as JND. Similarly, minimally noticeable distortion (MND) defines a supra-threshold generalization of it, which is not transparent, but still perceptually optimum.

11.6.3 How much watermark signal to add and where?

Watermarking an object is a communication process where the watermark signal should be inserted reliably into the host, the media object (Ramkumar, 2000;

Ramkumar and Akansu, 2001; Servette *et al.*, 1998; Sugihara, 2001). It was already proven that the communication is reliably possible theoretically but when the information rate does not go beyond the threshold. The threshold value in this scenario is called the channel capacity. The information theory can be used here to derive the rules which will be helpful in deciding the strength of the watermark used and the location on the original media object.

The watermark should be imperceptible and SNR is also highly limited. Communication will be reliable with the increase in bandwidth compensation mechanisms for a poor SNR. So when the watermarking process is being performed, a number of transform domain coefficients should be high and are easy to exploit for hiding information in the object media. An application for the spread spectrum communications can be considered to be digital watermarking. The Shannon limit may be considered by applying the error control codes. If necessary, more robust error correction methods may also be used here. This discusses how much watermark strength to be added to an image or media object. For an answer to the question, where to embed, the watermark should be embedded in a place where the local noise variance is small compared to the threshold value.

11.6.4 Spread spectrum communications

From the discussion until now, it has become more clear that the watermark should not be embedded into a place that is perpetually insignificant region on the image. A watermark that is embedded into a high-frequency domain can be removed using a little degradation of the quality directly or indirectly. Then the problem here is how a watermark can be embedded into the most significant part of the image without noticeable alterations to the image quality. Small changes can be made and spectral coefficient can be altered but the small changes are susceptible to high amounts of noise.

To avoid this issue, the frequency domain of the media object is considered as a communication channel and the watermark is the signal that is transmitted through it. The signal distortions and the attacks are considered as noise that affects the signal. So in this way, a watermarking can be considered as an application for spread spectrum communication (Sequeira and Kundur, 2001). In this case of a spread spectrum communications, a narrow band signal is transmitted over a broad bandwidth so that the signal energy is imperceptible. In the same manner, a watermark is spread over a large image such that it is imperceptible. Location of the watermark is known to the watermark detection systems.

11.7 Hardware-based approaches for watermarking

The watermarking research has advanced in the past few decades and the maturity attained during this period has advanced the software-based digital watermarking schemes. But when it comes to hardware watermarking schemes, the development is not at an exponential rate compared to the software watermarking. There are already implementation of hardware modules for watermarking. But in designing the watermarking modules based on the hardware, there are a lot of limitations such as power consumption, chip area, latency, and so on (Kougianos *et al.*, 2009; Krishnamurthy, 2018; Mohanty, 2015, 2017; Mohanty *et al.*, 2018; Pexaras *et al.*, 2017; Roy *et al.*, 2013; Sengupta *et al.*, 2017). There are also many advantages of hardware modules for watermarking, such as low-power consumption, optimization, high reliability, real-time performance, and easy integrability. The watermarking schemes hardware implementation can be design of custom-made VLSI chips or can be designed on an FPGA or can be built around a DSP board. It will become a part of still or motion camera, digital TV, DVD player, or recorder and MP3 player or recorder. For example, an encryption based mechanism called "Signcryption" has been integrated for protection of data in smart cameras in IoT environment (Figure 11.9) (Ullah *et al.*, 2017). This section presents different hardware watermarking schemes and the designs. A comparative analysis of various designs of watermarking chips that are presented in this section is provided in Table 11.1 (Kougianos *et al.*, 2009).

11.7.1 Image watermarking hardware systems

This section discusses the hardware-based watermarking schemes that are used for watermarking the images. This kind of hardware modules can be integrated into the still cameras and is useful in watermarking the images on the fly. The hardware modules embedded into the camera or other production systems will help a lot in reducing the extra production time of creating watermarked final products.

11.7.1.1 Invisible-robust wavelet domain VLSI architecture

A progressive watermarking module is presented by Hsiao *et al.* (2000a, 2000b), which is a framework of image encoder. The design of encoder and decoder is used for high-speed image and video compression using the "embedded zero wavelet" (EZW) tree algorithm. The EZW encoder–decoder is embedded with a progressive digital watermarking algorithm which can increase the robustness of the entire design. Embedder is a part of the architecture of compressor encoder and the detector is part of the decompressor.

A progressive invisible robust algorithm is presented which is able to detect the watermark if the received data is partial but not complete. This algorithm embeds the pseudorandom numbers into the wavelet coefficient. Then the detector will compute the correlation between the original watermark sequence that is already present in the memory and the received image. The number generated for the watermark is embedded in the position next to the nonzero bit of each significant wavelet coefficient. A minimal hardware is required to implement this algorithm.

Figure 11.10(a) presents the insertion algorithm that is used in this design. This module consists of the LFSR, linear feedback shift register, which will generate the pseudo random number for the watermarking scheme. Comparator will then make sure that the watermark bit is embedded into the significant coefficients of the original image. Figure 11.10(b) presents the detection module for the watermarking scheme. The detection module consists of LFSR, comparator, top index register, and a threshold table. The total number of significant coefficients that is received by the detector is stored in the top index register. Then the threshold correlation value is stored in the threshold table and are compared during the detection process.

Table 11.1 Watermarking chips proposed in the current literature

Research work	Design type	Watermarking type	Multimedia object	Working domain	Chip statistics
Pexaras *et al.* (2017)	FPGA Board	Invisible	Image	NA	88.69 MHz, 4,399 Logic elements utilized
Tsai and Wu (2003)	Custom IC	Robust	Raw video	Spatial	NA
Mathai *et al.* (2003a)	Custom IC	Invisible robust	Raw video	–	100 MHz
Roy *et al.* (2013)	FPGA Board	Semi-fragile	Video	MJPEG	40 MHz @ 607 fps (256×256) 40 MHz @ 130 fps (340×480)
Mohanty *et al.* (2018)	Simulink	Invisible Robust blind	Image	DCT	2,400 MHz @ 25 fps
Roy and Yadid-Pecht (2012)	FPGA Board	Visible Semi-fragile	Video	DCT	100 MHz @ 43 fps
Mohanty and Kougianos (2011)	FPGA Board	Invisible Semi-fragile	Image	DCT	40 MHz @ 591 fps
Strycker *et al.* (1999, 2000)	DSP processor board	Invisible Robust	Video	Spatial Fourier	100 MHz
Hsiao *et al.* (2000a, 2000b)	Custom IC	Invisible Robust	Image	Wavelet	NA
Maes *et al.* (2000)	FPGA board Custom IC	Invisible Robust	Video	Spatial Fourier	17 kG Logic 14 kG Logic
Tsai and Lu (2001)	Custom IC	Invisible Robust	Image	DCT	0.35 μ, 3.064×3.064 mm^2, 3.3 V, 50 MHz, 62.78 mW

(Continues)

Table 11.1 (Continued)

Research work	Design type	Watermarking type	Multimedia object	Working domain	Chip statistics
Petitjean *et al.* (2002)	FPGA board	Invisible	Image	Fractal	50 MHz takes 6 µs
	DSP processor	Robust	Video		250 MHz takes 118 µs
Garimella *et al.* (2003)	Custom IC	Invisible Fragile	Image	Spatial	0.13 µ, $3,453 \times 3,453$ mm^2 1.2 V, 37.6 µW
Mathai *et al.* (2003a, 2003b)	Custom IC	Invisible Robust	Video	Wavelet	0.18 µ, 3.53 mm^2, 1.8 V, 75 MHz 75 MHz, 160 mW (approx.)
Mohanty *et al.* (2003)	Custom IC	Invisible Robust-fragile	Image	Spatial	0.35 µ, 3.3 V, 545 MHz, 2.0547 mW
Seo and Kim (2003)	FPGA board	Invisible Robust	Image	Wavelet	82 MHz, 4037 LABs, 85 ESBs Altera APEX20KC
Mohanty *et al.* (2004, 2005)	Custom IC	Visible	Image	Spatial	0.35 µ, 3.34×2.89 mm^2 3.3 V, 292 MHz, 6.93 mW
Mohanty *et al.* (2005)	Custom IC	Invisible-robust Visible	Image	DCT	0.25 µ, 16.2 mm^2, 1.5, 2.5 V 70,280 MHz, 0.3 mW
Mohanty *et al.* (2004)	FPGA board	Invisible Robust	Image	Spatial	Xilinx, Virtex XCV50-BG256-6, 50.398 MHz
Garimella *et al.* (2004)	Custom IC	Invisible Fragile	Image	Spatial	0.13 µ, 545 µm × 525 µm 1.14 V, 166.6 MHz, 9.19 mW
Fan *et al.* (2005)	Custom IC	Visible	Image	Wavelet	

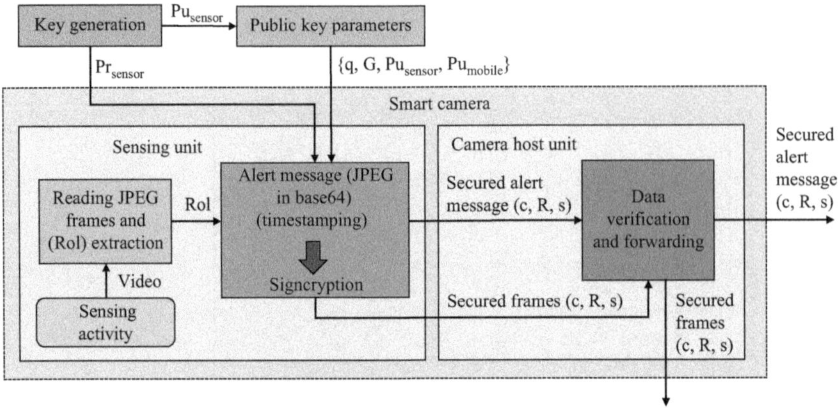

Figure 11.9 Signcryption

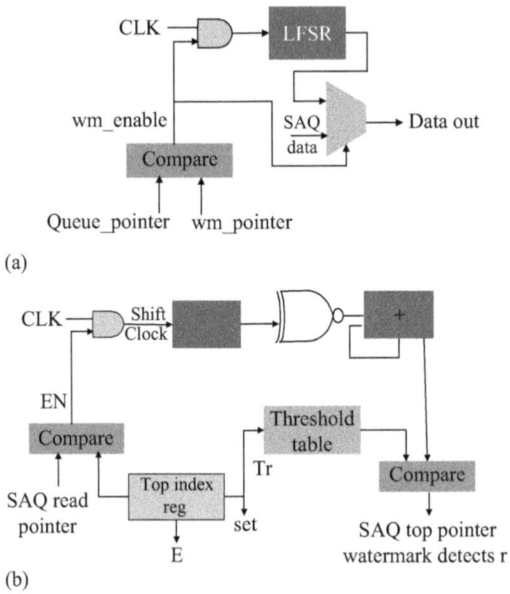

Figure 11.10 Watermarking embedder–detector modules: (a) embedder and (b) detector (Hsiao et al., 2000a, 2000b)

11.7.1.2 Invisible-robust DCT domain chip

The image watermarking chip should be able to insert the invisible-robust watermarking in DCT domain and was presented by Tsai and Lu (2001). This system will embed a pseudo random sequence into a previously selected set of DCT domain coefficients. When extracting the watermark, the original image is not needed. This can be a great advantage of the hardware watermarking scheme. The chip presented by Tsai and Lu (2001) used a TSMC 0.35 μm technology and occupies a die size of $3.064 \times 3.064\ \text{mm}^2$ for 46,374 gates. The power consumption of the chip is 62.78 mW and is operated at 50 MHz with a 3.3 V supply voltage. This is very memory efficient as the original image is not stored in the memory of the chip for extraction or detection.

A Gaussian random number generator is used for the watermark sequence and is represented as $N(0,1)$. The length of the original DCT coefficients will be the length of the sequence of watermark. Three DCT coefficients are embedded with the watermark with each nonoverlapping 8×8 block. The strength of watermark comes from the texture complexity and the average brightness. The image blocks are classified as:

- Bright and strong texture
- Dark and weak texture (strongest watermark)
- Rest (weakest watermark)

If the watermark sequence is denoted by w_i, the DCT coefficient of the kth image block is represented by $F_k(u, v)$, and the scaling factor is α_k, the watermarked DCT coefficients can be:

$$F_k^i(u,v) = \begin{cases} F_k(u,v) + \alpha_k w_i, & 3k < i < 3(k+1) \\ (u,v) \in \{(0,1),(1,0),(1,1)\} & \\ F_k(u,v), & \text{otherwise} \end{cases} \tag{11.6}$$

The scaling factor α_k is 2, 6, and 9 for class 1, class 2, and class 3 image blocks, respectively. Extraction of watermark can be dependent on the original watermark and the extracted watermark's correlation.

The chip is designed to be integrated with the JPEG encoder which has the DCT/IDCT, hash, quantization, zigzag (ZZ) scan, and VLC encoder modules. Figure 11.11 shows the architecture for watermarking functionality. The coefficients

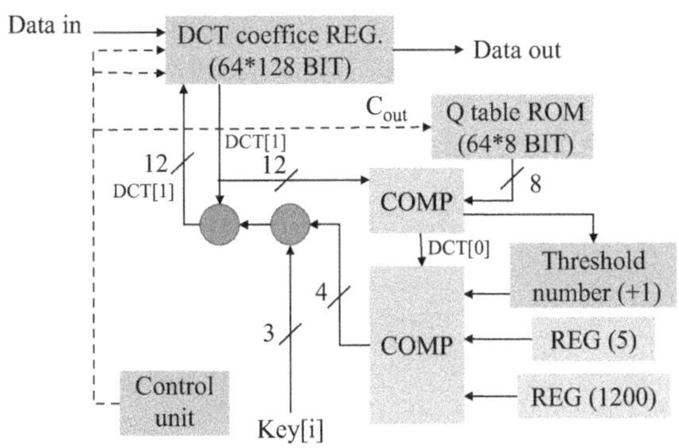

Figure 11.11 Architecture for watermarking embedding module (Tsai and Lu, 2001)

of DCT are stored in the register, watermark sequence in the ROM, and the comparison is carried out by the comparator. The FSM is used to implement the controller which gets the watermarking done using the data path.

11.7.1.3 Invisible-fragile spatial domain chip

A watermarking VLSI design for the implementation of watermark in the spatial design is presented by Garimella *et al.* (2003). Here, the differential error is encrypted and along the first sample, it is interleaved. The digital watermark extraction can be done by the accumulation of consecutive LSBs of the pixels and then decrypting. The watermark that is extracted will then be compared with the original watermark for authentication purposes. This is designed with a 0.13 μ technology. Chip area occupied by the module is $3,453 \times 3,453\ \mu m^2$ and power consumption 37.6 μW at an operating voltage of 1.2 V. The critical path delay of the circuit is 5.89 ns. This design has an advantage of low-power consumption but only works for grayscale images.

The watermark used here is a text-based one that is converted into an ASCII code. This code will undergo an encryption process using a differential pulse code modulation (DPCM) technique for increased security. The final size of the watermark will be 2,048 bits. The LSB of the original image is replaced with the encrypted ASCII code. In the process of extraction, the LSBs of the image are accumulated and decrypted. This gives the original watermark from the image.

Figure 11.12 presents the block diagram of the watermarking chip. Two CMOS clocked RAM modules, RAM 1 and RAM 2, are integrated, which are 40,000 and 2,018 bytes, respectively. The original and the watermarked images are handled in the RAMs. Watermark data is stored in the internal registers permanently. The extraction and insertion process is handled by the controller.

Color image watermarking for the same scheme is presented by Garimella *et al.* (2004). RGB-YUY and YUV-RGB converters are designed along with the

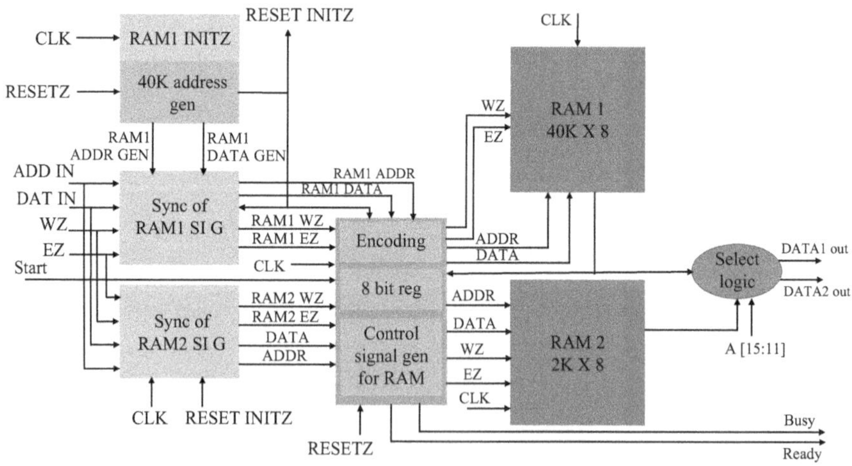

Figure 11.12 Functional block diagram of watermarking module (Garimella et al., 2003)

watermark encryption, embedding, extracting, detection, and authentication modules. Similar to the previous design, the LSB of the pixels is integrated with the watermark and the pixel array of 8-bit grayscale image. The color image is initially transformed into the YUV color space and the encrypted watermark is inserted into it. After the embedding process, the YUV space is retransformed into the RGB color space. The luminance component Y is the watermarking channel where the RGB image transforms into the YUV color space. LSBs from the Y are used for the watermarking process in which the watermark is the encrypted stream of the ASCII code text file. Each bit is swapped with first LSB of Y and second LSB is the complemented bit of the first LSB. DPCM technique is used for the encryption process. During the detection process, the LSBs are accumulated and then decrypted to be compared with the original ASCII code text.

11.7.1.4 Invisible-robust-fragile spatial domain chip

A hardware watermarking scheme that can embed both the invisible robust and invisible fragile watermarking is presented by Mohanty *et al.* (2003). A ternary watermark is inserted into the original media object in case of an invisible-robust watermarking. This can be done with an encoding function with an addition of scaled gray value of the neighboring pixels. The watermark is a binary sequence of numbers which are pseudo random and are XORed with the original image bit plane. The chip is implemented using 0.35 μ technology and consumes 2.0547 mW when operated at 3.3 V and 545 MHz frequency.

An invisible-fragile algorithm proposed by Mohanty *et al.* (1999) and invisible-robust algorithm proposed by Tefas and Pitas (2001) are used in the chip presented above. This design can be implemented with a JPEG encoder. In the invisible-robust watermarking, W, the watermark is the ternary image of three pixel values that are generated with a key K. The original image pixels, I, are altered to obtain the watermarked image I_W and are as follows:

$$I_W(i,j) = \begin{cases} I(i,j) & \text{if } W(i,j) = 0 \\ E_1(I(i,j), I_N(i,j)) & \text{if } W(i,j) = 1 \\ E_2(I(i,j), I_N(i,j)) & \text{if } W(i,j) = 2 \end{cases} \tag{11.7}$$

Assuming $\alpha_1 > 0$ and $\alpha_2 > 0$, the encoding functions E_1 and E_2 are defined as follows:

$$\begin{aligned} E_1(I, I_N) &= (1 - \alpha_1)I_N(i,j) + \alpha_1 I(i,j) \\ E_2(I, I_N) &= (1 - \alpha_1)I_N(i,j) - \alpha_2 I(i,j) \end{aligned} \tag{11.8}$$

The neighborhood image pixel gray value I_N is calculated as the average gray value of neighboring pixels of the original image as shown below:

$$I_N(i,j) = \left\{ \frac{\dfrac{I(i+1,j) + I(i+1,j+1)}{2} + I(i,j+1)}{2} \right\} \tag{11.9}$$

In the invisible-fragile insertion process, a pseudo random binary-sequence {0,1} is inserted in the kth bit plane of the original image using the XOR function:

$$\begin{aligned} I_W[0 \rightarrow k-1](i,j) &= I[0 \rightarrow k-1](i,j) \\ I_W[k](i,j) &= I[k](i,j) \text{ XOR } W(i,j) \\ I_W[k+1 \rightarrow](7i,j) &= I[k+1 \rightarrow 7](i,j) \end{aligned} \quad (11.10)$$

Finding of the candidate bit plane for watermark insertion is an iterative process, but the second ($k=2$) bit plane is chosen as the candidate for watermark insertion (for LSB, $k=0$). The watermarked image I_W is obtained by merging all the bit planes.

The data path architecture is presented in Figure 11.13. In this design, there are two RAM modules, watermark RAM and image RAM. They are used to temporarily store the image and watermark in them. Shift registers are used to generate the sequence of watermark. Address of the ram can be decoded using the address decoders. Main functional units are used for the adder/subtractor, adder, multipliers, and XOR. When a selection is needed, the multipliers are used. Data path that is designed as an FSM is used as a controller.

11.7.1.5 Invisible-robust wavelet domain module

In case of a blind watermarking scheme, the original image is not needed. An FPGA implementation of the blind watermarking scheme is presented by Seo and Kim (2003).

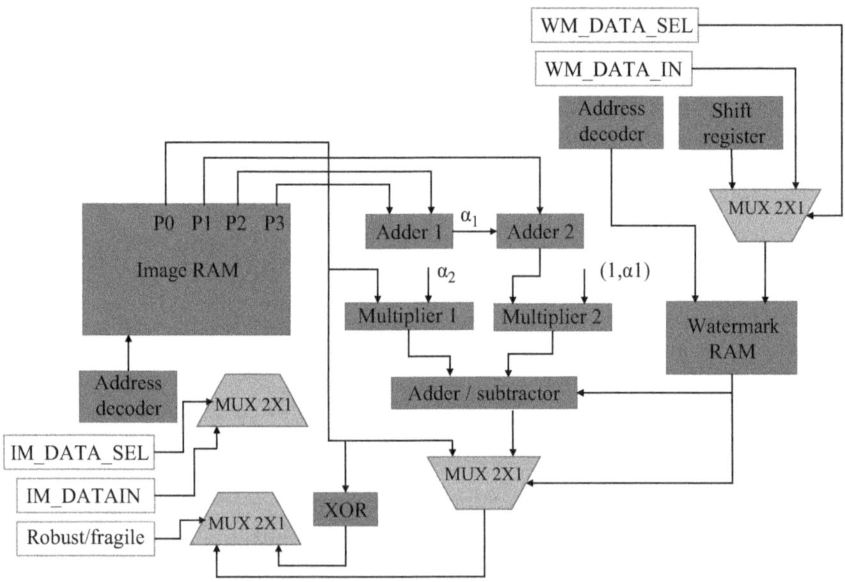

Figure 11.13 Architecture for invisible-robust/fragile watermarking chip (Mohanty et al., 2003)

In the two-dimensional discrete wavelet transform (2D-DWT), one or two image bit planes are replaced with the watermark. The hardware module is designed to be operated at 82 MHz, and 24% of logic array blocks and 3% of embedded system blocks are used of an Altera APEX20KC EP20K400CF672C7 device. JPEG 2000 CODEC was integrated with the watermarking scheme.

The watermark used in this scheme is created from the binary numbers. The candidate for the watermark embedding is the DC which makes it more robust to the attacks. The bit plane selected is done with more care such that the human eye cannot see the watermark because the DC subband has most important information regarding the image. During the embedding, the watermark is divided into 4-level 2D-DWT. Binary watermark is embedded in the second plane, the third plane from LSB plane. Selection of bit plane p is chosen from a set of k bit planes. Secret key is the initial value of the LFSR. Then, k parallel outputs of LFSR $(I_{k-1}, I_{k-2}, \ldots, I_0)$ are chosen followed by computation of p as $p = \left[I_{k-1}.2^{k-1} + I_{k-2}.2^{k-2} + \cdots + I_0.2^0\right] \bmod k$. The same procedure is performed in reverse to extract the watermark from the image.

Figure 11.14(a) presents the encoder for JPEG 2000 architecture. Decoder is also similar to the encoder architecture. There are six functional units in the architecture:

- Image-side controller to control the image input
- Wavelet kernel
- Quantizer/entropy encoder
- Storage-side controller/buffer to control the output of the compressed data
- Memory controller to control the data access with external memories
- Global controller/scheduler to determine and control the whole processing

Quantizer/entropy encoder modules consist of the watermarking scheme. Figure 11.14(b) shows the architecture of the watermarking module.

11.7.1.6 Invisible-robust spatial domain module

Invisible-robust insertion algorithm is presented by Mohanty *et al.* (2004). This algorithm was implemented on a Xilinx FGPA board and VIRTEX technology with XCV50-BG256-6 target device was synthesized.

Figure 11.15 shows the functional diagram of the encoder chip. The encoder of this design consists of watermark generator, insertion, address decoder, controller, and registers. External storage is used to load the images. An LFSR is used for the watermark generation, which is the most important part of the design. LFSR should be avoided to power on randomly, which may become permanently stuck with some prohibited states. Insertion module consists of adder/subtractor, adders, subtractors, multipliers, and shifters. Controller is an eight-state FSM.

11.7.1.7 Visible spatial domain chip

A watermarking VLSI architecture that can insert two visible watermarks in images is presented by Mohanty *et al.* (2004, 2005b). In this case, either of the watermarks

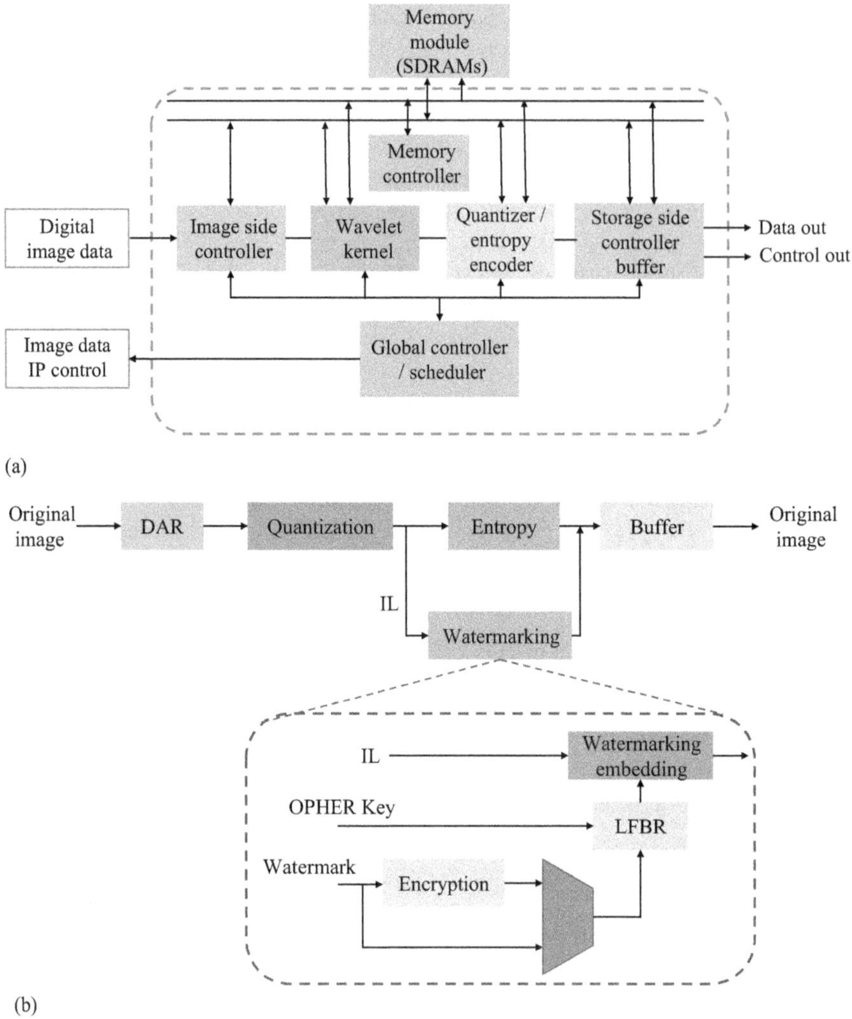

Figure 11.14 (a) JPEG2000 encoder and (b) watermarking embedding module (Seo and Kim, 2003)

can be embedded into the image based on the requirements. This design is implemented using a 0.35 μ technology and occupies an area of 3.34 × 2.89 mm^2. The power consumption of the module is 6.9286 mW with an operational voltage of 3.3 V and 292.27 MHz frequency. Two spatial domain visible watermarking algorithms were implemented in this design. The first one is the visible watermarking that is proposed by Braudaway *et al.* (1996) and the second one is the spatial domain watermarking presented by Mohanty *et al.* (1999).

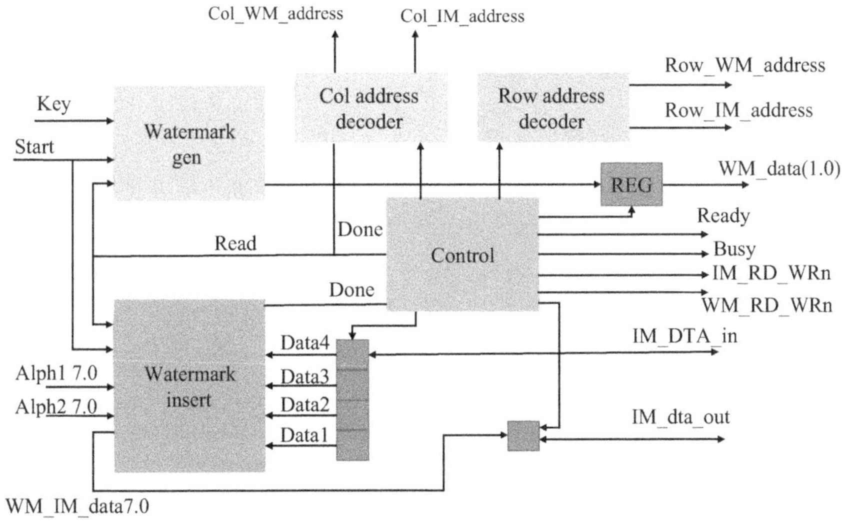

Figure 11.15 Invisible-robust spatial domain watermarking module (Mohanty et al., 2004)

Let i be the original image, w be the watermark image, and i_w be the watermarked image. The watermarked image for the first algorithm is obtained as follows:

$$i_w = \begin{cases} i(m,n) + \left(\dfrac{\alpha_i}{903.3}\right) i^*(m,n), & \text{for } i(m,n) \leq 2 \\ i(m,n) + \left(\dfrac{\alpha_i C_1}{6.0976}\right) i^*(m,n), & \text{for } 2 < i \leq 64 \\ i(m,n) + \left(\dfrac{\alpha_i C_2}{6.0976}\right) i^*(m,n), & \text{for } 64 < i \leq 128 \\ i(m,n) + \left(\dfrac{\alpha_i C_3}{6.0976}\right) i^*(m,n), & \text{for } 128 < i \leq 192 \\ i(m,n) + \left(\dfrac{\alpha_i C_4}{6.0976}\right) i^*(m,n), & \text{for } 192 < i < 256. \end{cases} \quad (11.11)$$

where α_i is a global scaling factor, $i^*(m,n) = W(m,n) \times i(m,n)$, and C_1, C_2, C_3, and C_4 are linear regression coefficients used to linearize cubic root. In the second algorithm the watermarked image blocks are obtained by modifying the kth original image block i_k as shown below:

$$i_{Wk} = \alpha_k i_k + \beta_k w_k, \text{ where } k = 1, 2 \ldots \quad (11.12)$$

where the kth watermarked image block is represented by i_{Wk}, the scaling factor is α_k, and embedding factor is β_k. The scaling and embedding factors for the kth image block are obtained using the following equation:

$$\begin{aligned} \alpha_k &= \alpha_{\min} + (\alpha_{\max} - \alpha_{\min}) \\ &\quad \times \frac{1}{\hat{\sigma}_{kI}} \exp(-(\hat{\mu}_{kI} - \hat{\mu}_I)^2) \\ \beta_k &= \beta_{\min} + (\beta_{\max} - \beta_{\min}) \\ &\quad \times \hat{\sigma}_{kI}(1 - \exp(-(\hat{\mu}_{kI} - \hat{\mu}_I)^2)) \end{aligned} \tag{11.13}$$

where $\hat{\mu}_{kI}$, $\hat{\mu}_I$, and $\hat{\sigma}_{kI}$ are, respectively, the normalized mean gray value of kth image block, the normalized standard deviation of the gray values of the kth image block, and normalized mean gray value of the image.

Figure 11.16 shows the architecture of the watermarking scheme. There are various units in the design such as the edge detection unit, α_k and β_k calculation unit, comparator, register files, multipliers, and adders. Scaling and embedding factors are calculated by the α_k and β_k calculation unit. The watermarking process uses the constants which are saved in the registers. Pixel processing is performed in the adders and multipliers. Both the α_k and β_k calculation unit and the edge detection unit are constructed using adder, adder/subtractor, multiplier, divider, comparator, accumulator, and exponential unit. The controller is designed using a five-stage FSM.

11.7.1.8 Visible wavelet domain chip

A discrete WT-based watermarking design is proposed by Fan *et al.* (2005). This reduces the operational and resource sharing which can increase the overall efficiency of the design. The watermark and the original image are transformed into three-level multiresolution structures. Original image is then divided into two different sequences of the same pattern length. A two-path parallel processing is used for calculating the processing time. De-multiplexers will then send the signal to different elements for processing. The watermark image is embedded by modifying the coefficients of the image.

11.7.1.9 Low-power chip with DCT domain-invisible-robust and visible functionalities

A low-power watermarking chip is presented by Mohanty *et al.* (2005a). This is capable of both the invisible and visible watermarking techniques. The power consumption of this chip is significantly low and there is also the integration of dynamic frequency clocking and clock gating. A parallel architecture and three-stage pipelining is used to significantly improve the performance of the watermarking. TSMC 0.25 μ technology was used for design and verification. The dual frequency of the chip is 280 and 70 MHz and operates at voltages of 2.5 and 1.5 V. The average power consumption of the chip is estimated to be 0.3 mW.

DCT domain-visible watermarking algorithm presented by Mohanty (1999) and Mohanty *et al.*, (2000a, 2000b) and the spectrum-invisible watermarking algorithm presented by Cox *et al.* (1996) are used for the implementation of the chip. Original image DCT coefficients are computed and considered as a single block. After that, 1,000 largest coefficients are identified as the perceptually

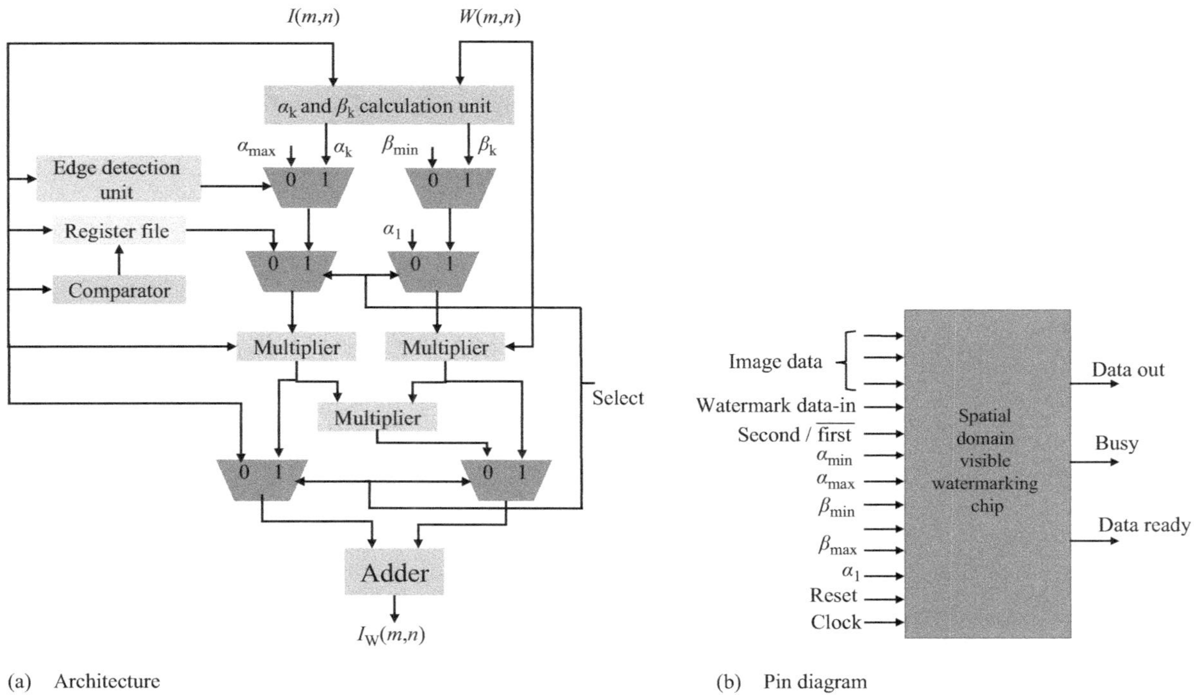

(a) Architecture (b) Pin diagram

Figure 11.16 Architecture for spatial domain visible watermarking (Mohanty et al., 2004, 2005b)

significant for the original media object. The watermark $X = x_1, x_2, ..., x_{1,000}$ is computed in which each x_i is chosen according to $N(0,1)$, where $N(0,1)$ denotes a normal distribution with mean 0 and variance 1. Then the watermark is embedded in the DCT domain of the original object, with the frequency components in C_I using the following:

$$C_{I_W}(m,n) = C_I(m,n) * (1 + \alpha x_i) \tag{11.14}$$

The values of m and n correspond to the pixels locations for the 1,000 largest DCT coefficients and $\alpha = 0.1$. The visible insertion algorithm is as follows: The original image I (one to be watermarked) and the watermark image W are divided into blocks. The DCT coefficient C_I for all the blocks of the original image is calculated. The scaling and embedding factors for each block are computed as follows:

$$\begin{aligned} \alpha_k &= \sigma^*_{ACI_k} \exp(-(\mu^*_{DCI_k} - \mu^*_{DCI})^2) \\ \beta_k &= \frac{1}{\sigma^*_{ACI_k}} (1 - \exp(-(\mu^*_{DCI_k} - \mu^*_{DCI})^2)) \end{aligned} \tag{11.15}$$

The α_k and β_k are scaled to the range $(\alpha_{\min}, \alpha_{\max})$ and $(\beta_{\min}, \beta_{\max})$, respectively. The edge blocks are determined, and the α_k and β_k for edge blocks are taken to be $\alpha_{\max}$ and $\beta_{\min}$, respectively. Then the calculation of the DCT coefficient, C_W, is performed and the visible watermark is inserted in the original image block-by-block and watermarked image block is obtained as follows:

$$c_{I_{Wk}} = \alpha_k c_{Ik} + \beta_k c_{Wk} \tag{11.16}$$

Figure 11.17(a) presents the architecture of the watermarking scheme and Figure 11.17(b) presents how different modules are run with different supply voltages. DCT modules are used for the insertion along with the random number generator and invisible insertion. Insertion module will add the random numbers to the DCT coefficients which are calculated from the original image. Insertion module input is the α parameter. The modules that are used for visible watermarking are edge detection, DCT, perceptual, scaling and embedding factor, and visible watermark insertion. Scaling and embedding modules will compute α_k and β_k. Visible watermark insertion module will embed the watermark into the host image.

11.7.2 Video watermarking hardware systems

Until now, this section discussed the implementation of hardware-based watermarking schemes for the image files. But this part of the section presents different modules for the implementation of the watermarking schemes for video systems.

11.7.2.1 Invisible-robust spatial domain module

A real-time embedder–detector modules are developed for the implementation in broadcasting in the framework of VLIW DSP processor and are presented by

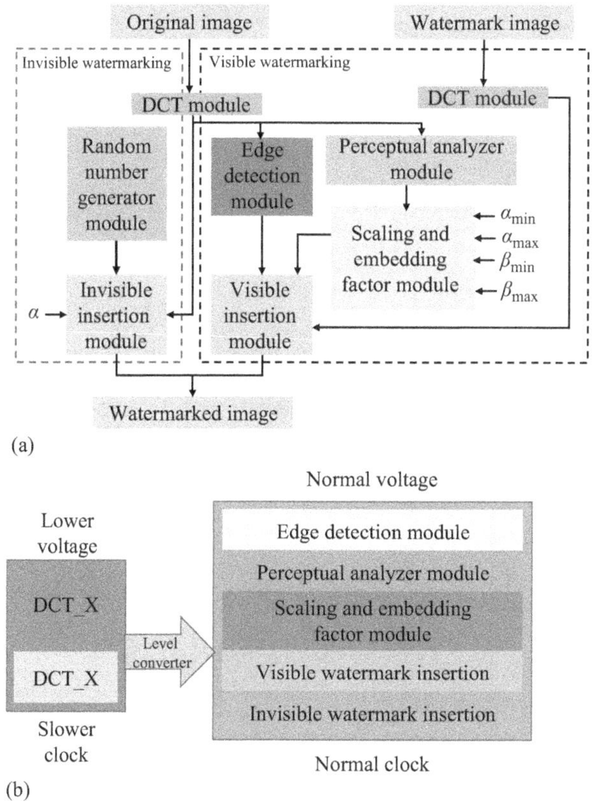

Figure 11.17 Low-power DCT domain-invisible and domain-visible watermarking chip: (a) datapath architecture and (b) dual voltage and frequency operation (Mohanty et al., 2005a)

Depovere *et al.* (1999), Strycker *et al.* (1999, 2000), and Termont *et al.* (1999). During the insertion of the watermark, the pseudorandom numbers are added to the received video stream and it is based on the luminance value of the frame. Detection of the watermark will depend on the correlation value calculations. Spatial and Fourier domains are used for watermark insertion and detection. Insertion process consists of adding pseudo random numbers to the stream. The initial sequence is a user key and then it is shifted and sum is calculated to generate n basic pseudo random patterns. Then one pattern out of those n is chosen thus creating a key. The size of this created number is $N \times M$, which is same as that of the size of video frame. Depth of insertion will determine the imperceptibility and the robustness of the watermark. Larger values make the watermark more secure but are likely to be visible than the smaller numbers. The algorithm is designed to increase the security while reducing the visibility. The detection module will retrieve the key if a watermark is present. A group of G frames are selected and then folded. Sum of it is

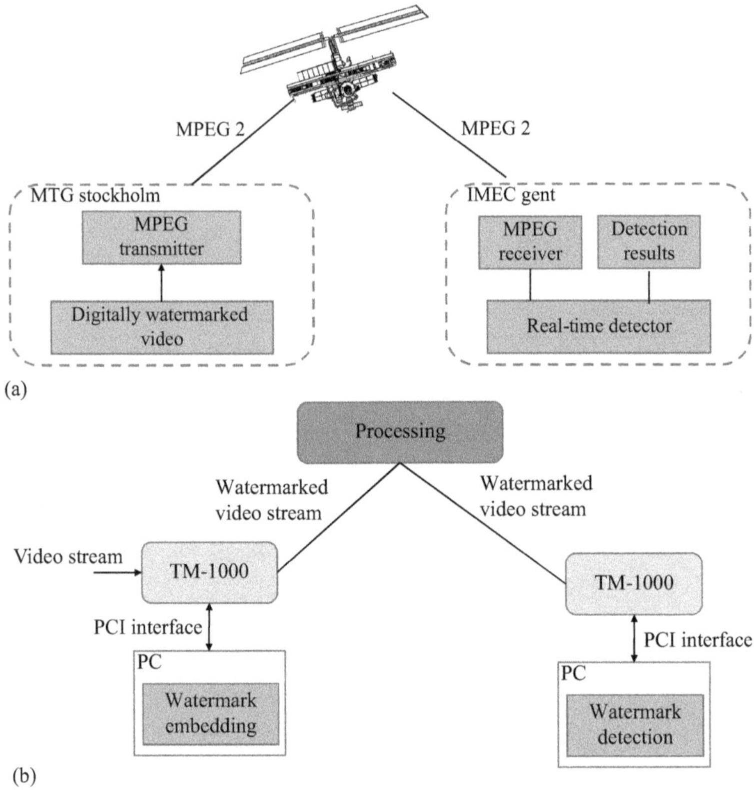

Figure 11.18 Simulation of broadcast environment: (a) TV broadcast and (b) simulation (Depovere et al., 1999; Strycker et al., 1999, 2000; Termont et al., 1999)

calculated with a buffer of size $N \times M$ and correlated with each of n pseudo random numbers. This is equivalent to the 2D cyclic convolution which can be efficiently performed in the Fourier domain. If the correlation so obtained, known as decision variable, exceeds the predefined threshold, the video is declared as watermarked. Figure 11.18(a) shows the commercial TV broadcasting where two Trimedia processors are used. Figure 11.18(b) shows the setup of two PCs, each with a Trimedia PCI card inside.

11.7.2.2 Invisible-robust spatial domain chip

A millennium watermarking scheme system is presented by Maes *et al.* (2000) for the implementation in DVD video system. The watermark detector location and the generation of copies of the DVD are discussed by Maes *et al.* (2000). The insertion phase of the watermark in this algorithm consists of addition of scaled watermark to the original media object signal in the spatial domain. The detection phase consists of calculating the correlation in Fourier domain. The algorithms are presented by Strycker *et al.* (1999, 2000). The memory usage in this scheme for DVD video watermarking should be minimal.

A real-time detection module is designed and implemented using three different modules such as Trimedia processor board, high-end silicon graphics workstation, and an FPGA. FPGA platform implementation has been the most relevant implementation for the DVD watermarking (Maes *et al.*, 2000). A random number generator is implemented which increases the robustness of the entire modules and the watermark.

Implementation of watermarking for the raw video encoding is presented by Tsai and Wu (2003). A standard video encoder is used for implementing the watermarking algorithm like MPEG-1/2/4, H.26x, and Motion JPEG for a real-time IP protection. The insertion algorithm has three steps. Initially the watermark is read and the binary transformation is performed with a chaotic function. Then pseudo-noise is generated, which is of lengths 63 and 31, for both spatial and chaotic transformed domains. Then insertion is performed with a human visual characteristics that will dictate the weighting factor of α for watermark insertion to be 16 for edge blocks and 4, otherwise.

In case of detection of the watermark, the correlation of the original pseudo-random noise is calculated. For implementing the algorithm, verilog was used for the watermark function. This has an advantage of using an MPEG encoder for the compression and increasing the robustness of the video. Watermark function has three parts, pseudo-noise generator, data path, and inserting component and watermarking information generator.

11.7.2.3 Invisible-robust fractal domain module

A real-time video watermarking scheme was presented by Petitjean *et al.* (2002). This actually embeds the watermark with the help of fractal approximation. Extraction of the watermark is done by performing the process in reverse to the insertion phase. C programming language was used for the implementation of the encoder and decoder. A general-purpose PC implementation of this algorithm will yield better results than that of the others.

For an increased security of the watermark, embedding exploits the affine transformation invariance properties of fractal coding. Embedding the watermark has three different phases:

- Formatting and encryption
- Cover generation
- Embedding

During formatting and encryption, the message bits that are to be hidden will be oversampled and duplicated. This gives out the watermark of the size of object or host image video frame. Pseudo-random noise that is generated will be XORed to the watermark and is secured. The difference of original image and fractal approximation gives the cover signal. Watermark is then modulated to the cover which includes forcing some of the cover pixels to zero based on the watermark bit and sign. Modulated cover is added to the fractal approximation which gives the watermarked image. The process is reversed to extract the watermark from the frame.

11.7.2.4 Invisible-robust wavelet domain chip

A hardware implementation of video watermarking algorithm is presented by Mathai *et al.* (2003a, 2003b). The algorithm is based on the implementation from Depovere *et al.* (1999), Strycker *et al.* (1999, 2000), and Termont *et al.* (1999). In the paper, both the embedder and the detector are presented with a floating point data path architecture. For the implementation of algorithm, 0.18 μ technology is used. The power consumption of the embedder module is 60 mW and detector is 100 mW at an operational voltage of 1.8 V.

The embedder consists of payload modulator, high-pass filter, and a watermark superposition unit. A video frame that is distorted without the degradation of quality is selected by the high-pass filter. The payload modulator modulates the pseudo random noise pattern using input payload to generate a base watermark which is then multiplied with the high-pass filter output to create the desired watermark. Then the raw images of the video will be added to the watermark that is created. Detection process contains the filter that is similar to the high-pass filter which is used for removing unnecessary components. The fold loop consisting of fold buffer and the adder components will accumulate the frames that are filtered. After accumulation, the input is stalled and detection will continue with the FFT computation. Then the pixel-wise multiplication is performed. The scan hardware is a set of registers used to keep track of highest correlation peaks.

11.7.2.5 Visible watermarking hardware

A visible watermarking hardware for video copyright protection in compressed domain is shown in Figure 11.19 (Cai, 2006; Mohanty and Kougianos, 2011; Mohanty *et al.*, 2009). The watermark-embedding IBP embeds a watermark image into IBP video frame, where I is intra-frame and B and P are inter-frames. The watermark embedding I module embeds a watermark to intra-frame only. The "Frame Buffer" buffers the frames for every block procedure. Its size capacity is enough for one input group of picture (GOP), output motion vectors, and output stream. DCT module performs DCT. Motion estimate (ME) module performs motion estimation among the frames. The quantization module quantizes DCT coefficients. The ZZ module performs scanning procedure. The entropy module performs entropy coding procedure using a Huffman coding. The controller which is an FSM generates addressing and control signals with clock for each individual component module in the system to synchronize the system working functions. The IE module performs the inverse entropy coding or decoding. The inverse zigzag (IZZ) module performs IZZ scanning. The inverse quantization (IQuan) module performs inverse quantization. The motion compensation (MC) module performs motions compensation. A new frame is rebuilt which resembles the original one in terms of reference frame and motion vectors, prediction errors. If it is intra-frame, this block will be skipped. The input and output are buffered to frame buffer.

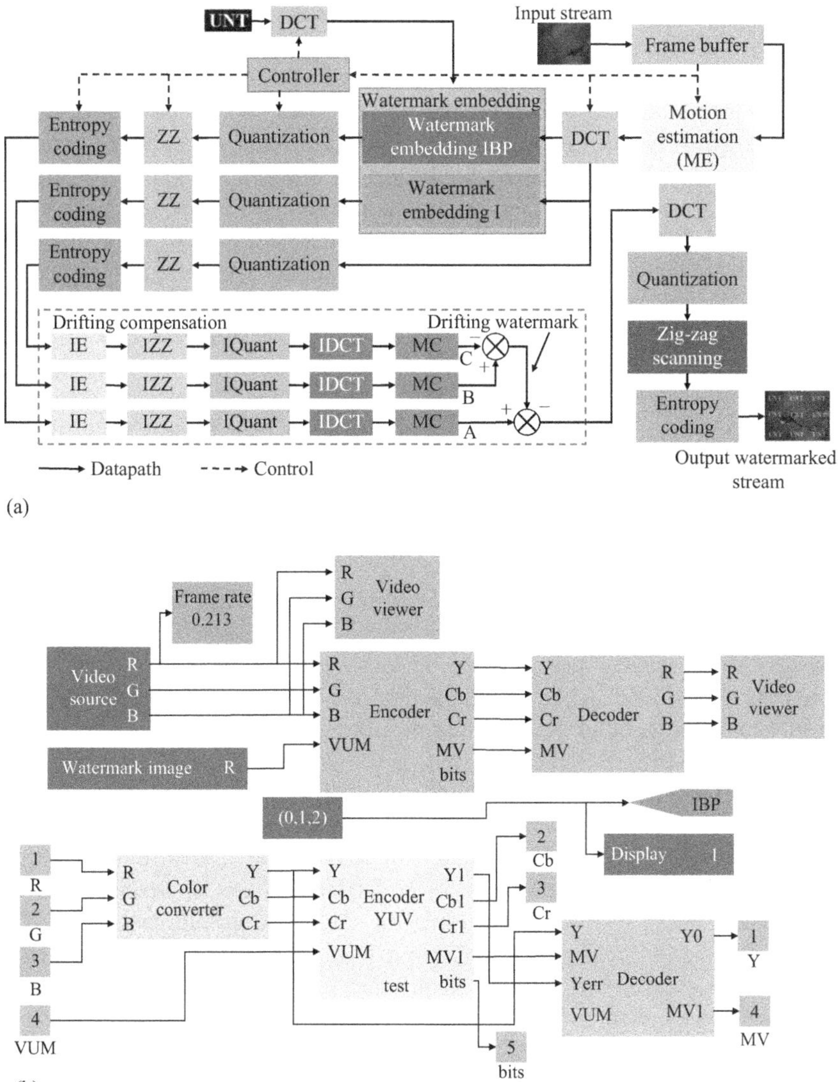

Figure 11.19 Copyright protection hardware DCT domain: (a) video watermarking architecture datapath and (b) video watermarking architecture Simulink design

The drifting compensation block has three processing branches as shown in the figure. In branch A, the watermark is embedded to all I, B, and P frames. In effect making two watermarks for inter-frames B and P, one is predicted from intra-frame, and one is embedded. In branch B, the watermark is inserted to intra-frame I only. However, inter-frames B and P have the same watermark by prediction. This watermark is the drift one and needs to be canceled in inter-frames B and P. In branch C, the frames are compressed without any watermark. So after decompressing, branch A has two watermarks, one is stable, another is drifting; branch B has one drifting watermark; branch C has no watermark. So, by subtracting branch B from branch C, the drifting watermark is extracted, and by subtracting branch A from the extracted drifting watermark, the drifting watermark effect in inter-frames is canceled. Overall a good visible watermark is inserted in the video frames. It may be noted that visible watermarking in compressed domain is bit tricky as compared to the watermarking in uncompressed domain. The host video frames are not independent as still images, but they are correlated frames with each other in temporal mode, that is, inter-frames (P or B) predicated from intra-frame. So, every object in base intra-frame is inherited by predicted inter-frames (P or B) such that the watermark in intra-frame appears in inter-frames (P or B) even though they are not embedded with the watermark. However, if it overlaps with any moving objects in the video scene, the watermark drifts around with the moving objects. The drift compensation is needed to cancel the side effect to obtain a stable watermark.

11.7.3 Secure better portable graphics (SBPG)

An SDC is a device that consists of all the features of a normal digital camera with additional onboard security features like watermarking. This can help in reducing the more time and resources that are used for watermarking the images taken with a normal camera. A hardware-based design also gives a low-power design with a high performance. Design of an Secure better portable graphics (SBPG) is presented by Mohanty *et al.* (2018). When compared to a JPEG compression, BPG compression gives a better and low image size with a comparatively higher image quality. This makes BPG more suitable for implementations in the real-time bandwidth restricted applications. SDC with SBPG is usually implemented as a SoC. Encryption and watermarking are two different layers of protection that can be attained from this setup. Hence, it can help in safeguarding the DRM, ownership, and so on. This makes it suitable for the implementation in IoT applications and smart health care.

Figure 11.20(a) shows the big picture and the architectures of secure BPG camera implementation. BPG encoding involves two different stages, the pre-encoding and the encoding. In the initial stage, the metadata like the bit-depth, color-space are read. The number of bits used for the indication of color are the bit depth: 8, 10, 12 are the commonly used bit depths. The bandwidth requirements for a higher bit rate images is higher and not all the displays are capable of displaying at higher bit rates. So, the bit depth of 8 is considered for BPG. During encoding, the high-efficiency video coding (HEVC) reduces the spatial and temporal redundancies. This is performed by the inter- and intra-frame prediction. After that an 8×8 block as the basic coding unit transforms the image data to the frequency domain, which uses the DCT or discrete sine transform (DST). The standard sizes that are used in this scenario are 4×4, 8×8, 16×16, and 32×32. The image that is transformed is then quantized and compressed additionally with the help of entropy coding. This is done to eliminate the redundancies that are not removed in

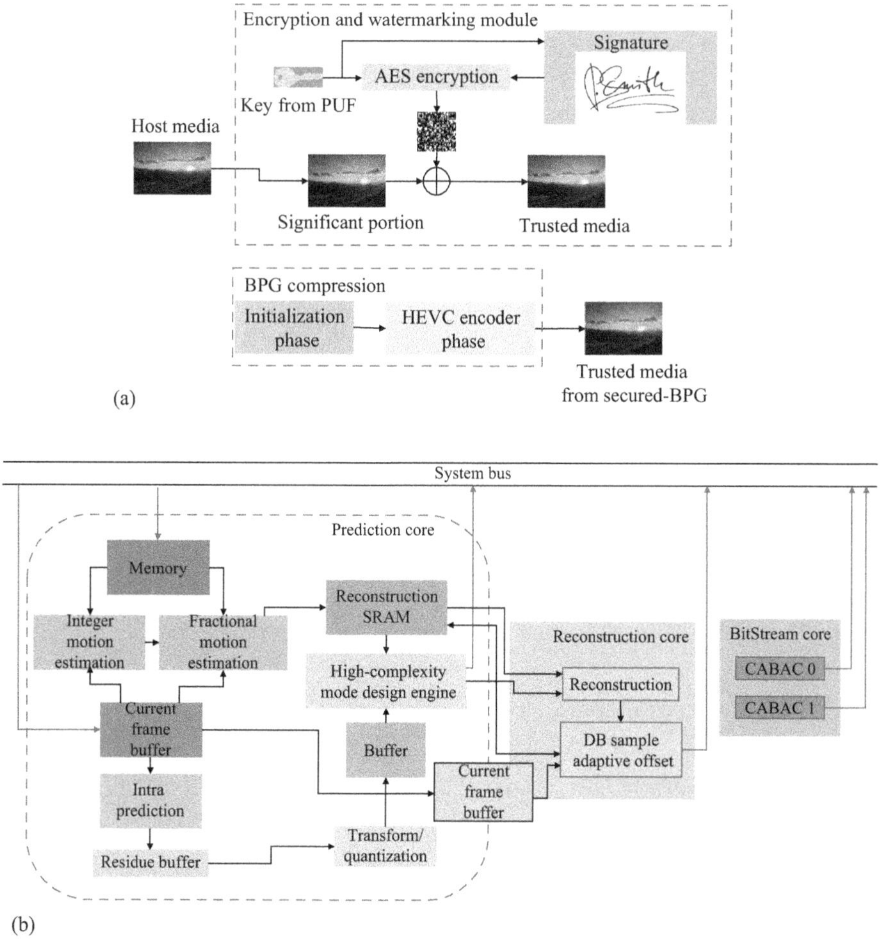

Figure 11.20 Secure BPG–DRM hardware: (a) BPG big picture and (b) SBPG architecture

the prediction mechanisms. Image is further compressed with the help of the entropy coding. Figure 11.20(b) shows the architecture of HEVC CODEC that is part of BPG.

11.7.4 Trust cam

Figure 11.21 presents the design outline of a Trust Cam (Winkler and Rinner, 2010). This is a camera that is able to achieve the ROI detection when the raw images are taken by the sensor. Various privacy levels are present in the camera by performing edge detection for ROI. Then the ROI images are compressed to JPEG and a key is used to encrypt them with AES algorithm.

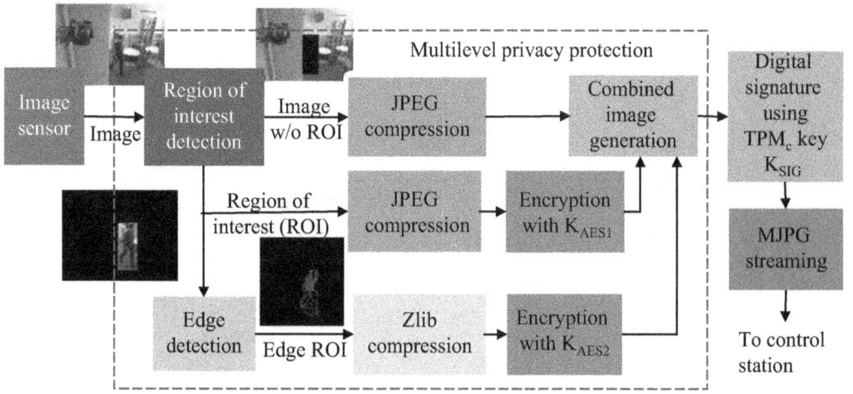

Figure 11.21 Trust cam

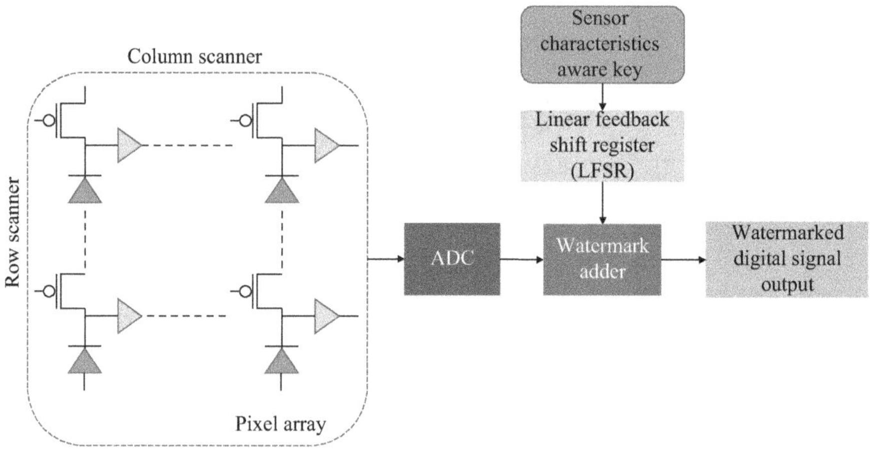

Figure 11.22 Secure image sensor

Figure 11.22 shows the processing flow of the Trust Cam. Image from the sensor is obtained by the image acquisition block and is sent to the privacy block. The sensor is able to send RAW images or JPEG compressed. Then detection is performed on the ROI which are randomly computed every time. After selecting the ROI, they are extracted from the input images captured by the sensor. Then it is compressed using JPEG compression using libjpeg. The resultant image is then sent as an output which can be used to monitor the ROI.

Figure 11.23 shows the designs of secure image sensors and the PUF-based image sensors. They are used to secure the images that are captured by the sensors. A PUF is a module that generates random numbers of cryptographic purposes. The PUF-based sensor attests the sensed data that is coming from the sensor for integrity and authenticity. Sensor controller will also undergo a secure boot to increase the robustness of the entire design.

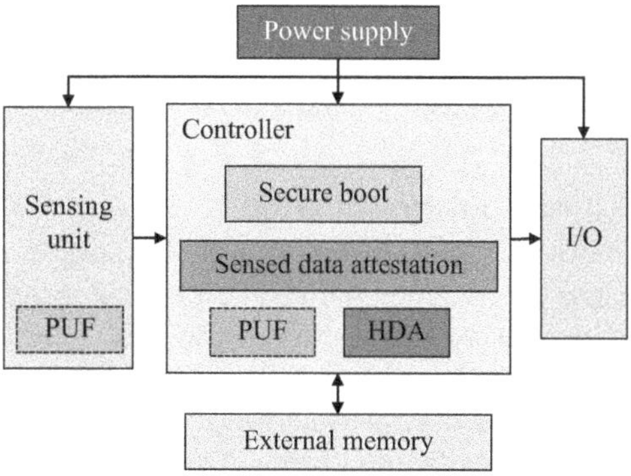

Figure 11.23 PUF-based trusted sensor

11.8 Dynamic watermarking in smart car or UAV

IoT has become a major focus of research nowadays. With all the new technologies coming out everyday, designing new THINGS has been possible and sensor integration into various devices has become an important task in every applications. Smart transportation and UAV technology have developed significantly since their advent (Mohanty, 2015). Autonomous vehicles and UAVs are being used for various purposes. Self-driving cars are not so far ahead in the future. With all these coming into picture, security is one of the main concerns that is bothering the research community. Both UAV and smart car will have video-recording capabilities. Watermarking that video will help in determining if the video that is recorded in the camera is tampered or not.

The video that is recorded in the vehicles is in some cases processed real time and in some cases, stored in the memory to be analyzed at a later time. Both these cases are important when it comes to a smart car and UAV. When the video is recorded, the cars and UAV might be subject to different kinds of attacks that let the attacker to tamper with the video surveillance footage of the vehicle. So, the watermark that is introduced in the vehicle should be tamper resistant (Fallahpour *et al.*, 2014). Many video editing software lets the user to tamper with the video and make the stream unreliable which defeats the purpose of the recording (Su *et al.*, 2011). So, detection and distinguishing of the tampered video is also necessary in such scenarios.

Hence the watermark should be able to provide transparency, robustness, and higher capacity so that it can be embedded into different kinds of video footage. A fragile watermarking is used in most cases for detecting the tampering of a video (Fridrich *et al.*, 2000). The attacker will try to change the watermarked video or object in the case of tampering but without touching the watermark itself. This poses a huge threat as distinguishing the original untampered video will become difficult. A fragile watermarking will be able to protect the footage against such tampering attacks. In case of a fragile watermarking algorithm, the video footage will become highly sensitive to the modifications and this will make it difficult to perform any kinds of modifications to the video like re-compression. A semi-fragile scheme was also proposed by Sang and Alam (2008).

11.9 Medical signals authentication

As mentioned in the previous sections, the advancements of technology have allowed us to explore high-bandwidth communications which led to transmitting data globally (Singh *et al.*, 2017). There has become no bounds for devices used for transmission, especially for medical use, such as the Internet, mobiles phones, wireless communications, and so on. There are different kinds of threats that are posed on the communication networks of such kind. The information that is being shared is also highly sensitive, which contains medical records of patients. This demands for higher levels of confidentiality and security. Medical image watermarking can provide a better service which can address these issues (Acharya *et al.*, 2001; Coatrieux *et al.*, 2006). There have been few implementations of such security initiatives. For example, the use of watermarking in tele-medical applications in order to enhance the security of the transmitted sensitive medical data has been implemented. Tele-medicine watermarking will demand more care when the data is encrypted within the medical images as the information itself and its quality cannot be affected. The reliability, confidentiality, and availability are the major concerns of the security in the electronic patient record (EPR) exchange. Modern telecommunication infrastructure supports the possibility of delivering quality health care without the physical presence of medical experts. The integrity of biomedical signals being transmitted through communication channels must be established before their utilization. Investigations have been performed where digital watermarking techniques for signal integrity verification in an electroencephalogram (EEG) monitoring application were performed.

A digital watermark which uses both the ECG and the patient demographic text as two-level watermarks was proposed by Annadurai *et al.* (2012). In the insertion process, DWT is applied to the object, and it is decomposed into three sub-bands. After that the texture matrix will be calculated for each of the sub-bands. The wavelet coefficients are then selected for the watermarks which are used as threshold values. Authenticating the watermarked object is the main goal of this watermarking scheme. A wavelet spread-spectrum watermarking mechanism for both text and image is presented by Singh *et al.* (2015). The watermark size is varied and the gain factor is obtained in this scenario for the experimental results. BCH codes are used for an increased robustness.

Researchers across the globe are using the hybrid watermarking schemes to improve the performance of the watermarking systems (Fatima and Tuptewar, 2016; Vo *et al.*, 2017). DCT and DWT can be used to embed the watermark into an image and the watermark can be selected at random. After a third level of decomposition, a DCT can be applied on them. A watermark with a zero mean and one variance is inserted in this process. The extraction process does not require the original image to be present in the memory.

Machine learning can also be used for the watermarking techniques. There are many variants of watermarking techniques that use machine learning and offer more imperceptibility and security compared to the other but demand high computational power. A blind watermarking technique which uses the multiwavelet and support vector machine (SVM) is presented by Peng *et al.* (2010). Initially the multiwavelet is performed on the each of the blocks of image. After this stage is complete, the lowest frequency sub-band is selected and the watermark is embedded into that using modulation. The watermark information contains a reference to the owner and the signature of binary logo image. The reference information of the owner will be helpful in training the SVM when extracting the watermark.

11.10 Side-channel information leakage attacks and countermeasures

Watermarking and cryptography hardware modules have an advantage of the low-power consumption along with a high performance and throughput. They can also be designed at a low cost per module and are integrated into different applications with ease. Besides these advantages, there are also some disadvantages to the hardware modules. They are made out of CMOS technology and are susceptible to various kinds of attacks such as side-channel attacks. The side-channel attacks collect the information from the hardware modules like the hardware watermarking chip directly unlike the software attacks.

During World War II, an engineer observed that the oscilloscope signal had a spike in it when a key was pressed in the secure typewriter Model 131-B2 at Bell Labs in 1943 (Mohanty, 2015). Based on that the first publicly known side-channel attack was reported in 1965. Any attack that uses the physical implementation of cryptosystem and gains information is called a side-channel attack. There are two types of attacks that are possible on a crypto system: nondestructive and destructive. The target of the first type is analyzing the target module and gain information from it without destroying or tampering the device itself. Side-channel attacks are included in this category of attacks. Different side-channel attacks are presented in Figure 11.24 (Parameswaran, 2018). This section discusses in detail the CPA and DPA attacks on the modules. There are also fault attacks, acoustic noise attacks, elapsed time attacks, EM radiation attacks, and so on, which exploit the advantages of different components of a device, like the time taken to complete a process, injecting faults into the device algorithms, and so on.

11.10.1 An encryption hardware

Figure 11.25 shows the process of AES encryption. AES is widely used for encryption and decryption in communication between two devices or nodes. It is considered to be the most secure cipher as the variable data block is supported by this standard with a variable key size (Mohanty *et al.*, 2018). The key size and the block size can be chosen as any multiple of 32 and between 128 and 256. AES has no practical attacks against it and is more secure. Brute-forcing an AES encryption

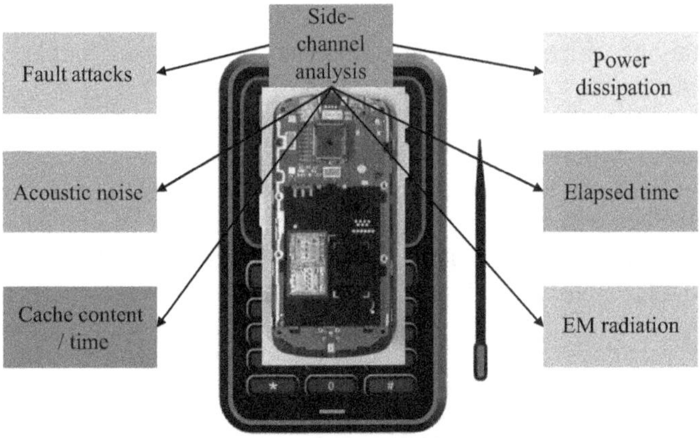

Figure 11.24 Different side-channel attacks (Parameswaran, 2018)

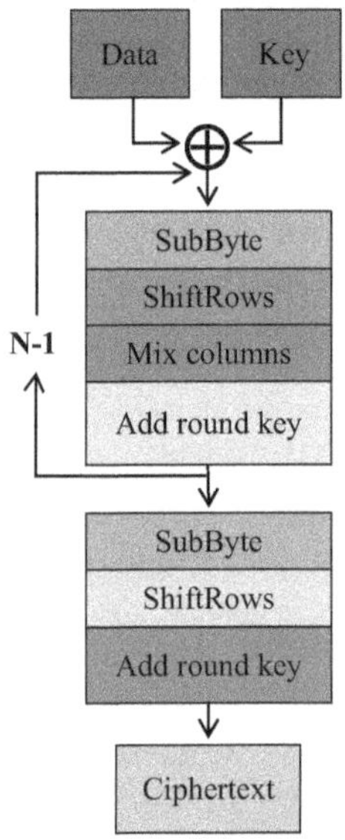

Figure 11.25 Advanced encryption standard (AES)

is highly difficult and requires high computational power. Fundamental algebraic transformations are present in AES like shift row, mix column, round key, and substitution. They are present for each of the standard round but there is an exception of the least standard round.

11.10.2 Side-channel analysis attacks

Two major side-channel analysis attacks are considered in this section, CPA and DPA (Jayasinghe *et al.*, 2015). CPA is the analysis of relationship between the

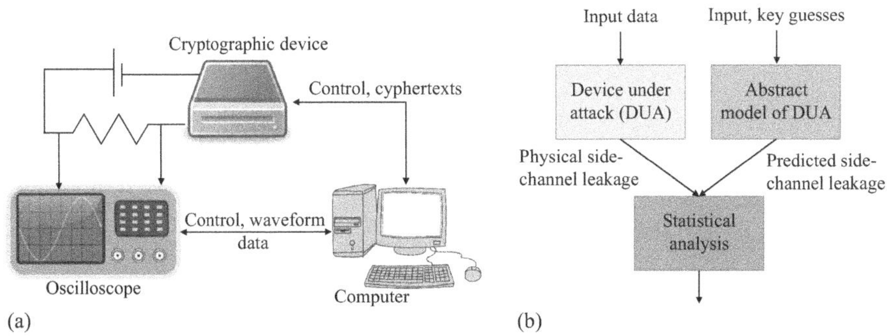

Figure 11.26 DPA and CPA on HMAC: (a) oscilloscope and (b) detector (Zhang and Shi, 2011)

cipher-text and the plain-text and it's correlation as well as the instantaneous power consumption of the cryptographic device. When compared to DPA, CPA is a more effective attacking method. DPA uses the relation between the power and the data. Using that DPA will try to look at the difference of the category averages for all the keys. As it uses the averages, more power correlation data is needed for successfully launching this attack. So it consumes more time and is less efficient compared to the CPA. Figure 11.26 presents a big picture of DPA and CPA (Zhang and Shi, 2011).

A DPA attack is a side-channel attack which statistically analyzes the consumed power by the respective module in a cryptosystem. The module can be a watermarking chip or a cryptographic processor which performs encryption and decryption. Figure 11.27 shows the overall process of DPA (Parameswaran, 2018). The biases varying power consumption is taken. Here, the microcontroller is monitored and the power analysis is performed when various operations are performed on the chip, which involve the secret keys. Signal processing and error correction properties are necessary for performing DPA, which are used for the extraction of secrets from the power consumption measurements. There is also noise that comes with the measurements so it must be filtered before the analysis is performed. A secret key can be obtained by the attacker by analyzing the power from the cryptographic operations on the chip using DPA.

A CPA uses the relation between the data and the power as mentioned before. Figure 11.28 shows the overall picture of the CPA (Parameswaran, 2018). This uses a correlation between the guesses of the keys. Hence, low power is required compared to the DPA. CPA is used to find a key that is stored in memory of the victim device. Launching a CPA attack involves the following four steps.

- The power consumption model of the victim device should be noted.
- The victim should encrypt different files of plain text.
- Record the trace of the power consumption on the victim device.
- Attack the small parts of the secret key from the observed power analysis.

Every possible combination for the subkey should be considered. When each guess is performed, it is compared to the power that is analyzed on the main board.

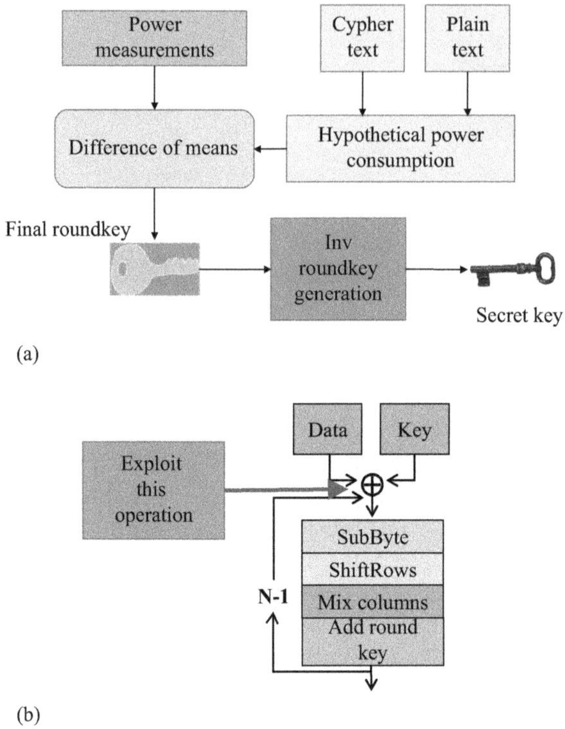

Figure 11.27 Differential power analysis: (a) differential power analysis big picture and (b) differential power analysis attack overview (Parameswaran, 2018)

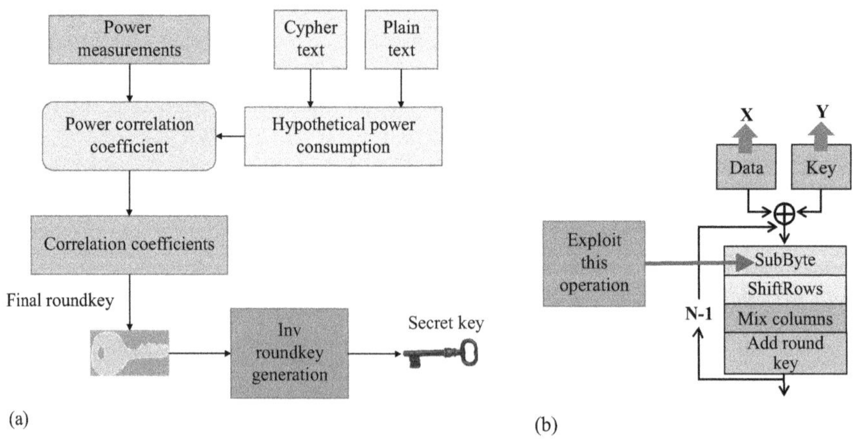

Figure 11.28 Correlation power analysis: (a) correlation power analysis big picture and (b) correlation power analysis attack overview (Parameswaran, 2018)

Repeating this process will give a guess of the subkey which will equal to the main key and power of the key used on the plain text will match the power observed from the module.

11.10.3 Side-channel attack countermeasures

Side-channel attacks are highly efficient in stealing information from the micro-controller or memory. The attacker can use various kinds of attacks such as CPA, DPA, timing analysis, and so on on the hardware to get the information stored on it. Many techniques are available to mitigate the issues of power analysis attacks (Ambrose *et al.*, 2015; Jayasinghe *et al.*, 2015). There are many ways to avoid the side-channel attacks on the hardware modules. Figures 11.29 and 11.30 show different ways of safeguarding the hardware modules from CPA and DPA attacks.

One of the main drawbacks of the power analysis attacks is that the hardware should be powered on during the time of the attacks in most of the cases. So, if the power supply to the module is turned off, the attacker cannot perform the attack anymore. Introducing some interrupts to the power can also mitigate the security

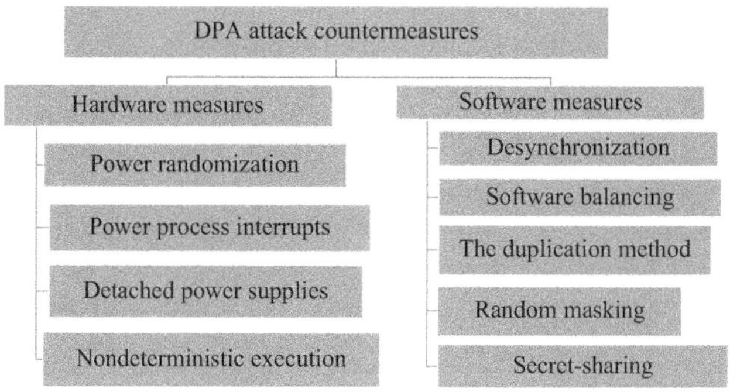

Figure 11.29 Differential power analysis attacks countermeasures

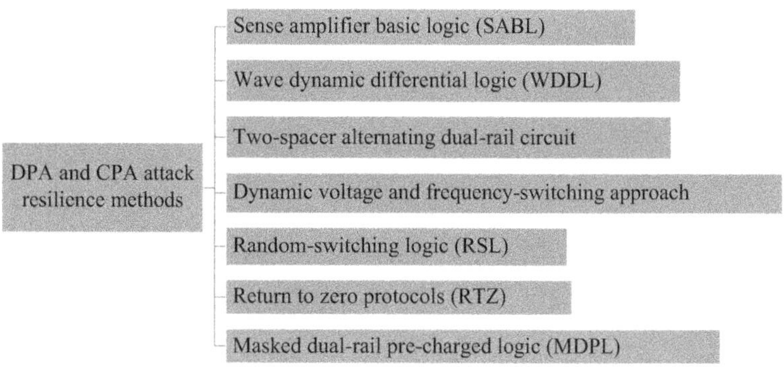

Figure 11.30 Correlation power analysis attacks countermeasures

threats to some extent. Power supply to the module can be detached when unused but this will only apply when the module is not solely in the hands of the attacker. During the execution process, there can be a nondeterministic execution which will not give out the exact power statistics of the device during the processing of secret keys. These are some of the hardware solutions that can be implemented. De-synchronization, software balancing, random masking, and so on are the software countermeasures to the DPA attacks on the security modules.

11.11 Attacks on watermarks and watermarking systems

When a watermark is embedded into a media object, there is a great chance that it might be subjected to manipulative attacks before the media reaches the receiver. Some of the manipulations are analog-to-digital and digital-to-analog conversions, resampling, quantization, dithering, low-pass and high-pass filtering, and so on. The media can be printed and rescanned in some cases intended to remove the watermark from the media. Any processing of the watermarked image that is intended to remove or tamper the watermark and fail the detection process is an attack on the watermark (Cox and Linnartz, 1998; Mintzer *et al.*, 1998; Sequeira and Kundur, 2001; Swanson *et al.*, 1998; Voloshynovskiy *et al.*, 2001). The characteristics of the watermark and its performance against the attacks which are mutually conflicting FoM. Comparing metrics for different algorithms is needed to come to a decision in selecting a suitable watermarking scheme for a particular application. Benchmarks of the watermark are different attacks in a unified framework which allow the owner to check the scheme for applicability.

This section discusses different attacks that are possible on a watermarked image. There are different types of attacks on a watermarked image to remove the watermark. They are classified into four different types (Liu *et al.*, 2017; Pereira *et al.*, 2001; Vellaisamy and Ramesh, 2014; Voloshynovskiy *et al.*, 2001; Wang *et al.*, 2017a, 2017b; Xiang *et al.*, 2014) as shown in Figure 11.31:

- Geometric attacks
- Removal and interference attacks
- Cryptographic attacks
- Protocol attacks

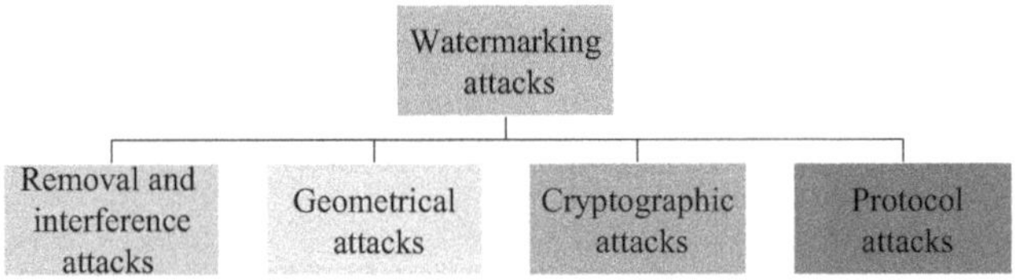

Figure 11.31 Attacks on watermarks (Pereira et al., 2001; Voloshynovskiy et al., 2001)

11.11.1 Removal and interference attacks

Removal attack on the watermark is targeted to remove the watermark from the object. A watermark is the noise that is added to the original image and this is taken as an advantage by the attacker to remove the watermark. Interference attacks target the watermark by adding more noise to the watermarked image. Quantization, lossy compression, collusion, remodulation, denoise, averaging, and noise storm are the examples of such kinds of attacks. JPEG, MPEG, and so on are the lossy compression styles that can degrade the quality of watermark that is embedded in the media object. When many legit owners of the watermarked copy of the media come together to create a nonwatermarked image, it is called a collude attack. This can be achieved by averaging different objects in the media that is already watermarked.

11.11.2 Geometric attacks

The geometric attacks are more targeted toward the watermarked images and videos. These kinds of attacks do not remove the watermark present in the media but will manipulate the watermarked media so that the detector will not be able to extract or retrieve the watermark. Affine transformations such as translation, rotation, and scaling are used to successfully launch this attack. Line removal, wrapping, and cropping can also be done to remove the watermark from the media object. Mosaic attack is another example of this category (Koubaa *et al.*, 2007). In such attacks, the media object is divided into different parts and again rearranged into one using HTML, which makes the watermark difficult to be detected by the detector.

11.11.3 Cryptographic attacks

The two attacks that are discussed above will not be able to breach the algorithm security. A cryptographic attack is the one that precisely attacks the security of the watermarking algorithm and cracks it. When the watermark is secured with a secret key, as discussed in the previous sections, a brute-force attack can be performed and broken, which can be considered as a cryptographic attack. Oracle attack is another kind of cryptographic attack which uses a public watermark detector to create a nonwatermarked media object.

11.11.4 Protocol attacks

This kind of attacks takes advantage of the loopholes that are present in the watermarking algorithm (Craver *et al.*, 1996, 1998; Wang, 2009; Zhang and Liu, 2017). This kind of attack is also known as inversion, deadlock, or fake–original attack. One or many new watermarks are introduced on an already watermarked media which makes it difficult to detect the original watermark and the owner. Some kinds of attacks will produce a fake original media object with a watermark so that the detector will detect it as an original watermarked legit copy. Watermark from an already watermarked object is detected and transferred to another media object which is called copy attack, which is also very popular.

11.12 Limitations of watermarks and watermarking

Limitations of the watermarking schemes are discussed in this section. When a watermark is inserted into the original media object, it should be detected with ease if not tampered with. Even in the case of attacks, the watermark should be resistant. In this chapter, we discussed a lot about different kinds of watermarks and the characteristics. With the hardware watermarking coming into picture, the IoT integration is becoming easy. Different hardware modules that were discussed above can be integrated into IoT applications. But IoT devices need to be connected to the Internet all the time and some of them will be battery operated. So, the latency and the power consumption of the watermarking modules is a great challenge.

When it comes to the visible watermarking schemes, the copyright is clearly mentioned and need not be worried (Mintzer *et al.*, 1996; Mohanty *et al.*, 2000a). The watermark should occupy a significant portion of the original media object. Besides that the quality of the image should not be degraded. When removing the watermark it costs same as a new media object. This kind of security is sufficient for use in different applications. This also has some drawbacks where the watermark should be enforced with laws.

Invisible watermarking, unlike the visible watermarking techniques, is imperceptible and is more complicated for that reason. Robustness of the watermark is considered the highest characteristic that a watermarking algorithm has to meet. But this is not true in case of an invisible watermarking. Because, every watermark can be attacked using various techniques and invisible watermark is also not left out from that category. Any media can be subject to counterfeit production by attacking the watermark. So, invertible or quasi-invertible invisible watermark schemes are not able to resolve the rightful ownership. Without a standardized watermarking scheme, anyone can claim ownership.

Due to the advancements in the technology scaling and development of high-performance devices, signal manipulations are being introduced into various artworks and the indistinguishable noises are added to the media object. This makes it difficult to differentiate the manipulations and prove original ownership of the object. The additional data in the object is not known to occur due to the watermark or any new noise. In such scenarios, extracting watermark and authenticating the ownership of the object is a challenge. With the use of highly sophisticated DSP, this problem is further deteriorating. A watermarked image that is not protected properly cannot be used for authentication and hence cannot be used in a court of law. To avoid this situation, watermark can be embedded into the object during the acquisition phase directly using hardware modules.

For a satisfactory robustness of watermark image, in case of a c traitors out of N, the length of the codewords to be inserted into the watermark is given as: $O\left(c^4 \log\left(\frac{N}{\varepsilon}\right)\right)$, where the probability of error is denoted by ε. The discussed amount of added noise embedded into the image will surely degrade the quality of the image which in turn reduces the commercial value of the object. Digital watermarking will not be able to prevent illegal activity but can only provide evidence that such an activity has been performed by a party. But cryptographic techniques can protect the data from such attacks. So digital watermarking should be used as a secondary protection layer.

11.13 Conclusion

This chapter discusses the watermarking schemes, some issues with watermarking and its limitations, and attacks and solutions on different watermarking algorithms. A big picture of watermarking from basics to deployment is presented in this chapter. Different hardware watermarking schemes along with some very-low-power chips which can be integrated into battery operated devices were discussed. With the introduction of a new watermarking scheme or a new cryptographic algorithms, the attackers are soon coming up with ways to attack them and breach the security. There are many counterfeit devices being produced and media content being shared without proper legalities being met. This has become a major issue that has to be addressed designing of high sophisticated and low-power security modules in both cryptography and watermarking.

11.14 Exercises

1. What is watermarking? What is the purpose of watermarking an image?
2. What is intellectual property protection? Describe with a broad overview.
3. What is digital rights management (DRM)?
4. What are the different parts in watermarking process? Discuss a broad picture of the three parts and their functions.
5. Why hardware modules can be used for watermarking schemes? What are its advantages?
6. Describe the general framework of the digital watermarking techniques.
7. What are different categories of a watermark? Discuss in detail two of the categories.
8. What are the differences between visible and invisible watermarking schemes?
9. What is source watermarking? What is destination watermarking? Distinguish them and describe briefly.
10. Describe five applications of watermarking in detail.
11. What are the major characteristics of a watermark? Discuss any five of them in detail.
12. What are the characteristics of visible watermarking?
13. Discuss invisible-robust watermark and invisible-fragile watermarking techniques.
14. Discuss different challenges faced during the process of watermarking and designing an algorithm.
15. Discuss the properties of human visual system.
16. How much watermark strength should be added to an object and where?
17. Discuss briefly about different attacks on watermarking.

18. What are the key differences in image watermarking and video watermarking techniques?
19. Discuss the invisible-robust DCT domain chip in detail.
20. What is invisible-robust-fragile watermarking? Discuss the spatial domain chip for the invisible-robust-fragile watermarking.
21. What are the different functional units in invisible-robust wavelet domain module?
22. Discuss the visible spatial domain chip.
23. Discuss the invisible-robust spatial domain chip in video watermarking schemes.
24. What is secure better portable graphics? How is it compared to the regular JPEG compression?
25. Discuss the importance of watermarking in medical domain.
26. What are different watermarking schemes for smart car and UAV?
27. How machine learning is used for watermarking in telemedical communication?
28. What are side-channel attacks? How are they categorized?
29. What are different types of side-channel attacks on a module?
30. What is correlation power analysis? Describe briefly the steps involved in performing a CPA attack.
31. What is differential power analysis? Discuss it briefly.
32. What are the countermeasures for side-channel attacks?

References

Acharya U. R., Acharya D., Bhat P. S., and Niranjan U. C.,. 'Compact storage of medical images with patient information,' *IEEE Transactions on Information Technology in Biomedicine*, 2001, vol. 5 (4), pp. 320–323.

Ahmaderaghi B., Kurugollu F., Rincon J. M. D., and Bouridane A., 'Blind image watermark detection algorithm based on discrete Shearlet transform using statistical decision theory,' *IEEE Transactions on Computational Imaging*, 2001, vol. 4 (1), pp. 46–59.

Ambrose J. A., Ragel R. G., Jayasinghe D., Li T., and Parameswaran S., 'Side channel attacks in embedded systems: A tale of hostilities and deterrence,' *Proceedings of the Sixteenth International Symposium on Quality Electronic Design*, 2015, pp. 452–459.

American Society of Composers, Authors and Publishers (ASCAP): URL https://www.ascap.com/. Last Accessed: 06-20-2018.

Annadurai K., Pavithra K., and Subharani S., 'Double watermarking of dicom medical images using wavelet decomposition technique,' *European Journal of Scientific Research*, 2012, vol. 70 (1), pp. 46–55.

Barnett. R., 'Digital watermarking: Application, techniques, and challenges,' *IEE Electronics and Communication Engineering Journal*, 1999, vol. 11 (4), pp. 173–183.

Bender W., Gruhl D., and Morimoto N., 'Techniques for data hiding,' *IBM Systems Journal*, 1996, vol. 35 (3), pp. 313–336.

Bi W., Zhang W., He W., and Li C., 'A modified decoding algorithm involving priori characteristics bits for LDPC,' *6th IEEE/International Conference on Advanced Infocomm Technology (ICAIT)*, 1996, pp. 245–246.

Braudaway G. W., Magerlein K. A., and Mintzer F., 'Protecting publicly available images with a visible image watermark,' *Proceedings of the SPIE Conference on Optical Security and Counterfiet Deterrence Technique* (vol. SPIE-2659), 1996, pp. 126–132.

Busch C., Funk W., and Wolthusen S., 'Digital watermarking: From concepts to real-time video applications,' *IEEE Computer Graphics and Applications*, 1999, vol. 19(1), pp. 25–35.

Cai W., 'FPGA prototyping of a watermarking algorithm,' 2006, Master's thesis, Department of Electrical Engineering Technology, University of North Texas, Denton, TX.

Chan H. T., Hwang W. J., and Cheng C. J., 'Digital hologram authentication using a Hadamard-based reversible fragile watermarking algorithm,' *Journal of Display Technology*, 2015, vol. 11 (2), pp. 193–203.

Chen S. T., Hsu C. Y., and Huang H. N., 'Wavelet-domain audio watermarking using optimal modification on low-frequency amplitude,' *IET Signal Processing*, 2015, vol. 9 (2), pp. 166–176.

Cheng C. J., Hwang W. J., Zeng H. Y., and Lin Y. C., 'A fragile watermarking algorithm for hologram authentication,' *Journal of Display Technology*, 2014, vol. 10 (4), pp. 263–271.

Coatrieux G., Lecornu L., Sankur B., and Roux C., 'A review of image watermarking applications in healthcare,' *2006 International Conference of the IEEE Engineering in Medicine and Biology Society*, 2006, pp. 4691–4694.

Copy Protection Technical Working Group (CPTWG): URL http://www.cptwg.org/. Last Accessed: 06-20-2018

Cox I. J. and Linnartz J. P. M. G., 'Some general methods for tampering with watermarks,' *IEEE Journal on Selected Areas in Communications*, 1998, vol. 16 (4), pp. 587–593.

Cox I. J. and Miller M., 'A review of watermarking and importance of perceptual modelling,' *Proceedings of SPIE Human Vision and Imaging*, 1997, vol. 3016, pp. 92–99.

Cox I. J. and Miller M. L., 'Electronic watermarking: The first 50 years,' *EURASIP Journal of Applied Signal Processing*, 2002, vol. 2002(2), pp. 126–132.

Craver S., Memon N., Yeo B. L., and Yeung M., 'Can invisible watermarks resolve rightful ownerships?,' Tech. rep., 1996, IBM Research Report, RC 20509.

Craver S., Memon N., Yeo B. L., and Yeung M. M., 'Resolving rightful ownerships with invisible watermarking techniques: Limitations, attacks and implications,' *IEEE Journal on Selected Areas in Communications*, 1998, vol. 16 (4), pp. 573–586.

Delaigle J. F., Devleeschouwer C., Macq B., and Langendijk I., 'Human visual systems features enabling watermarking,' *Proceedings of the IEEE International Conference Multimedia and Expo*, 2002, pp. 489–492.

Delaigle J. F., Vleeschouwer C. D., and Macq B., 'Watermarking algorithm based on human visual model,' *Signal Processing*, 1998, vol. 66 (3), pp. 319–335.

Depovere G., Kalker T., Haitsma J., *et al.*, 'The VIVA project : Digital watermarking for broadcast monitoring,' *Proceedings of the IEEE International Conference on Image Processing* (vol. 2), 1999, pp. 202–205.

Devillier N., 'Aging, well-being, and technology: From quality of life improvement to digital rights management – A French and European perspective,' *IEEE Communications Standards Magazine*, 2017, vol. 1 (3), pp. 46–49.

Fallahpour M., Shirmohammadi S., Semsarzadeh M., and Zhao J., 'Tampering detection in compressed digital video using watermarking,' *IEEE Transactions on Instrumentation and Measurement*, 2014, vol. 63 (5), pp. 1057–1072.

Fan Y. C., Van L. D., Huang C. M., and Tsao H. W., 'Hardware-efficient architecture design of wavelet-based adaptive visible watermarking,' *Proceedings of 9th IEEE International Symposium on Consumer Electronics*, 2005, pp. 399–403.

Fatima N. and Tuptewar D. J., 'Comparison of hybrid watermarking technique on different color spaces,' *Conference on Advances in Signal Processing (CASP)*, 2016, pp. 13–17.

Frattolillo F., 'Watermarking protocols: Problems, challenges and a possible solution,' *The Computer Journal*, 2015, vol. 58 (4), pp. 944–960.

Fridrich J., Goljan M., and Baldoza A. C., 'New fragile authentication watermark for images,' *Proceedings 2000 International Conference on Image Processing*, 2000, vol. 1, pp. 446–449 vol. 1.

Gafsi M., Ajili S., Hajjaji M. A., and Mtibaa A., 'XSG for hardware implementation of a robust watermarking system,' *17th International Conference on Sciences and Techniques of Automatic Control and Computer Engineering (STA)*, 2016, pp. 117–122.

Garimella A., Satyanarayan M. V. V., Kumar R. S., Murugesh P. S., and Niranjan U. C., 'VLSI implementation of online digital watermarking techniques with difference encoding for the 8-bit gray scale images,' *Proceedings of the International Conference on VLSI Design*, 2003, pp. 283–288.

Garimella A., Satyanarayana M. V. V., Murugesh P. S., and Niranjan U. C., 'ASIC for digital color image watermarking,' *Proceedings of 11th IEEE Digital Signal Processing Workshop*, 2004, pp. 292–295.

Hsiao S. F., Tai Y. C., and Chang K. H., 'VLSI design of an efficient embedded zerotree wavelet coder with function of digital watermarking,' *Proceedings of the IEEE International Conference on Consumer Electronics*, 2000a, pp. 186–187.

Hsiao S. F., Tai Y. C., and Chang K. H., 'VLSI design of an efficient embedded zerotree wavelet coder with function of digital watermarking,' *IEEE Transactions on Consumer Electronics*, 2000b, vol. 46 (3), pp. 628–636.

Hua X. S., Feng J. F., and Shi Q. Y., 'Public multiple watermarking resistant to cropping,' *Proceedings of the 6th International Conference on Pattern Recognition and Information Processing*, 2001, pp. 263–268.

Cox I. J, Kilian J., Shamoon T., and Leighton T., 'Secure spread spectrum watermarking of images, audio and video,' *Proceedings of IEEE International Conf on Image Processing*, 1996, vol. 3, pp. 243–246.

Jayasinghe D., Ignjatovic A., Ambrose J. A., Ragel R., and Parameswaran S., 'QuadSeal: quadruple algorithmic symmetrizing countermeasure against power based side-channel attacks,' *Proceedings of the International Conference on Compilers, Architecture and Synthesis for Embedded Systems (CASES)*, 2015, pp. 21–30.

Kahn D., 'The history of steganography,' *Proceedings of First International Workshop on Information Hiding*, 1996, vol. 1174, pp. 1–7.

Kamaruddin N. S., Kamsin A., Por L. Y., and Rahman H., 'A review of text watermarking: Theory, methods, and applications,' *IEEE Access*, 2018, vol. 6, pp. 8011–8028.

Kankanhalli M. S., Rajmohan, and Ramakrishnan, K. R., 'Content based watermarking for images,' *Proceedings of the 6th ACM International Multimedia Conference*, 1998, pp. 61–70.

Kaur R., Kalra G. S., and Kansal M., 'A user friendly GUI based benchmark for image watermarking,' *International Conference on Computing Sciences*, 2012, pp. 70–76.

Khan M. I., Jeoti V., and Malik A. S., 'Designing a joint perceptual encryption and blind watermarking scheme compliant with JPEG compression standard,' *International Conference on Computer Applications and Industrial Electronics*, 2010, pp. 688–691.

Kirovski, D., and Malvar, H., 'Spread-spectrum audio watermarking: Requirements, applications, and limitations,' *IEEE Fourth Workshop on Multimedia Signal Processing*, 2001, pp. 219–224.

Koubaa M., Amar C. B., and Nicolas H., 'Adaptive video watermarking using mosaic images,' *IEEE International Conference on Signal Processing and Communications*, 2007, pp. 1143–1146.

Kougianos E., Mohantya S. P., and Mahapatra R. N., 'Hardware assisted watermarking for multimedia,' *Computers & Electrical Engineering*, 2009, vol. 35 (2), pp. 339–358.

Krishnamurthy R. K., Humble T., Cheung S. C., Lyke J., Mohanty S. P., and Casto M., 'Energy and cybersecurity constraints on consumer electronics,' http://www.icce.org/expert-panels/ (January 13, 2018). Last visited on 20th Nov 2017

Lah U., and Lewis J. R., 'How expertise affects a digital-rights-management-sharing application's usability,' *IEEE Software*, 2016, vol. 33 (3), pp. 76–82.

Li R., Xu S., and Yang H., 'Spread spectrum audio watermarking based on perceptual characteristic aware extraction,' *IET Signal Processing*, 2016, vol. 10 (3), pp. 266–273.

Liu S., Pan Z., and Song H., 'Digital image watermarking method based on DCT and fractal encoding,' *IET Image Processing*, 2017, vol. 11 (10), pp. 815–821.

Loan N. A., Hurrah N. N., Parah S. A., Lee J. W., Sheikh J. A., and Bhat G. M., 'Secure and robust digital image watermarking using coefficient differencing and chaotic encryption,' *IEEE Access*, 2018, vol. 6 (19), pp. 876–19,897.

Ma Z., 'Digital rights management: Model, technology and application,' *China Communications*, 2017, vol. 14 (6), pp. 156–167.

Ma Z., Huang J., Jiang M., and Niu X., 'A novel image digital rights management scheme with high-level security, usage control and traceability,' *Chinese Journal of Electronics*, 2016, vol. 25 (3), pp. 481–494.

Maes M., Kalker T., Linnartz J. P. M. G., Talstra J., Depovere G. F. G., and Haitsma J., 'Digital watermarking for DVD video copyright protection,' *IEEE Signal Processing Magazine*, 2000, vol. 17 (5), pp. 47–57.

Makbol N. M., Khoo B. E., and Rassem T. H., 'Block-based discrete wavelet transform-singular value decomposition image watermarking scheme using human visual system characteristics,' *IET Image Processing*, 2016, vol. 10 (1), pp. 34–52.

Marcinak M. P. and Mobasseri B. G., 'Digital video watermarking for metadata embedding in UAV video,' *MILCOM 2005 – 2005 IEEE Military Communications Conference*, 2005, vol. 3–5, pp. 1637.

Mathai N. J., Kundur D., and Sheikholeslami A., 'Hardware implementation perspectives of digital video watermarking algortithms,' *IEEE Transanctions on Signal Processing*, 2003a, vol. 51 (4), pp. 925–938.

Mathai N. J., Sheikholeslami A., and Kundur D., 'VLSI implementation of a real-time video watermark embedder and detector,' *Proceedings of the IEEE International Symposisum on Circuits and Systems (vol. 2)*, 2003b, pp. 772–775.

Mathivanan, P., and Ganesh, A.B., 'Colour image steganography using XOR multibit embedding process,' *International Conference on Energy, Communication, Data Analytics and Soft Computing (ICECDS)*, 2017, pp. 1980–1988.

Memon N., and Wong P. W., 'Protecting digital media content,' *Communications of the ACM*, 1998, vol. 41 (7), pp. 35–43.

Mintzer F., Braudaway G., and Yeung M., 'Effective and ineffective digital watermarks,' *IEEE International Conference on Image Processing (ICIP-97)*, 1997, vol. 3, pp. 9–12.

Mintzer F., Braudaway G. W., and Bell A. E., 'Opportunities for watermarking standards,' *Communications of the ACM*, 1998, vol. 41 (7), pp. 57–64.

Mintzer F. C., Boyle L. E., Cazes A. N., *et al.*, 'Towards online worldwide access to Vatican library materials,' *IBM Journal of Research and Development*, 1996, vol. 40 (2), pp. 139–162.

Mohanty S. P., 'Everything you wanted to know about internet of things (IoT),' https://cesoc.ieee.org/images/files/pdf/Mohanty_IEEE-DL_IoT.PDF (16th November, 2017). Last visited on 20th Nov 2017

Mohanty, S. P., 'Digital watermarking of images,' Master's thesis, 1999, Department of Electrical Engineering, Indian Institute of Science, Bangalore, India.

Mohanty S. P., 'A secure digital camera architecture for integrated real-time digital rights management,' *Journal of Systems Architecture – Embedded Systems Design*, 2009, vol. 55 (10–12), pp. 468–480.

Mohanty S. P., '*Nanoelectronic Mixed-Signal System Design*,' McGraw-Hill Education, New York, 2015, ISBN: 9780071825719.

Mohanty S. P., Kumara C. R., and Nayak S., 'FPGA based implementation of an invisible-robust image watermarking encoder,' *Lecture Notes in Computer Science*, 2004, vol. 3356, pp. 344–353.

Mohanty S. P. and Kougianos E., 'Real-time perceptual watermarking architectures for video broadcasting,' *Journal of Systems and Software*, 2011, vol. 84 (5), pp. 724–738.

Mohanty S. P., Kougianos E., Cai W., and Ratnani M., 'VLSI architectures of perceptual based video watermarking for real-time copyright protection,' *Proceedings of the 10th International Symposium on Quality of Electronic Design (ISQED)*, 2009, pp. 527–534.

Mohanty S. P., Kougianos E., and Guturu P., 'SBPG: Secure better portable graphics for trustworthy media communications in the IoT,' *IEEE Access*, 2018, vol. 6, pp. 5939–5953.

Mohanty S. P., Ramakrishnan K. R., and Kankanhalli M. S., 'A dual watermarking technique for images,' *Proceedings of the 7th ACM International Multimedia Conference* (vol. 2), 1999, pp. 49–51.

Mohanty S. P., Ramakrishnan K. R., and Kankanhalli M. S., 'A DCT domain visible watermarking technique for images,' *Proceedings of the IEEE International Conference on Multimedia and Expo*, 2000a, pp. 1029–1032.

Mohanty S. P., Ramakrishnan K. R., and Kankanhalli M. S., 'An adaptive DCT domain visible watermarking technique for protection of publicly available images,' *Proceedings of the International Conference on Multimedia Processing and Systems*, 2000b, pp. 195–198.

Mohanty S. P. and Ranganathan N., 'A framework for energy and transient power reduction during behavioral synthesis,' *IEEE Transactions on Very Large Scale Integration Systems*, 2004, vol. 12 (6), pp. 562–572.

Mohanty S. P. and Ranganathan N., 'Energy efficient datapath scheduling using multiple voltages and dynamic clocking,' *ACM Transactions on Design Automation of Electronic Systems*, 2005, vol. 10 (2), pp. 330–353.

Mohanty S. P., Ranganathan N., and Balakrishnan K., 'Design of a low power image watermarking encoder using dual voltage and frequency,' *Proceedings of 18th IEEE International Conference on VLSI Design*, 2005a, pp. 153–158.

Mohanty S. P., Ranganathan N., and Namballa R. K., 'VLSI implementation of invisible digital watermarking algorithms towards the development of a secure JPEG encoder,' *Proceedings of the IEEE Workshop on Signal Processing Systems*, 2003, pp. 183–188.

Mohanty S. P., Ranganathan N., and Namballa R. K., A VLSI 'Architecture for visible watermarking in a secure still digital camera (S^2 DC) design,' *IEEE Transactions on Very Large Scale Integration Systems*, 2005b, vol. 13 (8), pp. 1002–1012.

Mohanty S. P., Rangnathan N., and Namballa R. K., 'VLSI implementation of visible watermarking for a secure digital still camera design,' *Proceedings of the 17th International Conference on VLSI Design*, 2004, pp. 1063–1068.

Mohanty S. P., Sengupta A., Guturu P., and Kougianos E., 'Everything you want to know about watermarking,' *IEEE Consumer Electronics Magazine*, 2017, vol. 6 (3), pp. 83–91.

Parameswaran S., 'QUADSEAL: A hardware countermeasure against side channel attacks on AES,' *IEEE International Symposium on Nanoelectronic and*

Information Systems, Bhopal, India (19th December, 2017). Last visited on 29th June 2018.

Peng H., Wang J., and Wang W., 'Image watermarking method in multiwavelet domain based on support vector machines,' *Journal of Systems and Software*, 2010, vol. 83 (8), pp. 1470–1477. URL http://www.sciencedirect.com/science/article/pii/S0164121210000701

Pereira S., Voloshynovskiy S., Madueo M., Marchand-Maillet S., and Pun T., 'Second generation benchmarking and application oriented evaluation,' *Proceedings of 3rd International Workshop on Information Hiding*, 2001.

Petitjean G., Dugelay J. L., Gabriele S., Rey C., and Nicolai J., 'Towards real-time video watermarking for systems-on-chip,' *Proceedings of the IEEE International Conference on Multimedia and Expo (vol. 1)*, 2002, pp. 597–600.

Pexaras K., Tsiourakis C., Karybali I. G., and Kalligeros E., 'Optimization and hardware implementation of image watermarking for low cost applications,' *24th IEEE International Conference on Electronics, Circuits and Systems (ICECS)*, 2017, pp. 347–350.

Pfitzmann B., 'Information hiding terminology,' *Proceedings of First International Workshop on Information Hiding*, 1996, vol. 1174, pp. 347–350.

Ramkumar M., 'Data hiding in multimedia – theory and applications,' Ph.D. thesis, Department of Electrical and Computer Engineering, 2000, New Jersey Institute of Technology, Newark, NJ, USA.

Ramkumar M., and Akansu A. N., 'Capacity estimates for data hiding in compressed images,' *IEEE Transactions on Image Processing*, 2001, vol. 10 (8), pp. 1252–1263.

Ross P. E., 'Commandments,' *IEEE Spectrum*, 2003, vol. 40 (12), pp. 30–35.

Roy S. D., Li X., Shoshan Y., Fish A., and Yadid-Pecht O., 'Hardware implementation of a digital watermarking system for video authentication,' *IEEE Transactions on Circuits and Systems for Video Technology*, 2013, vol. 23 (2), pp. 289–301.

Roy S. D. and Yadid-Pecht O., 'Design and implementation of hardware based watermarking solutions for CMOS image sensors,' *IEEE International NEWCAS Conference*, 2012, pp. 341–344.

Sang J. and Alam M. S., 'Fragility and robustness of binary-phase-only-filter-based fragile/semifragile digital image watermarking,' *IEEE Transactions on Instrumentation and Measurement*, 2008, vol. 57 (3), pp. 595–606.

Sarreshtedari S., Akhaee M. A., and Abbasfar A., 'A watermarking method for digital speech self-recovery,' *IEEE/ACM Transactions on Audio, Speech, and Language Processing*, 2015, vol. 23 (11), pp. 1917–1925.

Sengupta A. and Roy D., 'Antipiracy-aware IP chipset design for CE devices: A robust watermarking approach [Hardware matters],' *IEEE Consumer Electronics Magazine*, 2017, vol. 6 (2), pp. 118–124.

Seo Y. H. and Kim D. W., 'Real-time blind watermarking algorithm and its hardware implementation for motion JPEG2000 image CODEC,' *Proceedings of the 1st Workshop on Embedded Systems for Real-Time Multimedia*, 2003, pp. 88–93.

Sequeira A. and Kundur D., 'Communications and information theory in watermarking: A survey,' *Proceedings of SPIE Multimedia Systems and Application IV*, 2001, vol. 4518, pp. 216–227.

Servette S. D., Podilchuk C., and Ramchandran K., 'Capacity issues in digital watermarking,' *IEEE International Conference on Image Processing*, ICIP-98, 1998, vol. 1, pp. 445–449.

Singh A., Kumar B., Dave M., and Mohan A., 'Multiple watermarking on medical images using selective DWT coefficients,' *Journal of Medical Imaging and Health Informatics*, 2015, vol. 5 (3), pp. 1–8.

Singh A. K., Kumar B., Singh G., and Mohan A. (eds.), '*Medical Image Watermarking*,' Springer International Publishing, Delhi, ISBN: 978-3-319-57699-2, pp. XXIV, 244.

Strycker L. D., Termont P., Vandewege J., *et al.*, 'An implementation of a real-time digital watermarking process for broadcast monitoring on a trimedia VLIW processor,' *Proceedings of the IEE International Conference on Image Processing and Its Applications* (vol. 2), 1999, pp. 775–779.

Strycker L. D., Termont P., Vandewege J., *et al.*, 'Implementation of a real-time digital watermarking process for broadcast monitoring on trimedia VLIW processor,' *IEE Proceedings on Vision, Image and Signal Processing*, 2000, vol. 147 (4), pp. 371–376.

Su P. C., Wu C. S., Chen I. F., Wu C. Y., and Wu Y.C., 'A practical design of digital video watermarking in H.264/AVC for content authentication,' *Image Communication*, 2011, vol. 26 (8–9), pp. 413–426. URL http://dx.doi.org/10.1016/j.image.2011.07.004

Sugihara R., 'Practical capacity of digital watermark as constrained by reliability,' *Proceedings of the International Conference on Information Technology: Coding and Computing*, 2001, pp. 85–59.

Swanson M., Kobayashi M., and Tewfik A., 'Multimedia data embedding and watermarking technologies,' *Proceedings of the IEEE*, 1998, vol. 86 (6), pp. 1064–1087.

Tefas A. and Pitas I., 'Robust spatial image watermarking using progressive detection,' *Proceedings of the IEEE International Conference on Acoustics, Speech, and Signal Processing*, vol. 3, 2001, pp. 1973–1976.

Termont P., Strycker L. D., Vandewege J., *et al.*, 'Performance measurements of a real-time digital watermarking system for broadcast monitoring,' *Proceedings of the IEEE International Conference on Multimedia Computing and Systems*, 1999, vol. 2, pp. 220–224.

The Broadcast Music Inc., URL https://www.bmi.com/. Last Accessed: 06-20-2018

The Motion Picture Licensing Corporation: URL https://www.mplc.org/. Last Accessed: 06-20-2018

Tsai T. H. and Lu C. Y., 'A systems level design for embedded watermark technique using DSC systems,' *Proceedings of the IEEE International Workshop on Intelligent Signal Processing and Communication Systems*, 2001.

Tsai T. H. and Wu C. Y., 'An implementation of configurable digital watermarking systems in MPEG video encoder,' *Proceedings of the IEEE International Conference on Consumer Electronics*, 2003, pp. 216–217.

Ullah S., Rinner B., and Marcenaro L., 'Smart cameras with onboard signcryption for securing IoT applications,' *Proceedings of the Global Internet of Things Summit (GIoTS)*, Geneva, 2017, pp. 1–6.

Vellaisamy S. and Ramesh V., 'Inversion attack resilient zero-watermarking scheme for medical image authentication,' *IET Image Processing*, 2014, vol. 8 (12), pp. 718–727.

Vijayan R. and Sreedivya R. S., 'Cryptographic-steganography network communication,' *IEEE International Conference on Power, Control, Signals and Instrumentation Engineering (ICPCSI)*, 2017, pp. 694–697.

Vo P. H., Nguyen T. S., Huynh V. T. and Do T. N., 'A robust hybrid watermarking scheme based on DCT and SVD for copyright protection of stereo images,' *4th NAFOSTED Conference on Information and Computer Science*, 2017, pp. 331–335.

Voloshynovskiy S., Pereira S., Pun T., Eggers J. J., and Su J. K., 'Attacks on digital watermarks: Classification, estimation based attacks, and benchmarks,' *IEEE Communications Magazine*, 2001, vol. 39 (8), pp. 118–126.

Voyatzis G. and Pitas I., 'The use of watermarks in the protection of digital multimedia products,' *Proceedings of the IEEE*, 1999, vol. 87 (7), pp. 1197–1207.

Wang H., Peng D., Wang W., Sharif H., and Chen H. H., 'Energy-aware adaptive watermarking for real-time image delivery in wireless sensor networks,' *IEEE International Conference on Communications*, 2008, pp. 1479–1483.

Wang J-P., Sun S-F., Jiang M., Xie D-G., and Lei, B-J., 'Anti-protocol attacks digital watermarking based on media-hash and SVD,' *Fifth International Conference on Information Assurance and Security*, 2009, vol. 1, pp. 364–367.

Wang S., Cui C., and Niu X., 'A novel DIBR 3D image watermarking algorithm resist to geometrical attacks,' *Chinese Journal of Electronics*, 2017a, vol. 26 (6), pp. 1184–1193.

Wang Y., Liu J., Yang Y., Ma D., and Liu R., '3D model watermarking algorithm robust to geometric attacks,' *IET Image Processing*, 2017b, vol. 11 (10), pp. 822–832.

Wang Y. G., Xie D., and Gupta B. B., 'A study on the collusion security of LUT-based client-side watermark embedding,' *IEEE Access*, 2018, vol. 6 (15), pp. 816–15,822.

Winkler T. and Rinner B., 'TrustCAM: Security and privacy-protection for an embedded smart camera based on trusted computing,' *7th IEEE International Conference on Advanced Video and Signal Based Surveillance*, 2010, pp. 593–600.

Xiang Y., Natgunanathan I., Guo S., Zhou W., and Nahavandi, S., 'Patchwork-based audio watermarking method robust to de-synchronization attacks,' *IEEE/ACM Transactions on Audio, Speech, and Language Processing*, 2014, vol. 22 (9), pp. 1413–1423.

Yeung M. M., 'Digital watermarking,' *Communications of the ACM*, 1998, vol. 41 (7), pp. 31–33 .

Zhang F., and Shi Z. J., 'Differential and correlation power analysis attacks on HMAC-Whirlpool,' *Eighth International Conference on Information Technology: New Generations*, 2011, pp. 359–365.

Zhang J., and Liu L., 'Publicly verifiable watermarking for intellectual property protection in FPGA design,' *IEEE Transactions on Very Large Scale Integration (VLSI) Systems*, 2017, vol. 25 (4), pp. 1520–1527.

Zhang Y., Xu Z., and Huang B., 'Channel capacity analysis of the generalized spread spectrum watermarking in audio signals,' *IEEE Signal Processing Letters*, 2015, vol. 22 (5), pp. 519–523.

Zhao J., Koch E., O'Ruanaidh J., and Yeung M., 'Digital watermarking: What will it do for me? and what it won't!,' *Proceedings of SIGGRAPH Conference*, 1999, pp. 153–155.

Zhao J., Zhang N., Jia J., and Wang H., 'Digital watermarking algorithm based on scale-invariant feature regions in non-subsampled contourlet transform domain,' *Journal of Systems Engineering and Electronics*, 2015, vol. 26 (6), pp. 1309–1314.

Chapter 12
Physical unclonable functions (PUFs)

It has been the practice to store information under lock and key for safeguarding it. Even today, we use cryptographic primitives to store information securely under lock and key, encryption, and decryption. For the process of cryptography, keys are necessary for any operation. But these keys should be stored in the memory so that they can be used whenever necessary. When a key is stored in the memory, it can be stolen by the adversary using various methods. Therefore, storing it in a non-volatile memory is not an option in this age of security threats. Physical unclonable functions (PUFs) are the promising security primitives used for generating the keys instead of storing them in the memory. These modules use the naturally occurring manufacturing variations in the fabrication process for generating the keys for cryptographic purposes. This chapter discusses different types of PUFs. Section 12.1 gives a brief introduction of PUFs. Section 12.2 discusses working principles of PUFs. Section 12.3 discusses different characteristics of a PUF design. Section 12.4 presents different classifications of PUFs. Various designs of PUF based on ring oscillators (ROs) are presented in Section 12.5, based on multiplexers and reconfigurability are presented in Section 12.6. Static random access memory (SRAM)-based PUFs are presented in Section 12.7, memristor-based PUFs are presented in Section 12.8, and diode-based PUFs are presented in Section 12.9. There are also non-silicon-based PUF designs, such as carbon PUFs, presented in Section 12.10. Microprocessors can also be used for implementing the PUF designs which are presented in Section 12.11. Magnetic material-based PUFs are presented in Section 12.12 and the FPGA implementations of PUF and security measures for FPGA are presented in Section 12.13. Some case study applications are presented in Section 12.14. Further the issues and challenges faced during the design of PUF modules are presented in Section 12.15. The conclusion and future directions are presented in Section 12.16.

12.1 Introduction

The first transistor was invented in 1948. With the introduction of the transistor, vacuum tubes occupying huge space disappeared and led to the invention of the IC in 1958 (Kilby, 1976). Along came the Moore's Law which states that the number of transistors on an IC will double every two years. This came true until recently and the number of transistors on a single chip has exponentially increased since its first introduction. With the introduction of the new 10 nm technology, each square millimeter of an IC can be packed with 100 million transistors (Courtland, 2017). With technology improving at such a drastic pace, it has penetrated into almost every sphere of human life. Day-to-day activities such as communications, business, financial transactions, and so on depend on technology. Digital footprint of human being has become an inevitable and unavoidable parameter (Gencoglu *et al.*, 2015). IoT has become an integral part of our lives. In an IoT environment, every device is connected to the network and will have a unique IP address (Mohanty *et al.*, 2016). Smart cities, smart grid, smart healthcare, and so on are various forms of IoT.

With so many devices connected to each other, there are security concerns that are raised. The data that is being collected by various applications, sensors, and other devices should be stored securely. There are many attacks reported recently where security needs to be given the highest importance (Charette, 2018a, 2018b). To secure the data that is being generated, cryptography is used. It has been around for a very long time. Cryptography is the process of concealing a message or information by converting it into unreadable text and only the end user will be able to extract the original information from the text. It was found that in the year 1900 BC Egyptians used techniques such as steganography (2001). Various algorithms are available currently for encryption and decryption which can transfer the data securely. The National Bureau of Standards proposed DES algorithm which lasted for 20 years. Then came AES algorithm which is very popular now but there have been reports of AES being successfully broken (Mangard *et al.*, 2005). An amount of $7.5 billion is estimated to be lost by the US semiconductor industry because of IC counterfeiting (Rose *et al.*, 2013). In addition to all of this, in the case of many devices, sensors, or applications, the device itself will not be monitored all the time. It will be in a remote location transmitting and receiving necessary data to the cloud or a remote server. In such cases, the current practice of encryption and decryption is to store the necessary keys in a nonvolatile memory and use it (Herder *et al.*, 2014) when required. The memory is subject to side-channel attacks where the attacker can get access to the physical device and retrieve data from it.

A *PUF* is a new primitive that is being introduced into various applications. Without storing a key in nonvolatile memory, it can provide keys for encryption and decryption (as shown in Figure 12.1) (Yanambaka *et al.*, 2017). The cryptography key will be generated by PUF whenever necessary, based on the imperfections that are introduced into an IC while fabrication. There are various advantages to PUF. Fabrication of PUF components is much easier compared to the memory modules necessary to store the key. This also means the reduction in cost and chip area necessary for the memory modules like EEPROM. Side-channel attacks on the device itself will become more difficult. With some types of PUF, expensive cryptographic processors can be avoided. The number of keys generated by the PUF increases exponentially with the complexity of the architecture and design used. The same keys generated by a PUF module can also be used for IP protection where the input and output pair of PUF can be used to authenticate the ownership of a device. PUF uses the manufacturing variability of the IC to generate the output keys. Hence, the output keys generated by a PUF module should also be unique. The concept of

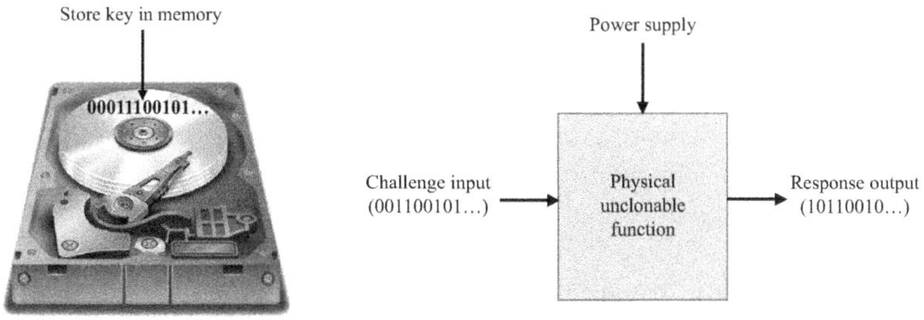

(a) Conventional method of storing keys in memory (b) Generating keys with PUF symbols

Figure 12.1 The concept of a PUF over traditional key stored in a nonvolatile memory

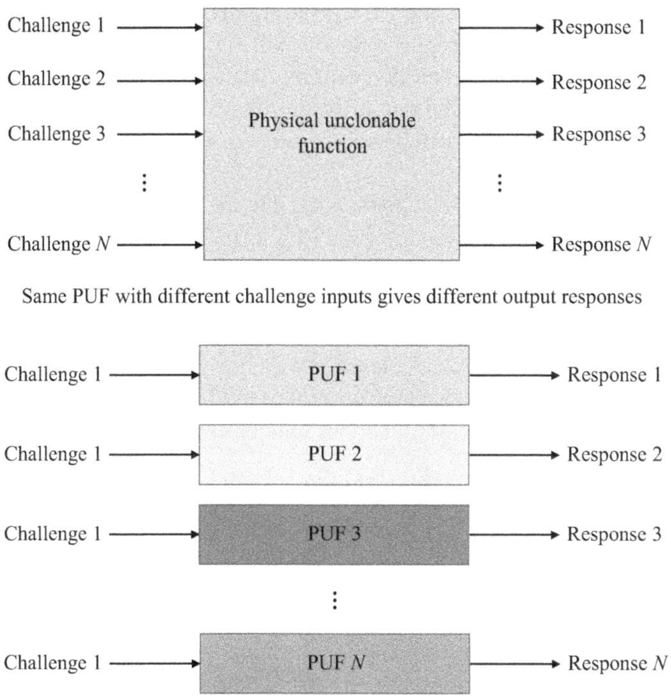

(a) Same PUF with different challenge inputs gives different output responses

(b) Different instantiations of a PUF circuit with same input challenge gives different output responses symbols

Figure 12.2 Concept of PUF

PUF is shown in Figure 12.2. For the same PUF module, different inputs yield different outputs as shown in Figure 12.1. For different instantiations of a PUF circuit (logically same) on different chips, the same input yields different outputs as shown in Figure 12.1.

12.2 PUF: Principle

When an IC is fabricated, there is a process that will be followed from the beginning to the end. From the time silicon becomes an ingot to the final IC, these series deals with various fabrication steps such as masking, ion implantation, photolithography, chemical diffusion, polishing, electrolysis, and so on (Mohanty, 2015). During this process of fabrication, the devices that are being fabricated on the wafer will be affected due to various factors. Figure 12.3 shows the manufacturing variability of devices. The geometry of the devices such as length, gate oxide thickness, doping concentration, and so on will also be affected. Due to these variations, no two devices that are fabricated will look the same.

Even when the devices are fabricated on the same die, there will be variations in their geometry that will make them distinct from the others on the same wafer. A PUF will take advantage of these variations to generate the keys. The manufacturing variations that are introduced into the IC are unpredictable and unavoidable. A naturally occurring phenomenon will introduce unwanted changes in the geometry of the device. When two different transistors on the same die have variations in their geometry, the output of the two will also be different. The output from two different devices will be compared with each other to generate a single bit key of PUF.

The design of a simple PUF is shown in Figure 12.4. The input to a PUF is called the "Challenge" and the output from a PUF is called the "Response." The pair of input and output of a PUF is called "Challenge Response Pair" (CRP). Figure 12.4 shows the design of arbiter PUF (Lee *et al.*, 2004). As shown in the figure, the PUF uses two sets of multiplexers connected in series. The output from each set is connected to the D-input and clock signal of a D-flip-flop. The select lines of the multiplexers are the challenge input the user gives to PUF. Based on the challenge input, the multiplexers through which the input signal traverses will change. Once the input

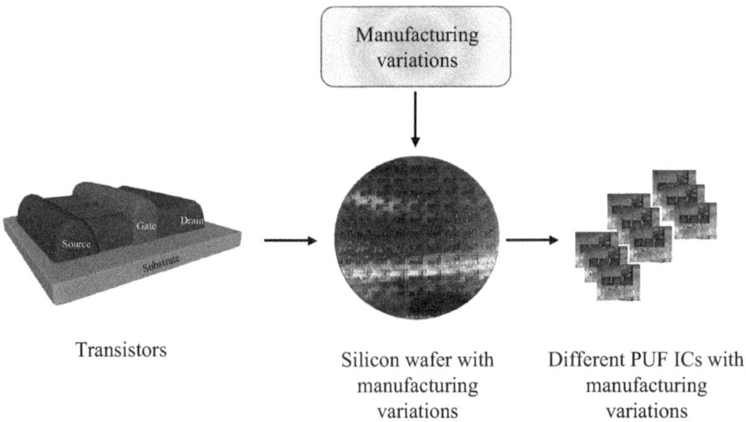

Figure 12.3 Working characteristics of the PUF

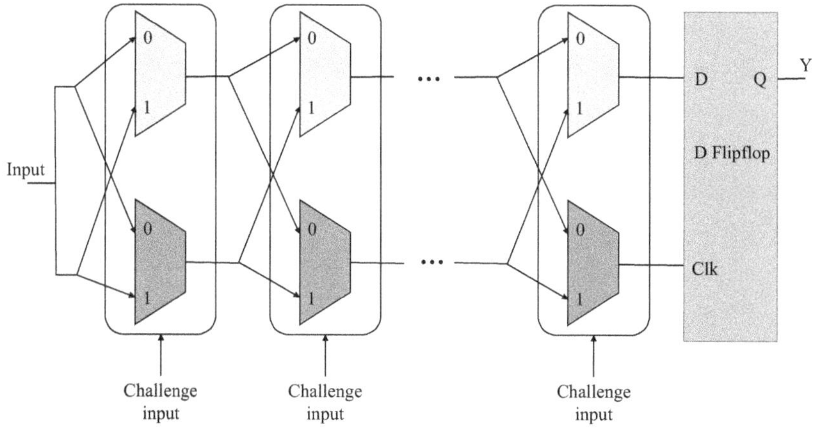

Figure 12.4 Arbiter-based simple PUF

signal is given to the module, based on the challenge input, the delay will be added before it reaches the D-flip-flop. The output from the flip-flop will depend on the amount of delay added to the input signal by the devices it passes through. In this scenario, due to manufacturing variations, each transistor in the multiplexers will add various amounts of delay. This delay added is unpredictable. Thus, the PUF key generated is also unique to the module and secure. However, predicting the CRPs in the arbiter PUF is also not difficult. To avoid this, much complex designs of PUF are designed, which are introduced in the later sections.

12.3 Properties or characteristics of PUFs

A PUF is used to generate keys for cryptographic purposes. Various properties should be satisfied by the outputs of PUF to be used for various applications. Figure 12.5 shows the different characteristics of PUF that should be satisfied for it to be ideal (Böhm, 2013; Chang *et al.*, 2017; Maiti *et al.*, 2013; Yanambaka *et al.*, 2017) (Table 12.1).

12.3.1 Uniqueness

A PUF module generates an N-bit key which should be unique. PUF uses the manufacturing variations on an IC to generate the output key. Hence, the same PUF key should not be generated by the same module for a different challenge input. The uniqueness can be considered to be calculated over different chips. Let us consider a pair of chips, i and j, where i is not same as j. The responses generated be R_i and R_j. Then the uniqueness can be given as (Chatterjee *et al.*, 2017; Maiti and Schaumont, 2011):

$$\text{Uniqueness} = \frac{2}{K(K-1)} \sum_{i=1}^{k-1} \sum_{j=i+1}^{k} \frac{HD(R_i, R_j)}{n} \times 100\% \qquad (12.1)$$

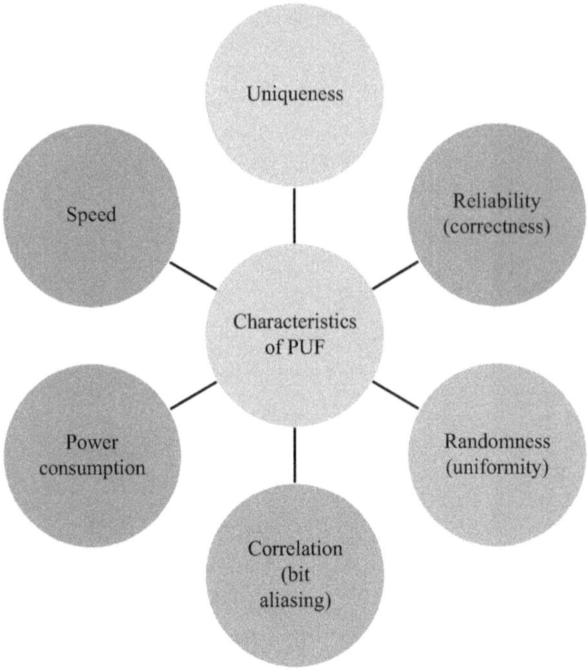

Figure 12.5 Properties of PUFs

Table 12.1 Comparison of various publications

Research works	**Technology**	**Architecture used**	**Power consumption**	**Uniqueness (Hamming distance %)**
Rahman *et al.* (2014)	90 nm CMOS		–	50
Maiti and Schaumont (2011)	180 nm CMOS	Ring oscillator	–	50.72
Sahoo *et al.* (2016)	90 nm CMOS	Ring oscillator	–	45.78
Maiti *et al.* (2010)	–	Ring oscillator	–	47.31
Yanambaka *et al.* (2017)	14 nm	Current-starved ring oscillator	250.15 μW	49.6
Yanambaka *et al.* (2018)	14 nm DLFET	Current-starved ring oscillator	121.34 μW	48.0
Tanaka *et al.* (2018)	CMOS	Bistable rings		50
Yu *et al.* (2017)		Interconnect PUF		48.3
Kumar *et al.* (2008)	FPGA	SRAM		≈50
Giterman *et al.* (2018)	28 nm	GC-eDRAM		50
Jeloka *et al.* (2017)	28 nm	Bistable SRAM	97% efficient	38
Su *et al.* (2008)	130 nm	SRAM	0.93 μW	64.7
Rosenblatt *et al.* (2013)	32 nm	Embedded DRAM		50
Hashemian *et al.* (2015)	65 nm	Embedded DRAM		50.15
Bai *et al.* (2016)	0.13 μm	Thyristor-based PUF	380 μW	16.012
Wang and Tehranipoor (2010)	90 nm	Ring oscillator		41

HD is the hamming distance of the two response bits and n is the length of the response. The number of different elements that two strings of the same length have is called the HD. Initially, XOR is performed on the two strings and the result is added to get the HD. For example, consider the two strings "11001100" and "11100111." The XOR of the string will be "00101011" and adding all bits will get us 4, which is the HD (Böhm, 2013). HD is also represented in percentage, where the result is divided by the length of the string and multiplied by 100. In the previous example, it is $\frac{4}{8} * 100$ is 50%. The ideal HD for a PUF to be securely used for cryptographic purposes is 50%.

12.3.2 Reliability (correctness)

When necessary PUF will generate the keys for various purposes. But once it is implemented, for a single challenge input, the output response should not change. Various factors such as aging, voltage fluctuations, noise, and other factors should not affect the output of a PUF. This is the reliability or correctness characteristic of PUF where it will be able to generate the same key for a given challenge input. The reliability of a PUF can also be calculated using the HD. If the same key is given to a single PUF over a long period of time, the ideal HD should be 0%. No errors in the output bits should occur. As no design is ideal, various techniques were employed to avoid them, especially to avoid the aging effects of a design.

Let us consider a chip i from which an "n" bit response R_i is extracted for reference at optimal conditions of environment. Now the same n-bit response is extracted from the same chip i at different environmental conditions, such as supply voltage and power supply, and let the value be $R_{i'}$, where x different samples are taken from the same i at different conditions. The reliability can be given as (Maiti and Schaumont, 2011):

$$S = \frac{1}{x}\sum_{y=1}^{x}\frac{\mathrm{HD}(R_i, R'_{i,y})}{n} \times 100\% \tag{12.2}$$

12.3.3 Randomness (uniformity)

For the output of a PUF to be random, the key should contain equal number of 1's and 0's. This ensures that the output of a PUF is random and secure to be deployed in an application. Apart from the HD, the randomness parameter validates the keys generated by the PUF. In some cases, it is also useful in determining the error rate in the output.

12.3.4 Correlation (bit aliasing)

The correlation characteristics of a PUF are of two different types: (1) bit and (2) chip. There should be no correlation between any of the bits generated by the PUF module. If a different challenge is given to the PUF, the response should be different. And various chips on the same wafer should not affect the working of a PUF module and generate the same keys. In some cases, the manufacturing variations will be uniform across the wafer. In such cases, there is a chance that with the CRPs of one PUF chip, the other can be predicted.

12.3.5 Power consumption

A PUF is designed to be deployed in devices like RFID tags which consume very low power. In an IoT environment, there is a chance that the devices are battery operated and need to work with low power consumption. PUF is also being deployed in various smart devices which require a low-power operation. Hence designing the PUF with low-power consumption is vital.

12.3.6 Speed

There are various IoT environments depending on the place they are deployed and the application tasks they are supposed to perform. Based on the various tasks, there is a need that the environment has to send and receive the data in real time. In such cases, there is a need for security but the same cannot be a bottleneck to the processing speed or increase the transfer latency. Hence the speed with which the keys are generated is also going to play a major role in where the respective design is going to be deployed.

12.4 Classification of PUFs

PUFs can be classified based on the fabrication method and security strength as shown in Figure 12.6 (Chang *et al.*, 2017; Joshi *et al.*, 2017; Škorić *et al.*, 2005). In a broad sense, there are two categories of PUFs. They are categorized based on the types of devices used to design the PUF and the other category is based on the level of security provided by the module itself. Further, each of the categories has different types of PUFs. Various researchers around the world have proposed different types of PUF: these are silicon-based, non-silicon-based, electronic, and non electronic. Based on the level of security provided, there are three types of PUFs, strong, controlled, and weak.

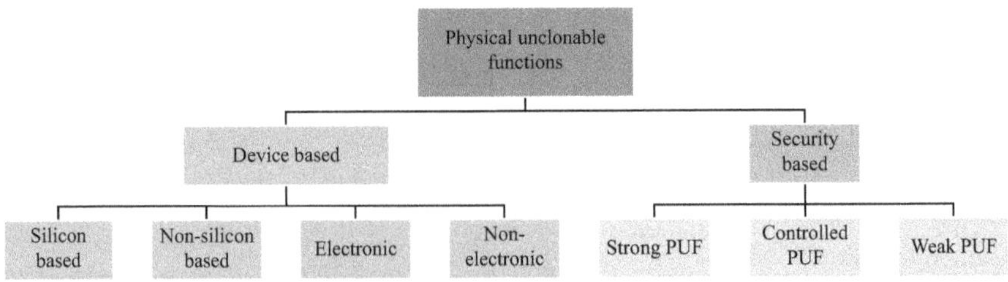

Figure 12.6 Different types of PUFs

12.4.1 Device-based PUFs

Various devices are used by researchers to manufacture PUF. A PUF uses the manufacturing variability present in the devices to generate the outputs. Not all the devices in the world are ideal. Therefore, different varieties of devices can be implemented at the core of design to generate the PUF keys.

12.4.1.1 Silicon-based PUFs

Almost all the ICs are made from silicon and other materials. This category of PUFs is made on the silicon and fabricated on the same die as part of the IC. Various designs of PUF are proposed by researchers around the world depending on the implementation and the area of deployment. Examples of this type of PUFs are SRAM (Schrijen and Leest, 2012), butterfly (Kumar *et al.*, 2008), arbiter (Lao and Parhi, 2011), RO (Rahman *et al.*, 2014), and lightweight secure (Majzoobi *et al.*, 2008). These PUFs will be extracting the manufacturing variations on the IC which can generate the keys for cryptographic purposes.

12.4.1.2 Non-silicon PUFs

PUFs that are not classified as silicon PUFs are referred as non-silicon PUFs. They are fabricated in silicon systems but require special fabrication techniques which are not part of a generic CMOS fabrication technology. The response is generated from the challenge set obtained from the random physical variation in the physical system rather than the ICs. Such types of PUFs are optical, acoustic, magnetic, radio frequency (RF), coating, etc.

12.4.2 Security-based PUFs

Once a PUF is fabricated, the design itself cannot be modified. Hence, various designs of PUFs will generate different number of CRPs. The CRPs will decide where the particular design of PUF can be deployed and used. Based on the category, PUF is of three types: strong, controlled, and weak.

12.4.2.1 Strong PUF

The strong PUF is the types of designs that will be able to generate more number of CRPs. The CRPs can also be not easily extracted from the device. A protection is provided through a complex input–output relations and an exponential number of CRPs. This type of PUF operates in the range of few MHz, so even a short duration of eavesdropping can collect many CRPs. However, all known strong PUFs are vulnerable to one or more types of attacks. This type of PUFs is frequently used for device authentication purposes. Some common example of circuits belonging to this class are arbiter, feed-forward arbiter, XOR arbiter, lightweight secure, and optical PUFs. This class of PUFs is also frequently referred to as super-high information content (SHIC) (Lugli *et al.*, 2013; Ruhrmair *et al.*, 2011).

12.4.2.2 Weak PUF

This category of PUFs will not be able to generate more number of CRPs. The least number of CRPs that can be generated is "1" in a worst-case scenario. Therefore, this design can be easily vulnerable to attacks. This is the reason that the design of this PUFs cannot be revealed publicly. These are mainly used to derive a secret key in cryptographic algorithms and are the least susceptible to modeling attacks. Some common examples of circuits belonging to this class are SRAM, coating, and butterfly PUFs. These are also known as physically obfuscated keys (POKs).

12.4.2.3 Controlled PUFs

This class of PUFs has a strong PUF as an underlying core additionally protected by control logic. Since the challenges are preprocessed first before passing to the strong PUF and the responses are preprocessed before coming out of the strong PUF, the CRPs of the strong PUF are never accessed directly. This preprocessing of challenges and responses significantly improves the security of controlled PUFs. A controlled PUF (Yu and Devadas, 2010) is only accessed via an algorithm which is fabricated in the same chip as the PUF such that the PUF and the algorithm are inseparable. This implies that the actual PUF outputs are never given to the outside world directly and the outputs that are given to the outside world are the processed outputs from the algorithm.

12.5 Ring oscillator–based PUFs

The design of a ring oscillator PUF is shown in Figure 12.7 (Suh and Devadas, 2007). There are two different sets of ROs that are connected to two multiplexers

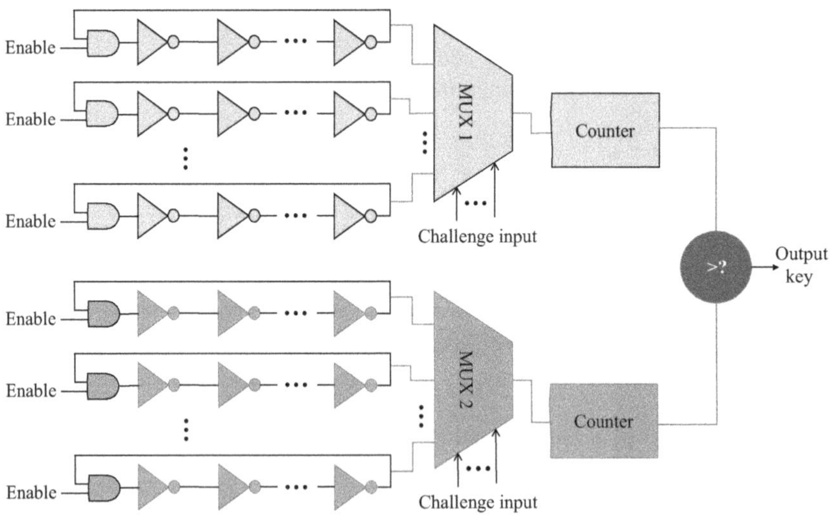

Figure 12.7 Ring oscillator PUF

"Mux1" and "Mux2." The two multiplexers are connected to two counters which are connected to a comparator, which gives the final output. The multiplexers will select different combinations of ROs and feed the oscillations to the counters. The number of oscillations over a selected period of time are given to the input and clock of the comparator. The comparator compares the two input values and gives out "1" or "0." The input to the multiplexers is the challenge input given to PUF and the output from the comparators is the response of the PUF.

Due to the manufacturing variations in the IC, all the ROs will not oscillate at the same frequency. This causes the counters to give out different number of oscillations for a same period of time. When different inputs are selected by the multiplexers, different ROs will give distinct oscillation frequencies, which generate different bits at the output of the comparator. Thus the keys generated by the RO PUF will be unique. If there are N different ROs, only $N \times (N - 1)/2$ different comparisons will be possible. The main component in the RO PUF design is the inverter. Due to the manufacturing variations in the inverter, the delay added is different for each inverter. The delay of an inverter can be calculated as (Wang and Tehranipoor, 2010):

$$t_{\text{inv}} = \frac{0.52 C_{\text{L}} V_{\text{DD}}}{\dfrac{W}{L_{\text{eff}}} \dfrac{\mu \varepsilon_{\text{ox}}}{t_{\text{ox}}} V_{\text{D}_{\text{SAT}}} \left(V_{\text{DD}} - V_{\text{T}} - \dfrac{V_{\text{D}_{\text{SAT}}}}{2} \right)}, \tag{12.3}$$

where V_{DD} is the supply voltage, C_{L} is the load capacitance, W and L are the effective width and effective length of the channel, respectively, μ is the mobility of carriers, $V_{\text{D}_{\text{SAT}}}$ is the saturation source to drain voltage, t_{ox} is the oxide thickness, ε_{ox} is the oxide dielectric permittivity, and V_{T} is the threshold voltage. The oscillation frequency of the ROs with N inverters is given as:

$$f_o = \frac{1}{2 N_{\text{inv}} t_{\text{inv}}} \tag{12.4}$$

There are some advantages to the implementation of ring oscillator PUFs. They are easier to build, evaluate, highly reliable, and relaxed layout and routing constraints in the design. There are some disadvantages also to the design. It is comparatively slow; to increase the number of bits, the chip area needs to be increase, which results in consumption of more power. In addition, aging is also one of the main issues that can affect the reliability of the ring oscillator PUF. Various factors like positive bias temperature instability (PBTI), negative bias temperature instability (NBTI), time-dependent dielectric breakdown (TDDB), hot carrier injection (HCI), electro-migration (EM), and thermal cycling (TC) also affect PUF. Due to the aging effect, the change in the output oscillations of the ROs will affect the output bits of PUF module itself. This can become a major issue when implemented in the cryptographic applications. In a PMOS transistor, when a negative voltage is applied at gate-to-source terminal, an interface trap is generated at Si–SiO_2 interface. Over a period of time, this results in the change of transistor parameters like transconductance, drain current, and threshold voltage. This effect is called NBTI. The HCI occurs due to the switching activity of the transistor

(Mohanty, 2015). NBTI and HCI will depend on various factors, including workload of the transistor and technology.

TDDB is occurred due to the rapid scaling of the device. The high operating voltages at the gate terminal of the transistor will lead to the structural degradation of the device, like the gate oxide and its interface. With the continuity, the dielectric layer will lose the property to insulate and as a result, a conduction path will form between the gate oxide and the substrate. This effect has been addressed and reduced significantly with the implementation of high-κ material. Other parameters that affect the performance of the device are the temperature and supply voltage. With the variations in the supply and temperature, the output oscillations of the ROs will change. When the temperature of the chip increases, the threshold voltage of the device is reduced. The increase in temperature can also decrease the drain saturation current and the carrier mobility. But the operating temperatures of the chip and the supply voltages can be regulated and maintained. If the temperature increases, it can be decreased using various methods. But the effects caused by aging cannot be reversed and have a permanent damage on the device and hence the chip itself. This affects the performance and reliability of the chip over time.

To avoid these, aging-resistance designs are introduced. Figure 12.8 shows the design of an aging-resistant PUF (Rahman *et al.*, 2014). The NBTI effect is decreased by introducing an NMOS transistor to connect two inverters in the RO as shown in the figure. The supply voltage V_{DD} is connected to the M_{C_i} transistors which are activated using the ENABLE signal. These are introduced to avoid the logical "0" at the input of the inverter's PMOS transistor. When the RO is not active, there is a chance that a "0" might be at the input of PMOS in the inverter and can cause power consumption. This design of PUF will reduce the NBTI effect in the off state. Various other designs of aging-resistant RO PUF designs were also introduced (Chowdhury *et al.*, 2017; Rahman *et al.*, 2015, 2016).

There are various designs of PUF that are implemented for different applications (Yanambaka *et al.*, 2017, 2018). One of the properties of PUF is the speed of key generation as discussed in Section 12.3. A PUF can be deployed in various applications, including network switches and routers. The speed with which the key is generated by the PUF is important in such cases. Figure 12.9 shows the design of

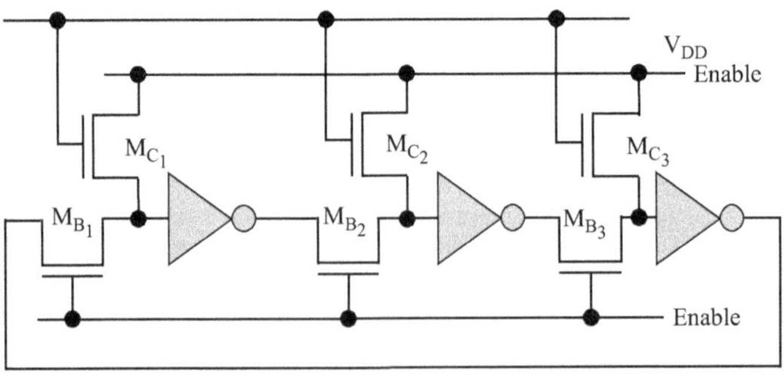

Figure 12.8 Aging-resistant RO PUF

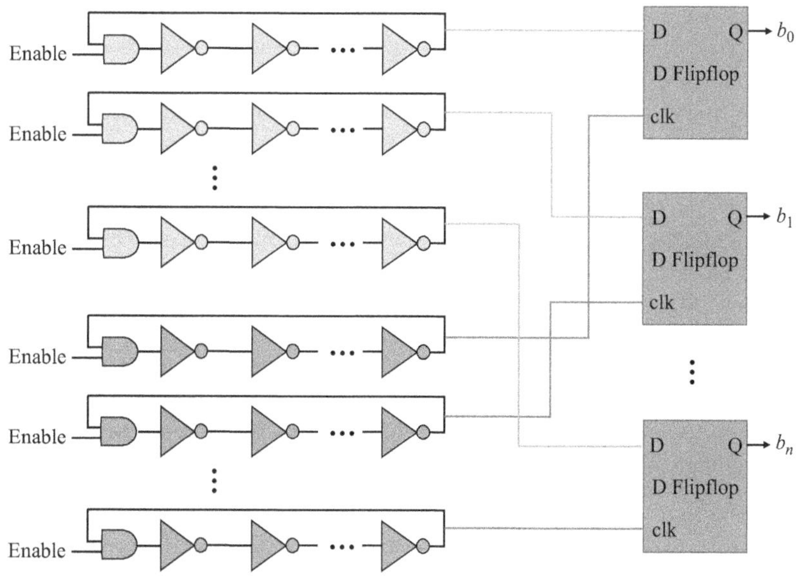

Figure 12.9 Speed-optimized hybrid oscillator arbiter PUF

hybrid oscillator arbiter PUF intended to be deployed in applications where the speed of key generation is pivotal. As shown in the figure, there are two sets of ROs in the design. Unlike the conventional ROs, multiplexers, counters, and the comparator are replaced with the D-flip-flops. The D-input and the clock input of D-flip-flops are connected with two ROs, one from each set as shown in the figure. Due to the manufacturing variations in the inverters of ROs, the delay added in each of the oscillator will be different and so will be the oscillation frequency. Based on the oscillation frequency of the ROs, the output bit from the flip-flop changes. Bits b_o to b_n give the key generated by the PUF.

12.6 Reconfigurable or dynamic PUFs

The previous section presented a ring oscillator PUF and different designs which can be implemented in various applications. Some of the designs that are presented in the previous section are reconfigurable but the others generate a constant set of CRPs and the design itself is not reconfigurable. This section concentrates on the reconfigurability of the PUF design. Reconfigurability is an important aspect of PUF design. There is a chance that if the attacker gets hold of the device and perform attacks such as machine learning, emulation, man-in-the-middle and RE, he/she will be able to regenerate the CRPs. The reconfigurable or dynamic PUFs are introduced in order to increase the security of PUF against these attacks and to be able to generate new set of CRPs when necessary. A feed-forward structure is introduced in the arbiter PUF to avoid the attacks using linear modeling. This introduces some nonlinearity in the design and increases the security and robustness of the design trading-off the design complexity. This can also be achieved using other techniques like preprocessing the challenge input before feeding it to the PUF, postprocessing the output response from the PUF, and adding some components to the output of module which can be used for reconfiguring the response bits. The preprocessing challenge bits can be done by introducing an additional circuitry which can implement linear feedback shift register (LFSR) or by introducing some hashing. The postprocessing of response bits can mean the introduction of hashing to the bits. Figure 12.10 shows the design of arbiter PUF with an LFSR at the input. The challenge inputs given to the PUF are not directly fed to the arbiter PUF. Initially they are passed through the LSFR, preprocessed and then they are given as input to the PUF, "Real Challenge" as shown in Figure 12.10 (Lao and Parhi, 2011). With the introduction of the input-reconfigurable circuitry, the robustness of the overall design and also the reconfigurability can be increased.

To avoid various attacks on the arbiter PUF, different configurations of the design are implemented (Lao and Parhi, 2014):

1. Feed-forward MUX PUF overlap structure (FFO): Design of FFO is shown in Figure 12.11. In this design, there is an overlap between the two feed-forward connections.
2. Feed-forward MUX PUF cascade structure (FFC): Design of FFC is shown in Figure 12.12. In this design, the stage which got the input from feed-forward path will be the start of a new feed-forward path.

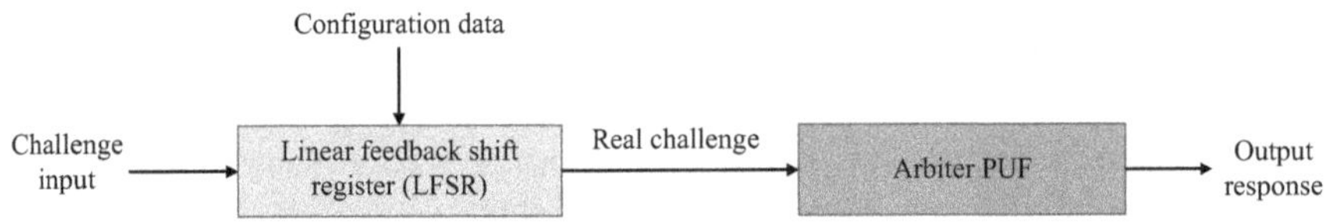

Figure 12.10 Arbiter PUF with LFSR at input

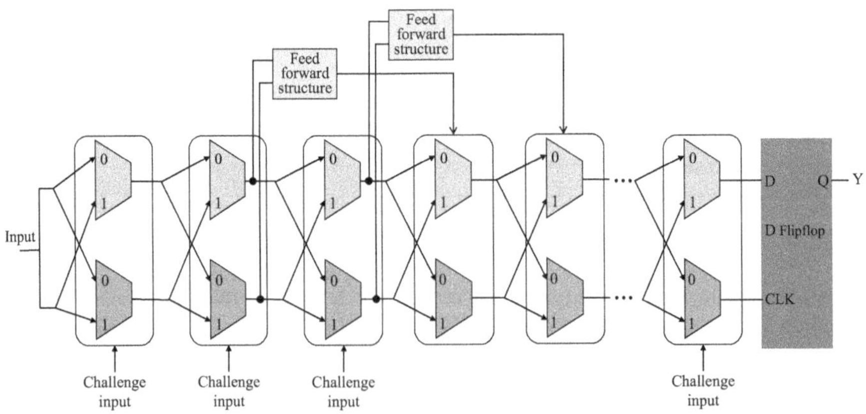

Figure 12.11 Arbiter feedforward MUX-based overlap structure PUF

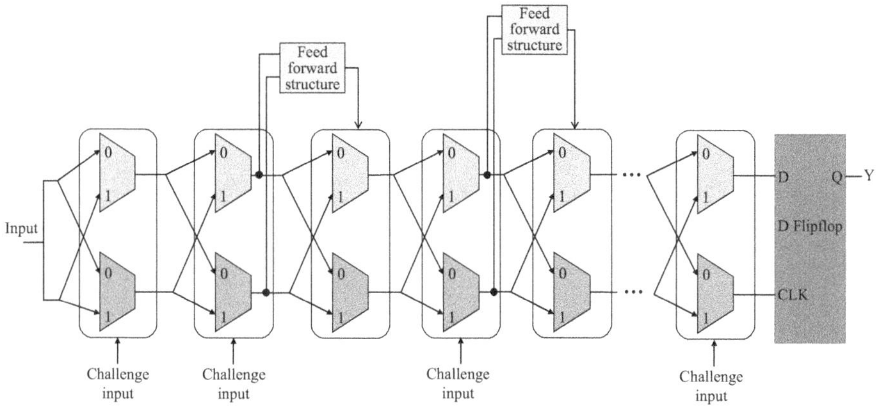

Figure 12.12 Arbiter feedforward MUX-based cascade structure PUF

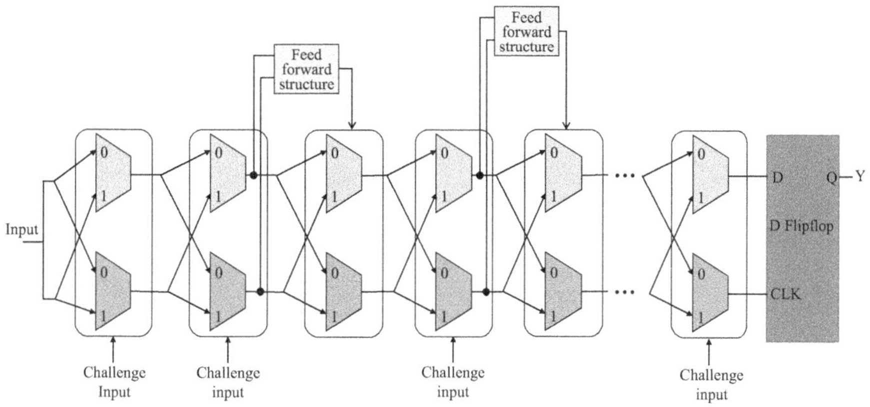

Figure 12.13 Arbiter feedforward MUX-based separate structure PUF

3. Feed-forward MUX PUF separate structure (FFS): Design of FFS is shown in Figure 12.13. In this design, there will be no overlap between the feed-forward paths of the multiplexers.

Figure 12.14 shows the design of a lightweight PUF. The design is presented by Majzoobi *et al.* (2008) with an improved robustness and increased security. In this PUF design, there are two networks at the input and the output, "Input Network," which takes in the challenge inputs, and "Output Network," which gives out the response outputs. The module itself will contain a number of parallel PUF structures which are run by the input and the output networks. For connecting the challenge inputs to a number of parallel PUF structures, an interconnect network is implemented. A single PUF will be satisfying the strict avalanche criterion (SAC). Any cryptographic primitive is said to satisfy the SAC if a complement in the single input

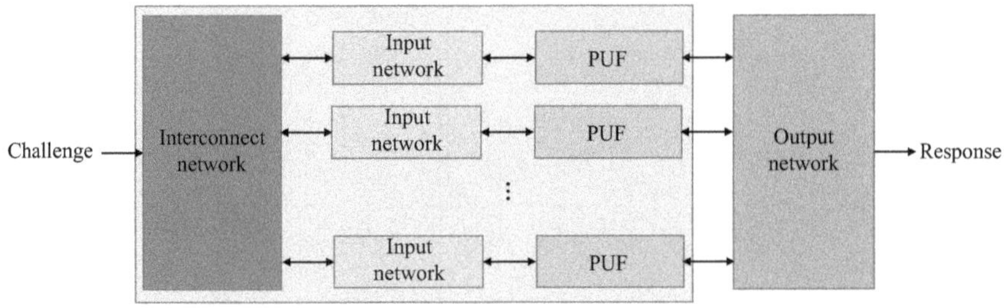

Figure 12.14 Lightweight PUF

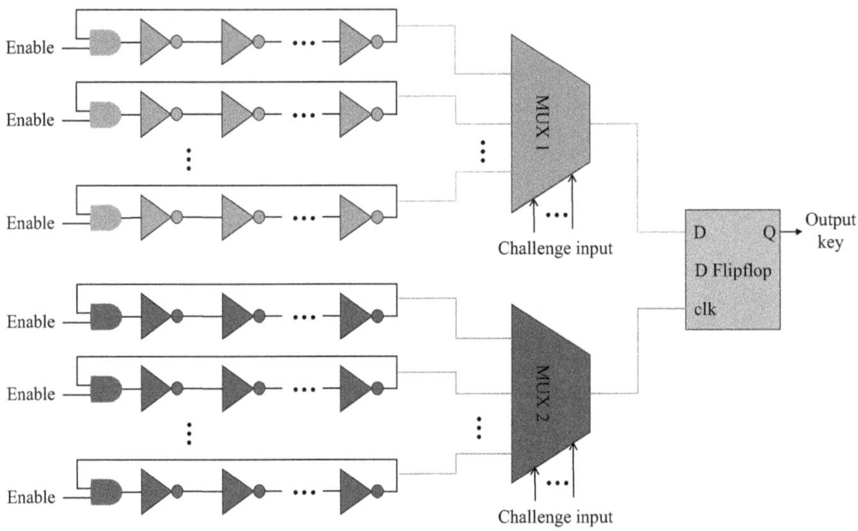

Figure 12.15 Power-optimized hybrid oscillator arbiter PUF

bit induces a change in each of the output bits with a 50% probability. But when the overall parallel network of the design is considered, the SAC is not satisfied.

Figure 12.15 shows a design of hybrid oscillator arbiter PUF which uses RO as a main component in the module and can also be reconfigured (Yanambaka *et al.*, 2018). As shown in the figure, there are two different sets of ROs which are connected to two multiplexers "Mux 1" and "Mux 2." The ROs will generate the oscillations which are fed to the multiplexers. Challenge input given to the PUF module acts as the select lines to the multiplexers. Based on the challenge input given, the multiplexers select different ROs. The output from Mux 1 is given as D-input to the D-flip-flop and output from Mux 2 is given as clock signal to the D-flip-flop. Due to the variations in the ROs, the oscillations produced will not be of the same frequency. Therefore, the output of D-flip-flop will change based on the input challenge. Different configurations of inputs can be achieved by changing the select line inputs. The robustness and security offered by the design can be increased by increasing the number of ROs in the design, but it also increases the complexity and chip are occupied.

12.7 SRAM-based PUF

Many devices and development board have SRAM which can be used to design a PUF. Figure 12.16 shows the circuit design of a 6T SRAM cell (Mohanty, 2015). The start-up value of the SRAM cell is used by the PUF to generate the output bits. When compared to the other designs of PUF, the reliability and uniqueness of the SRAM PUF are less. As shown in Figure 12.16, an SRAM cell can be designed using cross-coupled inverters and will be able to store a single bit of binary data, "1" or "0." When the SRAM cell is powered up, its initial condition is taken as the advantage by the PUF to generate the unique output. But there are issues with the behavior of SRAM cells at the start-up. Different SRAM memory modules were compared and analyzed by Schrijen and Leest (2012), operating at different temperatures and supply voltages. The analysis was performed by calculating the reference bits at the starting, at various temperatures of 20 °C, −40 °C, and 80 °C by Schrijen and Leest (2012), which showed the stability of SRAM. The supply voltage of the SRAM was also varied by 10%. This does not affect the performance of the PUF as claimed (Schrijen and Leest, 2012) because the start-up conditions are attained during the initial few nanoseconds.

In the case of an SRAM memory, one of the important challenges is to get a sufficient entropy into a key that is of a fixed length. In the case of an SRAM PUF, the generated keys will have noise and the entropy will be lost during the generation of helper data and the error correction (Koeberl *et al.*, 2014; Maes *et al.*, 2012). In some cases, the loss of entropy during the helper data will be considered where it is uniformly distributed. The entropy and the length of output are equal when the entropy density is multiplied by the error correction. In that scenario, the entropy of the raw PUF outputs, even when low, can be used for generation of keys. The entropy of the output of PUF is usually unknown and so the estimation of entropy should be performed using a more rigorous method in worst-case scenarios. In the syndrome constructions and the code-offset, when a binary linear code $C(n, k, t)$, where the

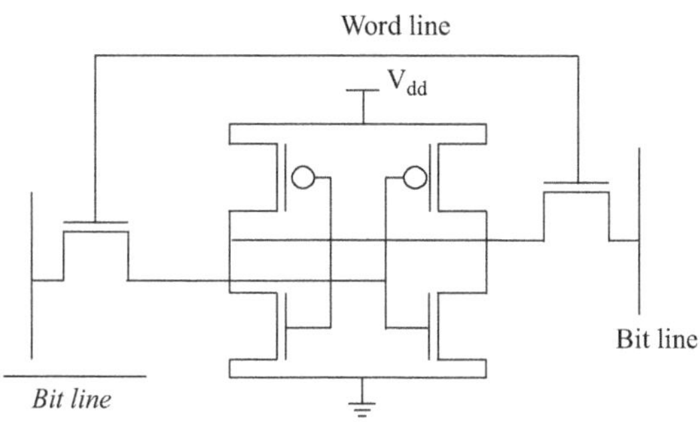

Figure 12.16 6T SRAM cell

length of the dimension of code is represented by n and k and the maximum number of errors that can be corrected is represented by t, is considered. When a worst-case scenario is taken into account, the entropy lost can be calculated as $n - k$ bits which can be considered satisfactory (Maes *et al.*, 2012). When the original bit entropy is less than $\frac{n-k}{n}$/bit, entropy will not be present in the output in case of a worst-case condition. Therefore, PUFs are not suited for generating the keys under a worst-case condition. For calculating the min-entropy, NIST specification 800-90 (Barker and Kelsey, 2015) is usually used. Let p_0 and p_1 be the probabilities of binary source of outputs where $p_0 + p_1 = 1$, the maximum probability of p_0 and p_1 which can be used for the calculation of min-entropy is as follows (Liu *et al.*, 2018):

$$H_{\min} = -\log_2(p_{\max}) \tag{12.5}$$

where $p_{\max} = \max(p_0, p_1)$.

For increasing the entropy of the source bit, Von Neumann extractor is used to compress the bits (Leest *et al.*, 2010). In the case of Von Neumann extractor, the process is as follows: a sample of the pattern is considered and is divided into successive nonoverlapping bits. When the pair of bits differ from each other, the second bit that is considered is discarded. If both the bits are equal, both of them are discarded. After the completion of processing all the bits, the remaining bit will be in order. A bias to "1" is considered for simplicity. Pair of bits A and B having a maximum probabilities of $p_{\mathrm{A}_{\max}}$ and $p_{\mathrm{B}_{\max}}$ are the probabilities due to the bit "1" occurrence, with the probabilities $p_{\mathrm{A}_{\max}}(1 - p_{\mathrm{B}_{\max}})$ and $(1 - p_{\mathrm{B}_{\max}})$, respectively. The ratio between the lengths of the output and input of extractor is called "Extraction Rate." This can be given as (Liu *et al.*, 2018):

$$r = \frac{(p_{\mathrm{A}_{\max}} = p_{\mathrm{B}_{\max}} - 2p_{\mathrm{A}_{\max}} p_{\mathrm{B}_{\max}})}{2} \tag{12.6}$$

12.8 Memristor-based PUFs

Technology scaling has led to designing high-performance, low-power chips and SoCs (Mohanty, 2015). Transistor sizes have reached 12 nm (AMD, 2018) and are going low still. There are speculations of 7 nm chips being released commercially. But there is a threshold that can be reached and alternatives to the transistors are being explored to replace them. Scaling has also led to many issues, including power consumption, fabrication, and switching speeds. High switching speeds and low power consumption have been achieved using the RRAM, resistive random access memory technology, and 3D arrays (Mohanty, 2015). These RRAMs are also called "Memristors." It is a two-terminal device, whose resistance depends on the amount of voltage and the direction of current that passes through it. So taking this property of a memristive device as an advantage, multiple resistive levels can be achieved. A continuous variable resistance can also be achieved by switching states from one resistive to other states. The resistance change will also be dependent on the state that it is already in, that is, the voltage/current that has already passed through the device in the previous iteration. This proves that a Memristor has the properties of a "Resistor" and "Memory." Hence the name "Memristor" (Memory + Resistor). The memristance of the memristor is given as (Rose *et al.*, 2013):

$$M(t) = R_0\sqrt{1 - \frac{2\eta\Delta R\varphi(t)}{D^2R_0^2}\mu R_{\text{on}}} \tag{12.7}$$

where the maximum resistance ($\sim R_{\text{off}}$) offered is represented as R_0, the difference between the maximum and minimum memristance offered is represented as ΔR, and applied voltage polarity is represented by η. μ is the mobility and D represents the device thickness.

In case of an analog model, the memristance offered by the device will change gradually from minimum to maximum and takes place from varying W and D. Figure 12.17 shows the device structure and working of a memristor. The doped and undoped regions of the device will change based on the direction and amount of current that is passing through it. The memristor contains two titanium dioxide layers, one of which contains excess amount of oxygens sandwiched between two metal electrodes. When a current is applied in the forward direction, the oxygen ions will be distributed all over the device, which increases the doped region as shown in Figure 12.17. Then the resistance offered by the device will reduce. Similarly, when current is applied in the opposite direction, the oxygen atoms will be concentrated at one place, which reduces the doped region as shown in Figure 12.17. Then resistance offered by the device will increase significantly. Thus, the memristance can be varied in case of a memristor.

The memristor-based PUFs use the variation of resistance in the device, which depends on the thickness of device, "D." Due to the manufacturing variations, the

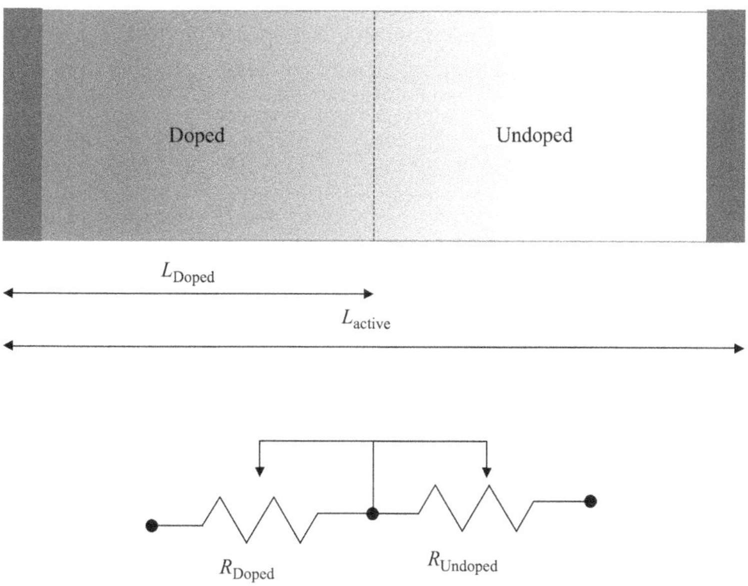

Figure 12.17 A memrister device

device thickness will vary which will affect the read and write times of the device in a memory cell. In case of the memristor, a logical "1" can be written to it when it is set from high resistive state (HRS) (R_{ON}) to a low resistive state (LRS) (R_{OFF}) for a write time above a minimum threshold of time $t_{wr,min}$; else, a logical "0" will be written. T is called the SET time and when it is close to $t_{wr,min}$, the probability of having a "1" in the cell will be equal to that of a "0." This uncertainty will make memristor much suitable for the design of PUF. Also, in this case of a binary memristor, where two states are considered, "1" and "0," transition from one state to the other will take place very abruptly. In such cases, the tamper resistance of memristor-based PUFs will be very high. For the initialization process, a high voltage should be applied so that it will be in a SET state and then switching should be performed between the HRS and LRS at a lower voltage. Below this SET voltage threshold, the device will be operating as a normal resistor. This property can be used as a tamper resistive mechanism. Figure 12.18 shows a 1-bit memristor memory-based PUF (McDonald *et al.*, 2011, 2012). This uses the write time variability of the devices for design of a PUF.

As shown in Figure 12.18, there are two control signals in the design, where ($\overline{R}$/W) is used to determine the write/read operation of the device and the NEG is used to determine the polarity of the writing operation. When a RESET operation should be performed on the device, $\overline{R}$/W and NEG should be supplied with a "1." Memristor will go into an HRS with the $\overline{R}/W$signal. After this, SET signal can be applied for writing a "1" or a "0" by supplying "1" to $\overline{R}/W$ and "0" to NEG. Reading can be done by supplying a "0" to $\overline{R}/W$. An XOR operation is performed on the output with the challenge input resultant of which will be the output of PUF.

In case of a lateral switching PUF cell designed using memristor (McDonald *et al.*, 2011, 2012), forming step is highly important for the operations. During the forming step, the conductive filament of the device is supplied with a large voltage in the dielectric sheet. This gives an advantage of values being hidden even during the

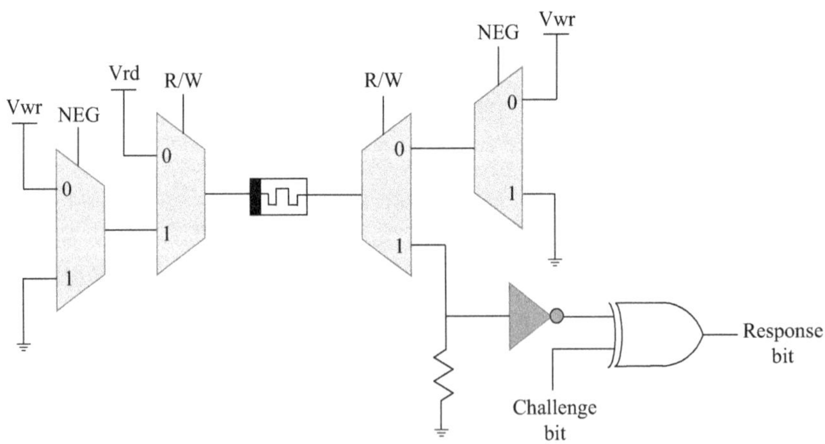

Figure 12.18 A 1-bit memristive memory-based PUF

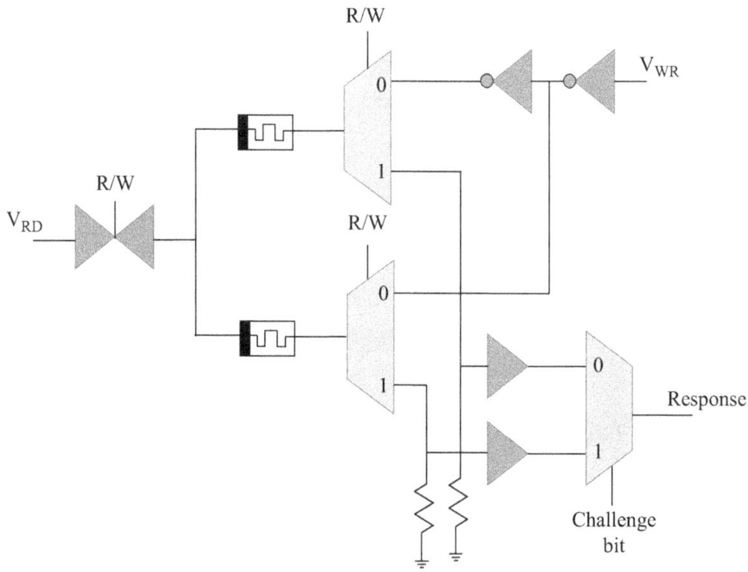

Figure 12.19 A 1-bit filament-based memristive PUF

fabrication state in the foundry. Therefore, if a filament is not present in the dielectric of the device, then it can be assumed that the device has not been tampered with and is secure. In the case of a lateral switching configuration, when the device is SET, during the RESET operation, only one device can go to HRS while other will stay in LRS. Which devices stays in LRS and which goes to HRS is completely random.

Figure 12.19 shows the design of a lateral switching filament-based memristor PUF. Like the previous design, there are control signals that are used for various operations and $R/\overline{W}$ is used for selecting formation step or operation modes of memristor devices, M1 or M2. When the signal "0" is applied to $R/\overline{W}$, either RESET $(-V_{WR})$ or SET (V_{WR}) is applied to the devices, during which the nodes between them are floating. When a "1" is applied to $R/\overline{W} = 1$, both the devices are driven with V_{RD} and memristors are loaded. The RESET operation will be done when the filament is formed and one of the memristors will go to HRS. Select signals of MUX will be given the challenge bit and response is read. Noise immunity of a binary memristor PUF is high compared to the others as it is forced either to LRS or HRS. Noise margin is high provided the states are far from the thresholds. But the operation of the memristor is near the threshold voltages which reduces the noise margin.

Figure 12.20 shows the design of a hybrid memristor CMOS PUF (Chatterjee *et al.*, 2016). As shown in the figure, the memristor array is connected to an arbiter at the end from which the response is collected. The delay produced by the memristors will decide the output bit that comes from the arbiter where a D-flip-flop can be implemented. Both the memristor arrays are given the same input at the same time. The signal that is present at the input of the arbiter at time of response collection will decide the output bit of PUF.

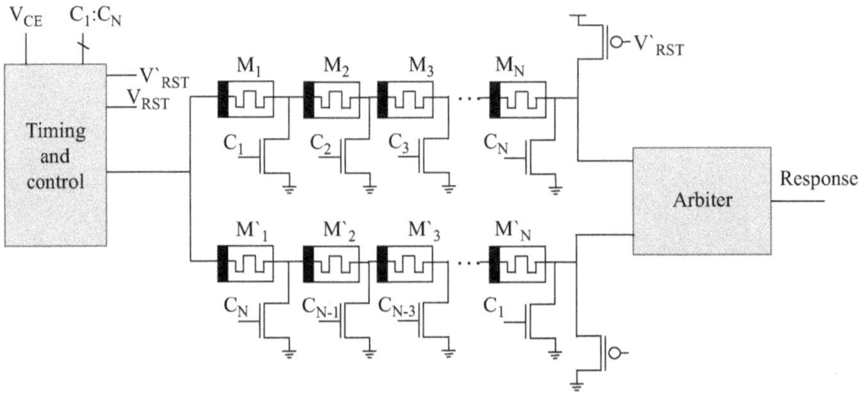

Figure 12.20 Hybrid memristor CMOS PUF

12.9 Diode-based PUF

There are various advantages to a crossbar array compared to other circuit designs which make them more suitable to be used in a PUF design. Figure 12.21 shows the design of a memristor crossbar array. Regular structure, low energy consumption, high packing density, and simple implementation of the designs are some of the advantages. Besides these, there are also some disadvantages to this design, like sneak current path and crosstalk. The sneak current is dependent on the reverse junction current.

There are two sets of wires in the crossbar array, horizontal and vertical, as shown in Figure 12.21. The intersection of the wires is separated by a thin material or by a molecular compound. The compound will be resistive to any voltage variations among the wires. Therefore, achieving different resistive states can be done using the intersection which is used for storing the data. Horizontal wires are the word lines and vertical are the bit lines with the voltages "V_{word}" and "V_{bit}." The voltage applied for reading is high compared to that of the reading.

When the memristors in the figure are replaced with diodes, it becomes a diode-based crossbar array. In this case, there are three schemes: in floating schemes, the bit and word lines that are not accessed are kept floating. In the single-supply biasing scheme, the word and bit lines that are not accessed are kept at a voltage V_b. When it is "0," it is considered grounded scheme. In case of a double supply biasing scheme, the word lines that are not accessed are biased at a voltage V_{b_1} and bit lines at V_{b_2}. When they are equal, it will be in a single-supply scheme and when they are opposite polarity, they are called reverse biasing scheme. A crossbar array cell is a bistable element consists of two resistive states R_{on} and R_{off}, which are used to store "1" and "0." The noise margin in this design can be calculated as (Ruhrmair *et al.*, 2011):

$$N_{\text{m}} = \frac{1}{2}\left(\frac{I_{\text{on}} - I_{\text{off}}}{I_{\text{on}} - I_{\text{off}}}\right) \tag{12.8}$$

where I_{on} and I_{off} are the junction currents.

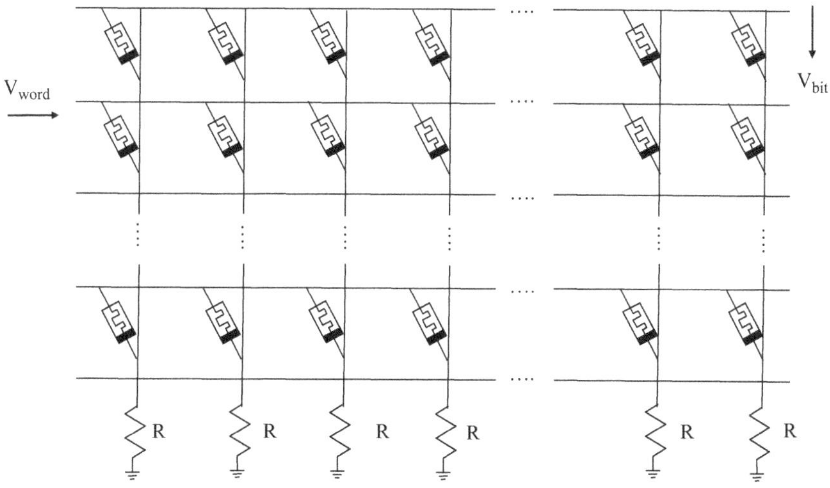

Figure 12.21 Memristor crossbar array

In the case of a memristor-based crossbar array, the forward voltage applied will reduce the resistance offered by the device and the input voltage in opposite direction will increase the resistance of the device (Lugli *et al.*, 2013) thus writing a "1" or "0" to the array. Another design of PUF is a super-high information content PUF (SHIC-PUF), which can be characterized by slow readout time and high information content (Ruhrmair *et al.*, 2011). For storing information, a crossbar array is used in this case. But for using this, there are some requirements that should be met: the readout time should be low and this will be controlled by the construction of memory cell or slow-process modules. For higher security, the slow readout times should be due to the intrinsic construction of memory cell. Another requirement is that the circuit should be characterized completely and should be very large. The bits of information that are stored on the memory should also be random. Characterization time for a $N = 10^{10}$ bits (~10 GB) and $k = 100$ bits/s circuit is $N/k \sim 3$ years. This array should also be able to fit in the small area which reduces the occupied chip area.

The bit and word lines are used for the addressing of bits at the intersection of vertical and horizontal lines. This is where the current will flow through the crossing. The word and bit lines will be biased at $V_{\mathrm{bit}}^{\mathrm{read}}$ and $V_{\mathrm{word}}^{\mathrm{read}}$ and during this period, and the word and bit lines that are not accessed will be biased at $V_{\mathrm{bit}}^{\mathrm{unaccessed}}$ and $V_{\mathrm{word}}^{\mathrm{unaccessed}}$ voltages. So, the junctions that are at the accessed word and bit lines are only made to go into forward bias and others are reverse biased in case of a diode-based crossbar array. Word lines are connected with a sense resistor and depending on the voltage at the word line, there are two states that are shown, high or low resistance. In this case, the information in the bits is random and is also important to see that the structural features are irregular. To achieve this, aluminum-induced layer exchange is considered. This gives an advantage of randomizing the distribution of conductivity which will result in an inhomogeneous conductivity. This can be taken as an advantage in case of PUF. When the array size is increased, reverse bias junction current will increase, which in turn will increase the power dissipation of the module. But the parasitic paths will avoid the accessed bit and word line. This reduces the noise margin and makes it suitable for PUF design. And large generator resistance values are required for readout cycles to be slow and for limiting the capacitive peak current. As an alternative, high-κ material can be used.

A crossbar array can also be used and a weak PUF and SHIC PUF can be implemented as presented by Lugli *et al.* (2013). In this case, rewriting and erasing the PUF response can be performed without the other responses changed. This is only possible in case of a crossbar array but not in the case of other PUF designs. An increased crossbar array will be a huge issue as the sneak current path becomes a problem and this also decreases the noise margin and degrades power dissipation. A rectifying element can be implemented for reducing the crosstalk and leakage between the accessed and unaccessed cells. The design of a crossbar can also contain a memristor in series with a diode which can reduce the noise margin and power dissipation. There is a compatibility issue in case of a traditional silicon process and transitional metal oxide memristors (Kim *et al.*, 2010).

When the floating biased scheme was implemented, the size of the array was reduced and noise margin was also reduced because of the increase in sneak current. This is also same in case of power dissipation. Due to this, the size of memory is limited to a few megabits. External resistance effect is minimal on the noise margin and reverse-biased scheme is suitable for the noise margin and power dissipation. But when compared to the floating point scheme, the power dissipation is high due to the higher off currents. It can be inferred that the memory size can be increased if power consumption is not a problem, which also decreases the noise margin. Scaling also helps in reducing the diode current and wire resistance.

In case of an ALILE, the diodes that come from a weak-doped wafer will have high rectification ratio and high randomness. But noise margin is low when compared to a ZnO junction because of low rectification of about 10^8 at ± 2 V.

An ALILE diodes of small size is suited for use in case of a weak PUF when compared to a ZnO Schottky diodes, as the randomness is high (Ruhrmair *et al.*, 2010). The stability of ALILE diodes is also high over a long period and resistive to invasive readouts. Temperature dependency of the diode's performance is high, which can be improved with the use of differential measurement techniques.

12.10 Carbon-based PUFs

In the current research trend, silicon devices are used widely for the development of PUF designs and are more focusing on the randomness, reliability, and robustness. But the power consumption of the devices is not being monitored or focused. Scaling of devices has led to a place where low-power consuming devices are more required and low-voltage operational devices are required, which becomes a challenge to the PUF design. Scaling of devices also makes it difficult to control the gate and charge carriers in the channel. This makes it necessary to explore post silicon device which can be used for scaling. Carbon derivative options such as the carbon nanotube (CNT) and graphene are promising candidates.

12.10.1 CNT-based PUF

CNT FET-based PUFs can be implemented for various advantages such as low-power consumption, high reliability, and low energy (Konigsmark *et al.*, 2014). In case of a CNTFET, a CNT is used as a channel in the transistor. The configuration of a CNTFET changes depending on various properties of the CNT, such as diameter, chirality, growth density, misalignment, and doping concentration. Chirality is used for the generation of secret bits. The direction in which the graphene sheet is rolled decides the chirality of a CNT which decides the metallic or semiconductor nature of the device. Figure 12.22 shows the design of a CNPUF, which contains the CNTFETs connected in series. The parallel elements of the transistors, CNPUF-PEs, share a common gate voltage. In this case, there can be a large number of semiconducting CNTs and a small number of metallic CNTs. The ratio in which they are used is 10%–30% where the semiconducting current will dominate the metallic current. Each parallel element consists of a distinct state for high- and low-input gate voltages. Due to the process variation, this states will vary, which affects the output. The improvement of area can be achieved by 300% when compared to design of silicon-based arbiter PUF (Konigsmark *et al.*, 2014). The interchip HD of 49.67% has been presented by Konigsmark *et al.* (2014). When the simulation was performed in different conditions, by giving random values to the temperature and the supply voltage of the devices, the performance of CNPUF-PE has been more reliable and stable compared to that of an RO PUF, current-based PUF, and SRAM PUF. For a higher security, a trade-off between security and power consumption should be made. An extended CNPUF adds an area overhead but is comparable to the Si PUF with higher security.

Even though the CNTFET has various advantages over the silicon transistors and allows more scalability, fabricating the CNTFET gives high difficulties. The planar geometry of the graphene makes it easier compared to the CNTs during fabrication and is also comparatively cheaper to silicon devices. Graphene is also compatible with the CMOS technology, which gives additional advantage of scaling and performance. Therefore, high-density SoC and higher performance can be achieved by combining the CMOS fabrication technologies with the graphene's unique properties.

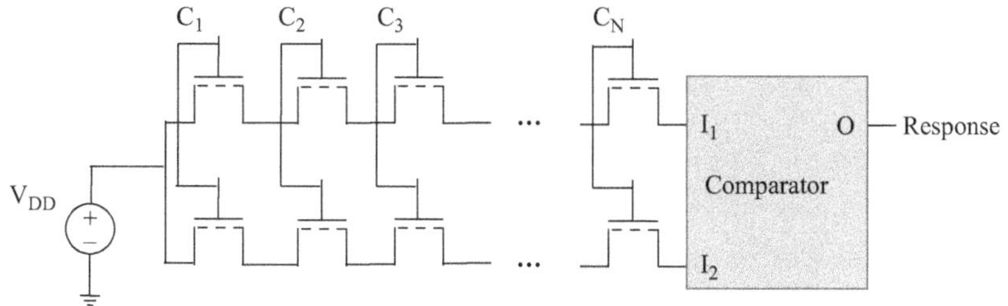

Figure 12.22 CNT-based PUF design

12.10.2 Graphene-based PUF

Figure 12.23 shows the design of a graphene nanoribbon transistor. Graphene has high electron mobility which has an added advantage to be used as a channel in case of FETs (Khan *et al.*, 2014). It is a two-dimensional carbon atom network which is arranged like a honeycomb lattice. Graphene is a single layer of graphite structure which is interconnected in a lattice structure. It is highly strong in all the directions. A CNT structure can be formed by rolling a graphene nanoribbon and one with 12 pentagon defects will give a fullerene structure. Properties of graphene which makes it more suitable than silicon for future technology are two-dimensional one atom crystal, light and thin material with high flexibility and strength. It is transparent and nonflammable and highly conductive.

Besides these advantages, there are also some challenges in this graphene technology: it is highly sensitive to environmental changes and does not work easily with discrete electronic signals as there is a no band gap in the semi-conducting graphene structure. Unlike a silicon transistor, the graphene-based transistor cannot be turned off as the bandgap is low. But the high bandgap makes it more suitable for use in RF applications, which controls the gate current in digital logic environment. Off-state leakage currents are high, which results in smaller I_{ON}/I_{OFF} ratio. Saturation of drain voltage current is not readily available.

A measurement circuit is usually present in a PUF-based security modules. This can be used for measurement of the property of graphene devices such as resistance, capacitance, and voltages for deriving the output of PUF which can be used for cryptographic applications (Dimitrakopoulos *et al.*, 2012). There are some points that should be used for graphene-based PUF designs:

1. Properties of graphene-based PUFs should be stable over a wide range of time, temperature, and other environmental variables.
2. During the fabrication process under extremely controlled conditions, there will be variations in the properties of the devices because of chemical processing, the

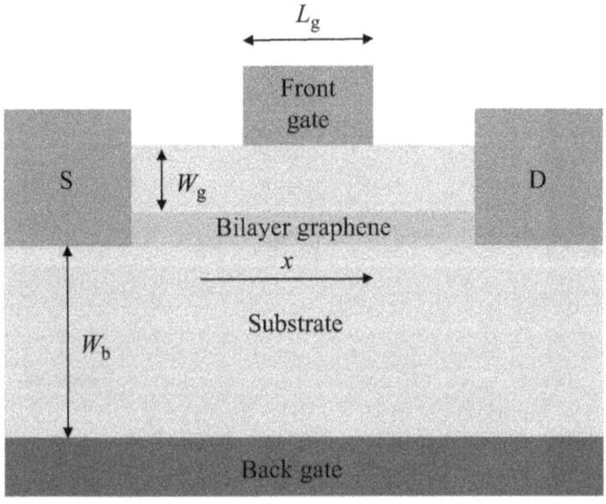

Figure 12.23 Graphene transistor

presence of terrace edges, and growth process during growth of graphene. This results in device-to-device variations on the wafer. For the introduction of additional variations, different methods can be employed:

   (a) Molecular doping, hydrogen inhalation, width/length scaling, contact treatments, and contact metal variation.
   (b) Laser annealing can be introduced during the fabrication where beam width, energy, pulse duration, and wavelength can be varied.
   (c) Contact treatment and contact metal variations.
   (d) Different ambient conditions such as varied dopant quantities and types, oxygen levels, and so on can also introduce variations.
   (e) For increasing the variability, a single graphene device can be packed with more layers. Which will vary the electrical properties of device.

3. For an improved stability, passivation layers can be employed, which also does not affect the security of the device during attacks based on scanning electron microscopy (SEM). A higher yield can be achieved using the passivated graphene. Resistance is inversely related to temperature in case of an unpassivated graphene device, which can reduce the resistance with the increase in temperature.
4. Array of devices that can be used for the implementation of PUF has a very low variability range almost close to threshold values of the core device used. In this case, reliability decreases with time and so will the temperature variation. This can cause to read errors while reading the output bits from the PUF module. There are also various solutions that can be implemented in this case. Some of them are:
   (a) Error correction schemes can be employed but by sacrificing security.
   (b) Average values of device in the array can be considered for setting the threshold value for digital fingerprint.
   (c) Temperature-compensating algorithms can be used along with the temperature sensors for mitigating the temperature variance issues.
5. Robustness against attacks can be increased by implementing the following:
   (a) The device location in a large array can be changed randomly which makes it difficult for the attacker to correctly give the challenge input to get an appropriate response.
   (b) A measurement circuit can be employed on the circuitry to make it difficult for the attacker to use probes for accessing the devices.
   (c) Delayering the measurement circuits can also be implemented, which makes the circuit more robust. This makes it difficult for the attacker to convert the read values to digital outputs. Attempting to remove the delayering will render the device useless.

12.11 Microprocessor-based PUF

Until now, this chapter discussed different PUF circuits that were designed for security and need to be fabricated. This section discusses different components such as microprocessor, power rail, and so on for designing PUF circuits. Figure 12.24 shows the design of a microprocessor-based PUF (Kumar *et al.*, 2017). Two 32-bit

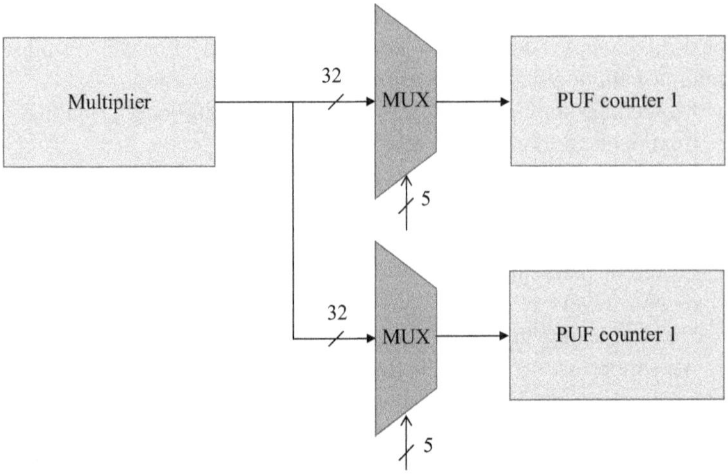

Figure 12.24 Microprocessor-based PUF

multiplexers and two PUF counters are used for designing the module. The output of the multiplier is connected to MUX as shown in the figure. The MUX will choose which PUF module will be driven by the bits of the multiplier. PUF counter is 32-bit long and the general-purpose registers of the microprocessor will store the 9th bit to 24th bit output from PUF counters. A 32-bit long output is generated from the two different PUF counters as shown in the figure (Kumar *et al.*, 2017).

There are five select lines for each of the MUX for selecting the output of the multiplier. Part of the challenge input to the PUF module will be the select lines that go into the MUX. Different paths of the multiplier are selected using the assembly program, which becomes the later part of the challenge input to the PUF module. Response from the counters gets stored in the registers of the microprocessor. For an efficient CRP generation, the assembly program development is an important task. The interconnects will add delays with the manufacturing variations introduced during the fabrication stage.

In the design presented by Kumar *et al.* (2017), the openmsp430 design has been modified into two different modes of operation, normal and PUF. I/O pins of the microprocessor are given the 10 select lines and "En" pin. Multiplier output has been modified for generating the addition of multiplexer and PUF counter. Additional circuitry in the module will become active when the PUF mode is selected and during the normal operation, they will be separated from the microprocessor.

12.12 Magnetic PUF

Sections 12.10 and 12.8 presented two different types of non-silicon material-based PUF designs. This section presents another type of non-silicon-based PUF designs, magnetic PUFs. There are various devices which use magnetic properties of materials for performing operations (Gao *et al.*, 2016).

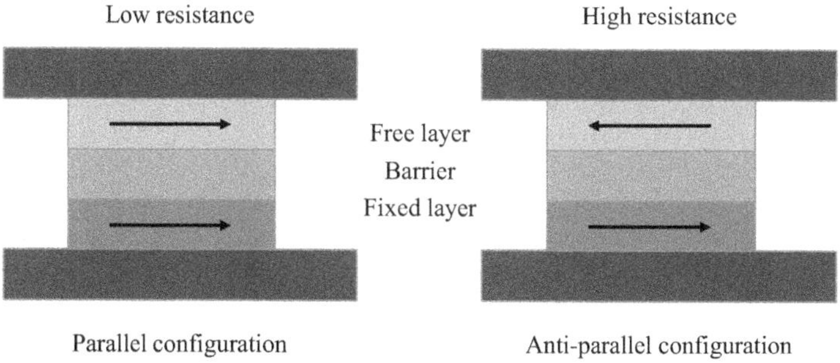

Figure 12.25 Spin transfer torque magnetic random access memory PUF

Spin transfer torque magnetic random access memory (STT-MRAM) devices are one of them. Figure 12.25 shows the design of an STT-MRAM (Huai, 2010). When current passes through the fixed layer, a spin-polarized current is produced. If the current that is produced is directed toward the free layer, the angular momentum can be transferred to the layer which can change the orientation of the magnet. When the two layers are parallel (P) to each other, the resistance is low. When the two layers are anti-parallel (AP) to each other, the resistance is high.

A magnetic tunnel junction (MTJ)-based PUF was proposed by Marukame *et al.* (2014) which can extract the fingerprint of the spin transfer-switching properties. The characterized STS characteristics were measured experimentally by Marukame *et al.* (2014) and a method to extract the PUF signature was also presented. It was also observed that different programming voltages for AP to P state and P to AP state transitions were required by the MTJ. Extrinsic properties such as the process damage and junction resistance lead to the MTJ. Due to these variations, an operation sequence for the extraction of bits from PUF-based 4×4 array was implemented. Initially, the reset voltage is applied to initialize them into a state of parallelism. Then the precalibrated voltage V_{PUF} is applied for inducing the switching with a probability of 50% for majority of the MTJs. The switching will result in a random distribution of the devices in an array in various states, AP and P. Signature extraction circuit will take care of the information resistance. After extracting the information, the entire array is set into P state. For increasing the reliability, the operation sequence is repeated for extraction of the bits.

Figure 12.26 shows a phase change material (PCM) which is similar to the previous one and takes advantage of the magnetic properties for designing the PUF. A PCM has amorphous and crystalline natures in the PCM. The difference in the resistance offered by the two states is largely different. To change the material to amorphous state, a pulse with high amplitude and fast fall time is supplied to heater to initially melt it and then quench it rapidly. So a region of high resistive amorphous material is formed in the PCM. A long fall time moderate amplitude pulse is supplied to the cell for converting it into a crystalline state which will anneal the active region at a temperature between crystallization and melting temperature which forms a crystalline material in the cell which offers low resistance.

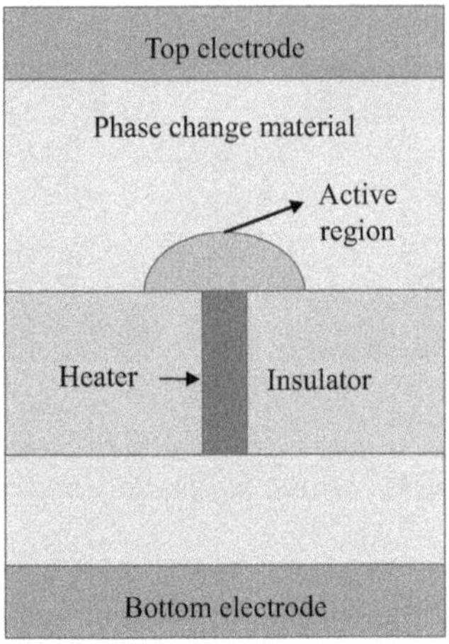

Figure 12.26 Phase change material

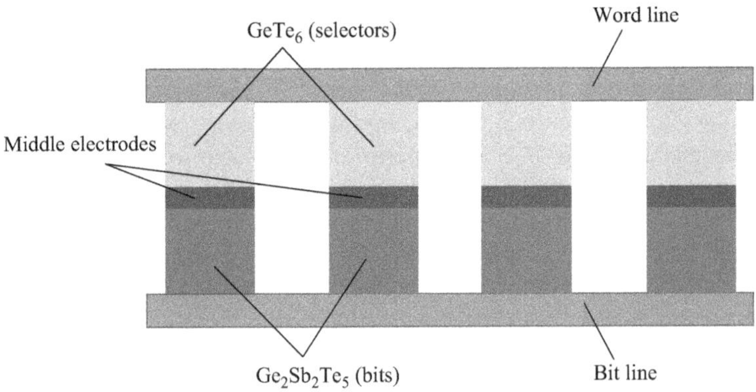

Figure 12.27 Phase change material memory array PUF

Figure 12.27 shows the structure of a PCM array. During the reading operation, the resistance of the cell is read by sending a small current to it. This pulse will not be disturbing the state of the cell (Wong *et al.*, 2010). The resistance offered by the PCM cell can also be controlled by passing a small amount of current, a programming pulse with intermediate amplitude rise/fall time.

12.13 Practical implementation of PUF

Designing a PUF module can be half the problem but the implementation and deploying the same in an environment can pose additional challenges. Any design to be validated has to be implemented in a real-time scenario where it is subject to different real-time conditions. FPGA implementations of PUF are being explored by researchers for validation of the design and implementation of their proposed designs into different applications. Deploying the design on an FPGA opens up a lot of opportunities to the PUF design and adds more to the list of applications (Chowdhury *et al.*, 2017; Ikezaki *et al.*, 2017; Johnson *et al.*, 2015).

The FPGAs are also evolving day-by-day and adding more features for better integration into the IoT world (Johnson *et al.*, 2015). One such capability is the dynamic partial reconfiguration (DPR) or runtime partial reconfiguration (RPR). With these features of an FPGA, the circuit design that is on the FPGA can be configured partially during the runtime. This ability is called "Partial Reconfiguration." The power consumption of the FPGA can be reduced by using the DPR which makes them ideal for use in IoT applications. But these DPR-enabled FPGAs are subjected to various attacks by the adversary which can get the entire network of devices into jeopardy. A protection technique that can be used for protecting the DPR-enabled FPGAs was proposed by Johnson *et al.* (2015). An XoR PUF was employed in the paper for securing the protocol. Because of the complex nature of the PUF that is deployed in the application, modeling an attack by the attacker will be difficult as there will be no access to the PUF responses. If that access is gained by the adversary, attacking the PUF with a modeling attack will become easy and successful in some cases.

There are also many different aging-resistant PUFs designs that are implemented on the FPGAs (Chowdhury *et al.*, 2017). The aging effects on the FPGA are presented by Maiti and Schaumont (2014). It was also presented that a change in the location of the output response bits and the devices that are affected by the aging are also random. There are also many designs of PUF that are implemented based on SRAM on the FPGA that use the aging effects on the design to an advantage. An aging resistant RO PUF is proposed for reliable key generation in (Rahman *et al.*, 2016). With the FPGA lacking the ability to change the circuit architecture, implementing this idea becomes more difficult.

12.14 PUF: case study applications

All the above sections discussed different designs of PUFs. This section discusses different applications of PUF in various domains. There are many applications and domains where PUF can be used and implemented. One such major issue that is becoming a problem is counterfeiting. Figure 12.28 shows a big picture of counterfeiting in the case of an IC. When an IC is designed by a manufacturer, there can be many clones designed by various third parties to be sold in the market (Chang *et al.*, 2017; Mohanty, 2015). But this is a major issue economically and also in terms of security. The ICs that are created by the counterfeiting party can be malicious and might contain trojans. A PUF can be implemented in the IC design so that it can be used for authenticating the original device (Yan *et al.*, 2017).

A PUF module that contains on-chip hash function can increase the cost of the entire chip but can also increase the robustness of the entire design. An index-based

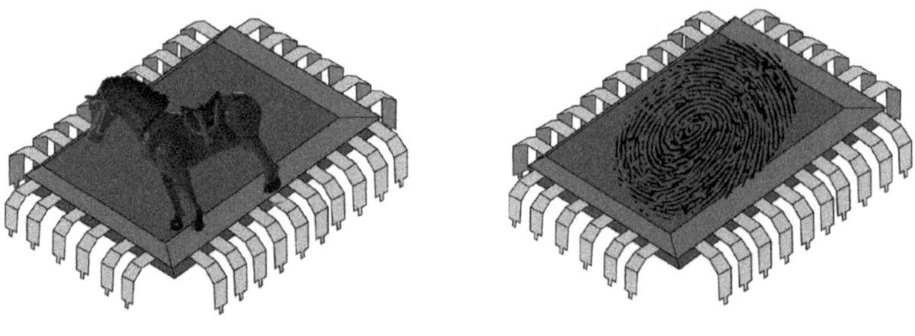

Counterfeit IC PUF-secured authentic IC

Figure 12.28 Securing against IPR PUFs

Table 12.2 Comparison of different types of hash functions in terms of chip area and clock (Majzoobi et al.*, 2008)*

Sample	Algorithm	Chip area	Clock cycles
	SHA-256	10,868	1128%
	SHA 1	8,120	1274%
	MD5	8,400	612
	MD4	7,350	456

syndrome encoding has been shown to be implemented with RO PUF for a better reliability (Yu and Devadas, 2010). There are also variations in the output bits that get flipped due to the environmental and operational conditions. So error correction mechanisms can also be implemented with a PUF so that the output can be read without errors. But error correction mechanism can increase the chip area and hence the total cost of the chip. There is also a power trade-off that has to be considered while adding additional circuitry. A comparison of built-in hash functions and their effects is shown in Table 12.2.

Man-in-the-middle attacks can be controlled by implementing PUF modules in various applications like distributed computing (Gassend *et al.*, 2002). Different steps involved in such an attack are as follows:

- "A" should interact with the PUF module by using the CRPs shared between A and PUF.
- "B" is an attacker who has access to the PUF module. Therefore, B will be able to gather the responses from the module by giving the challenge inputs that are chosen by B. B wants to use it with the identity of "A" and impersonates with fake results.
- A gives out PUF challenges from CRP list and compares responses that come from the output of PUF module with a private list.

- B has access to the challenges and PUF, so gets the response from PUF for that specific challenge.
- B will be able to fool A by pretending that A has PUF access and can impersonate the PUF as if it is A.

The applications of PUF are very broad as shown in Figure 12.29. They can also be used in applications or devices where randomness is needed. PUF modules are elegant solutions in various applications such as RFID tags, random number generation, secret key generation, device authentication, and so on (Gunlu and Iscan, 2014; O'Donnell *et al.*, 2004; Suh and Devadas, 2007; Suh *et al.*, 2005). PUF has also been used in consumer electronic devices where authentication is needed at a low cost (http://www.eecs.mit.edu/news-events/media/mit-spinoff-verayo-developed-srini-devadas-gains-increased-use-commercial).

In case of the cryptographic and security applications, the random number generation can be used for generating the encryption and decryption keys, initial seed values, and in some cases, where required, introduce some randomness to the padding schemes and protocols. Pseudo-random number generation (PRNG) is not truly random and in that case, the pattern will repeat after a certain number of iterations. Each of the number generated by the PRNG has to go through a very complex algorithm. Initial seed value should be kept highly secret in case of PRNG. Due to all these different issues with the PRNG, hardware random number generation (HRNG) has been implemented in various cases (O'Donnell *et al.*, 2004).

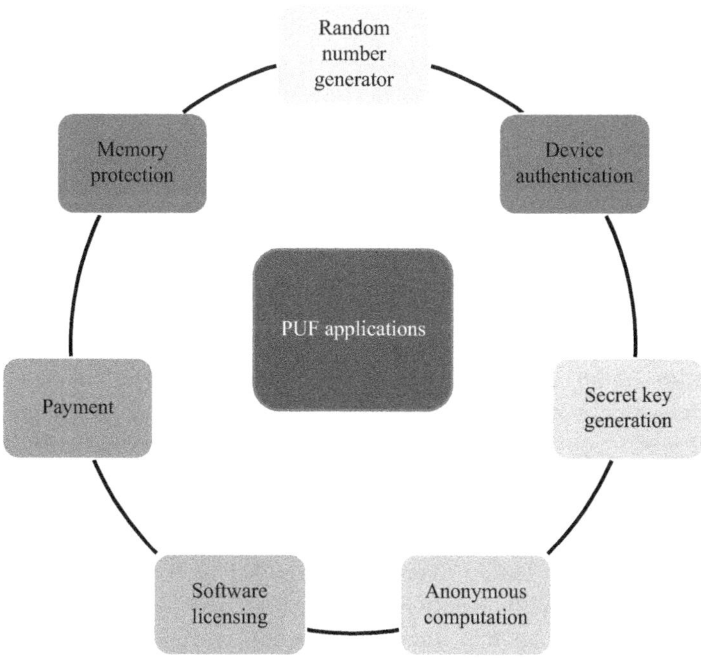

Figure 12.29 Typical applications of PUFs

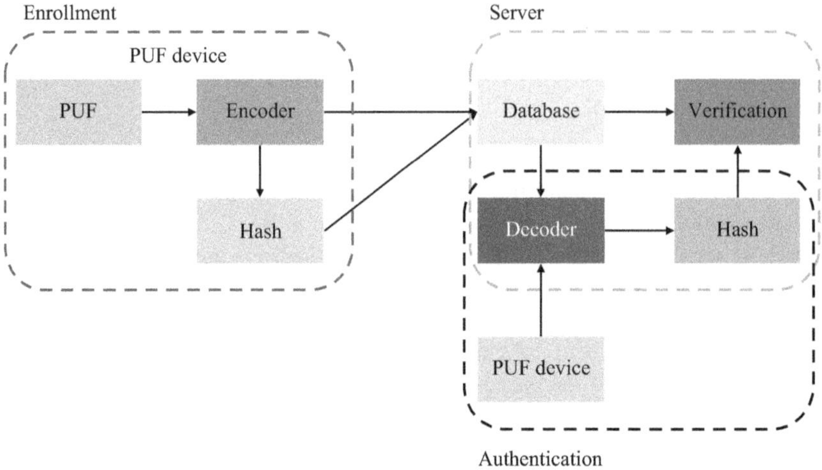

Figure 12.30 Secure key generation system using PUFs

In case of an HRNG, an initial seed is not required for generating a truly random number. A delay-based PUF circuit will have the rising edge of the input signal and challenge bits as select lines to the multiplexers. Input will propagate through the series of multiplexers and depending on the delay offered by each of them, a D-flip-flop will generate the output bit "1" or "0." An unbiased output response will be given by a meta-stable PUF. A Von Neumann correction gives a 32-bit output response from an *N*-bit input. Similarly, there are various designs of PUF which can be implemented based on the requirements of application. Figure 12.30 shows a key generation system using PUF modules.

For using the authentication schemes of PUF on a chip, a trusted party or the manufacturer should store the CRPs which are exponential in number in a secure database which can be matched for authentication (Suh and Devadas, 2007). For an input challenge, the output of PUF is the same, in that case, the device that has been challenged will be authentic. But when using the challenge inputs, the man-in-the-middle attack should be avoided by discarding the challenge once used.

A controlled PUF module can also be used for certified execution. It is a process of executing or running a program on a remote server where the client can rely on the results produced by the server (Gassend *et al.*, 2002). Figure 12.31 shows the process of authenticating a person over the network using PUF. In this case, the PUF device will generate a challenge input and has the response respectively. When authentication is needed, the response is sent to the server where it confirms the identity of the client. This can help avoid identity fraud and rendering the results of the client useless. This can also avoid an attacker from generating different malicious results and sending it to the user. This scenario is often seen in case of a distributed computing where one processor cannot handle the workflow. This type of PUF modules can also be used to restrict the usage of a specific software to a respective amount of processors.

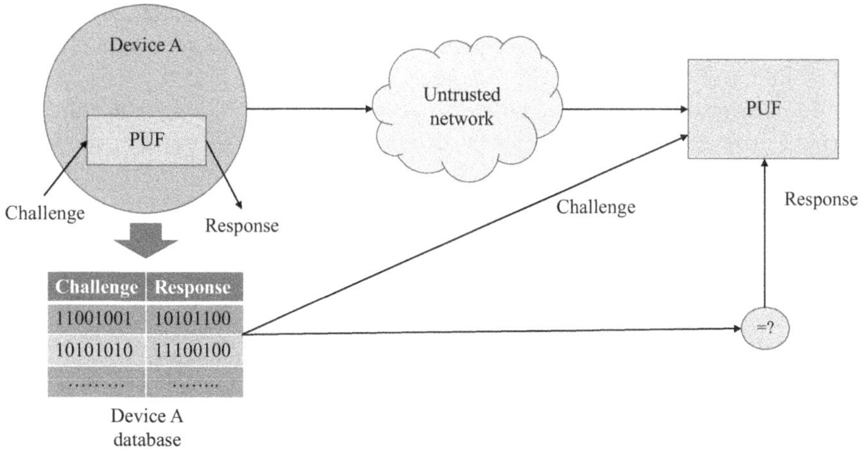

Figure 12.31 Authentication of user using PUFs

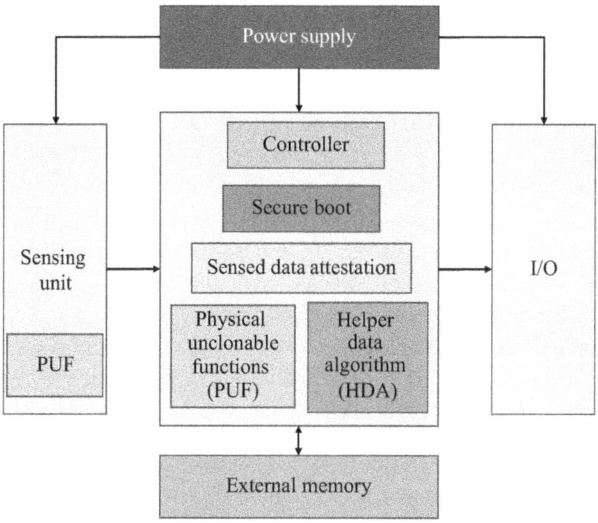

Figure 12.32 Secure camera using PUFs

Figure 12.32 shows the design of a secure digital camera which has a PUF module that is built into it (Haider *et al.*, 2016). Secure boot of sensor controller of the camera is used to ensure the integrity of the platform at booting. PUF-based secure key generation and storage module provides key for sensed data attestation to ensure integrity and authenticity. In system architecture of the secure camera, the PUF-based key generation and storage module has two cascaded components, a PUF source and a helper data algorithm (HDA). The choice of the HDA depends on the selection of the PUF source. The HDA is used to obtain: (1) a perfectly reproducible

PUF response under a range of environmental and operating conditions, in other words, the HD between any two responses from the same PUF is zero, and (2) an almost uniformly distributed PUF response.

IoT applications have become more necessary nowadays. Many situations are arising where the devices have to be deployed in places where they cannot be monitored or have human interaction. Such cases are more often prone to attacks by the attacker who has access to the device without security. PUF can be used to stop attacks in such situations where the entire network of devices can be secured (Mohanty *et al.*, 2016; Yanambaka *et al.*, 2017).

12.15 PUF: issues

The understanding of this chapter so far is that a PUF provides better security and robustness to the entire application. But there are many issues and challenges that have to be addressed while designing and implementing the PUF in an application (Chang *et al.*, 2017; Wan *et al.*, 2015; Wen and Lao, 2018; Ye *et al.*, 2017, 2018). There are various attempts for providing better security and robustness by implementation of PUF, but some of the common attacks are:

- There is a chance that the attacker can create a large number of chips with the similar design to the PUF module used in a respective application for obtaining the response. In cases where the design of PUF is simple, this can be very successful.
- It is also possible to predict the response of the PUF or calculate it using mathematical modeling techniques (Ma *et al.*, 2018; Mispan *et al.*, 2018; Tanaka *et al.*, 2018). Adding a nonlinearity to the PUF design can avoid these kinds of attacks.
- There are different types of attacks, invasive or semi-invasive, which can determine the time delay and power of the devices. Tamper-resisting packaging can avoid these attacks.
- Introducing errors into the module can also be performed by the attacker to tamper with it and render it useless.

These attacks are common and there are many different kinds of solutions that are suggested. Improvements can also be made to the PUF designs to make them difficult to attack by the attacker.

- A hash function can also be implemented before or after the PUF module which can prevent the attacker from gaining knowledge of the CRPs. This can prevent the modeling attacks on the modules.
- Error correction mechanisms can be implemented in the PUF modules which can make them more reliable and avoid bit-flips due to environmental variations and operational disturbances.
- Control of PUF response is often desired by the client who is using the module. Additional secret keys for the input of the modules can be implemented, which is totally in hands of the user or client.
- Tamper-resistant packaging should be used for the ICs and PUF modules. This will increase the cost of the entire application and will increase the robustness also.

Attacking the PUF modules is a kind of attack that has to be performed by an adversary with an ill intent. But there are also other challenges faced during the design of PUF modules (Ganta and Nazhandali, 2014; Sahoo *et al.*, 2018; Wang and Tehranipoor, 2010; Yanambaka *et al.*, 2017, 2018). Some of the challenges include crosstalk, noise in the signals, noise in power supply, and so on. Noise sources can also originate from an unusual input pattern which varies with a different input. Such a noise can be taken as an advantage by the user in designing different input challenges which will increase the robustness of the module against the attackers.

When a large switching activity is triggered due to the input pattern, a power supply noise can occur. The dynamic power is increased due to this and will be responsible for a voltage drop in the power line and an increase in voltage at the ground. Such changes in the supply voltages can cause the change in gate voltages and induce unusual delays in the circuitry. Then the entire circuit operation changes which affect the output response inducing more errors into it. Parasitics should also be taken into consideration during the design of a PUF module.

Aging is another issue that is present in the PUF designs. With the aging effects, the integrity of the circuit degrades over time and renders it useless. Over a period of time, the circuit health degrades and the output responses of a set of challenges can also be affected due to these changes (Chowdhury *et al.*, 2017; Ganta and Nazhandali, 2014; Meguerdichian and Potkonja, 2011; Rahman *et al.*, 2015, 2016). The aging effects on the PUF module depending on the frequency of turning the device on have been presented by Ganta and Nazhandali (2014). An RO PUF was subjected to the study and some of the observations presented by Ganta and Nazhandali (2014) are:

- The bit error rate (BER) in the case of PUF, which is always on is the highest. But when the device is turned on during the day only once, the BER is 2%. A 35% reduction in BER can be achieved by turning on the device once a day.
- The manufacturing variations of the devices are modeled during the simulation as threshold voltage variations. A mismatch was observed by Ganta and Nazhandali (2014) in identical circuit elements, and when the process variation is high, the PUF is stable.
- For the effects of temperature and the aging of the modules, and their effects on the stability of design, the PUFs were operated at various temperatures and ages. The effect of aging on the PUF design was not observed to be significant on the stability of the device due to the differential nature. The correlation on stability is shown as 70% of the aging effect and the temperature.
- Supply voltage and the noise in the supply voltage are also the challenges in the designs. The PUFs' performance is measured in the presence of these issues and aging by Ganta and Nazhandali (2014). The effect of supply noise is less than 0.2% in the presence of aging. But the BER of the PUF that is operating at a subthreshold voltage with aging is high.
- The stability of the PUF depends on the ring oscillation configuration highly where with an increase in the stages, there is also an increase in the susceptibility to the environmental changes.

12.16 Conclusion

This concept of a PUF was initially proposed in 2001 as physical one-way function. Since then, there are various concepts of PUF that were designed using different architectures, devices, and materials. Many of the PUFs are highly robust and others are not secure which can be easily broken with the implementation of a simple machine learning algorithm.

Research is being conducted for designing more robust and secure designs of PUFs for better security. But with the high implementations of IoT architectures, there is more necessity and scope for the development of low-power designs of PUF. There are low-power-consuming designs being implemented and proposed by many researchers around the world but the main focus of research has been the increase in the robustness and resistance to attacks. As a future research, low-power design of PUF can also be implemented for better wide range of implementation and integration.

There are also the threats and attacks on the PUF that are yet to be addressed. With the introduction of new applications every day, it is necessary to secure them against attacks that come with it. Side-channel resilient PUF designs are yet to be more saturated and better designs are yet to come into the consumer-level market. PUFs are being implemented in many enterprise domain applications like servers, but commercial market is still open to the implementations of PUF. This can also be considered as a future research of the PUFs.

12.17 Exercises

1. What is encryption and decryption?
2. How is implementing a PUF be better than using a traditional cryptographic primitive?
3. What is PUF?
4. Explain the principle of PUF.
5. Describe the working of an arbiter PUF.
6. Explain uniqueness of the PUF with an example. How hamming distance is calculated for uniqueness?
7. What is reliability? How is it measured?
8. Explain four characteristics of PUFs.
9. How are PUFs classified?
10. Describe in detail security-based PUF designs.
11. Explain the operation of an RO PUF.
12. Describe how the delay is calculated in an RO PUF.
13. Why are aging-resistant designs of PUF introduced? What are the aging effects on a circuit?
14. Describe NBTI and PBTI.
15. How can a reconfigurability of a PUF module be attained?
16. Explain in detail the arbiter PUF with an LFSR at the input.
17. Describe working of an SRAM.
18. Explain the working of an SRAM PUF.

19. Explain the working of a Von Neumann extractor. What is the extractor rate of the Von Neumann extractor?
20. What is memristor? Explain the working of a memristor.
21. Explain the working of a 1-bit filament-based memristor PUF.
22. How does a crossbar array PUF work? What are the three different schemes in the diode-based PUF design?
23. Explain the difference between CNT and GNR.
24. Explain the different properties of a graphene-based PUF design.
25. How robustness of GNRFET PUF can be increased?
26. Why microprocessor-based PUFs are used? How are they better compared to other designs?
27. What is an STT-MRAM? Explain its working.
28. Explain the working of a PCM-based PUF design.
29. Explain two different architectures of PUF in detail.
30. Explain five applications of PUFs.
31. What are different issues and challenges during the design of PUFs?

References

AMD, 'AMD Ryzen,' 2018. URL https://www.amd.com/en/products/cpu/amd-ryzen-7-2700x. Last Accessed: 7-1-2018

Bai C., Zou X., and Dai K., 'A novel thyristor-based silicon physical unclonable function,'*IEEE Transactions on Very Large Scale Integration (VLSI) Systems*, 2016, vol. 24 (1), pp. 290–300.

Barker E. B. and Kelsey J. M., 'Recommendation for random number generation using deterministic random bit generators,' National Institute of Standards and Technology, NIST, 2015.

Böhm C., 'Physical Unclonable Functions in Theory and Practice,' Springer New York, 2013, ISBN: 9781461450405.

Chang C. H., Zheng Y., and Zhang L., 'A retrospective and a look forward: Fifteen years of physical unclonable function advancement,' *IEEE Circuits and Systems Magazine*, 2017, vol. 17 (3), pp. 32–62.

Charette R., '2017 was a record year for ID theft in the U.S.' *IEEE Spectrum*, 2018a.

Charette R., 'Healthcare IT systems: Tempting targets for ransomware,' *IEEE Spectrum*, 2018b.

Chatterjee U., Chakraborty R. S., Mathew J., and Pradhan D. K., 'Memristor based arbiter PUF: Cryptanalysis threat and its mitigation,' *29th International Conference on VLSI Design and 2016 15th International Conference on Embedded Systems (VLSID)*, 2016, pp. 535–540.

Chatterjee U., Chakraborty R. S., and Mukhopadhyay D., 'A PUF-based secure communication protocol for IoT,' *ACM Trans. Embed. Comput. Syst.*, 2017, vol. 16 (3), pp. 67:1–67:25.

Chowdhury S., Xu X., Tehranipoor M., and Forte D., 'Aging resilient RO PUF with increased reliability in FPGA,' *2017 International Conference on ReConFigurable Computing and FPGAs (ReConFig)*, 2017, pp. 1–7.

Courtland R., 'Intel now packs 100 million transistors in each square millimeter,' *IEEE Spectrum*, 2017.

Dimitrakopoulos C. D., Pfeiffer D., and Smith J. T., 'Authentication using graphene based devices as physical unclonable functions,' 2012.

Ganta D. and Nazhandali L., 'Study of IC aging on ring oscillator physical unclonable functions,' *15th International Symposium on Quality Electronic Design (ISQED)*, 2014, pp. 461–466.

Gao Y., Ranasinghe D. C., Al-Sarawi S. F., Kavehei O., and Abbott D., 'Emerging physical unclonable functions with nanotechnology,' *IEEE Access*, 2016, vol. 4, pp. 61–80.

Gassend B., Clarke D., Dijk M. V., and Devada S., 'Controlled physical random functions,' *18th Annual Computer Security Applications Conference*, 2002, pp. 149–160.

Gencoglu O., Similä H., Honko H., and Isomursu M., 'Collecting a citizen's digital footprint for health data mining,' *37th Annual International Conference of the IEEE Engineering in Medicine and Biology Society (EMBC)*, 2015, pp. 7626–7629.

Giterman R., Weizman Y., and Teman A., 'Gain-cell embedded DRAM-based physical unclonable function,' *IEEE Transactions on Circuits and Systems I: Regular Papers*, 2018, pp. 1–11.

Gunlu O. and Iscan O., 'DCT based ring oscillator physical unclonable functions,' *IEEE International Conference o Acoustics, Speech and Signal Processing (ICASSP)*, 2014, pp. 8198–8201.

Haider I., Hoberl M., and Rinner B., 'Trusted sensors for participatory sensing and IoT applications based on physically unclonable functions,' *Proceedings of the 2nd ACM International Workshop on IoT Privacy, Trust, and Security*, 2016, pp. 14–21.

Hashemian M. S., Singh B., Wolff F., Weyer D., Clay S., and Papachristou C., 'A robust authentication methodology using physically unclonable functions in DRAM arrays,' *Design, Automation Test in Europe Conference Exhibition (DATE)*, 2015, pp. 647–652.

Herder C., Yu M. D., Koushanfar F., and Devadas S., 'Physical unclonable functions and applications: A tutorial,' *Proceedings of the IEEE*, 2014, vol. 102 (8), pp. 1126–1141.

Huai Y., 'Spin-transfer torque MRAM (STT-MRAM): Challenges and prospects,' *AAPPS Bulletin*, 2010, vol. 18 (6), pp. 33–40.

Ikezaki Y., Nozaki Y., Nagata H., and Yoshikawa M., 'FPGA implementation technique for power consumption aware tamper resistance accelerator of lightweight PUF,' *2017 IEEE 6th Global Conference on Consumer Electronics (GCCE)*, 2017, pp. 1–2.

Jeloka S., Yang K., Orshansky M., Sylvester D., and Blaauw D., 'A sequence dependent challenge-response PUF using 28 nm SRAM 6T bit cell,' *Symposium on VLSI Circuits*, 2017, pp. C270–C271.

Johnson A. P., Chakraborty R. S., and Mukhopadhyay D., 'A PUF-enabled secure architecture for FPGA-based IoT applications,' *IEEE Transactions on Multi-Scale Computing Systems*, 2015, vol. 1 (2), pp. 110–122.

Joshi S., Mohanty S. P., and Kougianos E., 'Everything you wanted to know about PUFs,' *IEEE Potentials*, 2017, vol. 36 (6), pp. 38–46.

Khan M. A., Mohanty S. P., and Kougianos E., 'Statistical process variation analysis of a graphene FET based LC-VCO for WLAN applications,' *Proceedings of the 15th IEEE International Symposium on Quality Electronic Design (ISQED)*, 2014, pp. 569–574.

Kilby J. S., 'Invention of the integrated circuit,' *IEEE Transactions on Electron Devices*, 1976, vol. 23 (7), pp. 648–654.

Kim G. H., Kim K. M., Seok J. Y., *et al* ., 'A theoretical model for Schottky diodes for excluding the sneak current in cross bar array resistive memory,' *Proceedings of IOP Science on Nanotechnology*, 2010, pp. 1–7.

Koeberl P., Li J., Rajan A., and Wu W., 'Entropy loss in PUF-based key generation schemes: The repetition code pitfall,' *IEEE International Symposium on Hardware-Oriented Security and Trust (HOST)*, 2014, pp. 44–49.

Konigsmark S. T. C., Hwang L .K., Chen D., and Wong M. D. F., 'CNPUF: A carbon nanotube-based physically unclonable function for secure low-energy hardware design,' *19th Asia and South Pacific Design Automation Conference (ASP-DAC)*, 2014, pp. 73–78.

Kumar S., Sahoo S., Mahapatra A., Swain A. K., and Mahapatra K. K., 'Microprocessor based physical unclonable function,' *IEEE International Symposium on Nanoelectronic and Information Systems (iNIS)*, 2017, pp. 246–251.

Kumar S. S., Guajardo J., Maes R., Schrijen G. J., and Tuyls P., 'Extended abstract: The butterfly PUF protecting IP on every FPGA,' *IEEE International Workshop on Hardware-Oriented Security and Trust, HOST*, 2008, vol. 11, pp. 67–70.

Lao Y. and Parhi K. K., 'Reconfigurable architecture for silicon physical unclonable function,' *IEEE International Conference on Electro/Information Technology (EIT)*, 2011, pp. 1–7.

Lao Y. and Parhi K. K., 'Statistical analysis of MUX-based physical unclonable functions,' *IEEE Transactions on Computer-Aided Design of Integrated Circuits and Systems*, 2014, vol. 33 (5), pp. 649–662.

Lee J., Lim D., Gassend B., Suh G. E., van Dijk M., and Devadas S., 'A technique to build a secret key in integrated circuits with identification and authentication applications,' *Digest of Technical Papers. 2004 Symposium on VLSI Circuits*, 2004, pp. 176–179.

Leest V. V. D., Jan S. G., Helena H., and Pim T., 'Hardware intrinsic security from D flip-flops,' *Proceedings of the Fifth ACM Workshop on Scalable Trusted Computing, STC '10*, 2010, pp. 53–62. ACM, New York, NY, USA.

Liu H., Liu W., Lu Z., Tong Q., and Liu Z., 'Methods for estimating the convergence of inter-chip min-entropy of SRAM PUFs,' *IEEE Transactions on Circuits and Systems I: Regular Papers*, 2018, vol. 65 (2), pp. 593–605.

Lugli P., Mahmoud A., Csaba G., Algasinger M., Stutzmann M., and Ruhrmair U., 'Physical unclonable functions based on crossbar arrays for cryptographic applications,' *Proceedings of the International Journal of Circuit Theory and Applications*, 2013, pp. 619–633.

Ma Q., Gu C., Hanley N., Wang C., Liu W., and O'Neill M., 'A machine learning attack resistant multi-PUF design on FPGA,' *23rd Asia and South Pacific Design Automation Conference (ASP-DAC)*, 2018, pp. 97–104.

Maes R., Van Herrewege A., and Verbauwhede I., 'PUFKY: A fully functional PUF-based cryptographic key generator,' E. Prouff, P. Schaumont (eds.) Cryptographic Hardware and Embedded Systems – CHES, 2012, pp. 302–319. Springer Berlin Heidelberg.

Maiti A. and Schaumont P., 'Improved ring oscillator PUF: An FPGA-friendly secure primitive,' *Journal of Cryptology*, vol. 24 (2), 2011, pp. 375–397.

Maiti A. and Schaumont P., 'The impact of aging on a physical unclonable function,' *IEEE Transactions on Very Large Scale Integration (VLSI) Systems*, 2014, vol. 22 (9), pp. 1854–1864.

Maiti A., Casarona J., McHale L., and Schaumont P., 'A large scale characterization of RO-PUF,' *IEEE International Symposium on Hardware-Oriented Security and Trust (HOST)*, 2010, pp. 94–99.

Maiti A., Gunreddy V., and Schaumont P., 'A systematic method to evaluate and compare the performance of physical unclonable functions,' 2013, pp. 245–267. Springer New York, New York, NY.

Majzoobi M., Koushanfar F., and Potkonjak M., 'Lightweight secure PUFs,' *IEEE/ACM International Conference on Computer-Aided Design*, 2008, pp. 670–673.

Mangard S., Pramstaller N., and Oswald E., 'Successful attacking masked AES hardware implementations,' *Proceedings of the Cryptographic Hardware and Embedded Systems CHES*, 2005, pp. 157–171.

Marukame T., Tanamoto T., and Mitani Y., 'Extracting physically unclonable function from spin transfer switching characteristics in magnetic tunnel junctions,' *IEEE Transactions on Magnetics*, 2014, vol. 50 (11), pp. 1–4.

McDonald N., Bishop S., Briggs B. D., Nostrand J. E. V., and Cady N. C., 'Influence of the plasma oxidation power on the switching properties of Al/Cu_xO/Cu memristive devices,' *International Semiconductor Device Research Symposium (ISDRS)*, 2011, pp. 1–2.

McDonald N., Bishop S., and Cady N. C., 'Experimentally demonstrated filament-based switching mechanism for Al/Cu_xO/Cu memristive devices,' *IEEE International Integrated Reliability Workshop Final Report (IRW)*, 2012, pp. 195–198.

Meguerdichian S. and Potkonja M., 'Device aging-based physically unclonable functions,' *48th ACM/EDAC/IEEE Design Automation Conference (DAC)*, 2011, pp. 288–289.

Mispan M. S., Su H., Zwolinski M., and Halak B., 'Cost-efficient design for modeling attacks resistant PUFs,' *Design, Automation Test in Europe Conference Exhibition (DATE)*, 2018, pp. 467–472.

Mohanty S. P., '*Nanoelectronic Mixed-Signal System Design*,' McGraw-Hill Education, New York, 2015, ISBN: 9780071825719.

Mohanty S. P., Choppali U., and Kougianos E., 'Everything you wanted to know about smart cities,' *IEEE Consumer Electronics Magazine*, 2016, vol. 5 (3), pp. 60–70.

O'Donnell C. W., Suh G. E., and Devadas S., 'PUF-based random number generation,' 2004, MIT CSAIL CSG Technical Memo.

Rahman M. T., Forte D., Fahmy J., and Tehranipoor M., 'ARO-PUF: An aging-resistant ring oscillator PUF design,' Design, *Automation and Test in Europe Conference and Exhibition (DATE)*, 2014, pp. 1–6.

Rahman M. T., Forte D., Rahman F., and Tehranipoor M., 'A pair selection algorithm for robust RO-PUF against environmental variations and aging,' *2015 33rd IEEE International Conference on Computer Design (ICCD)*, 2015, pp. 415–418.

Rahman M. T., Rahman F., Forte D., and Tehranipoor M., 'An aging-resistant RO-PUF for reliable key generation,' *IEEE Transactions on Emerging Topics in Computing*, 2016, vol. 4 (3), pp. 335–348.

Rose G. S., McDonald N., Yong L. K. W., Wysocki B., and Xu K., 'Foundations of memristor based PUF architectures,' *IEEE/ACM International Symposium on Nanoscale Architectures (NANOARCH)*, 2013, pp. 52–57.

Rosenblatt S., Fainstein D., Cestero A., *et al.*, 'Field tolerant dynamic intrinsic chip ID using 32 nm high-K/metal gate SOI embedded DRAM,' *IEEE Journal of Solid-State Circuits*, 2013, vol. 48 (4), pp. 940–947.

Ruhrmair U., Jaeger C., Hilgers C., Algasinger M., Csaba G., and Stutzmann M., 'Security applications of diodes with unique current-voltage characteristics,' *Financial Cryptography and Data Security*, 2010, pp. 328–335.

Ruhrmair U., Jaeger C., Bator M., Stutzmann M., Lugli P., and Csaba G., 'Applications of high-capacity crossbar memories in cryptography,' *Proceedings of the IEEE Transactions on Nanotechnology*, 2011, pp. 489–498.

Sahoo S. R., Kumar S., and Mahapatra K., 'A modified configurable RO PUF with improved security metrics,' *Proceedings of the 2nd IEEE International Symposium on Nanoelectronic and Information Systems*, 2016, pp. 320–324.

Sahoo S. R., Kumar S., and Mahapatra K., 'A novel configurable ring oscillator PUF with improved reliability using reduced supply voltage,' *Microprocessors and Microsystems*, 2018, vol. 60, pp. 40–52.

Schrijen G. J. and Leest V. V. D., 'Comparative analysis of SRAM memories used as PUF primitives,'*Design, Automation & Test in Europe Conference & Exhibition (DATE)*, 2012, pp. 1319–1324.

Škorić B., Tuyls P., and Ophey W., 'Robust key extraction from physical unclonable functions,' In: J. Ioannidis, A. Keromytis, and M. Yung (eds.) *Applied Cryptography and Network Security*, 2005, pp. 407–422. Springer Berlin Heidelberg, Berlin, Heidelberg.

Steganography, 'Past, present, future,' SANS Institute InfoSec Reading Room, 2001. URL http://www.sans.org/reading-room/whitepapers/stenganography/steganography-past-present-future-552

Su Y., Holleman J., and Otis B. P., 'A digital 1.6 pJ/bit chip identification circuit using process variations,' *IEEE Journal of Solid-State Circuits*, 2008, vol. 43 (1), pp. 69–77.

Suh G. E. and Devadas S., 'Physical unclonable functions for device authentication and secret key generation,' *44th ACM/IEEE Design Automation Conference*, 2007, pp. 9–14.

Suh G. E., O'Donnell C. W., Sachdev I., and Devadas S., 'Design and implementation of the AEGIS single-chip secure process using physical random functions,' *Proceedings of the 32nd International Symposium on Computer Architecture (ISCA)*, 2005, pp. 1–12.

Tanaka Y., Bian S., Hiromoto M., and Sato T., 'Coin flipping PUF: A novel PUF with improved resistance against machine learning attacks,' *IEEE Transactions on Circuits and Systems II: Express Briefs*, vol. 65 (5), 2018, pp. 602–606.

Wan M., He Z., Han S., Dai K., and Zou X., 'An invasive-attack-resistant PUF based on switched-capacitor circuit,' *IEEE Transactions on Circuits and Systems I: Regular Papers*, 2015, vol. 62 (8), pp. 2024–2034.

Wang X. and Tehranipoor M., 'Novel physical unclonable function with process and environmental variations,'*Design, Automation & Test in Europe Conference & Exhibition (DATE)*, 2010, pp. 1065–1070.

Wen Y. and Lao Y., 'PUF modeling attack using active learning,' *2018 IEEE International Symposium on Circuits and Systems (ISCAS)*, 2018, pp. 1–5.

Wong H. S. P., Raoux S., Kim S., *et al.*, 'Phase change memory,' *Proceedings of the IEEE*, 2010, vol. 98 (12), pp. 2201–2227.

Yan W., Tehranipoor F., and Chandy J. A., 'PUF-based fuzzy authentication without error correcting codes,' *IEEE Transactions on Computer-Aided Design of Integrated Circuits and Systems*, 2017, vol. 36 (9), pp. 1445–1457.

Yanambaka V. P., Mohanty S. P., and Kougianos E., 'Making use of semiconductor manufacturing process variations: FinFET-based physical unclonable functions for efficient security integration in the IoT,' *IEEE Potentials*, 2017, vol. 93 (3), pp. 429–441.

Yanambaka V. P., Mohanty S. P., and Kougianos E., 'Making use of manufacturing process variations: A dopingless transistor based-PUF for hardware-assisted security,' *IEEE Transactions on Semiconductor Manufacturing*, 2018, **PP** (99), pp. 1–1.

Ye J., Gong Y., Hu Y., and Li X., 'Polymorphic PUF: Exploiting reconfigurability of CPU + FPGA SoC to resist modeling attack,' *IEEE 23rd International Symposium on On-Line Testing and Robust System Design (IOLTS)*, 2017, pp. 205–206.

Ye J., Guo Q., Hu Y., Li H., and Li X., 'Modeling attacks on strong physical unclonable functions strengthened by random number and weak PUF,' *IEEE 36th VLSI Test Symposium (VTS)*, 2018, pp. 1–6.

Yu L., Wang X., Rahman F., and Tehranipoor M., 'iPUF: Interconnect PUF with self-masking circuit for performance enhancement,' *18th International Workshop on Microprocessor and SOC Test and Verification (MTV)*, 2017, pp. 45–50. DOI 10.1109/MTV.2017.14

Yu M. D. and Devadas S., 'Secure and robust error correction for physical unclonable functions,' *Proceedings of the IEEE Design & Test of Computers*, 2010, pp. 48–65.

Appendix A

This appendix describes the derivation process of Haar wavelet transformation (HWT) CODEC data flow graph using *Sengupta–Roy HAAR Computation Reduction Functions*

HAAR WAVELET TRANSFORMATION—LOSSY CODEC

$$Image\ A = \begin{bmatrix} m_1^1 & m_1^2 & m_1^3 & \cdots & m_1^{256} & m_1^{257} & \cdots & m_1^{512} \\ m_2^1 & m_2^2 & m_2^3 & \cdots & m_2^{256} & m_2^{257} & \cdots & m_2^{512} \\ m_3^1 & m_3^2 & m_3^3 & \cdots & m_3^{256} & m_3^{257} & \cdots & m_3^{512} \\ \vdots & \vdots & \vdots & & \vdots & \vdots & & \vdots \\ m_{256}^1 & m_{256}^2 & m_{256}^3 & \cdots & m_{256}^{256} & m_{256}^{257} & \cdots & m_{256}^{512} \\ m_{257}^1 & m_{257}^2 & m_{257}^3 & \cdots & m_{257}^{256} & m_{257}^{257} & \cdots & m_{257}^{512} \\ \vdots & \vdots & \vdots & & \vdots & \vdots & & \vdots \\ m_{512}^1 & m_{512}^2 & m_{512}^3 & \cdots & m_{512}^{256} & m_{512}^{257} & \cdots & m_{512}^{512} \end{bmatrix}_{512\text{x}512} \tag{A.1}$$

Equation (A.1) shows a generic 512 × 512 input image in the form of a matrix (A). The subscript and superscript of each element indicate the row and the column numbers of the elements, respectively. For example, the pixel intensity of the third row and 512th column of matrix "A" is represented by element m_3^{512}.

$$\begin{matrix} W_{512\times512} = \\ \text{(HWT matrix)} \end{matrix} \left[\begin{array}{cccccccccc} \frac{1}{2} & \frac{1}{2} & 0 & 0 & 0 & 0 & \cdots & & 0 & 0 \\ 0 & 0 & \frac{1}{2} & \frac{1}{2} & 0 & 0 & \cdots & & 0 & 0 \\ 0 & 0 & 0 & 0 & \frac{1}{2} & \frac{1}{2} & \cdots & & 0 & 0 \\ \vdots & \vdots & \vdots & \vdots & \vdots & \vdots & & & \vdots & \vdots \\ 0 & 0 & 0 & 0 & 0 & 0 & \cdots & & \frac{1}{2} & \frac{1}{2} \\ \hdashline -\frac{1}{2} & \frac{1}{2} & 0 & 0 & 0 & 0 & \cdots & & 0 & 0 \\ 0 & 0 & -\frac{1}{2} & \frac{1}{2} & 0 & 0 & \cdots & & 0 & 0 \\ 0 & 0 & 0 & 0 & -\frac{1}{2} & \frac{1}{2} & \cdots & & 0 & 0 \\ \vdots & \vdots & \vdots & \vdots & \vdots & \vdots & & & \vdots & \vdots \\ 0 & 0 & 0 & 0 & 0 & 0 & \cdots & & -\frac{1}{2} & \frac{1}{2} \end{array}\right] \begin{array}{l} \\ \\ \\ \\ \leftarrow \text{256th row} \\ \\ \\ \\ \\ \leftarrow \text{512th row} \end{array} \tag{A.2}$$

$X = W \cdot A$ (First phase of transformation on image A – Column transformation)

$$X = W \cdot A \text{ (Column-wise)} \tag{A.3}$$

$$X = \begin{bmatrix} \frac{m_1^1 + m_2^1}{2} & \frac{m_1^2 + m_2^2}{2} & \frac{m_1^3 + m_2^3}{2} & \cdots & \frac{m_1^{512} + m_2^{512}}{2} \\ \frac{m_3^1 + m_4^1}{2} & \frac{m_3^2 + m_4^2}{2} & \frac{m_3^3 + m_4^3}{2} & \cdots & \frac{m_3^{512} + m_4^{512}}{2} \\ \frac{m_5^1 + m_6^1}{2} & \frac{m_5^2 + m_6^2}{2} & \frac{m_5^3 + m_6^3}{2} & \cdots & \frac{m_5^{512} + m_6^{512}}{2} \\ \vdots & \vdots & \vdots & \vdots & \vdots \\ \frac{m_{511}^1 + m_{512}^1}{2} & \frac{m_{511}^2 + m_{512}^2}{2} & \frac{m_{511}^3 + m_{512}^3}{2} & \cdots & \frac{m_{511}^{512} + m_{512}^{512}}{2} \\ \hline \frac{m_2^1 - m_1^1}{2} & \frac{m_2^2 - m_1^2}{2} & \frac{m_2^3 - m_1^3}{2} & \cdots & \frac{m_2^{512} - m_1^{512}}{2} \\ \frac{m_4^1 - m_3^1}{2} & \frac{m_4^2 - m_3^2}{2} & \frac{m_4^3 - m_3^3}{2} & \cdots & \frac{m_4^{512} - m_3^{512}}{2} \\ \frac{m_6^1 - m_5^1}{2} & \frac{m_6^2 - m_5^2}{2} & \frac{m_6^3 - m_5^3}{2} & \cdots & \frac{m_6^{512} - m_5^{512}}{2} \\ \vdots & \vdots & \vdots & \vdots & \vdots \\ \frac{m_{512}^1 - m_{511}^1}{2} & \frac{m_{512}^2 - m_{511}^2}{2} & \frac{m_{512}^3 - m_{511}^3}{2} & \cdots & \frac{m_{512}^{512} - m_{511}^{512}}{2} \end{bmatrix} \tag{A.4}$$

256th row → (last row of the Average SUM part); 512th row → (last row of the Average DIFF. part)

Average SUM

Average DIFF.

$$X = \begin{bmatrix} X_1^1 & X_1^2 & X_1^3 & \cdots & X_1^{256} & X_1^{257} & \cdots & X_1^{512} \\ X_2^1 & X_2^2 & X_2^3 & \cdots & X_2^{256} & X_2^{257} & \cdots & X_2^{512} \\ X_3^1 & X_3^2 & X_3^3 & \cdots & X_3^{256} & X_3^{257} & \cdots & X_3^{512} \\ \vdots & \vdots & \vdots & \vdots & \vdots & \vdots & \vdots & \vdots \\ X_{256}^1 & X_{256}^2 & X_{256}^3 & \cdots & X_{256}^{256} & X_{256}^{257} & \cdots & X_{256}^{512} \\ \hline X_{257}^1 & X_{257}^2 & X_{257}^3 & \cdots & X_{257}^{256} & X_{257}^{257} & \cdots & X_{257}^{512} \\ \vdots & \vdots & \vdots & \vdots & \vdots & \vdots & \vdots & \vdots \\ X_{512}^1 & X_{512}^2 & X_{512}^3 & \cdots & X_{512}^{256} & X_{512}^{257} & \cdots & X_{512}^{512} \end{bmatrix} \tag{A.5}$$

Average SUM

Average DIFF.

Sengupta–Roy HAAR level 1 forward transformation reduction functions

$$\therefore X_n^p = \frac{m_{2n-1}^p + m_{2n}^p}{2} \tag{A.6}$$

$$\therefore X_{\frac{N}{2}+i}^p = \frac{m_{\frac{N}{2}+i-j+1}^p - m_{\frac{N}{2}+i-j}^p}{2} \tag{A.7}$$

where N is the dimension of the square input image; n is the variable ranging from 1 to $N/2$, increases in every row; i is the variable ranging from 1 to $N/2$, increases in every row; p is the variable ranging from 1 to N, increases in every column; j is the variable ranging from $N/2$ to 1, decreases as "i" increases. Equation (A.6) is used for calculating 1 to $N/2$th row and Equation (A.7) is used for calculating $(N/2+1)$th to Nth row of the input image. The corresponding level 1 forward transformation image in the form of matrix "X" is shown in Equation (A.4), where the calculation of pixel intensity is performed based on the aforementioned equations.

Second phase of transformation

$$B = X \cdot W^{\mathrm{T}} \tag{A.8}$$

where $X = W \cdot A$

256th column ↓ (column 256), 512th column ↓ (column 512); 511th row →, 512th row →

$$W^{T}{}_{512\times512} = \begin{bmatrix} \frac{1}{2} & 0 & 0 & \cdots & 0 & -\frac{1}{2} & 0 & 0 & \cdots & 0 \\ \frac{1}{2} & 0 & 0 & \cdots & 0 & \frac{1}{2} & 0 & 0 & \cdots & 0 \\ 0 & \frac{1}{2} & 0 & \cdots & 0 & 0 & -\frac{1}{2} & 0 & \cdots & 0 \\ 0 & \frac{1}{2} & 0 & \cdots & 0 & 0 & \frac{1}{2} & 0 & \cdots & 0 \\ 0 & 0 & \frac{1}{2} & \cdots & 0 & 0 & 0 & -\frac{1}{2} & \cdots & 0 \\ 0 & 0 & \frac{1}{2} & \cdots & 0 & 0 & 0 & \frac{1}{2} & \cdots & 0 \\ \vdots & \vdots & \vdots & & \vdots & \vdots & \vdots & \vdots & & \vdots \\ 0 & 0 & 0 & \cdots & \frac{1}{2} & 0 & 0 & 0 & \cdots & -\frac{1}{2} \\ 0 & 0 & 0 & \cdots & \frac{1}{2} & 0 & 0 & 0 & \cdots & \frac{1}{2} \end{bmatrix} \tag{A.9}$$

$B = X \cdot W^{\mathrm{T}}$ (computing row-wise)

257th column ↓, 512th column ↓; 256th row →, 512th row →

$$B = \begin{bmatrix} \frac{X_1^1+X_1^2}{2} & \frac{X_1^3+X_1^4}{2} & \frac{X_1^5+X_1^6}{2} & \cdots & \frac{X_1^{511}+X_1^{512}}{2} & \frac{X_1^2-X_1^1}{2} & \cdots & \frac{X_1^{512}-X_1^{511}}{2} \\ \frac{X_2^1+X_2^2}{2} & \frac{X_2^3+X_2^4}{2} & \frac{X_2^5+X_2^6}{2} & \cdots & \frac{X_2^{511}+X_2^{512}}{2} & \frac{X_2^2-X_2^1}{2} & \cdots & \frac{X_2^{512}-X_2^{511}}{2} \\ \vdots & \vdots & \vdots & & \vdots & \vdots & & \vdots \\ \frac{X_{256}^1+X_{256}^2}{2} & \frac{X_{256}^3+X_{256}^4}{2} & \frac{X_{256}^5+X_{256}^6}{2} & \cdots & \frac{X_{256}^{511}+X_{256}^{512}}{2} & \frac{X_{256}^2-X_{256}^1}{2} & \cdots & \frac{X_{256}^{512}-X_{256}^{511}}{2} \\ \frac{X_{257}^1+X_{257}^2}{2} & \frac{X_{257}^3+X_{257}^4}{2} & \frac{X_{257}^5+X_{257}^6}{2} & \cdots & \frac{X_{257}^{511}+X_{257}^{512}}{2} & \frac{X_{257}^2-X_{257}^1}{2} & \cdots & \frac{X_{257}^{512}-X_{257}^{511}}{2} \\ \vdots & \vdots & \vdots & & \vdots & \vdots & & \vdots \\ \frac{X_{512}^1+X_{512}^2}{2} & \frac{X_{512}^3+X_{512}^4}{2} & \frac{X_{512}^5+X_{512}^6}{2} & \cdots & \frac{X_{512}^{511}+X_{512}^{512}}{2} & \frac{X_{512}^2-X_{512}^1}{2} & \cdots & \frac{X_{512}^{512}-X_{512}^{511}}{2} \end{bmatrix}$$

Average SUM (columns 1–256) — Average DIFF. (columns 257–512)

(A.10)

$$B = \begin{bmatrix} b_1^1 & b_1^2 & b_1^3 & \cdots & b_1^{256} & b_1^{257} & \cdots & b_1^{512} \\ b_2^1 & b_2^2 & b_2^3 & \cdots & b_2^{256} & b_2^{257} & \cdots & b_2^{512} \\ b_3^1 & b_3^2 & b_3^3 & \cdots & b_3^{256} & b_3^{257} & \cdots & b_3^{512} \\ \vdots & \vdots & \vdots & & \vdots & \vdots & & \vdots \\ b_{256}^1 & b_{256}^2 & b_{256}^3 & \cdots & b_{256}^{256} & b_{256}^{257} & \cdots & b_{256}^{512} \\ b_{257}^1 & b_{257}^2 & b_{257}^3 & \cdots & b_{257}^{256} & b_{257}^{257} & \cdots & b_{257}^{512} \\ \vdots & \vdots & \vdots & & \vdots & \vdots & & \vdots \\ b_{512}^1 & b_{512}^2 & b_{512}^3 & \cdots & b_{512}^{256} & b_{512}^{257} & \cdots & b_{512}^{512} \end{bmatrix} \tag{A.11}$$

Sengupta–Roy HAAR level 2 forward transformation reduction functions

$$B = \left(\frac{x_n^q \pm x_n^p}{2}\right) \tag{A.12}$$

$$B = \left(\frac{X_{\frac{N}{2}+i}^q \pm X_{\frac{N}{2}+i}^p}{2}\right) \tag{A.13}$$

where N is the dimension of the square level 1 transformed image; n is the variable ranging from 1 to N, increases in every row; i is the variable ranging from 1 to $N/2$, increases in every row; p is the odd variable ranging from 1 to N, increases in every column; q is the even variable ranging from 2 to N, increases in every column. The first and second parts of Equation (A.12) are used for calculating the upper-left and upper-right quarters of the level 1 transformed images, respectively. The first and second parts of Equation (A.13) are used for calculating the lower-left and lower-right quarters of the level 1 transformed images, respectively. The corresponding level 2 forward transformation image in the form of matrix "B" is shown in Equation (A.10) where the calculation of pixel intensity is performed based on the aforementioned equations.

Sengupta–Roy HAAR forward transformation reduction functions

$$B = \left(\left(\frac{m_{2n}^q + m_{2n-1}^q}{2}\right) + \left(\frac{m_{2n}^p + m_{2n-1}^p}{2}\right)/2\right) \tag{A.14}$$

$$B = \left(\left(\frac{m_{2n}^q + m_{2n-1}^q}{2}\right) - \left(\frac{m_{2n}^p + m_{2n-1}^p}{2}\right)/2\right) \tag{A.15}$$

$$B = \left(\left(\frac{m_{\frac{N}{2}+i-j+1}^q - m_{\frac{N}{2}+i-j}^q}{2}\right) + \left(\frac{m_{\frac{N}{2}+i-j+1}^p - m_{\frac{N}{2}+i-j}^p}{2}\right)/2\right) \tag{A.16}$$

$$B = \left(\left(\frac{m_{\frac{N}{2}+i-j+1}^q - m_{\frac{N}{2}+i-j}^q}{2}\right) - \left(\frac{m_{\frac{N}{2}+i-j+1}^p - m_{\frac{N}{2}+i-j}^p}{2}\right)/2\right) \tag{A.17}$$

To generate the compressed image matrix "B," Equation (A.14) is used for calculating the intensity of pixels corresponding to both row and column index 1 to $N/2$; Equation (A.15) is used for calculating the intensity of pixels corresponding to row index 1 to $N/2$ and column index greater than $N/2$ to N; Equation (A.16) is used for calculating the intensity of pixels corresponding to the row index greater than $N/2$ to N and column index 1 to $N/2$; Equation (A.17) is used for calculating the intensity of pixels corresponding to both row and column index greater than $N/2$ to N.

DFG of HWT-based image compression and scheduled DFG based on four adder–subtractor units and four multipliers

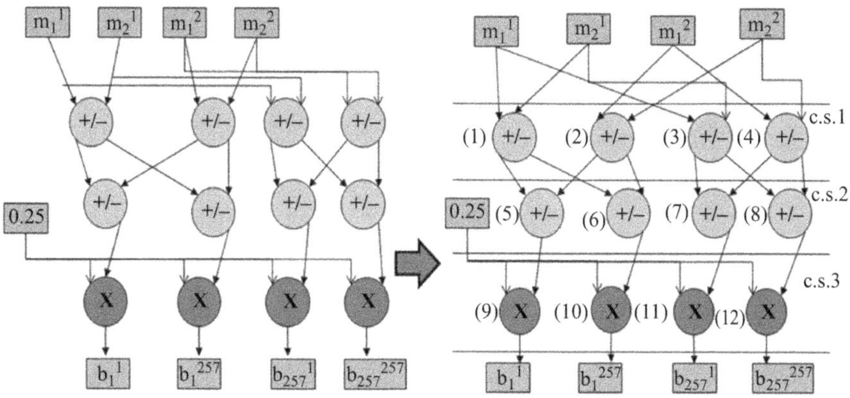

$$B' = f(B,T) = \begin{cases} B, & b_n^p > T \\ 0, & b_n^p \le T \end{cases} \tag{A.18}$$

where B' are the compressed image pixels after applying threshold based on the aforementioned function and "T" is the hard threshold value.

Image reconstruction

$$L = W^{-1} \cdot B' \tag{A.19}$$

$$W^{-1}_{512\times512} = 2 \cdot W^T \tag{A.20}$$

256th column; 512th column; 512th row

$$W^{-1}_{512\times512} = \begin{bmatrix} 1 & 0 & 0 & \cdots & 0 & -1 & 0 & 0 & \cdots & 0 \\ 1 & 0 & 0 & \cdots & 0 & 1 & 0 & 0 & \cdots & 0 \\ 0 & 1 & 0 & \cdots & 0 & 0 & -1 & 0 & \cdots & 0 \\ 0 & 1 & 0 & \cdots & 0 & 0 & 1 & 0 & \cdots & 0 \\ 0 & 0 & 1 & \cdots & 0 & 0 & 0 & -1 & \cdots & 0 \\ 0 & 0 & 1 & \cdots & 0 & 0 & 0 & 1 & \cdots & 0 \\ \vdots & \vdots & \vdots & & \vdots & \vdots & \vdots & \vdots & & \vdots \\ \vdots & \vdots & \vdots & & \vdots & \vdots & \vdots & \vdots & & \vdots \\ \vdots & \vdots & \vdots & & \vdots & \vdots & \vdots & \vdots & & \vdots \\ 0 & 0 & 0 & \cdots & 1 & 0 & 0 & 0 & \cdots & -1 \\ 0 & 0 & 0 & \cdots & 1 & 0 & 0 & 0 & \cdots & 1 \end{bmatrix}_{512\text{x}512} \tag{A.21}$$

$$L = \begin{bmatrix} \left(b_1'^1 - b_{257}'^1\right) & \left(b_1'^2 - b_{257}'^2\right) & \cdots & \left(b_1'^{512} - b_{257}'^{512}\right) \\ \left(b_1'^1 + b_{257}'^1\right) & \left(b_1'^2 + b_{257}'^2\right) & \cdots & \left(b_1'^{512} + b_{257}'^{512}\right) \\ \left(b_2'^1 - b_{258}'^1\right) & \left(b_2'^2 - b_{258}'^2\right) & \cdots & \left(b_2'^{512} - b_{258}'^{512}\right) \\ \left(b_2'^1 + b_{258}'^1\right) & \left(b_1'^2 + b_{258}'^2\right) & \cdots & \left(b_2'^{512} + b_{258}'^{512}\right) \\ \vdots & \vdots & \ddots & \vdots \\ \left(b_{256}'^1 - b_{512}'^1\right) & \left(b_{256}'^2 - b_{512}'^2\right) & \cdots & \left(b_{256}'^{512} - b_{512}'^{512}\right) \\ \left(b_{256}'^1 + b_{512}'^1\right) & \left(b_{256}'^2 + b_{512}'^2\right) & \cdots & \left(b_{256}'^{512} + b_{512}'^{512}\right) \end{bmatrix} \tag{A.22}$$

$$L = \begin{bmatrix} l_1^1 & l_1^2 & l_1^3 & \cdots & l_1^{256} & l_1^{257} & \cdots & l_1^{512} \\ l_2^1 & l_2^2 & l_2^3 & \cdots & l_2^{256} & l_2^{257} & \cdots & l_2^{512} \\ l_3^1 & l_3^2 & l_3^3 & \cdots & l_3^{256} & l_3^{257} & \cdots & l_3^{512} \\ \vdots & \vdots & \vdots & & \vdots & \vdots & & \vdots \\ l_{256}^1 & l_{256}^2 & l_{256}^3 & \cdots & l_{256}^{256} & l_{256}^{257} & \cdots & l_{256}^{512} \\ l_{257}^1 & l_{257}^2 & l_{257}^3 & \cdots & l_{257}^{256} & l_{257}^{257} & \cdots & l_{257}^{512} \\ \vdots & \vdots & \vdots & & \vdots & \vdots & & \vdots \\ l_{512}^1 & l_{512}^2 & l_{512}^3 & \cdots & l_{512}^{256} & l_{512}^{257} & \cdots & l_{512}^{512} \end{bmatrix} \tag{A.23}$$

Sengupta–Roy HAAR level 1 decompression reduction functions

$$L = b_x'^z - b_{\frac{N}{2}+x}'^z \tag{A.24}$$

$$L = b_x'^z + b_{\frac{N}{2}+x}'^z \tag{A.25}$$

where N is the dimension of the square-compressed image matrix; z is the variable ranging from 1 to N, increases in every column; x is the variable ranging from 1 to $N/2$, increases in every alternate row. Equation (A.24) is used for calculating the pixel value of odd rows and Equation (A.25) is used for calculating the pixel value of even rows of the compressed image. The corresponding level 1 decompressed image in the form of matrix "L" is shown in Equation (A.22) where calculation of each pixel is performed based on the aforementioned equations.

$$C = L \cdot \left(W^{-1}\right)T \tag{A.26}$$

$$C = L \cdot 2W \tag{A.27}$$

$$2W_{512\times 512}=\begin{bmatrix} 1 & 1 & 0 & 0 & 0 & 0 & . & . & . & 0 & 0\\ 0 & 0 & 1 & 1 & 0 & 0 & . & . & . & 0 & 0\\ 0 & 0 & 0 & 0 & 1 & 1 & . & . & . & 0 & 0\\ . & . & . & . & . & . & . & & & 0 & 0\\ . & . & . & . & . & . & & . & & 0 & 0\\ . & . & . & . & . & . & & & . & 0 & 0\\ 0 & 0 & 0 & 0 & 0 & 0 & . & . & . & 1 & 1\\ -1 & 1 & 0 & 0 & 0 & 0 & . & . & . & 0 & 0\\ 0 & 0 & -1 & 1 & 0 & 0 & . & . & . & 0 & 0\\ 0 & 0 & 0 & 0 & -1 & 1 & . & . & . & 0 & 0\\ . & . & . & . & . & . & . & & & . & .\\ . & . & . & . & . & . & & . & & . & .\\ . & . & . & . & . & . & & & . & 0 & 0\\ 0 & 0 & 0 & 0 & 0 & 0 & 1 & 0 & 0 & -1 & 1 \end{bmatrix}\begin{matrix} \\ \\ \\ \\ \\ \leftarrow \text{256th row} \\ \\ \leftarrow \text{257th row} \\ \\ \\ \\ \\ \\ \\ \end{matrix} \tag{A.28}$$

$$C=\begin{bmatrix} \left(l_1^1-l_1^{257}\right) & \left(l_1^1+l_1^{257}\right) & \left(l_1^2-l_1^{258}\right) & \left(l_1^2+l_1^{258}\right) & . & . & . & \left(l_1^{256}-l_1^{512}\right) & \left(l_1^{256}+l_1^{512}\right)\\ \left(l_2^1-l_2^{257}\right) & \left(l_2^1+l_2^{257}\right) & \left(l_2^2-l_2^{258}\right) & \left(l_2^2+l_2^{258}\right) & . & . & . & \left(l_2^{256}-l_2^{512}\right) & \left(l_2^{256}+l_2^{512}\right)\\ \left(l_3^1-l_3^{257}\right) & \left(l_3^1+l_3^{257}\right) & \left(l_3^2-l_3^{258}\right) & \left(l_3^2+l_3^{258}\right) & . & . & . & \left(l_3^{256}-l_3^{512}\right) & \left(l_3^{256}+l_3^{512}\right)\\ \left(l_4^1-l_4^{257}\right) & \left(l_4^1+l_4^{257}\right) & \left(l_4^2-l_4^{258}\right) & \left(l_4^2+l_4^{258}\right) & . & . & . & \left(l_4^{256}-l_4^{512}\right) & \left(l_4^{256}+l_4^{512}\right)\\ . & . & . & . & . & & & . & .\\ . & . & . & . & & . & & . & .\\ . & . & . & . & & & . & . & .\\ \left(l_{511}^1-l_{511}^{257}\right) & \left(l_{511}^1+l_{511}^{257}\right) & \left(l_{511}^2-l_{511}^{258}\right) & \left(l+l_{511}^{258}\right) & . & . & . & \left(l_{511}^{256}-l_{511}^{512}\right) & \left(l_{511}^{256}+l_{511}^{512}\right)\\ \left(l_{512}^1-l_{512}^{257}\right) & \left(l_{512}^1+l_{512}^{257}\right) & \left(l_{512}^2-l_{512}^{258}\right) & \left(l_{512}^2+l_{512}^{258}\right) & . & . & . & \left(l_{512}^{256}-l_{512}^{512}\right) & \left(l_{512}^{256}+l_{512}^{512}\right) \end{bmatrix} \tag{A.29}$$

$$C=\begin{bmatrix} C_1^1 & C_1^2 & C_1^3 & . & . & . & C_1^{256} & C_1^{257} & . & . & . & C_1^{512}\\ C_2^1 & C_2^2 & C_2^3 & . & . & . & C_2^{256} & C_2^{257} & . & . & . & C_2^{512}\\ C_3^1 & C_3^2 & C_3^3 & . & . & . & C_3^{256} & C_3^{257} & . & . & . & C_3^{512}\\ . & . & . & . & & & . & . & . & & & .\\ . & . & . & & . & & . & . & & . & & .\\ . & . & . & & & . & . & . & & & . & .\\ C_{256}^1 & C_{256}^2 & C_{256}^3 & . & . & . & C_{256}^{256} & C_{256}^{257} & . & . & . & C_{256}^{512}\\ C_{257}^1 & C_{257}^2 & C_{257}^3 & . & . & . & C_{257}^{256} & C_{257}^{257} & . & . & . & C_{257}^{512}\\ . & . & . & . & & & . & . & . & & & .\\ . & . & . & & . & & . & . & & . & & .\\ . & . & . & & & . & . & . & & & . & .\\ C_{512}^1 & C_{512}^2 & C_{512}^3 & . & . & . & C_{512}^{256} & C_{512}^{257} & . & . & . & C_{512}^{512} \end{bmatrix} \tag{A.30}$$

Sengupta–Roy HAAR level 2 decompression reduction functions

$$C = I_x^z - I_x^{\frac{N}{2}+z} \tag{A.31}$$

$$C = I_x^z + I_x^{\frac{N}{2}+z} \tag{A.32}$$

where N is the dimension of the square level 1 decompressed image matrix; z is the variable ranging from 1 to $N/2$, increases in every alternate column; x is the variable ranging from 1 to N, increases in every row. Equation (A.31) is used for calculating the pixel value of odd columns and Equation (A.32) is used for calculating the pixel value of even columns of the level 1 decompressed image. The corresponding level 2 decompressed image in the form of matrix "C" is shown in Equation (A.29) where calculation of each pixel is performed based on the aforementioned equations.

512th column ↓

$$C = \begin{bmatrix} (b'^1_1 - b'^1_{257}) - (b'^{257}_1 - b'^{257}_{257}) & (b'^1_1 - b'^1_{257}) + (b'^{257}_1 - b'^{257}_{257}) & \cdots & (b'^{256}_1 - b'^{256}_{257}) + (b'^{512}_1 - b'^{512}_{257}) \\ (b'^1_1 + b'^1_{257}) - (b'^{257}_1 + b'^{257}_{257}) & (b'^1_1 + b'^1_{257}) + (b'^{257}_1 + b'^{257}_{257}) & \cdots & (b'^{256}_1 + b'^{256}_{257}) + (b'^{512}_1 + b'^{512}_{257}) \\ (b'^1_2 - b'^1_{258}) - (b'^{257}_2 - b'^{257}_{258}) & (b'^1_2 - b'^1_{258}) + (b'^{257}_2 - b'^{257}_{258}) & \cdots & (b'^{256}_2 - b'^{512}_{258}) + (b'^{256}_2 - b'^{512}_{258}) \\ (b'^1_2 + b'^1_{258}) - (b'^{257}_2 + b'^{257}_{258}) & (b' + b'^1_{258}) + (b'^{257}_2 + b'^{257}_{258}) & \cdots & (b'^{256}_2 + b'^{512}_{258}) + (b'^{256}_2 + b'^{512}_{258}) \\ \vdots & \vdots & \vdots & \vdots \\ (b'^1_{256} - b'^1_{512}) - (b'^{257}_{256} - b'^{257}_{512}) & (b'^1_{256} - b'^1_{512}) + (b'^{257}_{256} - b'^{257}_{512}) & \cdots & (b'^{256}_{256} - b'^{256}_{512}) + (b'^{512}_{256} - b'^{512}_{512}) \\ (b'^1_{256} + b'^1_{512}) - (b'^{257}_{256} + b'^{257}_{512}) & (b'^1_{256} + b'^1_{512}) + (b'^{257}_{256} + b'^{257}_{512}) & \cdots & (b'^{256}_{256} + b'^{256}_{512}) + (b'^{512}_{256} + b'^{512}_{512}) \end{bmatrix} \tag{A.33}$$

Sengupta–Roy HAAR decompression reduction functions

$$C = \left(b'^g_i - b'^g_{\frac{N}{2}+1}\right) - \left(b'^{\frac{N}{2}+g}_i - b'^{\frac{N}{2}+g}_{\frac{N}{2}+i}\right) \tag{A.34}$$

$$C = \left(b'^g_i - b'^g_{\frac{N}{2}+1}\right) + \left(b'^{\frac{N}{2}+g}_i - b'^{\frac{N}{2}+g}_{\frac{N}{2}+i}\right) \tag{A.35}$$

$$C = \left(b'^g_i - b'^g_{\frac{N}{2}+1}\right) - \left(b'^{\frac{N}{2}+g}_i - b'^{\frac{N}{2}+g}_{\frac{N}{2}+i}\right) \tag{A.36}$$

$$C = \left(b'^g_i - b'^g_{\frac{N}{2}+1}\right) - \left(b'^{\frac{N}{2}+g}_i - b'^{\frac{N}{2}+g}_{\frac{N}{2}+i}\right) \tag{A.37}$$

DFG of HWT-based image decompression and scheduled DFG based on four adder–subtractor units, two adders, and two subtractors

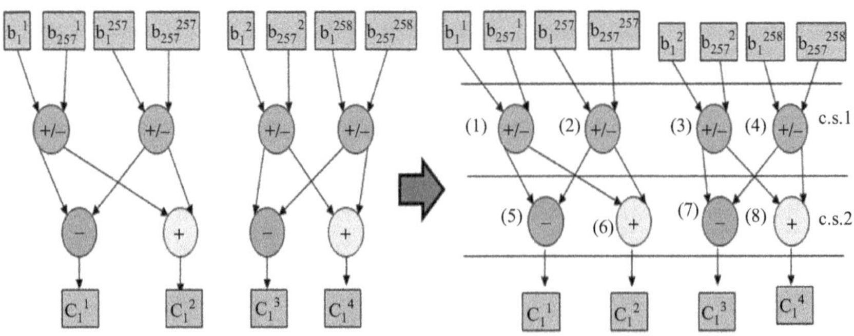

HAAR LOSSLESS CODEC

512th column

$$A^{Re} = \begin{bmatrix} (b_1^1 - b_{257}^1) - (b_1^{257} - b_{257}^{257}) & (b_1^1 - b_{257}^1) + (b_1^{257} - b_{257}^{257}) & \cdots & (b_1^{256} - b_{257}^{256}) + (b_1^{512} - b_{257}^{512}) \\ (b_1^1 + b_{257}^1) - (b_1^{257} + b_{257}^{257}) & (b_1^1 + b_{257}^1) + (b_1^{257} + b_{257}^{257}) & \cdots & (b_1^{256} + b_{257}^{256}) + (b_1^{512} + b_{257}^{512}) \\ (b_2^1 - b_{258}^1) - (b_2^{257} - b_{258}^{257}) & (b_2^1 - b_{258}^1) + (b_2^{257} - b_{258}^{257}) & \cdots & (b_2^{256} - b_{258}^{512}) + (b_2^{256} - b_{258}^{512}) \\ (b_2^1 + b_{258}^1) - (b_2^{257} + b_{258}^{257}) & (b + b_{258}^1) + (b_2^{257} + b_{258}^{257}) & \cdots & (b_2^{256} + b_{258}^{512}) + (b_2^{256} + b_{258}^{512}) \\ \vdots & \vdots & \vdots & \vdots \\ (b_{256}^1 - b_{512}^1) - (b_{256}^{257} - b_{512}^{257}) & (b_{256}^1 - b_{512}^1) + (b_{256}^{257} - b_{512}^{257}) & \cdots & (b_{256}^{256} - b_{512}^{256}) + (b_{256}^{512} - b_{512}^{512}) \\ (b_{256}^1 + b_{512}^1) - (b_{256}^{257} + b_{512}^{257}) & (b_{256}^1 + b_{512}^1) + (b_{256}^{257} + b_{512}^{257}) & \cdots & (b_{256}^{256} + b_{512}^{256}) + (b_{256}^{512} + b_{512}^{512}) \end{bmatrix} \tag{A.38}$$

$$B = \left(\left(\frac{m_{2n}^q + m_{2n-1}^q}{2} \right) + \left(\frac{m_{2n}^p + m_{2n-1}^p}{2} \right) /2 \right) \tag{A.39}$$

$$B = \left(\left(\frac{m_{2n}^q + m_{2n-1}^q}{2} \right) - \left(\frac{m_{2n}^p + m_{2n-1}^p}{2} \right) /2 \right) \tag{A.40}$$

$$B = \left(\left(\frac{m_{\frac{N}{2}+i-j+1}^q - m_{\frac{N}{2}+i-j}^q}{2} \right) + \left(\frac{m_{\frac{N}{2}+i-j+1}^p - m_{\frac{N}{2}+i-j}^p}{2} \right) /2 \right) \tag{A.41}$$

$$B = \left(\left(\frac{m_{\frac{N}{2}+i-j+1}^q - m_{\frac{N}{2}+i-j}^q}{2} \right) - \left(\frac{m_{\frac{N}{2}+i-j+1}^p - m_{\frac{N}{2}+i-j}^p}{2} \right) /2 \right) \tag{A.42}$$

To generate the compressed image matrix "*B*," Equation (A.39) is used for calculating the intensity of pixels corresponding to both row and column index 1 to $N/2$; Equation (A.40) is used for calculating the intensity of pixels corresponding to row index 1 to $N/2$ and column index greater than $N/2$ to N; Equation (A.41) is used for calculating the intensity of pixels corresponding to row index greater than $N/2$ to N and column index 1 to $N/2$; Equation (A.42) is used for calculating the intensity of pixels corresponding to both row and column index greater than $N/2$ to N.

Sengupta–Roy HAAR lossless image reconstruction reduction functions

$$C = \left(b_i^g - b_{\frac{N}{2}+i}^g \right) - \left(b_i^{\frac{N}{2}+g} - b_{\frac{N}{2}+i}^{\frac{N}{2}+g} \right) \tag{A.43}$$

$$C = \left(b_i^g - b_{\frac{N}{2}+i}^g \right) + \left(b_i^{\frac{N}{2}+g} - b_{\frac{N}{2}+i}^{\frac{N}{2}+g} \right) \tag{A.44}$$

$$C = \left(b_i^g + b_{\frac{N}{2}+i}^g \right) - \left(b_i^{\frac{N}{2}+g} + b_{\frac{N}{2}+i}^{\frac{N}{2}+g} \right) \tag{A.45}$$

$$C = \left(b_i^g + b_{\frac{N}{2}+i}^g \right) + \left(b_i^{\frac{N}{2}+g} + b_{\frac{N}{2}+i}^{\frac{N}{2}+g} \right) \tag{A.46}$$

Sengupta–Roy HAAR image CODEC validation functions

$$A^{Re} = \left[\left(\left(\frac{m^p_{2n-1} + m^p_{2n}}{2}\right) + \left(\frac{m^q_{2n-1} + m^q_{2n}}{2}\right)/2\right) - \left(\left(\frac{m^p_{\frac{N}{2}+i-j+1} - m^p_{\frac{N}{2}+i-j}}{2}\right) + \left(\frac{m^q_{\frac{N}{2}+i-j+1} - m^q_{\frac{N}{2}+i-j}}{2}\right)/2\right)\right] - \left[\left(\left(\frac{m^q_{2n-1} + m^q_{2n}}{2}\right) - \left(\frac{m^p_{2n-1} + m^p_{2n}}{2}\right)/2\right) - \left(\left(\frac{m^q_{\frac{N}{2}+i-j+1} - m^q_{\frac{N}{2}+i-j}}{2}\right) - \left(\frac{m^p_{\frac{N}{2}+i-j+1} - m^p_{\frac{N}{2}+i-j}}{2}\right)/2\right)\right] \quad (A.47)$$

$$A^{Re} = \left[\left(\left(\frac{m^p_{2n-1} + m^p_{2n}}{2}\right) + \left(\frac{m^q_{2n-1} + m^q_{2n}}{2}\right)/2\right) - \left(\left(\frac{m^p_{\frac{N}{2}+i-j+1} - m^p_{\frac{N}{2}+i-j}}{2}\right) + \left(\frac{m^q_{\frac{N}{2}+i-j+1} - m^q_{\frac{N}{2}+i-j}}{2}\right)/2\right)\right] + \left[\left(\left(\frac{m^q_{2n-1} + m^q_{2n}}{2}\right) - \left(\frac{m^p_{2n-1} + m^p_{2n}}{2}\right)/2\right) - \left(\left(\frac{m^q_{\frac{N}{2}+i-j+1} - m^q_{\frac{N}{2}+i-j}}{2}\right) - \left(\frac{m^p_{\frac{N}{2}+i-j+1} - m^p_{\frac{N}{2}+i-j}}{2}\right)/2\right)\right] \quad (A.48)$$

$$A^{Re} = \left[\left(\left(\frac{m^p_{2n-1} + m^p_{2n}}{2}\right) + \left(\frac{m^q_{2n-1} + m^q_{2n}}{2}\right)/2\right) + \left(\left(\frac{m^p_{\frac{N}{2}+i-j+1} - m^p_{\frac{N}{2}+i-j}}{2}\right) + \left(\frac{m^q_{\frac{N}{2}+i-j+1} - m^q_{\frac{N}{2}+i-j}}{2}\right)/2\right)\right] - \left[\left(\left(\frac{m^q_{2n-1} + m^q_{2n}}{2}\right) - \left(\frac{m^p_{2n-1} + m^p_{2n}}{2}\right)/2\right) + \left(\left(\frac{m^q_{\frac{N}{2}+i-j+1} - m^q_{\frac{N}{2}+i-j}}{2}\right) - \left(\frac{m^p_{\frac{N}{2}+i-j+1} - m^p_{\frac{N}{2}+i-j}}{2}\right)/2\right)\right] \quad (A.49)$$

$$A^{Re} = \left[\left(\left(\frac{m^p_{2n-1} + m^p_{2n}}{2}\right) + \left(\frac{m^q_{2n-1} + m^q_{2n}}{2}\right)/2\right) + \left(\left(\frac{m^p_{\frac{N}{2}+i-j+1} - m^p_{\frac{N}{2}+i-j}}{2}\right) + \left(\frac{m^q_{\frac{N}{2}+i-j+1} - m^q_{\frac{N}{2}+i-j}}{2}\right)/2\right)\right] + \left[\left(\left(\frac{m^q_{2n-1} + m^q_{2n}}{2}\right) - \left(\frac{m^p_{2n-1} + m^p_{2n}}{2}\right)/2\right) + \left(\left(\frac{m^q_{\frac{N}{2}+i-j+1} - m^q_{\frac{N}{2}+i-j}}{2}\right) - \left(\frac{m^p_{\frac{N}{2}+i-j+1} - m^p_{\frac{N}{2}+i-j}}{2}\right)/2\right)\right] \quad (A.50)$$

Corresponding DFG below represents the validation function as shown in Equations (A.47)–(A.50), where N is the dimension of the square input image matrix; n is the variable ranging from 1 to $N/2$, increases in every alternate row; i is the variable ranging from 1 to $N/2$, increases in every alternate row; p is the odd variable ranging from 1 to N, increases in every alternate column; q is the even variable ranging from 2 to N, increases in every alternate column; j is the variable

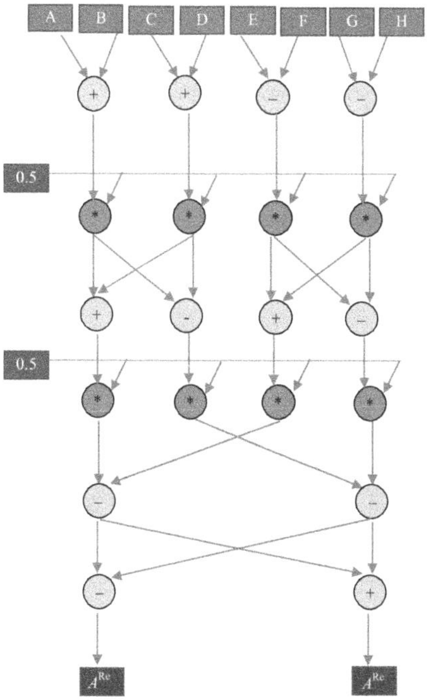

DFG of Equations (A.47) and (A.48)

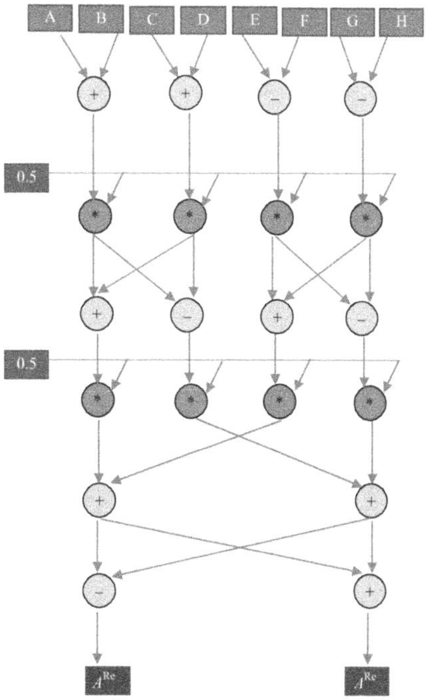

DFG of Equations (A.49) and (A.50)

ranging from $N/2$ to 1, decreases as 'i' increases. Equation (A.43) is used for calculating the intensity of pixels corresponding to odd rows and odd columns. Equation (A.44) is used for calculating the intensity of pixels corresponding to odd rows and even columns. Equation (A.45) is used for calculating the intensity of pixels corresponding to even rows and odd columns. Equation (A.46) is used for calculating the intensity of pixels corresponding to even rows and even columns.

$$A = m^p_{2n-1}, \qquad B = m^p_{2n}, \qquad C = m^q_{2n-1}, \qquad D = m^q_{2n}$$

$$E = m^p_{\frac{N}{2}+i-j+1}, \qquad F = m^p_{\frac{N}{2}+i-j}, \qquad G = m^q_{\frac{N}{2}+i-j+1}, \qquad H = m^q_{\frac{N}{2}+i-j}$$

Appendix B

Part A: Controller design details for JPEG codec IP core

	Control signals of M1					Control signals of M2				
Count	Lstr _M1	Enable _M1	Ostr _M1	Sel _M1	Desel _M1	Lstr _M2	Enable _M2	Ostr _M2	Sel _M2	Desel _M2
0	0	0	0	00000	00000	0	0	0	00000	00000
1	1	0	0			1	0			
2		1		00001			1		00001	
3	0	1				0	1			
4		1					1			
5		1					1			
6		1					1			
7		1	1				1	1		
8		1					1			
9		0					0			
10	1					1				
11		1	0	00010	00001		1	0	00010	00001
12	0	1				0	1			
13		1					1			
14		1					1			
15		1					1			
16		1	1				1	1		
17		1					1			
18		0					0			
19										
20										
21	1					1				
22		1	0	00011	00010		1	0	00011	00010
23	0	1				0	1			
24		1					1			
25		1					1			
26		1					1			
27		1	1				1	1		
28		1					1			
29		0					0			
30										
31										
32	1					1				
33		1	0	00100	00011		1	0	00100	00011
34	0	1				0	1			
35		1					1			

	Control signals of M1					Control signals of M2				
count	**Lstr _M1**	**Enable _M1**	**Ostr _M1**	**Sel _M1**	**Desel _M1**	**Lstr _M2**	**Enable _M2**	**Ostr _M2**	**Sel _M2**	**Desel _M2**
36		1					1			
37		1					1			
38		1	1				1	1		
39		1					1			
40		0					0			
41										
42										
43	1					1				
44		1	0	00101	00100		1	0	00101	00100
45	0	1				0	1			
46		1					1			
47		1					1			
48		1					1			
49		1	1				1	1		
50		1					1			
51		0					0			
52										
53										
54	1					1				
55		1	0	00110	00101		1	0	00110	00101
56	0	1				0	1			
57		1					1			
58		1					1			
59		1					1			
60		1	1				1	1		
61		1					1			
62		0					0			
63										
64										
65	1					1				
66		1	0	00111	00110		1	0	00111	00110
67	0	1				0	1			
68		1					1			
69		1					1			
70		1					1			
71		1	1				1	1		
72		1					1			
73		0					0			
74										
75										

	Control signals of M1					Control signals of M2				
count	**Lstr _M1**	**Enable _M1**	**Ostr _M1**	**Sel _M1**	**Desel _M1**	**Lstr _M2**	**Enable _M2**	**Ostr _M2**	**Sel _M2**	**Desel _M2**
76	1					1				
77		1	0	01000	00111		1	0	01000	00111
78	0	1				0	1			
79		1					1			
80		1					1			
81		1					1			
82		1	1				1	1		
83		1					1			
84		0					0			
85										
86										
87	1					1				
88		1	0	01001	01000		1	0	01001	01000
89	0	1				0	1			
90		1					1			
91		1					1			
92		1					1			
93		1	1				1	1		
94		1					1			
95		0					0			
96										
97										
98	1					1				
99		1	0	01010	01001		1	0	01010	01001
100	0	1				0	1			
101		1					1			
102		1					1			
103		1					1			
104		1	1				1	1		
105		1					1			
106		0					0			
107										
108										
109	1					1				
110		1	0	01011	01010		1	0	01011	01010
111	0	1				0	1			
112		1					1			
113		1					1			
114		1					1			
115		1	1				1	1		

	Control signals of M1					Control signals of M2				
count	**Lstr _M1**	**Enable _M1**	**Ostr _M1**	**Sel _M1**	**Desel _M1**	**Lstr _M2**	**Enable _M2**	**Ostr _M2**	**Sel _M2**	**Desel _M2**
116		1					1			
117		0					0			
118										
119										
120	1					1				
121		1	0	01100	01011		1	0	01100	01011
122	0	1				0	1			
123		1					1			
124		1					1			
125		1					1			
126		1	1				1	1		
127		1					1			
128		0					0			
129										
130										
131	1					1				
132		1	0	01101	01100		1	0	01101	01100
133	0	1				0	1			
134		1					1			
135		1					1			
136		1					1			
137		1	1				1	1		
138		1					1			
139		0					0			
140										
141										
142	1					1				
143		1	0	01110	01101		1	0	01110	01101
144	0	1				0	1			
145		1					1			
146		1					1			
147		1					1			
148		1	1				1	1		
149		1					1			
150		0					0			
151										
152										
153	1					1				
154		1	0	01111	01110		1	0	01111	01110
155	0	1				0	1			

	Control signals of M1					Control signals of M2				
count	**Lstr _M1**	**Enable _M1**	**Ostr _M1**	**Sel _M1**	**Desel _M1**	**Lstr _M2**	**Enable _M2**	**Ostr _M2**	**Sel _M2**	**Desel _M2**
156		1					1			
157		1					1			
158		1					1			
159		1	1				1	1		
160		1					1			
161		0					0			
162										
163										
164	1					1				
165		1	0	10000	01111		1	0	10000	01111
166	0	1				0	1			
167		1					1			
168		1					1			
169		1					1			
170		1	1				1	1		
171		1					1			
172		0					0			
173										
174										
175	1					1				
176		1	0	10001	10000		1	0	10001	10000
177	0	1				0	1			
178		1					1			
179		1					1			
180		1					1			
181		1	1				1	1		
182		1					1			
183		0					0			
184										
185										
186	1					1				
187		1	0	10010	10001		1	0	10010	10001
188	0	1				0	1			
189		1					1			
190		1					1			
191		1					1			
192		1	1				1	1		
193		1					1			
194		0					0			
195										

	Control signals of M1					Control signals of M2				
count	**Lstr _M1**	**Enable _M1**	**Ostr _M1**	**Sel _M1**	**Desel _M1**	**Lstr _M2**	**Enable _M2**	**Ostr _M2**	**Sel _M2**	**Desel _M2**
196										
197	1					1				
198		1	0	10011	10010		1	0	10011	10010
199	0	1				0	1			
200		1					1			
201		1					1			
202		1					1			
203		1	1				1	1		
204		1					1			
205		0					0			
206										
207										
208	1					1				
209		1	0	10100	10011		1	0	10100	10011
210	0	1				0	1			
211		1					1			
212		1					1			
213		1					1			
214		1	1				1	1		
215		1					1			
216		0					0			
217										
218	1					1				
219		1	0	10101	10100		1	0	10101	10100
220	0	1				0	1			
221		1					1			
222		1					1			
223		1					1			
224		1	1				1	1		
225		1					1			
226		0					0			
227										
228										
229	1					1				
230		1	0	10110	10101		1	0	10110	10101
231	0	1				0	1			
232		1					1			
233		1					1			
234		1					1			
235		1	1				1	1		

	Control signals of M1					Control signals of M2				
count	**Lstr _M1**	**Enable _M1**	**Ostr _M1**	**Sel _M1**	**Desel _M1**	**Lstr _M2**	**Enable _M2**	**Ostr _M2**	**Sel _M2**	**Desel _M2**
236		1					1			
237		0					0			
238										
239										
240	1					1				
241		1	0	10111	10110		1	0	10111	10110
242	0	1				0	1			
243		1					1			
244		1					1			
245		1					1			
246		1	1				1	1		
247		1					1			
248		0					0			
249										
250										
251	1					1				
252		1	0		10111		1	0		10111
253	0	1				0	1			
254		1					1			
255		1					1			
256		1					1			
257		1	1				1	1		
258		1					1			
259		0					0			
260										
261										
262										
263										
264										
265										
266										
267										
268										
269										
270										
271										
272										
273										
274										
275										

	Control signals of M3					Control signals of A1			
count	**Lstr _M3**	**Enable _M3**	**Ostr _M3**	**Sel _M3**	**Desel _M3**	**Lstr _A1**	**Ostr _A1**	**Sel _A1**	**Desel _A1**
0	0	0	0	00000	00000	0	0	00000	00000
1	1								
2		1		00001					
3	0	1							
4		1							
5		1							
6		1							
7		1	1						
8		1							
9		0							
10	1					1			
11		1	0	00010	00001				
12	0	1				0		00001	
13		1							
14		1							
15		1							
16		1	1				1		
17		1							
18		0							
19									
20									
21	1					1	0		
22		1	0	00011	00010				
23	0	1				0		00010	00001
24		1							
25		1							
26		1					1		
27		1	1						
28		1							
29		0							
30									
31									
32	1					1	0		
33		1	0	00100	00011				
34	0	1				0		00011	00010
35		1							
36		1							
37		1					1		
38		1	1						
39		1							
40		0							

	Control signals of M3					Control signals of A1			
count	**Lstr _M3**	**Enable _M3**	**Ostr _M3**	**Sel _M3**	**Desel _M3**	**Lstr _A1**	**Ostr _A1**	**Sel _A1**	**Desel _A1**
41									
42									
43	1					1	0		
44		1	0	00101	00100				
45	0	1				0		00100	00011
46		1							
47		1							
48		1					1		
49		1	1						
50		1							
51		0							
52									
53									
54	1					1	0		
55		1	0	00110	00101				
56	0	1				0		00101	00100
57		1							
58		1							
59		1					1		
60		1	1						
61		1							
62		0							
63									
64									
65	1					1	0		
66		1	0	00111	00110				
67	0	1				0		00110	00101
68		1							
69		1							
70		1					1		
71		1	1						
72		1							
73		0							
74									
75									
76	1					1	0		
77		1	0	01000	00111				
78	0	1				0		00111	00110
79		1							
80		1							

	Control signals of M3					Control signals of A1			
count	**Lstr _M3**	**Enable _M3**	**Ostr _M3**	**Sel _M3**	**Desel _M3**	**Lstr _A1**	**Ostr _A1**	**Sel _A1**	**Desel _A1**
81		1					1		
82		1	1						
83		1							
84		0							
85									
86									
87	1					1	0		
88		1	0	01001	01000				
89	0	1				0		01000	00111
90		1							
91		1							
92		1					1		
93		1	1						
94		1							
95		0							
96									
97									
98	1					1	0		
99		1	0	01010	01001				
100	0	1				0		01001	01000
101		1							
102		1							
103		1					1		
104		1	1						
105		1							
106		0							
107									
108									
109	1					1	0		
110		1	0	01011	01010				
111	0	1				0		01010	01001
112		1							
113		1							
114		1					1		
115		1	1						
116		1							
117		0							
118									
119									
120	1					1	0		

	Control signals of M3					Control signals of A1			
count	**Lstr _M3**	**Enable _M3**	**Ostr _M3**	**Sel _M3**	**Desel _M3**	**Lstr _A1**	**Ostr _A1**	**Sel _A1**	**Desel _A1**
121		1	0	01100	01011				
122	0	1				0		01011	01010
123		1							
124		1							
125		1					1		
126		1	1						
127		1							
128		0							
129									
130									
131	1					1	0		
132		1	0	01101	01100				
133	0	1				0		01100	01011
134		1							
135		1							
136		1					1		
137		1	1						
138		1							
139		0							
140									
141									
142	1					1	0		
143		1	0	01110	01101				
144	0	1				0		01101	01100
145		1							
146		1							
147		1					1		
148		1	1						
149		1							
150		0							
151									
152									
153	1					1	0		
154		1	0	01111	01110				
155	0	1				0		01110	01101
156		1							
157		1							
158		1					1		
159		1	1						
160		1							

	Control signals of M3					Control signals of A1			
count	**Lstr _M3**	**Enable _M3**	**Ostr _M3**	**Sel _M3**	**Desel _M3**	**Lstr _A1**	**Ostr _A1**	**Sel _A1**	**Desel _A1**
161		0							
162									
163									
164	1					1	0		
165		1	0	10000	01111				
166	0	1				0		01111	01110
167		1							
168		1							
169		1					1		
170		1	1						
171		1							
172		0							
173									
174									
175	1					1	0		
176		1	0	10001	10000				
177	0	1				0		10000	01111
178		1							
179		1							
180		1					1		
181		1	1						
182		1							
183		0							
184									
185									
186	1					1	0		
187		1	0	10010	10001				
188	0	1				0		10001	10000
189		1							
190		1							
191		1					1		
192		1	1						
193		1							
194		0							
195									
196									
197	1					1	0		
198		1	0	10011	10010				
199	0	1				0		10010	10001
200		1							

	Control signals of M3					Control signals of A1			
count	**Lstr _M3**	**Enable _M3**	**Ostr _M3**	**Sel _M3**	**Desel _M3**	**Lstr _A1**	**Ostr _A1**	**Sel _A1**	**Desel _A1**
201		1							
202		1					1		
203		1	1						
204		1							
205		0							
206									
207									
208	1					1	0		
209		1	0	10100	10011				
210	0	1				0		10011	10010
211		1							
212		1							
213		1					1		
214		1	1						
215		1							
216		0							
217									
218	1					1	0		
219		1	0	10101	10100				
220	0	1				0		10100	10011
221		1							
222		1							
223		1					1		
224		1	1						
225		1							
226		0							
227									
228									
229	1					1	0		
230		1	0	10110	10101				
231	0	1				0		10101	10100
232		1							
233		1							
234		1					1		
235		1	1						
236		1							
237		0							
238									
239									
240	1					1	0		

	Control signals of M3					Control signals of A1			
count	**Lstr _M3**	**Enable _M3**	**Ostr _M3**	**Sel _M3**	**Desel _M3**	**Lstr _A1**	**Ostr _A1**	**Sel _A1**	**Desel _A1**
241		1	0	10111	10110				
242	0	1				0		10110	10101
243		1							
244		1							
245		1					1		
246		1	1						
247		1							
248		0							
249									
250									
251						1	0		
252									
253						0		10111	10110
254									
255									
256							1		
257									
258									
259									
260									
261									
262						1	0		
263									
264						0		11000	10111
265									
266									
267							1		
268									
269									
270									
271									
272									
273	1								
274		1	0	11000	10111				
275	0	1							
276		1							
277		1							
278		1							
279		1	1						
280		1							

	Control signals of M3					Control signals of A1			
count	**Lstr _M3**	**Enable _M3**	**Ostr _M3**	**Sel _M3**	**Desel _M3**	**Lstr _A1**	**Ostr _A1**	**Sel _A1**	**Desel _A1**
281		0							
282									
283									
284						1	0		
285									11000
286						0			
287									
288									
289							1		
290									
291									
292									
293									
294									
295									
296									
297									
298									
299									
300									
301									
302									
303									
304									
305									
306									
307									
308									
309									
310									
311									
312									
313									
314									
315									
316									
317									
318									
319									
320									

	Control signals of M3					Control signals of A1			
count	**Lstr _M3**	**Enable _M3**	**Ostr _M3**	**Sel _M3**	**Desel _M3**	**Lstr _A1**	**Ostr _A1**	**Sel _A1**	**Desel _A1**
321									
322									
323									
324									
325									
326									
327									
328	1								
329		1	0		11000				
330	0	1							
331		1							
332		1							
333		1							
334		1	1						
335		1							
336		0							

	Control signals of A2				Control signals of A3			
count	**Lstr _A2**	**Ostr _A2**	**Sel _A2**	**Desel _A2**	**Lstr _A3**	**Ostr _A3**	**Sel _A3**	**Desel _A3**
0	0	0	00000	00000	0	0	0000	0000
1								
2								
3								
4								
5								
6								
7								
8								
9								
10								
11								
12								
13								
14								
15								
16								
17								
18								
19								
20								
21	1							
22								
23	0		00001					

(*Continued*)

	Control signals of A2				Control signals of A3			
count	**Lstr _A2**	**Ostr _A2**	**Sel _A2**	**Desel _A2**	**Lstr _A3**	**Ostr _A3**	**Sel _A3**	**Desel _A3**
24								
25								
26		1						
27								
28								
29								
30								
31								
32	1	0						
33								
34	0		00010	00001				
35								
36								
37		1						
38								
39								
40								

	Control signals of A2				Control signals of A3			
count	**Lstr _A2**	**Ostr _A2**	**Sel _A2**	**Desel _A2**	**Lstr _A3**	**Ostr _A3**	**Sel _A3**	**Desel _A3**
41								
42								
43	1	0			1			
44								
45	0		00011	00010	0		0001	
46								
47								
48		1				1		
49								
50								
51								
52								
53								
54	1	0			1	0		
55								
56	0		00100	00011	0		0010	0001
57								
58								
59		1				1		
60								
61								

(*Continues*)

(*Continued*)

	Control signals of A2				**Control signals of A3**			
count	**Lstr _A2**	**Ostr _A2**	**Sel _A2**	**Desel _A2**	**Lstr _A3**	**Ostr _A3**	**Sel _A3**	**Desel _A3**
62								
63								
64								
65	1	0						
66								
67	0		00101	00100				
68								
69								
70		1						
71								
72								
73								
74								
75								
76	1	0						
77								
78	0		00110	00101				
79								
80								

	Control signals of A2				**Control signals of A3**			
count	**Lstr _A2**	**Ostr _A2**	**Sel _A2**	**Desel _A2**	**Lstr _A3**	**Ostr _A3**	**Sel _A3**	**Desel _A3**
81		1						
82								
83								
84								
85								
86								
87	1	0			1	0		
88								
89	0		00111	00110	0		0011	0010
90								
91								
92		1				1		
93								
94								
95								
96								
97								
98	1	0			1	0		
99								

(*Continued*)

	Control signals of A2				Control signals of A3			
count	Lstr _A2	Ostr _A2	Sel _A2	Desel _A2	Lstr _A3	Ostr _A3	Sel _A3	Desel _A3
100	0		01000	00111	0		0100	0011
101								
102								
103		1				1		
104								
105								
106								
107								
108								
109	1	0			1	0		
110								
111	0		01001	01000	0		0101	0100
112								
113								
114		1				1		
115								
116								
117								
118								
119								
120	1	0						

	Control signals of A2				Control signals of A3			
count	Lstr _A2	Ostr _A2	Sel _A2	Desel _A2	Lstr _A3	Ostr _A3	Sel _A3	Desel _A3
121								
122	0		01010	01001				
123								
124								
125		1						
126								
127								
128								
129								
130								
131	1	0			1	0		
132								
133	0		01011	01010	0		0110	0101
134								
135								
136		1				1		
137								

(Continues)

(*Continued*)

	Control signals of A2				Control signals of A3			
count	**Lstr _A2**	**Ostr _A2**	**Sel _A2**	**Desel _A2**	**Lstr _A3**	**Ostr _A3**	**Sel _A3**	**Desel _A3**
138								
139								
140								
141								
142	1	0			1	0		
143								
144	0		01100	01011	0		0111	0110
145								
146								
147		1				1		
148								
149								
150								
151								
152								
153	1	0						
154								
155	0		01101	01100				
156								
157								
158		1						
159								
160								

	Control signals of A2				Control signals of A3			
count	**Lstr _A2**	**Ostr _A2**	**Sel _A2**	**Desel _A2**	**Lstr _A3**	**Ostr _A3**	**Sel _A3**	**Desel _A3**
161								
162								
163								
164	1	0						
165								
166	0		01110	01101				
167								
168								
169		1						
170								
171								
172								
173								
174								
175	1	0			1	0		

(*Continued*)

count	Control signals of A2				Control signals of A3			
	Lstr _A2	**Ostr _A2**	**Sel _A2**	**Desel _A2**	**Lstr _A3**	**Ostr _A3**	**Sel _A3**	**Desel _A3**
176								
177	0		01111	01110	0		1000	0111
178								
179								
180		1				1		
181								
182								
183								
184								
185								
186	1	0			1	0		
187								
188	0		10000	01111	0		1001	1000
189								
190								
191		1				1		
192								
193								
194								
195								
196								
197	1	0			1	0		
198								
199	0		10001	10000	0		1010	1001
200								

count	Control signals of A2				Control signals of A3			
	Lstr _A2	**Ostr _A2**	**Sel _A2**	**Desel _A2**	**Lstr _A3**	**Ostr _A3**	**Sel _A3**	**Desel _A3**
201								
202		1				1		
203								
204								
205								
206								
207								
208	1	0						
209								
210	0		10010	10001				
211								
212								
213		1						

(*Continues*)

(*Continued*)

	Control signals of A2				Control signals of A3			
count	**Lstr _A2**	**Ostr _A2**	**Sel _A2**	**Desel _A2**	**Lstr _A3**	**Ostr _A3**	**Sel _A3**	**Desel _A3**
214								
215								
216								
217								
218	1	0			1	0		
219								
220	0		10011	10010	0		1011	1010
221								
222								
223		1				1		
224								
225								
226								
227								
228								
229	1	0			1	0		
230								
231	0		10100	10011	0		1100	1011
232								
233								
234		1				1		
235								
236								
237								
238								
239								
240	1	0						

	Control signals of A2				Control signals of A3			
count	**Lstr _A2**	**Ostr _A2**	**Sel _A2**	**Desel _A2**	**Lstr _A3**	**Ostr _A3**	**Sel _A3**	**Desel _A3**
241								
242	0		10101	10100				
243								
244								
245		1						
246								
247								
248								
249								
250								
251	1	0						

(*Continued*)

	Control signals of A2				Control signals of A3			
count	**Lstr _A2**	**Ostr _A2**	**Sel _A2**	**Desel _A2**	**Lstr _A3**	**Ostr _A3**	**Sel _A3**	**Desel _A3**
252								
253	0		10110	10101				
254								
255								
256		1						
257								
258								
259								
260								
261								
262	1	0			1	0		
263								
264	0		10111	10110	0		1101	1100
265								
266								
267		1				1		
268								
269								
270								
271								
272								
273								
274								
275								
276								
277								
278								
279								
280								

	Control signals of A2				Control signals of A3			
count	**Lstr _A2**	**Ostr _A2**	**Sel _A2**	**Desel _A2**	**Lstr _A3**	**Ostr _A3**	**Sel _A3**	**Desel _A3**
281								
282								
283								
284								
285								
286								
287								
288								
289								

(*Continues*)

(Continued)

	Control signals of A2				Control signals of A3			
count	**Lstr _A2**	**Ostr _A2**	**Sel _A2**	**Desel _A2**	**Lstr _A3**	**Ostr _A3**	**Sel _A3**	**Desel _A3**
290								
291								
292								
293								
294								
295	1	0						
296								
297	0			10111				
298								
299								
300		1						
301								
302								
303								
304								
305								
306					1	0		
307								
308					0		1110	1101
309								
310								
311						1		
312								
313								
314								
315								
316								
317					1	0		
318								
319					0			1110
320								
321								
322						1		

Control signals of Registers Reg_1 to Reg_40:

Registers	**Enable at count #**
Reg_1	10
Reg_2	21
Reg_3	32
Reg_4	32
Reg_5	43
Reg_6	54
Reg_7	65
Reg_8	65
Reg_9	65
1 Reg_0	76
Reg_11	76
Reg_12	87
Reg_13	98
Reg_14	98
Reg_15	109
Reg_16	109
Reg_17	120
Reg_18	120
Reg_19	120
Reg_20	131
Reg_21	142
Reg_22	153
Reg_23	153
Reg_24	153
Reg_25	164
Reg_26	164
Reg_27	175
Reg_28	186
Reg_29	186
Reg_30	197
Reg_31	197
Reg_32	208
Reg_33	208
Reg_34	208
Reg_35	218
Reg_36	229
Reg_37	240
Reg_38	240
Reg_39	240
Reg_40	251

Part B: Transfer functions and control data flow graph of DSP cores

1. Four-point DCT

Transfer functions of four-point DCT:

$$X[0] = c_2 * x[0] + c_2 * x[1] + c_2 * x[2] + c_2 * x[3]$$
$$X[1] = c_1 * x[0] + c_3 * x[1] + (-c_3) * x[2] + (-c_1) * x[3]$$
$$X[2] = c_2 * x[0] + (-c_2) * x[1] + (-c_2) * x[2] + c_2 * x[3]$$
$$X[3] = c_3 * x[0] + (-c_1) * x[1] + c_1 * x[2] + (-c_3) * x[3]$$

The generic equation of four-point DCT to compute the first sample:

$$X[0] = k_1 * x[0] + k_2 * x[1] + k_3 * x[2] + k_4 * x[3]$$

DFG of four-point DCT based on generic representation to compute the first sample:

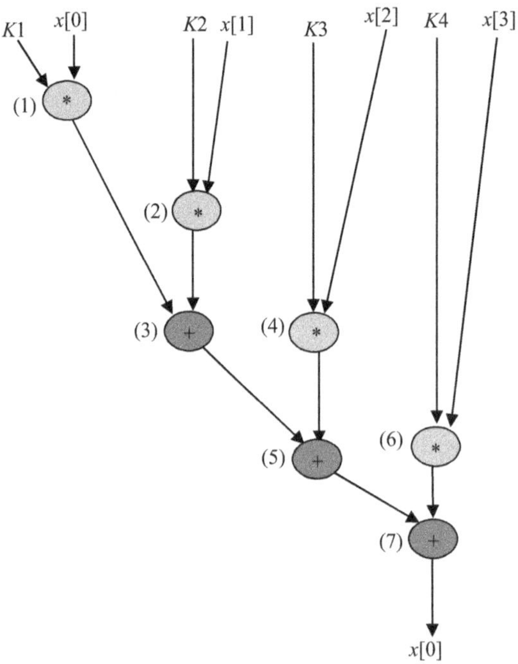

2. Four-point DIT-FFT

Transfer functions of four-point DIT-FFT:

$$X'[0] = H'[0] * W_4^0 + G'[0]$$
$$X'[1] = H'[1] * W_4^1 + G'[1]$$
$$X'[2] = H'[0] * W_4^2 + G'[0]$$
$$X'[3] = H'[1] * W_4^3 + G'[1]$$

DFG of four-point DIT-FFT:

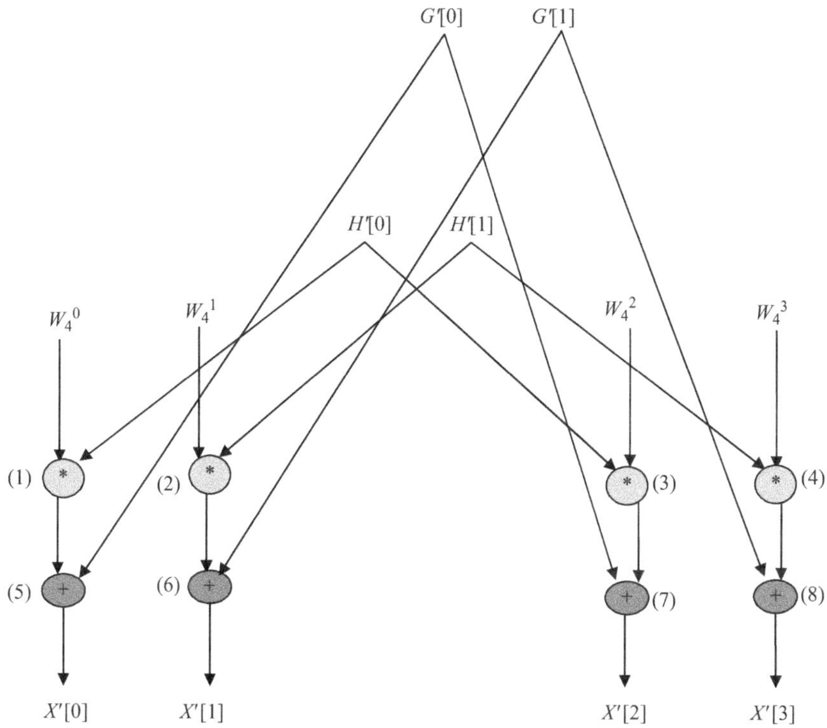

3. Four-point IDCT

Transfer functions of four-point IDCT:

$$x'[0] = c_2 * X'[0] + c_1 * X'[1] + c_2 * X'[2] + c_3 * X'[3]$$
$$x'[1] = c_2 * X'[0] + c_3 * X'[1] + (-c_2) * X'[2] + (-c_1) * X'[3]$$
$$x'[2] = c_2 * X'[0] + (-c_3) * X'[1] + (-c_2) * X'[2] + c_1 * X'[3]$$
$$x'[3] = c_2 * X'[0] + (-c_1) * X'[1] + c_2 * X'[2] + (-c_3) * X'[3]$$

The generic equation of four-point IDCT to compute the first sample:

$$x'[0] = k'1 * X'[0] + k'2 * X'[1] + k'3 * X'[2] + k'4 * X'[3]$$

DFG of four-point IDCT based on generic representation to compute the first sample:

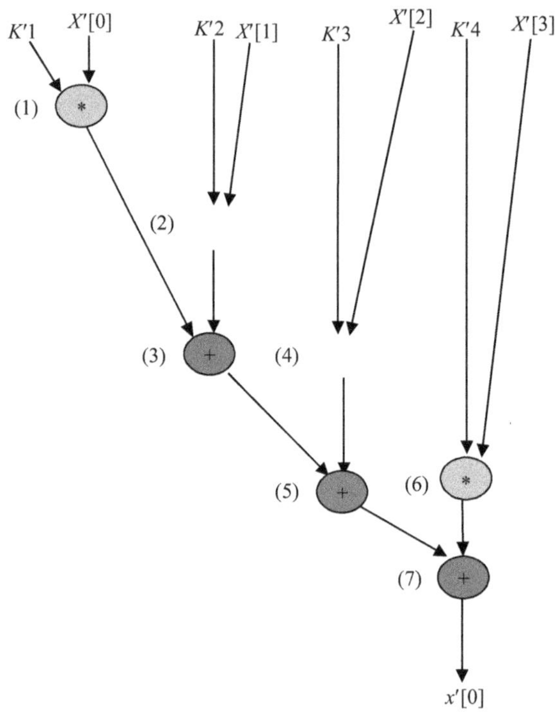

4. Eight-point DCT

   Functions of eight-point DCT

$$X[0] = c_4 * x[0] + c_4 * x[1] + c_4 * x[2] + c_4 * x[3] + c_4 * x[4] + c_4 * x[5] + c_4 * x[6] + c_4 * x[7]$$

$$X[1] = c_1 * x[0] + c_3 * x[1] + c_5 * x[2] + c_7 * x[3] + (-c_7) * x[4] + (-c_5) * x[5] + (-c_3) * x[6] + (-c_1) * x[7]$$

$$X[2] = c_2 * x[0] + c_6 * x[1] + (-c_6) * x[2] + (-c_2) * x[3] + (-c_2) * x[4] + (-c_6) * x[5] + c_6 * x[6] + c_2 * x[7]$$

$$X[3] = c_3 * x[0] + (-c_7) * x[1] + (-c_1) * x[2] + (-c_5) * x[3] + c_5 * x[4] + c_1 * x[5] + c_7 * x[6] + (-c_3) * x[7]$$

$$X[4] = c_4 * x[0] + (-c_4) * x[1] + (-c_4) * x[2] + c_4 * x[3] + c_4 * x[4] + (-c_4) * x[5] + (-c_4) * x[6] + c_4 * x[7]$$

$$X[5] = c_5 * x[0] + (-c_1) * x[1] + c_7 * x[2] + c_3 * x[3] + (-c_3) * x[4] + (-c_7) * x[5] + c_1 * x[6] + (-c_5) * x[7]$$

$$X[6] = c_6 * x[0] + (-c_2) * x[1] + c_2 * x[2] + (-c_6) * x[3] + (-c_6) * x[4] + c_2 * x[5] + (-c_2) * x[6] + c_6 * x[7]$$

$$X[7] = c_7 * x[0] + (-c_5) * x[1] + c_3 * x[2] + (-c_1) * x[3] + c_1 * x[4] + (-c_3) * x[5] + c_5 * x[6] + (-c_7) * x[7]$$

Generic representation of eight-point DCT to compute first sample:

$$X[0] = k_1 * x[0] + k_2 * x[1] + k_3 * x[2] + k_4 * x[3] + k_5 * x[4] + k_6 * x[5] + k_7 * x[6] + k_8 * x[7]$$

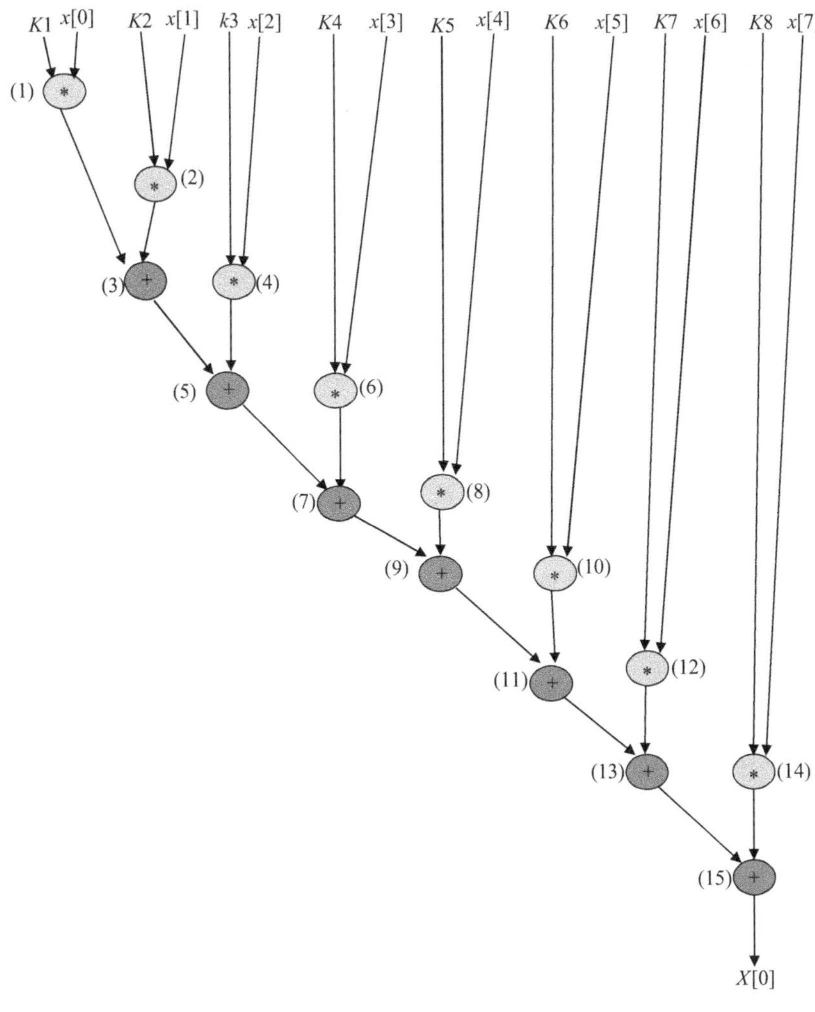

5. Eight-point DIT-FFT

Functions of eight-point DIT-FFT:

$$X[0] = H[0] * W_8^0 + G[0]$$
$$X[1] = H[1] * W_8^1 + G[1]$$
$$X[2] = H[2] * W_8^2 + G[2]$$
$$X[3] = H[3] * W_8^3 + G[3]$$
$$X[4] = H[0] * W_8^4 + G[0]$$
$$X[5] = H[1] * W_8^5 + G[1]$$
$$X[6] = H[2] * W_8^6 + G[2]$$
$$X[7] = H[3] * W_8^7 + G[3]$$

DFG of eight-point DIT-FFT:

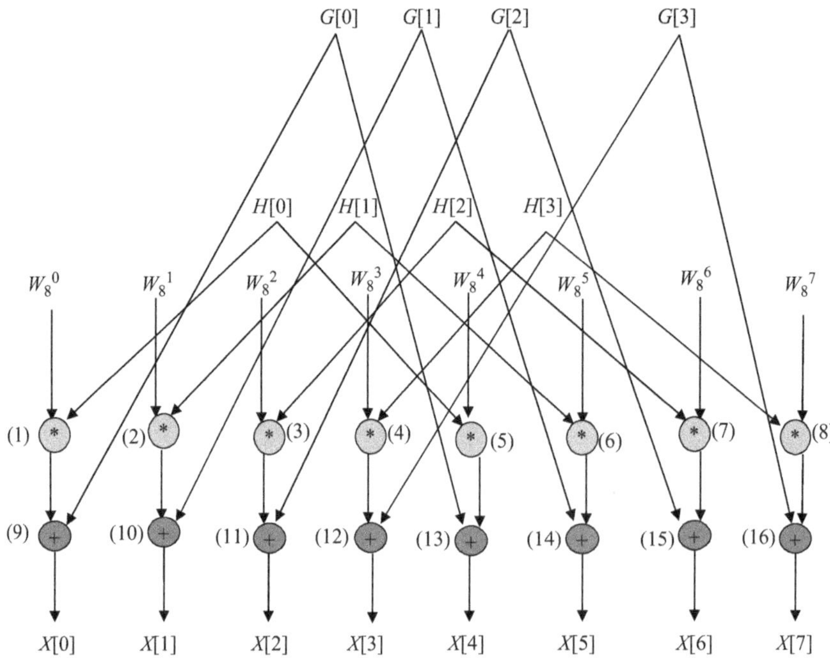

6. Eight-point IDCT

Functions of eight-point IDCT:

$$x'[0] = c_4 * X'[0] + c_1 * X'[1] + c_2 * X'[2] + c_3 * X'[3] + c_4 * X'[4] + c_5 * X'[5] + c_6 * X'[6] + c_7 * X'[7]$$
$$x'[1] = c_4 * X'[0] + c_3 * X'[1] + c_6 * X'[2] + (-c_7) * X'[3] + (-c_4) * X'[4] + (-c_1) * X'[5] + (-c_2) * X'[6] + (-c_5) * X'[7]$$
$$x'[2] = c_4 * X'[0] + c_5 * X'[1] + (-c_6) * X'[2] + (-c_1) * X'[3] + (-c_4) * X'[4] + c_7 * X'[5] + c_2 * X'[6] + c_3 * X'[7]$$

$$x'[3] = c_4 * X'[0] + c_7 * X'[1] + (-c_2) * X'[2] + (-c_5) * X'[3] \\ + c_4 * X'[4] + c_3 * X'[5] + (-c_6) * X'[6] + (-c_1) * X'[7]$$

$$x'[4] = c_4 * X'[0] + (-c_7) * X'[1] + (-c_2) * X'[2] + c_5 * X'[3] \\ + c_4 * X'[4] + (-c_3) * X'[5] + (-c_6) * X'[6] + c_1 * X'[7]$$

$$x'[5] = c_4 * X'[0] + (-c_5) * X'[1] + (-c_6) * X'[2] + c_1 * X'[3] \\ + (-c_4) * X'[4] + (-c_7) * X'[5] + c_2 * X'[6] + (-c_3) * X'[7]$$

$$x'[6] = c_4 * X'[0] + (-c_3) * X'[1] + c_6 * X'[2] + c_7 * X'[3] \\ + (-c_4) * X'[4] + c_1 * X'[5] + (-c_2) * X'[6] + c_5 * X'[7]$$

$$x'[7] = c_4 * X'[0] + (-c_1) * X'[1] + c_2 * X'[2] + (-c_3) * X'[3] \\ + c_4 * X'[4] + (-c_5) * X'[5] + c_6 * X'[6] + (-c_7) * X'[7]$$

Generic representation of eight-point IDCT to compute the first sample:

$$x'[0] = k'1 * X'[0] + k'2 * X'[1] + k'3 * X'[2] + k'4 * X'[3] \\ + k'5 * X'[4] + k'6 * X'[5] + k'7 * X'[6] + k'8 * X'[7]$$

DFG of generic representation of eight-point IDCT to compute the first sample:

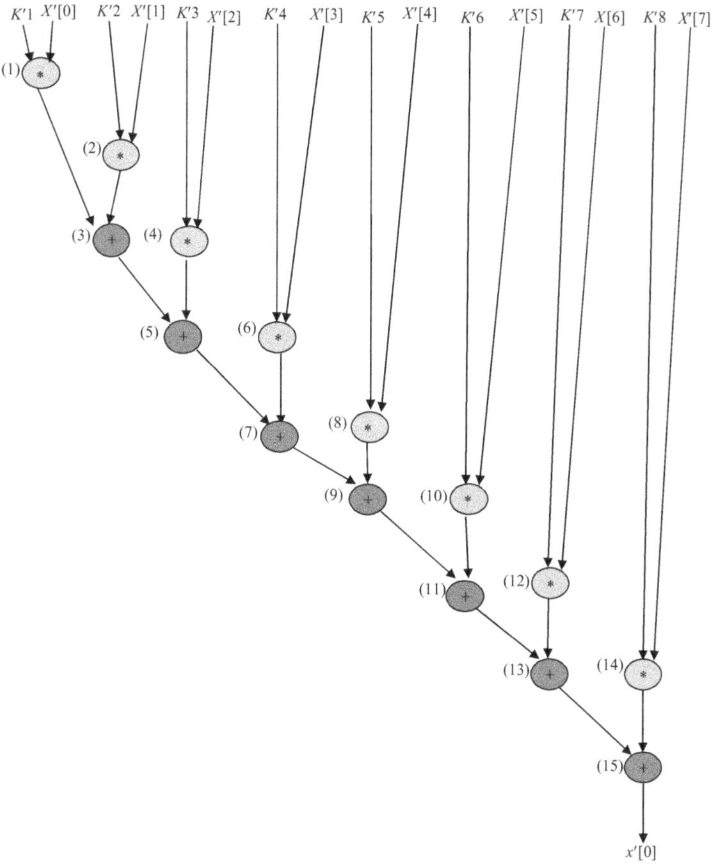

7. Third-order FIR filter

FIR transfer function:

$$Y'[n] = h_0 * X'[n] + h_1 * X'[n-1] + h_2 * X'[n-2] + h_3 * X'[n-3]$$

DFG of FIR:

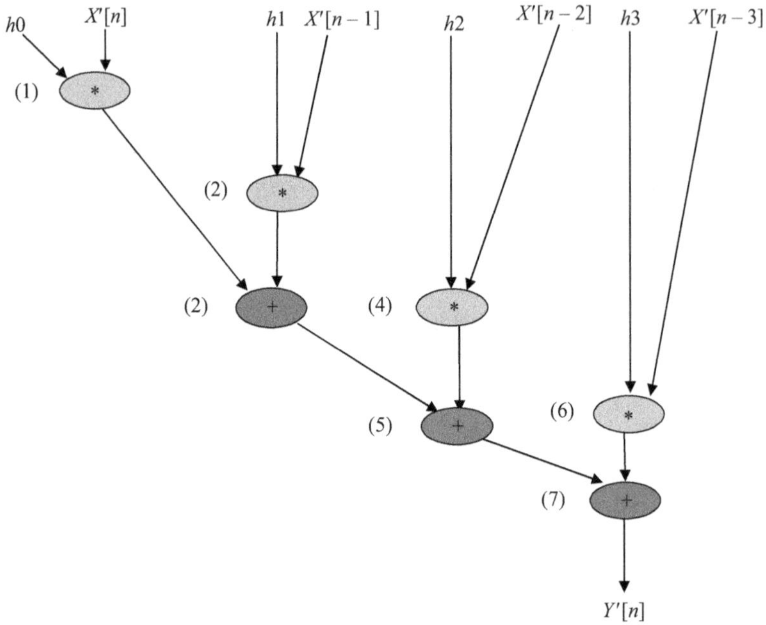

8. Third-order IIR filter

IIR transfer function:

$$Y[n] = b_0 * X[n] + b_1 * X[n-1] + b_2 * X[n-2] + b_3 * X[n-3] + (-a_1) * Y[n-1] + (-a_2) * Y[n-2] + (-a_3) * Y[n-3]$$

DFG of IIR:

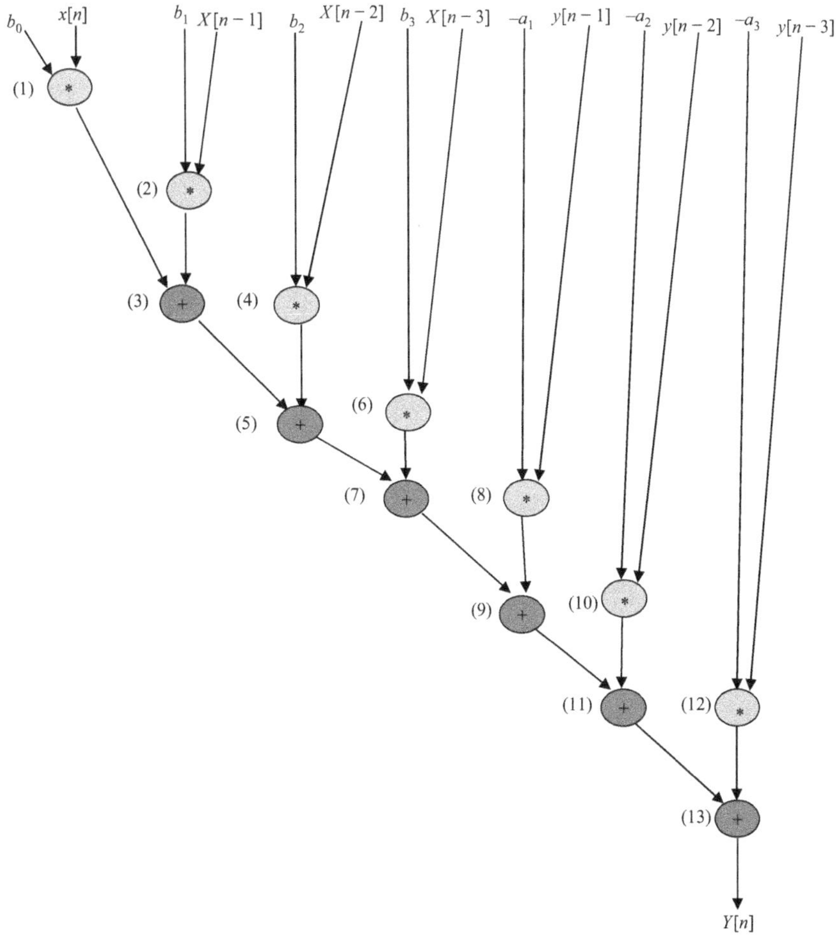

b_0
$x[n]$
b_1
$X[n-1]$
b_2
$X[n-2]$
b_3
$X[n-3]$
$-a_1$
$y[n-1]$
$-a_2$
$y[n-2]$
$-a_3$
$y[n-3]$
(1)
(2)
(3)
(4)
(5)
(6)
(7)
(8)
(9)
(10)
(11)
(12)
(13)
$Y[n]$

Index

active attack 60
additive technique 347
Address Space Layout Randomization (ASLR) 63
advanced encryption standard (AES) 12, 261, 317, 383–4
  AES algorithm 318
    byte substitution (S-box) layer 319–20
    key addition layer 323
    key scheduler for 128-bit AES 323–6
    MixColumn diffusion layer 320–2
    row-shifting diffusion layer 320
  case study 330
  digital watermarking 326
    process of embedding watermark in AES 329
    signature detection 329
    watermark encoding 327–8
analog/mixed-signal system on a chip (AMS-SoC) 2
AND/OR gates 232
application-specific integrated circuits (ASICs) 99
arbiter PUF 11, 415–16
architectural synthesis 13–15, 230–1, 233
As Late As Possible (ALAP) algorithm 204–5
As Soon As Possible (ASAP) algorithm 204–5
attack scenarios of functional obfuscation 256–9
audio watermarking 346
autoregressive filter (ARF) benchmark 248

bacterial foraging optimization algorithm (BFOA) control parameters 89
basal dose 60
Bitcoin-NG 69
blockchain 64
  application 66–7
  issues 68–9
  overview 65
  as security framework 67–8
Bluetooth 28
bolus dose 60
Boolean satisfiability (SAT) 232
buyer fingerprint 172, 176
byte substitution layer (S-box) 318–20

carbon-based PUFs 426
  CNT-based PUF 427
  graphene-based PUF 428
carbon nanotube (CNT)-based PUF 427
Challenge Response Pair (CRP) 406, 411–12
channel threats 44
code injection attack 63
code obfuscation techniques 231
code reuse attacks 63
Code-Word 232, 260
collude attack 389
colored interval graph (CIG) 178, 180
combinational-triggering-based hardware Trojan 81
comparator 344
compiler transformation-driven structural obfuscation 233
  evaluation models 235–6
  formulation 235
  multistage high-level transformation techniques 234, 236–42
compression-decompression (CODEC) 79, 288

computational forensic engineering (CFE) 9, 13, 175, 201–4
analysis on case studies 214–22
bus width support 202, 205–6
customized CFE approach 203
vs. hardware watermarking 13
resource type 202, 205
scheduling algorithm 202, 204–5
vs. watermarking 200
computer-aided design (CAD) software 176
concurrent error detection (CED) techniques 88
ConnX D2 DSP IP core 227
consumer electronics (CE) devices 79,123
and security perspectives 1–3
control data flow graph (CDFG) 212
digital signal processing applications 89
controlled PUFs 412
copy attack 389
Copy Protection Technical Working Group (CPTWG) 341
copyright, defined 12
copyright protection 38, 347–8
comparator 344
decoder 344
encoder 343–4
general framework for 343–4
of multimedia 339–42
CPA attack 384–5
Crypta tag architecture 50
cryptographic attacks 389
cryptographic primitives 45, 417
cryptography 317, 340, 404
customized CFE 203
important characteristics of 214
cypher text 318

data encoding 293, 306
data flow graph (DFG) 89, 177, 183
data pipelining 207–9
deadlock attack 389
deadlock in IP resolution 214
decoder 344
denial of service (DoS) payload-based hardware Trojan 84
denial of service attacks 56, 62
de-quantization 289
design area evaluation model 98, 235
design cost evaluation function 235–6
design delay evaluation model 235
design objective 209
design space exploration (DSE) module 6
desired characteristics of hardware-based watermarking systems 355
destination watermarking technique 347
device-based PUFs 411
non-silicon PUFs 411
silicon-based PUFs 411
differential pulse code modulation (DPCM) technique 364
diffusion 318
digital filters 6
digital fingerprint 11
digital rights management 338–9
digital signal processing (DSP) core 6
computational forensic engineering (CFE) vs. hardware watermarking 13
DSP benchmark suite 214
DSP kernels 227
eight-point DCT 486–7
eight-point DIT-FFT 488
eight-point IDCT 488–9
four-point DCT 484
four-point DIT-FFT 485
four-point IDCT 485–6
hardware metering vs. hardware watermarking 12–13
hardware obfuscation vs. hardware watermarking 13
HLS (architectural synthesis) as design backbone 13–15
intellectual property (IP) cores 171, 230
low-cost structural obfuscation for 242
movement of particle 242–3
PSO 242
resilience of, against RE 231
terminating condition of PSO 243
security and protection methodologies available for IP core/hardware 7
computational forensic engineering (CFE) for IP cores 9
hardware-based encryption 12
hardware metering 9–10
IP core/hardware fingerprinting 10

IP core/hardware obfuscation and encryption 10–11
IP core/hardware watermarking 8–9
patent and copyright 12
physically unclonable functions 11
Trojan security/detection of an IP core 11
third-order FIR filter 490
third-order IIR filter 490–1
transfer functions and control data flow graph of 484–91
digital watermarks 339–41
applications of 347
anti-counterfeiting 349
authentication and integrity verification 348
broadcast monitoring 348–9
content labeling 349
copyright protection 347–8
fingerprinting 348
medical signals authentication 350
misappropriation detection 349
ownership assertion 348
UAV safety 349–50
usage control 348
characteristics specific to 354
desired characteristics of hardware-based watermarking systems 355
invisible-fragile watermark 355
invisible-robust watermark 354
video watermark 355
visible watermark 354
watermarking for broadcasting applications 355
desired characteristics of 350
bit rate 351
blindness 352–3
cascadability 351
cost and complexity 353
energy consumption 353–4
integrability 354
modifiability 351
multiplicity 351
orthogonality 351
perceptibility 350
pixel alteration and human intervention 352
real-time operation 353
reliability 352
robustness 350–1
scalability 351
security 353
tamper resistance 351
unambiguity and universality 352
hardware-based approaches for 358–9
image watermarking hardware systems 359–72
secure better portable graphics (SBPG) 378–9
trust cam 379–81
video watermarking hardware systems 372–8
human visual system, properties of 357
brightness sensitivity 357
frequency sensitivity 357
JND and minimally noticeable distortion 357
texture sensitivity 357
spread spectrum communications 358
technical challenges for 355
nonstationarity 356
periodicity 356
power spectral density 356
properties of color signals 356–7
properties of visual signals 356
types of 344
based on embedding techniques 347
based on human perception 346
based on multimedia objects 345–6
from applications point of view 346–7
hardware-based watermarking systems 347
spatial versus frequency domain watermarking 345
watermark signal 357–8
diode-based PUF 424–6
discrete cosine transformation (DCT) algorithms 6, 79
DCT-based JPEG image compression process 288, 290
data encoding 293
performing DCT on input image 291–2
performing quantization 292
processing of the input image 290–1

DCT-based JPEG image decompression process 293
data decoding 293
performing IDCT on decompressed image 293
performing inverse quantization 293
reconstructing the decompressed image 293
2D-DCT coefficient matrix 290–1
discrete wavelet transformation (DWT) benchmark 154
distributed DoS (DDoS) attack 56, 62
DMR-based Trojan security system 213
DPA attack 385
dual modular redundant (DMR) schedule 81
dual modular redundant (DMR) system 96
dynamic attack 63
dynamic watermarking in smart car or UAV 381–2

eavesdropping 44, 48
E-Ink display 49
electronic patient record (EPR) exchange 382
embedded zero wavelet (EZW) 359
encoder 343–4
encryption hardware 383–4
end-to-end JPEG CODEC through designed hardware/IP core 302–8
EPIC 232
Express Benchmark Suite 214

fake–original attack 389
Faraday cages for security 45
fast Fourier transform (FFT) 6, 79
field programmable gate array (FPGA) 124, 146, 175, 347, 375, 433
financial transactions 64, 66
fingerprint encoding rules 178–9
fingerprinting constraints embedding process, demonstration of 183–6
finite impulse response (FIR) filters 79
finite state machine (FSM) 87, 231–2, 260
firmware 35, 62
attacks 62–3
solutions 63–4
forensic engineering (FE) 199–200
entity clustering 201–2
feature and statistical data collection 201
feature extraction 201
validation 202
functionality changeable payload-based hardware Trojan 83
functionally obfuscated finite impulse response (FO FIR)design 269, 271
functional obfuscation of DSP cores 200, 255
analysis of case studies 275
comparative analysis 279–82
overhead analysis 277–9
security analysis 275–7
attack scenarios 256–9
design of 262
formulation 262
low-cost obfuscation method 262–6
optimization engine for 273
particle encoding 273
particle fitness 273–4
updating particle 274
security of 266
countermeasures against attacks 267–73
keyspace 266–7
security analysis 267
selected functional obfuscation approaches 260–2
threat model 259–60

geometric attacks 389
global best particle 110
global system for mobile communication (GSM) modem 49
GPS spoofing 51–2, 56
graphene-based PUF 428

Haar wavelet transformation (HWT) algorithms 79
Haar wavelet transformation (HWT) CODEC data 447
DFG of HWT-based image compression 451
DFG of HWT-based image decompression 454–5
Image reconstruction 451–2

Hamming distance (HD) 232, 261, 409
hardware-assisted media protection 15–16
hardware-assisted security 4
  and IP core protection 3–4
  and protection 35–7
hardware-based encryption 12
hardware description language (HDL) 15, 127, 176
hardware hardening technique 256
hardware metering 9–10
  vs. hardware watermarking 12–13
hardware obfuscation 199–200
  vs. hardware watermarking 13
hardware random number generation (HRNG) 435–6
hardware Trojan 1, 5, 33–4, 80
  in 3PIP module 86
    example of a hardware Trojan 86–7
    Trojan detectability in a 3PIP module at RTL/lower levels 87–8
  types of 81
    combinational-triggering-based hardware Trojan 81
    functionality changeable payload-based hardware Trojan 83
    leaking confidential information and launching denial of service (DoS) payload-based hardware Trojan 84
    performance degradable payload-based hardware Trojan 83–4
    sequential-triggering-based hardware Trojan 82–3
hardware versus media protection 342–3
hardware watermarking
  computational forensic engineering (CFE) vs. 13
  hardware metering vs., 12–13
  hardware obfuscation vs. 13
hash cache 41
Health Information Privacy and Security Act (HIPAA) 59
Health Information Technology for Economic and Clinical Health (HITECH) Act 59
high-efficiency video coding (HEVC) 378–9
high-level synthesis (HLS) 13–15, 85, 171, 176
  based on license 230
  HLS tools 202, 204, 209, 215
high-level transformations 233–4
hill-climbing search-based attack 232

IBM visible watermarking techniques 341
identity management 33
ILBs 267–8, 272
  and AES removal attack 272–3
image compression and decompression (CODEC) 288
image watermarking 346, 359–72
infinite impulse response (IIR) filters 79
instrumentation 24
insulin delivery system 60–1
integrity attacks 56
Intel Cyclone II 302
intellectual property (IP), 38–9
intellectual property (IP) cores 79
  feature extraction algorithms
    analysis on case studies 214–22
    feature extraction rules 204–13
    important characteristics of customized CFE 214
  IP core-locking blocks 263–4
    insertion 265–6
    multi pair-wise security 264–5
    nonmutable convergent key-gates 265
    prohibiting key-gate isolation 265
    protection against run of key-gates 265
  protection, fundamentals of 174
    non-symmetric IP core protection techniques 174–5
    symmetric IP core protection techniques 175
  validation 213–14
intellectual property (IP) cores/hardware 4
  need for security and protection of hardware/IP cores 5–6
  traditional forms of IP protection 6
  utility of IP cores in CE devices 4–5
intellectual property (IP) metering 200
intellectual property (IP) piracy 256
  and Trojan attack 259, 268

intellectual property (IP) protection 337–8
copyright protection of multimedia 339–42
digital rights management 338–9
hardware versus media protection 342–3
intellectual property rights (IPR) 2, 32
intelligence 24
Intel Quartus Cyclone II FPGA 312
interconnection 24
interference attacks 389
Internet of Battlefield Things 29
Internet of Battle Things 29
internet of everything (IoE) 29
internet of health things (IoHT) 58
internet of medical things (IoMT) 58
internet of things (IoT) 23
applications 28–30
architecture 25–7
challenges 30–1
definition of 24
driving technology 27–8
intrusion detection systems (IDS) 57
inverse discrete cosine transformation (IDCT) 289, 293
2D-IDCT 298, 302
inverse quantization (IQuan) module 376
inverse zigzag (IZZ) module 376
inversion attack 389
invertible watermarking 346
invisible-fragile spatial domain chip 364–5
invisible-fragile watermark 355
invisible-robust DCT domain chip 362–4
invisible-robust fractal domain module 375–6
invisible-robust-fragile spatial domain chip 365–6
invisible-robust spatial domain chip 374–5
invisible-robust spatial domain module 367, 372–4
invisible-robust watermark 354
invisible-robust wavelet domain chip 376
invisible-robust wavelet domain module 366–7
invisible-robust wavelet domain VLSI architecture 359–62
invisible watermarking 346
isolated key-gates-based key-sensitization attack 258

Joint Picture Export Group (JPEG) CODEC IP core 287
analysis on case studies 308–12
controller design details for 459–83
DCT-based JPEG image compression process 290–3
DCT-based JPEG image decompression process 293
design process of structurally obfuscated JPEG IP core 293
constructing non-obfuscated DFG for JPEG compression 294–6
optimization framework 294
problem formulation 294
structurally obfuscated JPEG compression IP core, generating 296–8
structurally obfuscated JPEG decompression IP core, generating 298–9
threat model 293–4
implementation of 299
designing obfuscated JPEG compression IP core 299–302
designing obfuscated JPEG decompression IP core 302
end-to-end JPEG CODEC through designed hardware/IP core 302–8
JPEG compression and decompression 289–93
SoO of 310
JPEG encoder with watermarking modules 342
just noticeable distortion (JND) 350

key-sensitization 232, 234, 258, 261–2, 264–5, 267–8
keyspace 266–7

latch physical unclonable functions 11
linear feedback shift register (LFSR) 359, 416
list algorithm algorithm 204–5
logic encryption: *see* logic obfuscation techniques
logic obfuscation techniques 231–3
logic transformation process 239

long-range wide area network (LoRaWAN) 28
Lookup Table (LUT) 232
loop-invariant code motion process 240
loop support feature 211–12
loop unrolling (LU) process 240
low-cost JPEG CODEC IP core 308
low-power chip with DCT domain-invisible-robust and visible functionalities 370, 372

macro-IP 295, 297
magnetic PUF 430–2
mean square error (MSE) 311
media and information protection and authentication 337
  copyright protection, general framework for 343
    comparator 344
    decoder 344
    encoder 343–4
  desired characteristics of watermarks 350
    bit rate 351
    blindness 352–3
    cascadability 351
    cost and complexity 353
    desired characteristics of hardware-based watermarking systems 355
    energy consumption 353–4
    integrability 354
    invisible-fragile watermark 355
    invisible-robust watermark 354
    modifiability 351
    multiplicity 351
    orthogonality 351
    perceptibility 350
    pixel alteration and human intervention 352
    real-time operation 353
    reliability 352
    robustness 350–1
    scalability 351
    security 353
    tamper resistance 351
    unambiguity and universality 352
    video watermark 355
    visible watermark 354
    watermarking for broadcasting applications 355
  digital watermarks
    applications of 347–50
    types of 344–7
  dynamic watermarking in smart car or UAV 381–2
  hardware-based approaches for watermarking 358–9
    image watermarking hardware systems 359–72
    secure better portable graphics (SBPG) 378–9
    trust cam 379–81
    video watermarking hardware systems 372–8
  IP protection 337–8
    copyright protection of multimedia 339–42
    digital rights management 338–9
    hardware versus media protection 342–3
  medical signals authentication 382–3
  side-channel information leakage attacks and countermeasures 383
    encryption hardware 383–4
    side-channel analysis attacks 384–7
    side-channel attack countermeasures 387–8
  technical challenges for watermarking 355–6
    properties of the human visual system 357
    properties of visual signals 356
    spread spectrum communications 358
    watermark signal 357–8
  watermarks and watermarking, limitations of 390–1
  watermarks and watermarking systems, attacks on 388
    cryptographic attacks 389
    geometric attacks 389
    protocol attacks 389
    removal and interference attacks 389
medical signals authentication 350, 382–3
memory security attacks 39–40

memory security solutions 40–3
memristor-based PUFs 420–4
Memristors 420–6
MESA benchmark 183
micro-IP 148, 295, 297
microprocessor-based PUF 429–30
minimally noticeable distortion (MND) 357
MixColumn 319–29
motion compensation (MC) module 376
motion estimate (ME) module 376
multicycling 207
multiplicative technique 347
multi-variable signature embedding process 180–1
mutable key-gates-based key-sensitization attack 259
Muxes 84, 232

National Institute of Standards and Technology (NIST) 317
near-field communications (NFC) security 46
  attacks 47–8
  solutions 48–50
nonobfuscated control data flow graph 238, 243
non-repudiation 37
non-silicon physical unclonable functions 411
non-symmetric IP core protection techniques 174–5
  computational forensic engineering (CFE) 175
  hardware obfuscation 174
  watermark 175
nonvolatile memory 40

obfuscated JPEG compression IP core, designing 299–302
obfuscated JPEG decompression IP core, designing 299, 302
obfuscation 10–11, 199–200
  for IP core protection
    code obfuscation techniques 231
    logic obfuscation techniques 231–2
    structural obfuscation techniques 232–3

object-centric architecture 29
one-way random functions (ORF) 261
ownership right 15, 38
ownership right, 38

particle swarm optimization (PSO) 242
  PSO-DSE process 273
  terminating condition of PSO 243
passive attack 39, 60
patent and copyright 12
peak signal-to-noise ratio (PSNR) 311
performance degradable payload-based hardware Trojan 83–4
physically obfuscated keys (POKs) 412
physical unclonable functions (PUFs) 16, 403
  carbon-based PUFs 426
    CNT-based PUF 427
    graphene-based PUF 428
  case study applications 433–8
  classification of 410
    device-based PUFs 411
    security-based PUFs 411
  diode-based PUF 424–6
  issues 438–9
  magnetic PUF 430–2
  memristor-based PUFs 420–4
  microprocessor-based PUF 429–30
  practical implementation of PUF 432–3
  principle 406
  properties/characteristics of 407
    correlation (bit aliasing) 409–10
    power consumption 410
    randomness (uniformity) 409
    reliability (correctness) 409
    speed 410
    uniqueness 407–9
  reconfigurable/dynamic PUFs 415–19
  ring oscillator PUFs 412
  SRAM-based PUF 419–20
Power of Obfuscation 247–8
practical byzantine fault tolerance (PBFT) 67
proof-of-work 68
protocol attacks 389
pseudo-random number generation (PRNG) 435

quantization 289, 292, 295, 347
Quartus CAD tool 302
quasi-invertible watermarking 346, 390

radio-frequency identification (RFID) security 43
  attacks 43–4
  solutions 44–6
reconfigurable/dynamic PUFs 415–19
redundant operation elimination (ROE)-based obfuscated design 238
relay attack 44, 47
reliability 209–11, 352
removal attacks 259, 261, 389
replay attack 40, 44
reverse engineering (RE) 10, 36, 228, 256
ring of oscillators (RO) 10–11
ring oscillator PUFs 412–13, 415
ring oscillators (ROs) 403
robust functional locking (RFL) 275
robust functional obfuscation (RF-Ob) 275
run-length encoding (RLE) 293
run of key-gates-based key-sensitization attack 258

SAT attack 261, 268–72
S-box 319–20
scheduling algorithm 202, 204–5
  feature extraction methodology for 205
secure better portable graphics (SBPG) 378–9
secure digital camera (SDC) 378
Secure Spread Spectrum algorithm 346
security, defined 4
security-based PUFs 411
  controlled PUFs 412
  strong PUF 411
  weak PUF 412
seller watermark 172
Semiconductor Chip Protection Act of 1984 (SCPA) 228
Sengupta–Roy HAAR
  decompression reduction functions 454
  forward transformation reduction functions 450–1
  level 1 decompression reduction functions 452–3
  level 1 forward transformation reduction functions 448
  level 2 decompression reduction functions 454
  level 2 forward transformation reduction functions 450
  lossless image reconstruction reduction functions 455
  second phase of transformation 449
  validation functions 456–7
sensor module (SM) 41
sequential-triggering-based hardware Trojan 82–3
side-channel analysis attacks 384–7
side-channel attack countermeasures 387–8
side-channel information leakage attacks and countermeasures 383
  encryption hardware 383–4
  side-channel analysis attacks 384–7
  side-channel attack countermeasures 387–8
signal probability skew (SPS) attack 272
signature, desirable properties of 182
signature detection process 153–4, 181–2
signcryption 359, 362
silicon-based PUFs 411
6LowPAN 28
smart devices or THINGS 27, 29
smart healthcare 29
smart healthcare security 58
  attacks 59–60
  solutions 60–2
smart transportation security 50
  smart car security 51
    attacks 51–3
    solutions 53–5
  unmanned aerial vehicle (UAV)/drone security 55
    attacks 55–7
    solutions 57–8
software-based security solutions 36
software obfuscation 228
source code-based obfuscation 231
source watermarking 347
spatial versus frequency domain watermarking 345
spread spectrum communications 358

SRAM PUF 11
static attacks 62
static random access memory (SRAM) 403, 419–20
steganography 340
strength of obfuscation (SoO) 275
strong physical unclonable function 411
structurally obfuscated JPEG compression IP core, generating 296–8
structurally obfuscated JPEG decompression IP core, generating 298–9
structurally obfuscated JPEG IP core, design process of 293
  constructing non-obfuscated DFG for JPEG compression 294–6
  JPEG compression IP core, generating 296–8
  JPEG decompression IP core, generating 298–9
  optimization framework 294
  problem formulation 294
  threat model 293–4
structural obfuscation 174
structural obfuscation of DSP cores 227–8, 231–3
  analysis of case studies 246
    comparative study and discussion 248–50
    result of multistage structural obfuscation 246–8
  benefits of providing security at higher design abstraction level 230–1
  case study for multistage structural obfuscation 243–6
  compiler transformation-driven structural obfuscation 233
    evaluation models 235–6
    formulation 235
    multistage high-level transformation techniques 236–42
  low-cost structural obfuscation 242
    movement of particle 242–3
    PSO 242
    terminating condition of PSO 243
  threat model 230
super-high information content (SHIC) 411
support vector machine (SVM) 383
Swing-Pay system 48
symmetric IP core protection techniques 175–6
  case study 182, 187
    analysis of embedding cost, security metric on DSP Cores symmetrical protection 188–90
    comparative study between symmetrical and non-symmetrical technique 190–3
    demonstration of fingerprinting constraints embedding process 183–6
    demonstration of watermarking constraints embedding process 186–7
  desirable properties of signature 182
  encoding rules of buyer fingerprint and seller watermark 178–80
  multi-variable signature embedding process 180–1
  problem formulation 177
  signature detection process 181–2
  symmetrically protected design—area evaluation model 177
  symmetrically protected design—cost evaluation function 177
  symmetrically protected design—delay evaluation model 177
system-on-chip (SoC)
  architecture 79
  design process 123

text watermarking 346
ThingBot 62
third-party IP (3PIP) cores 3, 79
third-party IP (3PIP) vendors 85–6, 89, 255
threat model 88, 259–60
transport layer security (TLS) scheme 317
tree height transformation (THT) process 239–40
Trojan secured scheduling of DMR CDFG 108
  determining resource configuration 109
    particle initialization 109
    particle movement 110
    terminating criteria 110
    velocity clamping 110
  vendor allocation procedure 111–13

Trojan security aware DSP IP core and integrated circuits 79
benefit of, at higher abstraction level 84–5
case studies/test cases, analysis of 113
comparative perspectives 116–17
DSP applications and system setup for the case studies 113–14
non-loop CDFG DSP core 115
security analysis 114
single- and nested-loop CDFG DSP cores 115–16
definition 96
design area evaluation model 98
design cost evaluation model 99
design delay evaluation model 98
DFG and nested-loop-based CDFG 98–9
single-loop-based CDFG 98
design process 99
deriving the CDFG of an 8-point DCT function 99–102
deriving the CDFG of an 8-point DFT 102–7
deriving the CDFG of the FIR filter 107–8
generating the DMR of the CDFG 108
formulation 97
goal 96–7
hardware Trojan in a 3PIP module 86
example of a hardware Trojan 86–7
Trojan detectability in a 3PIP module at RTL/lower levels 87–8
hardware Trojans, types of 81
combinational-triggering-based hardware Trojan 81
denial of service (DoS) payload-based hardware Trojan 84
functionality changeable payload-based hardware Trojan 83
performance degradable payload-based hardware Trojan 83–4
sequential-triggering-based hardware Trojan 82–3
threat model 85–6
Trojan security approaches 88
for combinational/sequential circuits 94–5
for DSP cores 88–94
Trojan security/detection of an IP core 11
Trojan security feature 212–13
trust cam 379–81
trustworthy consumer electronic systems 34–5

unmanned aerial vehicle (UAV)/drone security 55
attacks 55–7
solutions 57–8

vendor allocation processes 81
Verilog of DSP applications 230
VHDL 230
video watermarking 346, 355, 372–8
visible spatial domain chip 367–70
visible watermarking 346, 351, 354, 376–8
visible wavelet domain chip 370
voice of internet protocol 317
voice-over internet protocol 227

watermark detection 344
watermarked IP core/hardware, design process of 128
case studies, analysis on 163
design cost analysis of triple-phase watermark for DSP IP cores 164–6
security analysis of triple-phase watermark for DSP IP cores 163–4
desired properties of IP core watermark 161–2
possible cases of dishonest claim of IP core/hardware ownership and its resolution 162–3
problem formulation 128–9
single-phase watermarked IP core/ hardware, design process of 129
didactic example of single-phase watermarking process 135–48
process of generating DFG from a DSP function 129–33
signature encoding scheme 134
watermark embedding process for DSP IP core 134–5
watermarking process for DSP IP core 133–4

triple-phase watermarked IP core/ hardware, design process of 148
signature detection process 153–4
signature encoding scheme 151–2
triple-phase watermarking process, didactic example of 154–61
watermark embedding process 152–3
watermark encoding rules 179–80
watermarking and fingerprinting 171
background 173
benefits of protection at higher abstraction 173
threat model 173
watermarking approaches 12, 38, 124–8
watermarking constraints embedding process, demonstration of 186–7
watermarks and watermarking systems
attacks on 388
cryptographic attacks 389
geometric attacks 389
protocol attacks 389
removal and interference attacks 389
for broadcasting applications 355
limitations of 390–1
weak PUF 412
Weber's law 357

XOR/XNOR gates 232

zigzag scanning 293, 306